Lecture Notes in Control and Information Sciences

Volume 491

This series reports new developments in the fields of control and information sciences—quickly, informally and at a high level. The type of material considered for publication includes:

1. Preliminary drafts of monographs and advanced textbooks
2. Lectures on a new field, or presenting a new angle on a classical field
3. Research reports
4. Reports of meetings, provided they are
    (a) of exceptional interest and
    (b) devoted to a specific topic. The timeliness of subject material is very important.

Indexed by EI-Compendex, SCOPUS, Ulrich's, MathSciNet, Current Index to Statistics, Current Mathematical Publications, Mathematical Reviews, IngentaConnect, MetaPress and Springerlink.

More information about this series at https://link.springer.com/bookseries/642

Vladimir Nikiforov · Dmitry Gerasimov

Adaptive Regulation

Reference Tracking and Disturbance Rejection

Springer

Vladimir Nikiforov
Faculty of Control Systems and Robotics
ITMO University
Saint Petersburg, Russia

Dmitry Gerasimov
Faculty of Control Systems and Robotics
ITMO University
Saint Petersburg, Russia

ISSN 0170-8643 ISSN 1610-7411 (electronic)
Lecture Notes in Control and Information Sciences
ISBN 978-3-030-96090-2 ISBN 978-3-030-96091-9 (eBook)
https://doi.org/10.1007/978-3-030-96091-9

This Springer imprint is published by the registered company Springer Nature Switzerland AG
The registered company address is: Gewerbestrasse 11, 6330 Cham, Switzerland

To the memory of my mentor,
Prof. Iliya Miroshnik
Vladimir Nikiforov

To my beloved children, Alexander and Egor
Dmitry Gerasimov

Preface

The main topic of this book is the design of adaptive control laws for asymptotic reference tracking and complete disturbance rejection in the presence of uncertainties. While there are a lot of approaches to such control problems, the book is focused on those ones that are based on the *adaptive implementation* of the *internal model principle* (IMP).

There are two fundamental problems in automatic control: attenuation (rejection) of external disturbances and tracking of reference signals. These both problems solved simultaneously are commonly referred to as the *regulation problem*, while the external signals (i.e., disturbances to be rejected and references to be tracked) are referred to as *exogenous signals* or simply *exosignals*.

There are a lot of control techniques proposed to solve the regulation problem. In the case when there is no a priori information about exogeneous signals, various control techniques provide prescribed attenuation of the external disturbances and prescribed residual tracking error only. At the same time, in many practical applications, more stringent requirement is imposed on controller and consists in zero steady-state control error. It can be enforced if only some information about exogenous signals is available a priori. In particular, this information can be extracted from an exosignal model. Such conjecture is used in the famous IMP.

According to the IMP, the exogeneous signal is modelled as the output of a dynamical autonomous system, called *exosystem*, excited by nonzero initial conditions. Then the disturbance influence on the plant response can be completely eliminated or the reference signal can be asymptotically tracked if the exosystem model is suitably reduplicated in the feedback loop of the control system.

From the practical point of view, the most important class of the exosignals is represented by periodic or quasi-periodic functions modelled as biased sinusoid or as a sum of biased sinusoids. Such exogeneous signals are typical for variety of practical applications: electromechanical drives, robotics, rotating machineries, closed-loop systems of active suppression of noise and vibration, audio signal processing, and so on.

The main question concerning practical implementation of the IMP consists in the following: what precise information about the plant and exosystem models should be known a priori? Initially, the IMP was proposed more than 50 years ago for a priori known linear plants and known linear exosystems, and later it was extended to different classes of a priori known nonlinear systems. In this case, the IMP results in control laws (controllers) with pre-determined and fixed structure and parameters. We will call this approach *nonadaptive implementation* of the IMP.

The main shortcoming of the nonadaptive implementation of the IMP lies in the fact that exact models of both a plant to be controlled and an exosystem generating exosignals must be known. In particular, it means that one must know exact waveform of the exogeneous signals. The case when the plant with uncertain model is affected by a priori uncertain external disturbance is more realistic and has a greater practical importance. This case implies *adaptive implementation* of the IMP.

During the last decades, adaptive implementation of the IMP was successfully developed for solutions to the problems of rejection/tracking of exogenous signals generated by linear exosystems with unknown parameters, for uncertain exogenous signals affecting linear and nonlinear plants with unknown parameters. Recently, new problem of adaptive implementation of the IMP for systems with input delay draws a serious attention. This is due to the fact that time delay is a common phenomenon observed in many real-world technical applications. A number of solutions of external disturbances rejection in systems with input delay were proposed on the basis of combination of prediction-like techniques with indirect (identification based) as well as with direct adaptation.

By now, there exists a great variety of different nonadaptive as well as adaptive implementations of the IMP that solve many important theoretical and practical problems. However, these solutions look unrelated and sometimes lead to similar results. The main purpose of this book is to give a unified and systematic presentation of the adaptive implementation of the IMP applied to the problems of asymptotic rejection and/or tracking of a priori uncertain exogenous signals. The key idea behind the approach, developed by the authors during the last two decades and presented in this book, is based on a special canonical form of the exosystem. This form enables one to design, in a unified and systematic way, appropriate observers of exosignals and obtain parameterizations of the external signals in the form of linear regressions with unknown constant parameters and measured regressors. Such parameterizations can be obtained using a unified and systematic way for different classes of linear and nonlinear plants, plants with measured state or output as well as for plants with unknown parameters. In own turn, the parameterizations enable one to employ well-known basic algorithms of adaptation and efficiently solve adaptive regulation problems for different classes of control systems.

Another focus important for practical applications is the transient performance of the disturbance rejection and/or reference tracking. In the case of adaptive control, high performance of the transients in the closed-loop system mainly implies the fast tuning of the controller adjustable parameters provided by an adaptation algorithm. Thus, an important part of this book is addressed to design and employment of adaptation algorithms with improved parametric convergence achieved by *dynamic* or *memory regressor extension* (DRE or MRE, respectively).

The book starts with the introductory Chap. 1, where the control problems addressed in this book are formulated; the main challenges related to the adaptive compensation of *signal uncertainties* are discussed; and the state of the art in the field of adaptive implementation of the IMP is surveyed.

Chapter 2 presents a unified approach to design observers for a class of uncertain exosignals considered in this book. This approach is based on the employment of the exosystem *canonical form* and results in representation of the uncertain exosignal in the form of linear regression with a constant vector of unknown parameters and a measured regressor. In own turn, this representation allows one to derive a suitable parameterization of the plants affected by uncertain external disturbances.

In Chap. 3, based on the plant parameterizations presented in Chap. 2, we define canonical *error models* and derive for them *basic* algorithms of adaptation. Constructing such algorithms and establishing their convergence properties, we provide theoretical background that will be used in the next chapters for design of different adaptive and/or robust controllers.

Chapter 4 presents the controllers of adaptive reference tracking and adaptive disturbance rejection for plants with known parameters. The employed design methods are based on combinations of the exosignal observers introduced in Chap. 2 and the basic adaptation algorithms presented in Chap. 3. The designed adaptive controllers provide complete rejection of the external disturbance and/or asymptotic reference tracking if the exosignals are modelled by a linear exosystem of known order but with unknown parameters.

Chapter 5 extends the results of Chap. 4 to linear and nonlinear plants with unknown parameters, while Chap. 6 presents controllers of adaptive reference tracking and adaptive disturbance rejection for LTI multi-input multi-output (MIMO) plants with known parameters and input delay.

Chapter 7 presents solutions of the *robust* adaptive regulation problems if the exosignals contain irregular components. In this chapter, we demonstrate how the adaptive controllers must be modified in order to provide desirable properties of the closed-loop control systems: boundedness of all the closed-loop signals in the presence of the irregular component and asymptotic achievement of the control objective when the irregular component vanishes.

The book is written in a self-contained fashion and contains appendices presenting some mathematical tools and relevant background concepts useful for understanding of the main contents of the book. It is assumed that a perspective reader is familiar with standard university courses of linear algebra, differential equations, and control system theory. The book is oriented to the senior undergraduate and graduate students in automatic control, researches, and industrial engineers interested in implementation of modern methods of adaptive control.

All chapters of the book are written in close collaboration by the authors. This book contains some results that originally appeared in Russian (“Никифоров В.О., *Адаптивное и робастное управление с компенсацией возмущений*, СПб: Наука, 2003, ISBN 5-02-025006-6.”) .

The work on the book was supported by the Government of Russian Federation (grant 08-08), the Ministry of Science and Higher Education of Russian Federation, project (Goszadanie) no.2019-0898 and the Russian Science Foundation grant (project 17-79-20341).

The book was typeset in LaTeX. Simulations are carried out in MATLAB/Simulink environment, the figures are made in MATLAB and CorelDRAW.

We are thankful to our colleagues I. Furtat, A. Pyrkin, A. Bobtsov, I. Kolmanovskiy, A. Paramonov, and O. Suzdalev for their helpful discussions, comments, and numerous valuable suggestions.

We are also grateful to Oliver Jackson for encouraging us to do this project with Springer.

Finally, we would like to testify that without encouragement and patience of Marina Vlasova and Liubov Urbanovich this book would not have been prepared.

Saint Petersburg, Russia

Vladimir Nikiforov
Dmitry Gerasimov

Contents

Abbreviations and Symbols

Abbreviations

AC	Alternating current
DC	Direct current
DRE	Dynamic regressor extension
HOTD	High-order time derivative
IMP	Internal model principle
ISS	Input-to-state stability (or input-to-state stable)
LTI	Linear time-invariant
MIMO	Multi-input multi-output
MRAC	Model reference adaptive control
MRE	Memory regressor extension
PASMP	Proper asymptotically stable and minimum phase
PMSM	Permanent magnet synchronous motors
PE	Persistent excitation (or persistently excited, or persistently exciting)
PDE	Partial differential equation
PWM	Pulse width modulation
SISO	Single-input single-output
SPR	Strictly positive real

Symbols

$\lvert x \rvert$	The absolute value of a scalar x or the Euclidean norm of a vector x given by $\sqrt{x^\top x}$
C_j^i	The binomial coefficients given by $C_j^i = \frac{i!}{j!(i-j)!}$
ς_i	The ith coordinate vector defined as $\varsigma_i = [0, \ldots, 0, \underset{i}{1}, 0, \ldots, 0]^\top$ (the vector dimension depends on a context)

j	The imaginary unit ($j = \sqrt{-1}$) in a complex number of the form $a + jb$
$\mathbb{N}_0$	Set of natural numbers including zero
$\mathbb{N}$	Set of natural numbers
$\mathbb{R}$ $\left(\mathbb{R}^{n\times m}\right)$	Space of real numbers ($n \times m$ matrices)
$\mathbb{R}_+$	Space of nonnegative real numbers or vectors with nonnegative elements
$\in$ ($\notin$)	Belongs (does not belong) to a space, a class, or a set
$\square$	The end of a proof
$\blacksquare$	The end of an example or remark
$\text{adj}\{A\}$	The adjoint or adjugate of a matrix A
$\det\{A\}$	Determinant of a matrix A
$\text{diag}\{A_i\}$	Block-diagonalized matrix with diagonal elements A_i given by matrices, vectors, or scalars
I or I_i	The identity matrix of an appropriate dimension indicated when it is necessary
O	The zero matrix of an appropriate dimension
$\|A\|_F$	The Frobenius norm of a matrix A given by $\sqrt{\text{tr}\left\{A^\top A\right\}}$
$\text{tr}\{A\}$	Trace of a matrix A
$\lambda\{A\}$	The set of eigenvalues of a matrix A
$\otimes$	The Kronecker product operator
$A \succ B$ ($A \succeq B$)	A matrix A is greater than (greater or equal to) a matrix B In particular, $A \succ 0$ denotes a positively defined matrix A
$f^{(i)}(t)$ or $f^{(i)}$	The ith time derivative of a time function $f(t)$
s	The time derivative operator or the Laplace transformation operator under zero initial conditions
$w(s)[f(t)]$ or $w(s)[f]$	The transfer function operator acting on a function of time $f(t)$
ϵ	Different exponentially decaying terms of appropriate dimensions with different indices and accents
$L_f h(x)$	Lie derivative of $h(x)$ along f defined as $\frac{\partial h(x)}{\partial x} f$
$\mathcal{L}$	Class of locally Lipschitz functions uniformly in time
$\mathcal{L}_1$	Space of integrally bounded functions of time
$\mathcal{L}_2$	Space of squared integrally bounded functions of time
$\mathcal{L}_\infty$	Space of globally bounded functions of time
$\mathcal{MS}(c)$	Space of time functions that are of the order of c in the mean square sense
$\mathcal{PC}_{[l,\infty)}$	Class of piecewise continuous functions with bounded discontinuities
$\mathcal{PE}$	Class of persistently exciting functions
$\nabla f(x)$	The gradient of a function $f(x)$ defined as $\frac{\partial f(x)}{\partial x}$
$\|f(t)\|_\infty$	The infinity norm of a function $f(t)$ given by $\sup\|f(t)\|$
$\sup_{t\geq 0} f(t)$	Essential supremum of a function $f(t)$ defined on the set $t \geq 0$

Chapter 1
Introduction and Problem Statement

Abstract The chapter formulates the control problems addressed in this book, discusses the main challenges related to the adaptive compensation of *signal uncertainties* and describes the state of the art in this field. The chapter starts in Sect. 1.1 with the motivation for *adaptive* control and introduces the adaptive control problem for *parametrically* uncertain plants. Then Sect. 1.2 formulates a general problem of adaptive compensation of *signal* uncertainties and discusses the main challenges caused by this type of uncertainty. Section 1.3 presents a nonadaptive version of the IMP which is a starting point for the design of adaptive controllers that can compensate signal uncertainties for linear and some classes of nonlinear plants with known and unknown parameters, with measured and partially measured states. The rigorous statements of *adaptive regulation* and *robust adaptive regulation* problems are given. Section 1.4 provides an overview of the adaptive regulation techniques, while Sect. 1.5 presents several real-world applications of adaptive tracking and/or disturbance rejection.

1.1 Adaptation Against Uncertainties

Conventional design tools of automatic control are based on the assumption that a model of a plant or a process to be controlled is completely known a priori. This enables one to use control laws (controllers) with predetermined fixed structure and parameters [20, 41, 69, 78, 136, 142].

However, in the most practical applications, the models of real plants as well as the properties of the environment, in which control systems operate, either are not exactly known a priori or are subjected to unpredictable variations. In this case, we will speak about *uncertain plant model* or, simply, *uncertainty*.[1] The following types of uncertainties can be distinguished [43, Sect. 1.3][2]:

[1] Often, a model of the environment is aggregated with a model of the plant and is considered as a part of the plant model.

[2] Alternative classifications of uncertainties can be found in [135] or [62, Sect. 8.2].

V. Nikiforov and D. Gerasimov, *Adaptive Regulation*, Lecture Notes in Control and Information Sciences 491, https://doi.org/10.1007/978-3-030-96091-9_1

- *parametric uncertainty* means that the plant model contains unknown constant parameters;
- *signal uncertainty* means that the plant model contains unknown functions of time which can represent time-varying parameters as well as external disturbances;
- *functional uncertainty* means that the plant model contains unknown functions of the state or output variables;
- *structural uncertainty* means that the structure of the plant model is not exactly known and contains unmodelled or so-called *parasitic* dynamics.

It is known that certain properties of closed-loop control systems are *robust*, i.e., they are preserved at least under sufficiently small (in a certain sense) perturbations of the plant model [43, Sect. 2.1.2], [71, Sect. 4.2]. For example, the property of asymptotic stability of linear time-invariant (LTI) systems is robust with respect to small perturbations of systems parameters. This robustness allows one to use control laws with predetermined fixed structure and parameters to guarantee satisfactory performance of the closed-loop systems under at least small perturbations or uncertainties of the plant model. However, if the perturbations or uncertainties are large, special control tools must be applied.

One of the efficient ways to cope with large uncertainties consists of exploiting *adaptation* in a control law. In the broadest sense, adaptation means tuning (adjustment) of the control law parameters or even its structure to overcome the influence of uncertainties and, as a result, to achieve acceptable performance of the closed-loop control system with uncertainties. There exist a few definitions of *adaptive control* (see, for example, [43, Sect. 1.3], [99, Sect. 1.3.1]). In this book, we consider a class of adaptive control laws (controllers) whose parameters are automatically tuned based on measured information about the closed-loop system behavior in a changing environment to ensure achievement of the control objective in the presence of pre-definite type of uncertainties.

Up to the present time, the case of parametric uncertainties has been extensively studied, and the tools of adaptive control for parametrically uncertain plants are the most developed [43, 62, 74, 99]. A wide class of parametrically uncertain plants can be described by an ordinary differential equation of the form[3]

$$\dot{x}(t) = f(x(t), u(t), \psi), \quad y(t) = h(x(t)), \tag{1.1}$$

where $x \in \mathbb{R}^n$ is the state with an initial condition $x(0)$, $y \in \mathbb{R}^m$ is the vector of output (measured and regulated) variables, $u \in \mathbb{R}^m$ is the vector of control signals, $\psi \in \mathbb{R}^p$ is a vector of unknown constant parameters, t is time, f and h are the vector functions to which additional requirements will be imposed if necessary.

In its most general form, the adaptive control objective is to design a control law in the form of *state feedback* (when the state x is available for measurements) or in

[3] For simplicity of notation, hereafter the argument t will be omitted except the cases in which it is necessary for understanding.

the form of *output feedback* (when only the output y is measured) so that for any values of unknown parameters ψ[4]:

(i) all signals in the closed-loop system are bounded;
(ii) the output (regulated) variables y satisfy a certain performance requirement; without loss of generality, let us assume that the control objective is to asymptotically track the reference signal $g(t) \in \mathbb{R}^m$ according to the equality

$$\lim_{t\to\infty} \varepsilon(t) = 0, \tag{1.2}$$

where $\varepsilon = y - g$ is the *tracking* or *output error*.

The process of adaptive control design typically consists of the following steps:

Step 1. Plant model parameterization and nonadaptive controller design. The *state error* $e = x - x^*$ (where x^* is some desired state trajectory) is defined, and the *open-loop error model* is derived as

$$\dot{e} = f_e(e, u, \psi, t), \quad \varepsilon = h_e(e). \tag{1.3}$$

Then on the basis of this model and under the temporary assumption that the vector ψ is known a control law ensuring the control objective is designed, for example, in the form of state feedback

$$u = U(e, \psi, t). \tag{1.4}$$

At this step, a designer can face a *structural obstacle* consisting in violation of the *matching condition*. The control u is said to be matched with uncertainty ψ if in the model (1.3) the terms containing ψ are in the span of control [74, Sect. 3.1]. In this case, the uncertainties can be directly canceled by the control u if they are known. Such a control law that relies on the direct cancellation of the matched uncertainties is called *certainty equivalence control*.

For example, in the model (1.3), the unknown vector ψ is matched with the control u if the function f_e can be represented in the form

$$f_e(e, u, \psi, t) = f_0(e, t) + g(e, t)\psi^\top \varphi(e, t) + g(e, t)u, \tag{1.5}$$

where f_0, g and φ are some known functions, and φ is available for measurements. Then the control $u = -\psi^\top \varphi(e, t)$ completely cancels the term with ψ.

In particular control tasks, the expression (1.5) can be derived with the use of techniques of *parameterization* and/or *change of coordinates*. In some cases, special auxiliary filters are used to obtain an appropriate parameterization. Auxiliary filters (observers) are also used when instead of the plant state error e only the output error ε is accessible for measurements. In such a case, the controller becomes a dynamic system that can be defined by the following equations:

[4] In some cases, it is assumed that unknown parameters belong to a compact a priori known set.

$$u = U(\varepsilon, \zeta, \psi, t), \tag{1.6}$$
$$\dot{\zeta} = Z(\varepsilon, u, \zeta, t), \tag{1.7}$$

where (1.7) represents an auxiliary filter (observer) with the state ζ.

However, a parameterization satisfying the matching condition does not exist for all plant models. To overcome the structural obstacle caused by violation of the matching conditions, special design tool called *adaptive backstepping* has been developed [74].

Step 2. Adjustable controller construction. The controller (1.4) ensures the control objective under the assumption that the vector ψ is known. The *adjustable* controller is obtained from (1.4) by replacing the vector of unknown parameters ψ by the vector of adjustable parameters $\hat{\psi}$:

$$u = U(e, \hat{\psi}, t). \tag{1.8}$$

The vector of adjustable parameters $\hat{\psi}$ is generated by an *adaptation algorithm*

$$\dot{\hat{\psi}} = \Psi(e, t) \tag{1.9}$$

with a function $\Psi(e, t)$ which will be defined in Step 4. In general, an adaptation algorithm represents an additional dynamic feedback converting measured signals into adjustable parameters $\hat{\psi}$. Therefore, the adjustable parameters $\hat{\psi}$ must be treated as additional state variables of a closed-loop adaptive system whose stability properties must be analyzed in the extended state space of dimension $n + p$ [99, Sect. 1.2].

Step 3. Error model derivation. To complete the design of an adaptive system, one needs to choose an adaptation algorithm ensuring the stability of the closed-loop system and achievement of the control objective (1.2). As suggested in [100] (see also [99, Sect. 1.6]), the problem of adaptation algorithm design can be formulated in terms of the *closed-loop error model*. Substituting (1.8) into (1.3) and introducing the *parametric error* $\tilde{\psi} = \psi - \hat{\psi}$, one can derive the *closed-loop error model* consisting of the *control error model*

$$\dot{e} = f_e(e, \tilde{\psi}, t), \quad \varepsilon = h_e(e) \tag{1.10}$$

and the *parametric error model*

$$\dot{\tilde{\psi}} = -\Psi(e, t). \tag{1.11}$$

Step 4. Adaptation algorithm design. The adaptation algorithm (1.9) is constructed so that the origin of the error model (1.10), (1.11) is stable and the control objective (1.2) is achieved. In this book, the *Lyapunov-based* approach is used to derive adaptation algorithms. According to this approach, the function $\Psi(e, t)$ is selected so that the time derivative of a Lyapunov function calculated along the solutions of the closed-loop error model has appropriate properties which ensure the specified stability requirements.

While designing an adaptation algorithm, when only the output error ε (rather than the state error e) is available for measurements, one may face one more structural obstacle consisting of a *high relative degree* of the error model. If the relative degree of the error model is higher than one, special design tools such as *augmented error* [43, Sect. 6.4.3], [99, Sect. 5.4] and *high-order tuners* [43, Sect. 6.4.4], [115] can be applied to ensure stability of the closed-loop adaptive system.

The next example illustrates two types of structural obstacles mentioned above.

Example 1.1 Consider two plants

$$\dot{x}_1 = x_2, \tag{1.12}$$

$$\dot{x}_2 = x_3, \tag{1.13}$$

$$\dot{x}_3 = \psi\phi(x_1, t) + u, \tag{1.14}$$

$$y = x_1 \tag{1.15}$$

and

$$\dot{x}_1 = x_2 + \psi\phi(x_1, t), \tag{1.16}$$

$$\dot{x}_2 = x_3, \tag{1.17}$$

$$\dot{x}_3 = u, \tag{1.18}$$

where ψ is an unknown parameter and $\phi(x_1, t)$ is a known smooth function.

In the plant (1.12)–(1.15), the uncertainty ψ and the control u are matched. However, the transfer function from ψ to $y = x_1$ has a relative degree equal to 3. This causes a structural obstacle for the design of stable adaptation algorithms using the output y as a feedback signal.

In the plant (1.16)–(1.18), the uncertainty ψ and the control u is not matched because they appear in different equations, and there are two integrators between the control and unknown parameter. Hence, there exists the structural obstacle for the design of a certainty equivalence control. ■

In Chap. 3, we present *basic adaptation algorithms* applicable to different classes of error models. Then, in the next chapters, the problems of adaptive control considered will be reduced to one of these classes of the error models with immediate application of an appropriate adaptation algorithm.

In the end of this section, we illustrate the design steps introduced above by the following example.

Example 1.2 Let us consider a problem of adaptive stabilization of the first-order plant

$$\dot{x} = \psi x + u, \tag{1.19}$$

where x, u and ψ are the scalar output, control and unknown parameter, respectively. Since, in this case, the desired state is zero, i.e., $x^*(t) \equiv 0$, then $e(t) \equiv x(t)$, and we will use x in subsequent calculations.

It is easy to see that ψ is in the span of the control u and the controller

$$u = -kx - \psi x,$$

where $k > 0$ is a feedback gain, would solve the control problem if the parameter ψ was known. Since ψ is unknown, we use the adjustable controller

$$u = -kx - \hat{\psi}x, \tag{1.20}$$

where the adjustable parameter $\hat{\psi}$ is generated by the adaptation law

$$\dot{\hat{\psi}} = \Psi(x) \tag{1.21}$$

with some function $\Psi(x)$ defined later. Introducing the parametric error $\tilde{\psi} = \psi - \hat{\psi}$, substituting (1.20) into (1.19) and calculating the time derivative of $\tilde{\psi}$, in view of (1.21), we obtain the following closed-loop error model:

$$\dot{x} = -kx + \tilde{\psi}x, \quad \dot{\tilde{\psi}} = -\Psi(x). \tag{1.22}$$

In order to define the function $\Psi(x)$, we use the Lyapunov function

$$V(x, \tilde{\psi}) = \frac{1}{2}x^2 + \frac{1}{2\gamma}\tilde{\psi}^2, \tag{1.23}$$

where $\gamma > 0$ is an arbitrary positive constant called the *adaptation gain*. Then the time derivative of $V(x, \tilde{\psi})$ along the solutions of (1.22) takes the form

$$\dot{V}(x, \tilde{\psi}) = -kx^2 + \tilde{\psi}x^2 - \frac{1}{\gamma}\tilde{\psi}\Psi(x).$$

If we choose $\Psi(x) = \gamma x^2$, then

$$\dot{V}(x, \tilde{\psi}) = -kx^2.$$

From this, it can be shown that the origin $(x, \tilde{\psi}) = (0, 0)$ is stable and $x(t) \to 0$ as $t \to \infty$ (for the details, see Proposition 3.5 and its proof). As a result, the adaptive controller

$$u = -kx - \hat{\psi}x, \quad \dot{\hat{\psi}} = \gamma x^2 \tag{1.24}$$

provides adaptive stabilization of the parametrically uncertain plant (1.19) for any values of the parameter ψ, gains $k > 0, \gamma > 0$ and any initial conditions $x(0)$, $\hat{\psi}(0)$. As it can be observed form (1.24), in this case, the adaptive controller uses the adaptation algorithm with the *"pure" integral action.* ■

1.2 Challenges of Signal Uncertainties

In Sect. 1.1, we assumed that the uncertainties of the plant (1.1) are represented by unknown constant parameters, and there are no external disturbances affecting the plant. However, such an assumption is a strong idealization, and in the most practical applications, the influence of external disturbances needs to be taken into account.

In this book, we focus primarily on the plants of the form

$$\dot{x}(t) = f(x(t), u(t), \psi, \upsilon(t)), \quad y(t) = h(x(t))$$

with both parametric uncertainties represented by the vector of constant unknown parameters ψ and signal uncertainties represented by an unknown external disturbance $\upsilon(t)$. The case of signal uncertainties is much more difficult and raises several challenges in design of closed-loop adaptive systems.

At first, the signal uncertainty υ can be treated as an *unmodelled* (i.e., not taken into account in the adaptive control design) disturbance. Unfortunately, as it was demonstrated by many authors [61, 62, 99, 106], unmodelled disturbances can destroy the stability of the closed-loop adaptive system. In particular, even an asymptotically vanishing unmodelled disturbance can cause unbounded *parametric drift* (i.e., unbounded growth of the adjustable parameters) in systems with standard adaptation algorithms with the "pure" integral action.

Example 1.3 Let us try to use the adaptive controller (1.24) for stabilization of the plant

$$\dot{x} = \psi x + u + \upsilon, \tag{1.25}$$

where υ is unmeasurable bounded disturbance. In this case, the closed-loop error model takes the form

$$\dot{x} = -kx + \tilde{\psi} x + \upsilon, \quad \dot{\tilde{\psi}} = -\gamma x^2,$$

and the time derivative of the Lyapunov function (1.23) along the solutions of the closed-loop system satisfies the inequality

$$\dot{V}(x, \tilde{\psi}) = -kx^2 + \upsilon x \leq -kx^2 + \|\upsilon\|_\infty |x|.$$

Since $\dot{V} < 0$ for $|x| > \frac{\|\upsilon\|_\infty}{k}$, it follows that x is bounded. However, the parametric error $\tilde{\psi}$, and therefore, the adjustable parameter $\hat{\psi}$ can grow unboundedly for sufficiently small x, when $|x| < \frac{\|\upsilon\|_\infty}{k}$ and $\dot{V} > 0$. This theoretical observation was confirmed by a number of numerical examples [61, 62, 99, 106]. In [61], it was shown that for

$$\upsilon(t) = (1+t)^{-\frac{1}{8}}\left[1 - \psi(1+t)^{-\frac{1}{4}} - \frac{3}{8}(1+t)^{-\frac{5}{4}}\right],$$

$k = 1$, $\gamma = \frac{1}{4}$, $\hat{\psi}(0) = 0$, $x(0) = 1$ we have

$$x(t) = (1+t)^{-\frac{3}{8}}, \quad \hat{\psi}(t) = (1+t)^{\frac{1}{4}} - 1.$$

Therefore, $\hat{\psi}(t) \to \infty$ in spite of the fact that $\upsilon(t) \to 0$ and $x(t) \to 0$ as $t \to \infty$. ■

To prevent unbounded growth of the closed-loop signals in adaptive systems with unmodelled uncertainties, different types of *robust modifications* of adaptation algorithms were proposed [61, 62, 99, 106]. However, in general, robust adaptation guarantees control error convergence to a bounded residual set rather than its asymptotic convergence to zero.

Example 1.4 To stabilize the plant (1.25), let us use the adjustable controller (1.20) and a modified algorithm of adaptation. Instead of the algorithm with "pure" integral action, we use the algorithm with parametric feedback

$$\dot{\hat{\psi}} = \gamma x^2 - \sigma\hat{\psi}, \tag{1.26}$$

where $\sigma > 0$ is a constant. The adaptation algorithm (1.26) is referred to as *fixed σ-modification* [61, 62]. The main idea of this modification, which introduces the parametric feedback in an adaptation algorithm, is to provide negative time derivative of an appropriate Lyapunov function in the case when adjustable parameters exceed certain bounds. This, in turn, prevents unbounded parametric drift.

For (1.25), (1.20), (1.26), the closed-loop error model takes the form

$$\dot{x} = -kx + \tilde{\psi}x + \upsilon, \quad \dot{\tilde{\psi}} = -\gamma x^2 - \sigma\tilde{\psi} + \sigma\psi,$$

and the time derivative of the Lyapunov function (1.23) satisfies the inequality

$$\dot{V}(x,\tilde{\psi}) = -kx^2 + x\upsilon - \frac{\sigma}{\gamma}\tilde{\psi}^2 + \frac{\sigma}{\gamma}\psi\tilde{\psi} \le -kx^2 - \frac{\sigma}{\gamma}\tilde{\psi}^2 + \|\upsilon\|_\infty|x| + \frac{\sigma}{\gamma}|\psi||\tilde{\psi}|.$$

Completing the squares, we obtain

$$\dot{V}(x,\tilde{\psi}) \leq -\frac{k}{2}x^2 - \frac{\sigma}{2\gamma}\tilde{\psi}^2 - \left(\sqrt{\frac{k}{2}}|x| - \frac{1}{\sqrt{2k}}\|\upsilon\|_\infty\right)^2 + \frac{1}{2k}\|\upsilon\|_\infty^2 -$$
$$\frac{\sigma}{2\gamma}\left(|\tilde{\psi}| - |\psi|\right)^2 + \frac{\sigma}{2\gamma}|\psi|^2 \leq -cV(x,\tilde{\psi}) + \frac{1}{2k}\|\upsilon\|_\infty^2 + \frac{\sigma}{2\gamma}|\psi|^2,$$

where $c = \min\{k, \sigma\}$. From the last inequality, we conclude that x and $\tilde{\psi}$ exponentially converge to the residual set

$$S = \left\{x, \tilde{\psi}:\ V(x,\tilde{\psi}) < \frac{1}{2c}\left(\frac{1}{k}\|\upsilon\|_\infty^2 + \frac{\sigma}{\gamma}|\psi|^2\right)\right\}. \tag{1.27}$$

Hence, in presence of unmodelled disturbance, the adjustable controller (1.20) with the robust adaptation algorithm (1.26) provides the boundedness of all the closed-loop signals and exponential convergence of x and $\tilde{\psi}$ to the compact residual set (1.27). ■

In many practically-oriented problems, a control objective is more stringent and requires zero steady-state control error, e.g., achievement of the asymptotic control objective (1.2). Achieving this becomes possible if the external disturbance can be represented by an appropriate model that is taken into account in the adaptive control design. In particular, for efficient application of adaptive control tools well-developed for parametric uncertainties, one can try to reduce the signal uncertainty of the disturbance υ to a parametric one. To this end, on the basis of prior information one can try to implement *direct parameterization* and represent the uncertain disturbance in the form of linear regression

$$\upsilon(t) = \theta^\top \xi_\upsilon(t), \tag{1.28}$$

where $\theta \in \mathbb{R}^q$ is a vector of unknown parameters and ξ_υ is a measured vector-function called *regressor*.

However, the direct parameterization leaves open the question of choosing the regressor ξ_υ. A more flexible approach is based on *indirect parameterization* when an unknown disturbance is modelled as the output of an autonomous system (called *exosystem*) with unknown parameters. In this book, we will consider linear exosystems of the form

$$\upsilon = h^\top \xi, \quad \dot{\xi} = \Gamma\xi, \tag{1.29}$$

where $\xi \in \mathbb{R}^q$ is an unmeasured state with some initial condition $\xi(0)$. The order q of the exosystem is assumed to be known a priori, while the parameters of Γ and h are unknown. As it will be discussed in Sect. 2.1, such an exosystem can generate a sufficiently wide class of functions υ.

It is worth noting that the disturbance representation (1.29) allows one to use the IMP which can be applied to both types of external signals: a disturbance to be rejected and a reference to be tracked [85, 142]. In accordance with this principle, in order to completely reject or asymptotically track the exosignal, the model of its exosystem must be suitably replicated in the structure of the closed-loop system.

A control problem of simultaneous disturbance rejection and reference tracking, when the exosignals are modelled by autonomous exosystems, is called the problem of *output regulation*, while a controller solving this problem sometimes is called *servocontroller*. This control problem is called *output* regulation, because the control objective is formulated for the output (regulated) variables: the influence of the external disturbances on the output variables must be compensated, while the output variables must asymptotically track the reference. At the same time, an appropriate control law can be designed in the form of a state feedback or an output feedback.

When parameters of the exosystem are unknown, the problems of tracking and rejection can be resolved via *adaptive implementation* of the IMP. The main topic of this book is *adaptive output regulation problem* solved for the linear exosystem (1.29) with unknown parameters and for different classes of linear and nonlinear plants with known and unknown parameters, with measured state or output. To solve this problem, we perform indirect parameterization using special *disturbance observers* that allow one to represent the uncertain disturbance in the form

$$\upsilon = \theta^{\top}\hat{\xi}_{\upsilon} + \epsilon, \tag{1.30}$$

where $\theta \in \mathbb{R}^q$ is a vector of unknown parameters depending on the parameters of Γ and h, $\hat{\xi}_{\upsilon}$ is the *regressor estimate* generated by a disturbance observer, and ϵ is an exponentially decaying signal (*disturbance estimation error*). We will call this parameterization *indirect* because we do not know a priori the exact waveform of the regressor estimate $\hat{\xi}_{\upsilon}$ generated online by the disturbance observer.

Once the parameterization (1.30) is derived, an adaptive controller with extended vector of adjustable parameters can be designed to counteract the plant parametric uncertainty as well as signal uncertainty of the disturbance.

Example 1.5 Let a bounded disturbance υ affecting the plant (1.25) admit the parameterization

$$\upsilon = \theta\hat{\xi}_{\upsilon} + \epsilon, \tag{1.31}$$

where θ is an unknown constant, and the regressor estimate $\hat{\xi}_{\upsilon}$ is generated by a special disturbance observer. The design of such observers is one of the main topics of this book (see Chap. 2) and is not considered in this example. Here we just assume that the parameterization (1.31) is implementable.

Then the plant (1.25) is given by

$$\dot{x} = \psi x + \theta\hat{\xi}_{\upsilon} + u + \epsilon$$

that motivates the following structure of the adjustable controller:

$$u = -kx - \hat{\psi}x - \hat{\theta}\hat{\xi}_\upsilon,$$

where $\hat{\psi}$ and $\hat{\theta}$ are the adjustable parameters. Choosing the adaptation algorithms

$$\dot{\hat{\psi}} = \gamma_1 x^2, \quad \dot{\hat{\theta}} = \gamma_2 \hat{\xi}_\upsilon x$$

with the adaptation gains $\gamma_1 > 0$ and $\gamma_2 > 0$, we can derive the following closed-loop error model:

$$\dot{x} = -kx + \tilde{\psi}x + \tilde{\theta}\hat{\xi}_\upsilon + \epsilon, \quad \dot{\tilde{\psi}} = -\gamma_1 x^2, \quad \dot{\tilde{\theta}} = -\gamma_2 \hat{\xi}_\upsilon x, \tag{1.32}$$

where $\tilde{\psi} = \psi - \hat{\psi}$ and $\tilde{\theta} = \theta - \hat{\theta}$ are the parametric errors.

Then, selecting the Lyapunov function

$$V(x, \tilde{\psi}, \tilde{\theta}, \epsilon) = \frac{1}{2}x^2 + \frac{1}{2\gamma_1}\tilde{\psi}^2 + \frac{1}{2\gamma_2}\tilde{\theta}^2 + \frac{1}{k}\int_t^\infty \epsilon^2(\tau)d\tau$$

and calculating its time derivative along solutions of (1.32), we have

$$\dot{V}(x, \tilde{\psi}, \tilde{\theta}, \epsilon) = -kx^2 + x\epsilon - \frac{1}{k}\epsilon^2.$$

Completing the squares, we obtain

$$\dot{V}(x, \tilde{\psi}, \tilde{\theta}, \epsilon) = -\frac{k}{2}x^2 - \left(\sqrt{\frac{k}{2}}x - \frac{\epsilon}{\sqrt{2k}}\right)^2 - \frac{1}{2k}\epsilon^2 \le -\frac{k}{2}x^2 - \frac{1}{2k}\epsilon^2 \le 0.$$

In the view of Theorems B.1, B.4, the latter means that the origin $(x, \tilde{\psi}, \tilde{\theta}, \epsilon) = (0, 0, 0, 0)$ is stable, and $x(t) \to 0$ as $t \to \infty$ for a disturbance υ admitting parameterization (1.31), arbitrary ψ, $k > 0$, $\gamma_1 > 0$, $\gamma_2 > 0$ and for any initial conditions $x(0)$, $\epsilon(0)$, $\hat{\psi}(0)$, $\hat{\theta}(0)$. ■

In the above example, the disturbance estimation error ϵ was omitted in the design of the adaptive controller since the plant (1.25) is linear, and the exponentially decaying term does not influence the stability of the closed-loop adaptive system. However, in the case of nonlinear plants, this term generally can cause instability of the closed-loop system. Therefore, there is a need in a control tool to guarantee boundedness of all the signals in closed-loop disturbed nonlinear system.

Example 1.6 Consider the nonlinear plant

$$\dot{x} = \psi x + u + \phi(x)\upsilon,$$

where $\phi(x)$ is a smooth function, while the disturbance υ admits the parameterization (1.31) and yields the plant model of the form

$$\dot{x} = \psi x + u + \theta\phi(x)\hat{\xi}_\upsilon + \phi(x)\epsilon.$$

First, we try to ignore the exponentially decaying term ϵ (as in the case of a linear plant) and construct a certainty equivalence control given by

$$u = -kx - \hat{\psi}x - \hat{\theta}\phi(x)\hat{\xi}_\upsilon,$$

which results in the following closed-loop error model:

$$\dot{x} = -kx + \tilde{\psi}x + \tilde{\theta}\phi(x)\hat{\xi}_\upsilon + \phi(x)\epsilon.$$

It is well known that for $\phi(x) = x^2$ even in the case of zero parametric errors, i.e., when $\tilde{\psi}(t) \equiv 0$ and $\tilde{\theta}(t) \equiv 0$ the equation

$$\dot{x} = -kx + x^2\epsilon \tag{1.33}$$

can have an unbounded solution [74, Sect. 2.5.1]. Indeed, if $\epsilon(t) = \mathrm{e}^{-at}\epsilon(0)$, where $a > 0$ is a constant, the explicit solution of (1.33) is given by

$$x(t) = \frac{(k+a)x(0)}{\mathrm{e}^{-at}\epsilon(0)x(0) + \mathrm{e}^{kt}(k + a - \epsilon(0)x(0))}.$$

If $k + a > \epsilon(0)x(0)$, $x(t)$ approaches zero as $t \to \infty$. However, if $k + a < \epsilon(0)x(0)$, the *finite-time escape phenomenon* is observed, i.e.,

$$x(t) \to \infty \quad as \quad t \to t_f = \frac{1}{k+a}\ln\left(\frac{\epsilon(0)x(0)}{\epsilon(0)x(0) - k - a}\right).$$

Therefore, in the case of nonlinear plants even exponentially decaying terms (caused, for example, by initial states of auxiliary filters and disturbance observers) must be taken into account and counteracted by means of special control tools. One such efficient design tools consists of employing the *nonlinear damping term* $-\mu\phi^2(x)x$, where $\mu > 0$, to augment the control law [74, Sect. 2.5]:

$$u = -kx - \hat{\psi}x - \hat{\theta}\phi(x)\hat{\xi}_\upsilon - \mu\phi^2(x)x. \tag{1.34}$$

Using the control law (1.34) together with the adaptation algorithms

$$\dot{\hat{\psi}} = \gamma_1 x^2, \quad \dot{\hat{\theta}} = \gamma_2\phi(x)\hat{\xi}_\upsilon x,$$

we obtain the following closed-loop error model:

$$\dot{x} = -kx - \mu\phi^2(x)x + \tilde{\psi}x + \tilde{\theta}\phi(x)\hat{\xi}_\upsilon + \phi(x)\epsilon, \tag{1.35}$$

$$\dot{\tilde{\psi}} = -\gamma_1 x^2, \quad \dot{\tilde{\theta}} = -\gamma_2\phi(x)\hat{\xi}_\upsilon x. \tag{1.36}$$

Then the time derivative of the Lyapunov function

$$V(x, \tilde{\psi}, \tilde{\theta}, \epsilon) = \frac{1}{2}x^2 + \frac{1}{2\gamma_1}\tilde{\psi}^2 + \frac{1}{2\gamma_2}\tilde{\theta}^2 + \frac{1}{\mu}\int_t^{\infty} \epsilon^2(\tau)d\tau$$

calculated along the solutions of (1.35), (1.36) is given by

$$\dot{V}(x, \tilde{\psi}, \tilde{\theta}, \epsilon) = -kx^2 - \mu\phi^2(x)x^2 + \phi(x)x\epsilon - \frac{1}{\mu}\epsilon^2.$$

Due to the nonlinear damping term we complete the square and obtain

$$\begin{aligned}\dot{V}(x, \tilde{\psi}, \tilde{\theta}, \epsilon) &= -kx^2 - \mu\left(\phi(x)x - \frac{1}{2\mu}\epsilon\right)^2 - \frac{3}{4\mu}\epsilon^2 \\ &\leq -kx^2 - \frac{3}{4\mu}\epsilon^2 \leq 0.\end{aligned}$$

The latter means that the origin $(x, \tilde{\psi}, \tilde{\theta}, \epsilon) = (0, 0, 0, 0)$ is stable and $x(t) \to 0$ as $t \to \infty$. ■

It is also worth noting another important challenge concerning adaptive compensation of the signal uncertainties. In most of the practical cases, it is impossible to expect that the exosignal (disturbance or reference) is modelled with absolute accuracy as a solution of the linear exosystem (1.29). The case when an exosignal contains both a dominant component modelled by the linear exosystem (1.29) as well as an irregular (potentially small) additive component is more realistic and has higher practical relevance. Such an exosignal υ_δ can be represented as

$$\upsilon_\delta = \upsilon + \delta, \tag{1.37}$$

where the *regular* component υ is modelled by the exosystem (1.29), while the *irregular* component δ is an unknown bounded function of time. The representation (1.37) covers three different types of the exosignals:

- **Case 1:** if $\delta(t) \equiv 0$, then the exosignal $\upsilon_\delta = \upsilon$ is exactly modelled by the linear exosystem (1.29) with unknown parameters; such an exosignal is of theoretical interest but limited practical relevance;
- **Case 2:** if $\delta(t) \not\equiv 0$, then the Eq. (1.37) describes, in particular, practically relevant case when the exosignal υ_δ consists of the dominant regular component υ exactly modelled by the exosystem (1.29) and a typically small irregular component δ representing unmodelled disturbances, e.g., noise, errors of the ideal model (1.29), etc.;
- **Case 3:** if $\upsilon(t) \equiv 0$, then the exosignal $\upsilon_\delta = \delta$ is just an unknown bounded function of time.

Therefore, the main challenge in adaptive compensation of signal uncertainties in the form (1.37) is to design a control law which simultaneously provides:

- **in Case 1**, the boundedness of all the closed-loop signals and asymptotic output regulation;
- **in Case 2**, the boundedness of all the closed-loop signals and a small (in a certain sense) residual control error for a small δ[5];
- **in Case 3**, the boundedness of all the closed-loop signals and a bounded residual control error.

The control problem stated for an uncertain exosignal represented in the form (1.37), i.e., when it contains both regular and irregular components, we will call *robust adaptive output regulation* problem. It will be demonstrated in this book that a control law resolving the robust adaptive output regulation problem can be designed with the use of the following special control tools:

- *disturbance observer* allowing one to obtain a disturbance parameterization of the form
$$\upsilon_\delta = \theta^\top \hat{\xi}_\upsilon + \epsilon_\delta, \tag{1.38}$$
where $\theta \in \mathbb{R}^q$ is a vector of unknown parameters, $\hat{\xi}_\upsilon$ is the regressor estimate generated by a disturbance observer, ϵ_δ is a bounded disturbance estimation error that converges to zero as $\delta \to 0$; if $\delta(t) \equiv 0$, then ϵ_δ exponentially decays;
- *nonlinear damping term* needed to counteract a bounded or an exponentially decaying disturbance estimation error ϵ_δ in nonlinear plants;
- *changes of coordinates* to match the disturbance with control in linear plants;
- *adaptive backstepping* to counteract unmatched disturbances;
- a special *robust modifications of adaptation algorithms* to prevent unbounded parametric drift for the case when $\delta(t) \not\equiv 0$ and simultaneously to drive the control error to zero if $\delta(t) \equiv 0$ or $\delta(t) \to 0$ as $t \to \infty$.

The key idea of a special robust modification of adaptation algorithms mentioned above is illustrated by the following example.

Example 1.7 Let us assume that the disturbance (1.37), that admits the parameterization (1.38), affects the plant (1.25) (where υ is replaced by υ_δ). Define the *extended regressor* as $\omega = [x, \hat{\xi}_\upsilon^\top]^\top$ and the *extended vector of unknown parameters* as $\vartheta = \left[\psi, \theta^\top\right]^\top$. Then the plant model takes the form

$$\dot{x} = \vartheta^\top \omega + u + \epsilon_\delta$$

and motivates the following structure of the adjustable controller:

$$u = -kx - \hat{\vartheta}^\top \omega,$$

[5] In particular, it means that the control error converges to zero as δ does so.

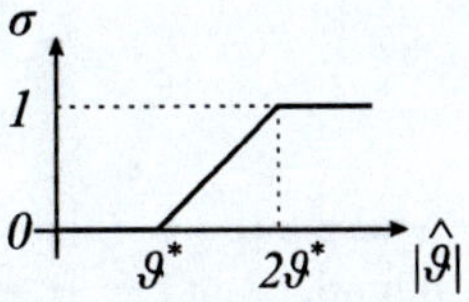

Fig. 1.1 Illustration of the switching mechanism (1.40) of the σ-modification

where $\hat{\vartheta}$ is the vector of adjustable parameters. To cope with the unmodelled disturbance ϵ_δ, which can be either exponentially decaying or just a bounded function, we use the following adaptation algorithm with a special *continuous switching σ-modification* [62, Sect. 8.4.1]:

$$\dot{\hat{\vartheta}} = \gamma\omega x - \sigma(\hat{\vartheta})\hat{\vartheta}, \tag{1.39}$$

where

$$\sigma(\hat{\vartheta}) = \begin{cases} 0, & |\hat{\vartheta}| < \vartheta^*, \\ \frac{|\hat{\vartheta}|}{\vartheta^*} - 1, & \vartheta^* \leq |\hat{\vartheta}| \leq 2\vartheta^* \\ 1, & |\hat{\vartheta}| > 2\vartheta^*, \end{cases}, \tag{1.40}$$

ϑ^* is a positive constant. In contrast to the adaptation algorithm (1.26) with fixed σ-modification, the function $\sigma(\hat{\vartheta})$ in (1.40) (see Fig. 1.1) switches off the parametric feedback if the adjustable parameters are within some prescribed bounds, i.e., when $|\hat{\vartheta}| < \vartheta^*$.

In this case, the closed-loop error model takes the form

$$\dot{x} = -kx + \tilde{\vartheta}^\top\omega + \epsilon_\delta, \quad \dot{\tilde{\vartheta}} = -\gamma\omega x - \sigma(\hat{\vartheta})\tilde{\vartheta} + \sigma(\hat{\vartheta})\vartheta. \tag{1.41}$$

To establish the properties of the closed-loop system, we consider the following two cases. First, consider the case when $\delta(t) \not\equiv 0$, and ϵ_δ is just a bounded function. Then the time derivative of the Lyapunov function

$$V(x, \tilde{\vartheta}) = \frac{1}{2}x^2 + \frac{1}{2\gamma}\tilde{\vartheta}^\top\tilde{\vartheta}$$

calculated along the solutions of (1.41) satisfies the inequality

$$\begin{aligned}\dot{V}(x, \tilde{\vartheta}) &\leq -kx^2 - \frac{\sigma(\hat{\theta})}{\gamma}|\tilde{\vartheta}|^2 + x \parallel \epsilon_\delta \parallel_\infty + \frac{\sigma(\hat{\theta})}{\gamma}|\tilde{\vartheta}||\vartheta| \\ &= -\frac{k}{2}x^2 - \frac{\sigma(\hat{\theta})}{2\gamma}|\tilde{\vartheta}|^2 - \frac{1}{2}\left(\sqrt{k}x - \frac{1}{\sqrt{k}} \parallel \epsilon_\delta \parallel_\infty\right)^2 + \frac{1}{2k} \parallel \epsilon_\delta \parallel_\infty^2 \\ &- \frac{\sigma(\hat{\theta})}{2\gamma}\left(|\tilde{\vartheta}| - |\vartheta|\right)^2 + \frac{\sigma(\hat{\vartheta})}{2\gamma}|\vartheta|^2\end{aligned}$$

$$\leq -\frac{k}{2}x^2 - \frac{\sigma(\hat{\vartheta})}{2\gamma}|\tilde{\vartheta}|^2 + \frac{1}{2k}|\epsilon_\delta|_\infty^2 + \frac{\sigma(\hat{\vartheta})}{2\gamma}|\vartheta|^2,$$

where $\sigma(\hat{\vartheta}) > 0$ if $|\hat{\vartheta}| > \vartheta^*$. The latter means that all the closed-loop signals are bounded, and x, $\tilde{\vartheta}$ exponentially converge to a residual set (for more details see Lemma B.1). In other words, for $\delta(t) \not\equiv 0$ the properties of the algorithm (1.39) are equivalent to the properties of the algorithm (1.26) with fixed σ-modification.

Second, consider the case when $\delta(t) \equiv 0$, ϵ_δ exponentially decays and ϑ^* is selected to be sufficiently large so that $\vartheta^* > |\vartheta|$.[6] Then the time derivative of the Lyapunov function

$$V(x, \tilde{\vartheta}, \epsilon_\delta) = \frac{1}{2}x^2 + \frac{1}{2\gamma}\tilde{\vartheta}^\top\tilde{\vartheta} + \frac{1}{k}\int_t^\infty \epsilon_\delta^2(\tau)d\tau,$$

calculated along the solutions of (1.41) takes the form

$$\begin{aligned}
\dot{V}(x, \tilde{\vartheta}, \epsilon_\delta) &= -kx^2 + \epsilon_\delta x - \frac{1}{k}\epsilon_\delta^2 + \frac{\sigma(\hat{\vartheta})}{\gamma}\tilde{\vartheta}^\top\hat{\vartheta} \\
&= -\frac{k}{2}x^2 - \left(\sqrt{\frac{k}{2}}x - \frac{1}{\sqrt{2k}}\epsilon_\delta\right)^2 - \frac{1}{2k}\epsilon_\delta^2 + \frac{\sigma(\hat{\vartheta})}{\gamma}\hat{\vartheta}^\top\tilde{\vartheta} \\
&\leq -\frac{k}{2}x^2 - \frac{1}{2k}\epsilon_\delta^2 + \frac{\sigma(\hat{\vartheta})}{\gamma}\hat{\vartheta}^\top\tilde{\vartheta}.
\end{aligned}$$

By assumption $\vartheta^* > |\vartheta|$, we have $\sigma(\hat{\vartheta})\hat{\vartheta}^\top\tilde{\vartheta} \leq 0$. Hence,

$$\dot{V}(x, \tilde{\vartheta}, \epsilon_\delta) \leq -\frac{k}{2}x^2 - \frac{1}{2k}\epsilon_\delta^2.$$

The latter means that the origin $(x, \tilde{\vartheta}, \epsilon_\delta) = (0, 0, 0)$ is stable and $x(t) \to 0$ as $t \to \infty$. Thus, for the case of zero unmodelled disturbance, i.e., when $\delta(t) \equiv 0$, the adaptation algorithm (1.39) with continuous switching σ-modification provides asymptotic zeroing of the regulated variable x much like the algorithm with "pure" integral action does. ■

In the next sections of this chapter, the control problems to be addressed in the book will be formulated and discussed, a brief overview of known approaches to these problems will be presented and practical implementations of the IMP in the problems of output regulation will be illustrated by examples of real engineering systems.

---

[6] In some practical cases, it can be done on the basis of prior information about ϑ.

1.3 Output Regulation

We start with the theoretical background developed for the idealized control problem of LTI plants with known parameters. Then we will extend the problem statements to different cases of practical significance.

1.3.1 *Nonadaptive Output Regulation: Problem Statement and the Francis Equation*

For methodological purposes, we first consider the important but mostly theoretical case when the full states of exosystem and plant are assumed to be accessible for measurements, while the plant and exosystem parameters are known [64].

Consider an LTI plant with input and output disturbances given by

$$\dot{x} = Ax + Bu + D\upsilon, \tag{1.42}$$

$$y = Cx + R\upsilon, \tag{1.43}$$

where $x \in \mathbb{R}^n$ is the state, $y \in \mathbb{R}^m$ is the vector of regulated variables, $u \in \mathbb{R}^m$ is the vector of control signals, $\upsilon \in \mathbb{R}^\rho$ is the vector of disturbances, $A \in \mathbb{R}^{n\times n}$, $B \in \mathbb{R}^{n\times m}$, $C \in \mathbb{R}^{m\times n}$, $D \in \mathbb{R}^{n\times \rho}$ and $R \in \mathbb{R}^{m\times \rho}$ are known constant matrices, $n \geq m$.

Remark 1.1 From the perspective of the adaptive control theory, in the plant (1.42), (1.43) the control u and the disturbance υ are not matched. Indeed, since, in general, the columns of B and D are linearly independent, and R is a nonzero matrix, the disturbance υ is not in the span of the control u. ■

We introduce the output error as

$$\varepsilon = y - g, \tag{1.44}$$

where $g \in \mathbb{R}^m$ is the reference signal. In this context, both the disturbance υ and the reference g are referred to as *exosignals*. A class of the considered exosignals is defined by the following assumption.

Assumption 1.1 The exosignals υ and g are defined as a class of all functions of time that can be generated as the outputs of the exosystem

$$\dot{\xi} = \Gamma\xi, \tag{1.45}$$

$$g = Q\xi, \tag{1.46}$$

$$\upsilon = H\xi \tag{1.47}$$

with the state $\xi \in \mathbb{R}^q$, known constant matrices $\Gamma \in \mathbb{R}^{q\times q}$, $Q \in \mathbb{R}^{m\times q}$ and $H \in \mathbb{R}^{\rho\times q}$ and all possible initial conditions $\xi(0)$.

In Sect. 2.1, we will impose additional requirements on the exosystem which will be used in the derivation of the technical results of this book. Now, on the basis of Assumption 1.1, we formulate a control problem which serves as a starting point for the main developments in the book.

Problem 1.1 *Output regulation problem* is to stabilize the plant (1.42), (1.43) affected by the disturbance υ and to achieve the asymptotic tracking of the reference g under Assumption 1.1.

When parameters of the plant (1.42), (1.43) and the exosystem (1.45)–(1.47) are known, the output regulation problem can be handled by a nonadaptive controller with fixed precalculated parameters [64, Chap. 4], [78, Sect. 46], [136, Chap. 9], [142, Chaps. 6–8]. In this case, we will speak about *nonadaptive* implementation of the output regulation problem.

Substituting the equality (1.47) into (1.42) and (1.43), and the equalities (1.43), (1.46) into (1.44), we obtain

$$\dot{x} = Ax + Bu + DH\xi, \tag{1.48}$$

$$\varepsilon = Cx + RH\xi - Q\xi. \tag{1.49}$$

Now, let us define the *state error*

$$e = x - M\xi,$$

where an $n \times q$ coordinate transformation matrix M will be defined later. Given that $x = e + M\xi$, (1.49) takes the form

$$\varepsilon = Ce + (CM + RH - Q)\,\xi. \tag{1.50}$$

Taking time derivative of e, in view of (1.45) and (1.48), we have

$$\begin{aligned}\dot{e} &= Ax + Bu + DH\xi - M\Gamma\xi \\ &= Ae + Bu + (AM - M\Gamma + DH)\xi.\end{aligned} \tag{1.51}$$

Let us assume that there exist matrices $M \in \mathbb{R}^{n\times q}$ and $\Theta_e \in \mathbb{R}^{m\times q}$ which solve the following matrix equation, referred to as the *Francis* or, also, the *regulator* equation:

$$AM - M\Gamma + DH = B\Theta_e, \tag{1.52}$$

$$CM = Q - RH. \tag{1.53}$$

Then the plant model expressed in the new variables e and ε takes the form

$$\dot{e} = Ae + B(u + \Theta_e\xi), \quad \varepsilon = Ce, \tag{1.54}$$

where the term $\Theta_e\xi$ is matched with the control u. Therefore, the model (1.54) enables one to exploit the certainty equivalence control for rejection of the matched disturbance, and it motivates the following structure of the controller:

$$u = -\Theta_e\xi + u_s, \tag{1.55}$$

where u_s is a feedback term solving the problem of plant stabilization. Substituting (1.55) into (1.54), we obtain the closed-loop error model

$$\dot{e} = Ae + Bu_s, \quad \varepsilon = Ce. \tag{1.56}$$

Therefore, to implement the controller (1.55), we need to find a solution pair (M, Θ_e) of the regulator matrix equation (1.52), (1.53) and to choose a stabilizing control u_s, for example, in the form of linear state feedback $u_s = -Ke$ with $K \in \mathbb{R}^{m\times n}$ selected so that the matrix $A - BK$ is Hurwitz. The latter can be done if the matrix pair (A, B) is stabilizable.

Thus, we have proved the sufficiency part of the following statement [46, 64].

Theorem 1.1 *The problem of output regulation is solvable for the plant* (1.42), (1.43) *and the exosystem* (1.45)–(1.47) *by the controller* (1.55) *iff:*

(1) the matrix pair (A, B) is stabilizable;

(2) there exists a solution pair (M, Θ_e) of the matrix equation (1.52), (1.53).

The necessity part of the proof can be found in [64, Sect. 4.2].

Remark 1.2 One can apply different approaches to select the stabilizing component u_s. In particular, instead of the state-error feedback $u_s = -Ke$, one can use the plant state feedback $u_s = -Kx$. In this case, the equality (1.52) takes the form

$$(A - BK)M - M\Gamma + DH = B\Theta_e.$$

■

The following result establishes a condition for the existence of a solution of the Francis equation [64, Lemma 4.1].

Lemma 1.1 *The Francis equation* (1.52), (1.53) *has a solution (M, Θ_e) for each DH and $(Q - RH)$ iff*

$$\operatorname{rank}\begin{bmatrix} A - \lambda I & B \\ C & 0 \end{bmatrix} = n + m \tag{1.57}$$

for each λ which is an eigenvalue of Γ. If this is the case and the control input u and the regulated output y have the same dimensions,[7] *the solution (M, Θ_e) is unique.*

---

[7] In this book, we consider the case of such systems only.

It was proved in [64, Sect. 4.1] that the condition of Lemma 1.1 means that none of the eigenvalues of Γ coincides with a transmission zero of the plant transfer matrix $W_u(s) = C(sI - A)^{-1}B$.

Example 1.8 Consider the single-input single-output (SISO) plant (1.42), (1.43) of the second order with

$$A = \begin{bmatrix} 0 & 1 \\ -2 & -3 \end{bmatrix}, \quad B = \begin{bmatrix} 0 \\ 1 \end{bmatrix}, \quad C = \begin{bmatrix} 0 & 1 \end{bmatrix}, \quad D = \begin{bmatrix} d_1 \\ d_2 \end{bmatrix}, \quad R = 0,$$

some constants d_1 and d_2, and the corresponding transfer function

$$W_u(s) = C(sI - A)^{-1}B = \frac{s}{s^2 + 3s + 2}.$$

Let the exosignal be a constant $\upsilon = a$. It can be modelled by (1.45), (1.47) with $\Gamma = 0$, $H = 1$ and $\xi(0) = a$. In this case, the zero of the transfer function $W_u(s)$ coincides with the eigenvalue of the exosystem, while the Francis equation (1.52), (1.53) takes the form

$$B\Theta_e = AM + D, \quad CM = 0, \tag{1.58}$$

where $\Theta_e \in \mathbb{R}^1$, $M = \begin{bmatrix} m_1, & m_2 \end{bmatrix}^{\mathrm{T}}$. Solving (1.58), we obtain

$$\begin{bmatrix} 0 \\ \Theta_e \end{bmatrix} = \begin{bmatrix} 0 & 1 \\ -2 & -3 \end{bmatrix} \begin{bmatrix} m_1 \\ m_2 \end{bmatrix} + \begin{bmatrix} d_1 \\ d_2 \end{bmatrix}, \quad m_2 = 0$$

or

$$d_1 = 0, \quad m_2 = 0, \quad \Theta_e = -2m_1 + d_2.$$

Therefore, the Francis equation is resolvable not for all D, but only for D with $d_1 = 0$. ■

Example 1.9 Consider the two-input and two-output plant (1.42), (1.43) of the third order with

$$A = \begin{bmatrix} 0 & 1 & 0 \\ -6 & -5 & 0 \\ 0 & 0 & -1 \end{bmatrix}, \quad B = \begin{bmatrix} 0 & 0 \\ 1 & 0 \\ 0 & 1 \end{bmatrix}, \quad D = \begin{bmatrix} 1 & 1 \\ 0 & 0 \\ 0 & 1 \end{bmatrix}, \tag{1.59}$$

$$C = \begin{bmatrix} 1 & 0 & 0 \\ 1 & 0 & 1 \end{bmatrix}, \qquad R = \begin{bmatrix} 1 & 0 \\ 0 & 1 \end{bmatrix},$$

and the corresponding transfer function matrices

$$W_u(s) = C(sI - A)^{-1}B = \begin{bmatrix} \frac{1}{s^2+5s+6} & 0 \\ \frac{1}{s^2+5s+6} & \frac{1}{s+1} \end{bmatrix},$$

$$W_v(s) = C(sI - A)^{-1}D + R = \begin{bmatrix} \frac{s^2+6s+11}{s^2+5s+6} & \frac{s+5}{s^2+5s+6} \\ \frac{s+5}{s^2+5s+6} & \frac{s^3+8s^2+22s+17}{s^3+6s^2+11s+6} \end{bmatrix}.$$

Let the exosystem (1.45)–(1.47) have the matrices

$$\Gamma = \begin{bmatrix} 0 & 1 & 0 \\ -1 & 0 & 0 \\ 0 & 0 & 0 \end{bmatrix}, \quad Q = \begin{bmatrix} 1 & 0 & 0 \\ 0 & 0 & 1 \end{bmatrix}, \quad H = \begin{bmatrix} 0 & 0 & 1 \\ 1 & 0 & 0 \end{bmatrix}. \tag{1.60}$$

In this case,

$$g(t) = \begin{bmatrix} g_1(t) \\ g_2(t) \end{bmatrix} = \begin{bmatrix} b\sin(t+\varphi) \\ a \end{bmatrix}, \tag{1.61}$$

and

$$v(t) = \begin{bmatrix} v_1(t) \\ v_2(t) \end{bmatrix} = \begin{bmatrix} a \\ b\sin(t+\varphi) \end{bmatrix}, \tag{1.62}$$

where the amplitudes a, b and the phase φ are determined by the initial condition $\xi(0)$ as $a = \xi_3(0)$, $b = \sqrt{\xi_1^2(0) + \xi_2^2(0)}$, $\varphi = \arctan\left(\frac{\xi_1(0)}{\xi_2(0)}\right)$.

Then the matrix equation (1.52), (1.53) gives the following solution:

$$M = \begin{bmatrix} 1 & 0 & -1 \\ -1 & 1 & -1 \\ -2 & 0 & 2 \end{bmatrix}, \quad \Theta_e = \begin{bmatrix} 0 & -4 & 11 \\ 3 & 2 & -2 \end{bmatrix}. \tag{1.63}$$

Since A is Hurwitz, the stabilization is unnecessary and the controller $u = -\Theta_e\xi$ with the precalculated matrix Θ_e (1.63) provides asymptotic tracking of the reference (1.61) with complete rejecting of the disturbance (1.62) for any initial conditions $x(0)$ and $\xi(0)$. It is worth noting that the initial condition $\xi(0)$ determines the amplitudes a, b and the phase φ, while the exosignal waveforms and the frequency of harmonic are determined by the structure and parameters of Γ.

Simulation results obtained for different initial conditions are presented in Fig. 1.2. As seen from the presented plots, the controller provides asymptotic tracking of the references $g_1(t)$, $g_2(t)$ independently of the initial conditions $\xi(0)$. ■

1.3.2 Decomposition on Tracking and Rejection

The regulation problem with a single exosystem (1.45)–(1.47) has elegant mathematical statement and solution. However, its practical significance is limited since

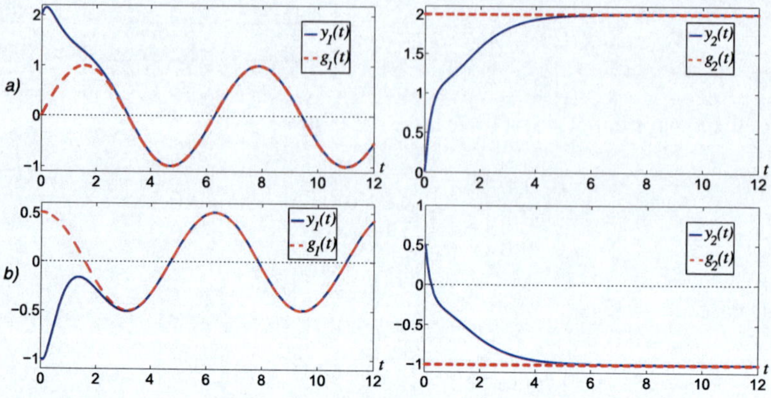

Fig. 1.2 Example 1.9: nonadaptive regulation of the system with the parameters (1.59), (1.60), (1.63), and the initial conditions $x(0) = [0, 0, 0]^\top$, and: **a** $\xi(0) = [0, 1, 2]^\top$ $(a = 2,\ b = 1,\ \varphi = 0)$; **b** $\xi(0) = [0.5, 0, -1]^\top$ $(a = -1, b = 0.5, \varphi = \frac{\pi}{2})$

the disturbance and the reference are composed of the same exosignals (see Example 1.9).

Let us consider a practically more relevant case when the reference g to be tracked and the disturbance υ to be rejected are described by the individual exosystems

$$\dot{\xi}_g = \Gamma_g \xi_g, \quad g = Q_g \xi_g, \tag{1.64}$$
$$\dot{\xi}_\upsilon = \Gamma_\upsilon \xi_\upsilon \quad \upsilon = H_\upsilon \xi_\upsilon, \tag{1.65}$$

where $\xi_g \in \mathbb{R}^{q_g}$ and $\xi_\upsilon \in \mathbb{R}^{q_\upsilon}$. Then the matrices Γ, H and Q of the single exosystem obtained by combining of the corresponding individual ones take the forms

$$\Gamma = \begin{bmatrix} \Gamma_g & O \\ O & \Gamma_\upsilon \end{bmatrix}, \quad Q = \begin{bmatrix} Q_g \ O \end{bmatrix}, \quad H = \begin{bmatrix} O \ H_\upsilon \end{bmatrix}.$$

In this case, the Eq. (1.53) can be rewritten as

$$CM = \begin{bmatrix} Q_g \ O \end{bmatrix} - \begin{bmatrix} O \ RH_\upsilon \end{bmatrix}$$

to partition the matrix M:

$$M = \begin{bmatrix} M_g \ M_\upsilon \end{bmatrix},$$

where $M_g \in \mathbb{R}^{n \times q_g}$ and $M_\upsilon \in \mathbb{R}^{n \times q_\upsilon}$. Hence, the equation (1.53) can be split into two independent ones:

$$CM_g = Q_g, \quad CM_\upsilon = -RH_\upsilon.$$

Let the matrix Θ_e be partitioned into the blocks:

$$\Theta_e = \begin{bmatrix} \Theta_g \ \Theta_v \end{bmatrix},$$

where $\Theta_g \in \mathbb{R}^{m \times q_g}$ and $\Theta_v \in \mathbb{R}^{m \times q_v}$. Then the Eq. (1.52) can be rewritten in the following partitioned form:

$$B \begin{bmatrix} \Theta_g \ \Theta_v \end{bmatrix} = A \begin{bmatrix} M_g \ M_v \end{bmatrix} - \begin{bmatrix} M_g \ M_v \end{bmatrix} \begin{bmatrix} \Gamma_g & O \\ O & \Gamma_v \end{bmatrix} + D \begin{bmatrix} O \ H_v \end{bmatrix}$$

or

$$\begin{bmatrix} B\Theta_g \ B\Theta_v \end{bmatrix} = \begin{bmatrix} AM_g \ AM_v \end{bmatrix} - \begin{bmatrix} M_g\Gamma_g \ M_v\Gamma_v \end{bmatrix} + \begin{bmatrix} O \ DH_v \end{bmatrix}.$$

Equating the corresponding entries of the matrices on the left-hand side and the right-hand side, we obtain two independent equations. Now, we are in position to formulate the following statement.

Lemma 1.2 *In the case of individual exosystems* (1.64), (1.65), *the Francis equation* (1.52), (1.53) *splits into two independent ones:*

- *for the reference tracking problem*

$$B\Theta_g = AM_g - M_g\Gamma_g, \quad CM_g = Q_g \tag{1.66}$$

with the solution pair (M_g, Θ_g)*;*

- *for the disturbance rejection problem*

$$B\Theta_v = AM_v - M_v\Gamma_v + DH_v, \quad CM_v = -RH_v \tag{1.67}$$

with the solution pair (M_v, Θ_v)*.*

Therefore, the controller (1.55) *takes the form*

$$u = -\Theta_g\xi_g - \Theta_v\xi_v + u_s, \tag{1.68}$$

where $u_t = -\Theta_g\xi_g$ *is the control component responsible for reference tracking, while* $u_r = -\Theta_v\xi_v$ *is the control component responsible for disturbance rejection.*

The latter means that in the case of individual exosystems the problems of reference tracking and disturbance rejection can be solved separately on the basis of independent linear matrix equations (1.66) and (1.67). In this book, we will consider the case of individual exosystems (1.64), (1.65) only.

Example 1.10 Consider again the plant (1.42), (1.43) with the matrices (1.59). Let the reference g and the disturbance v be modelled by the exosystems (1.64) and (1.65) with

$$\Gamma_g = \begin{bmatrix} 0 & 1 & 0 \\ -1 & 0 & 0 \\ 0 & 0 & 0 \end{bmatrix}, \quad Q_g = \begin{bmatrix} 1 & 0 & 1 \\ 0 & 0 & 2 \end{bmatrix}, \quad \Gamma_\upsilon = \begin{bmatrix} 0 & 1 & 0 \\ -9 & 0 & 0 \\ 0 & 0 & 0 \end{bmatrix}, \quad H_\upsilon = \begin{bmatrix} 0 & 0 & 1 \\ 1 & 0 & 2 \end{bmatrix}. \tag{1.69}$$

Hence, it can be shown that

$$g(t) = \begin{bmatrix} g_1(t) \\ g_2(t) \end{bmatrix} = \begin{bmatrix} a_1 + b_1 \sin(t + \varphi_1) \\ 2a_1 \end{bmatrix}, \tag{1.70}$$

and

$$\upsilon(t) = \begin{bmatrix} \upsilon_1(t) \\ \upsilon_2(t) \end{bmatrix} = \begin{bmatrix} a_2 \\ 2a_2 + b_2 \sin(3t + \varphi_2) \end{bmatrix}, \tag{1.71}$$

where the amplitudes a_1, a_2, b_1, b_2 and the phases φ_1, φ_2 are determined by the initial conditions $\xi_g(0)$, $\xi_\upsilon(0)$ as

$$a_1 = \xi_{g3}(0), \quad b_1 = \sqrt{\xi_{g1}^2(0) + \xi_{g2}^2(0)}, \quad \varphi_1 = \arctan\left(\frac{\xi_{g1}(0)}{\xi_{g2}(0)}\right),$$

$$a_2 = \xi_{\upsilon 3}(0), \quad b_2 = \sqrt{\xi_{\upsilon 1}^2(0) + \frac{\xi_{\upsilon 2}^2(0)}{9}}, \quad \varphi_2 = \arctan\left(\frac{3\xi_{g1}(0)}{\xi_{g2}(0)}\right).$$

The solution of the regulation problem for the plant and exosystems considered consists of the matrices

$$\begin{aligned} M_g &= \begin{bmatrix} 1 & 0 & 1 \\ 0 & 1 & 0 \\ -1 & 0 & 1 \end{bmatrix}, \quad \Theta_g = \begin{bmatrix} -5 & -5 & -6 \\ 1 & 1 & -1 \end{bmatrix}, \\ M_\upsilon &= \begin{bmatrix} 0 & 0 & -1 \\ -1 & 0 & -3 \\ -1 & 0 & -1 \end{bmatrix}, \quad \Theta_\upsilon = \begin{bmatrix} 5 & 1 & 21 \\ 2 & 1 & 3 \end{bmatrix}. \end{aligned} \tag{1.72}$$

As in Example 1.9, since A is Hurwitz, the stabilization is unnecessary, and the controller $u = -\Theta_g \xi_g - \Theta_\upsilon \xi_\upsilon$ with the precalculated matrices Θ_g and Θ_υ defined in (1.72) completely rejects the disturbance (1.71) and provides asymptotical tracking of the reference g (1.70) for any initial conditions $x(0)$, $\xi_g(0)$ and $\xi_\upsilon(0)$.

Simulation results obtained for the splitted exosystem and different initial condition are presented in Fig. 1.3 and demonstrate asymptotic tracking independently of the initial conditions $\xi_\upsilon(0)$ and $\xi_g(0)$. ■

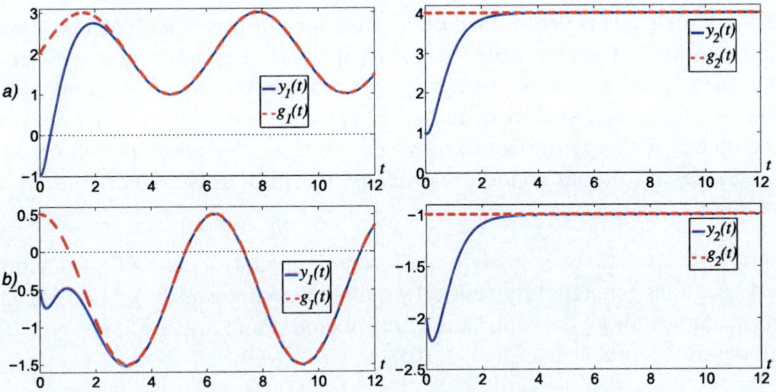

Fig. 1.3 Example 1.10: nonadaptive regulation in the system with parameters (1.59), (1.69), (1.72), $x(0) = [0, 0, 0]^\top$, and: **a** $\xi_g(0) = [0, 1, 2]^\top$, $\xi_\upsilon(0) = [3, 0, -1]^\top$ $(a_1 = 2,\ b_1 = 1,\ \varphi_1 = 0,\ a_2 = -1,\ b_2 = 3,\ \varphi_2 = \frac{\pi}{2})$; **b** $\xi_g(0) = [0, 1, -0.5]^\top$, $\xi_\upsilon(0) = [-1, 0, -0.5]^\top$ $(a_1 = -0.5,\ b_1 = 1,\ \varphi_1 = 0,\ a_2 = -0.5,\ b_2 = 1,\ \varphi_2 = -\frac{\pi}{2})$

1.3.3 Adaptive Output Regulation: Statements of the Problems

As mentioned above, the approach to the output regulation problem presented in Sects. 1.3.1, 1.3.2 has primarily theoretical rather than practical significance since it is based on assumptions that the exosignals are exactly modelled by a linear exosystem, the states of the exosystem and the plant are accessible for measurements, while the parameters of the exosystem and the plant are known a priori. To increase its practical applicability, the setting conditions of the control problem can be extended in several different directions.

First, in most of the practical cases, the states ξ_υ and ξ_g of the exosystems (1.64) and (1.65) as well as the disturbance υ are not available for measurements. This assumption, motivated by a variety of practical applications (see, for example, Sect. 1.5), is typical for *classical* approaches to the output regulation problem in which the parameters of the plant and exosystems are assumed to be known a priori [46, 64, 78, 142]. To cope with such exosignals, when parameters of the plant and exosystems are known, special reduced-order observers can be used to reconstruct the inaccessible state of the exosystem.

Second, in practical applications, the plant state x is often inaccessible for measurements that leads to design of *output-feedback* controllers. Again, in the case of known plant parameters, an output-feedback controller is easily designed with the use of state observers within the classical output regulation framework [64, 78, 142].

Third, a class of the exosignals to be coped with can be essentially extended if we assume that the parameters of the exosystem are unknown, and all prior information about the exosignal waveform is defined by the order of its exosystem. In this case, a class of exosignals is determined by all functions of time which can be generated by an exosystem of known order but with all possible parameters and all possible initial states. This case is the main subject treated in this book. To cope with such a class of exosignals, special disturbance/reference observers will be constructed and combined with appropriate adaptive controllers. This problem will be referred to as the *adaptive output regulation problem*. It will be solved under the following assumption.

Assumption 1.2 The exosignals υ and g are defined as a class of all functions of time that can be generated by neutrally stable exosystems (1.64), (1.65) of known orders q_υ and q_g for all possible parameters of constant matrices Γ_g, Γ_υ, Q_g, H_υ and for all possible initial states $\xi_g(0)$, $\xi_\upsilon(0)$.

A more detailed discussion of exosystem properties will be presented in Sect. 2.1. Here we only mention that the requirement of neutral stability of the exosystems implies, in particular, the boundedness of the exosignals. It is necessary for a rigorous proof of asymptotic properties of the adaptive controllers to be designed.

Fourth, we can further extend a class of systems considered if we assume that plant parameters are also unknown. In this case, an adaptation mechanism must tune the controller to compensate both the plant as well as the exosystem parametric uncertainties.

Fifth, we can try to extend the above treatment to different classes of nonlinear plants as well as to linear plants with input, state and output delays.

Finally, we can consider the most realistic case when the exosignal contains a dominant regular component, which is exactly modelled by a linear exosystem and an irregular component represented by an unknown bounded function of time. Such a situation is described by the following assumption, under which the control problem of *robust adaptive output regulation* will be formulated and solved.

Assumption 1.3 The exosignals (disturbance and reference) are each represented by a sum of a *regular* component satisfying Assumption 1.2 and an *irregular* component given by an unknown bounded function of time.

In this book, we will focus on the adaptive and robust adaptive output regulation problems addressed by the third, fourth, fifth and sixth (final) extensions mentioned above.

Now, we formulate the main control problems to be solved in this book. For methodological purposes, we will first consider the tasks of reference tracking and disturbance compensation separately and later we will combine them into a general problem of output regulation.

Problem 1.2 The control problem of *adaptive reference tracking* is to ensure the boundedness of all signals in the closed-loop disturbance-free system and to achieve the asymptotic tracking of the reference g satisfying Assumption 1.2.

Problem 1.3 The control problem of *adaptive disturbance rejection* is to ensure the boundedness of all signals in the closed-loop system affected by the disturbance υ

satisfying Assumption 1.2 and to achieve the asymptotic convergence of the regulated variables to zero.

Problem 1.4 The control problem of *adaptive output regulation* is to ensure, under Assumption 1.2, the boundedness of all signals in the closed-loop system affected by the disturbance υ and to achieve the asymptotic tracking of the reference g.

Problem 1.5 The control problem of *robust adaptive reference tracking* is to ensure, under Assumption 1.3: (i) the boundedness of all signals in the closed-loop system and a small[8] steady-state tracking error for a small irregular component of the reference; (ii) asymptotic tracking of the reference g when it does not contain an irregular component.

Problem 1.6 The control problem of *robust adaptive disturbance rejection* is to ensure, under Assumption 1.3: (i) the boundedness of all signals in the closed-loop system affected by the disturbance υ; (ii) a small steady-state control output for a small irregular component of the disturbance; (iii) asymptotic convergence of the regulated variables to zero when the disturbance υ does not contain an irregular component.

Problem 1.7 The control problem of *robust adaptive output regulation* is to ensure, under Assumption 1.3: (i) the boundedness of all signals in the closed-loop system affected by the disturbance υ; (ii) a small steady-state regulation error for small irregular components of the disturbance and reference; (iii) asymptotic tracking of the reference g when both the reference and the disturbance do not contain irregular components.

The control Problems 1.2–1.7 will be considered in this book for cases of accessible and inaccessible plant state, for linear and for some classes of nonlinear plants as well as for plants with unknown parameters and known input delay. However, in all the cases, the following common idea will be explored. Instead of the controller (1.68) with the fixed parameters Θ_g, Θ_υ and the measured states ξ_g, ξ_υ, we will apply the *adjustable controller* with the nonadaptive-controller-inspired structure

$$u = -\hat{\Theta}_g \hat{\xi}_g - \hat{\Theta}_\upsilon \hat{\xi}_\upsilon + u_s, \tag{1.73}$$

where $\hat{\Theta}_g$ and $\hat{\Theta}_\upsilon$ are the adjustable parameters generated by an *adaptation algorithm*, while the regressors $\hat{\xi}_g$ and $\hat{\xi}_\upsilon$ are the estimates of inaccessible states generated by a special *reference observer* and a special *disturbance observer*, respectively. Of course, in certain cases of accessible or inaccessible plant states, of known or unknown plant parameters, the structure of an adaptive controller can be significantly more complicated.

[8] Here and hereafter, we use the term *small* in a certain sense which will be defined later. As a rule, we will use the notion of the same order in the mean square sense (see Definition C.8).

1.4 Internal Model and Output Regulation: An Overview

The problem of control in the presence of external signals plays an important role in the modern control theory. Several conceptually different approaches to this problem have been proposed in literature during the last decades. One of them is based on the famous IMP which allows one to efficiently cope with external signals (exosignals) to be rejected or tracked. According to this principle, the reference to be tracked or the disturbance to be rejected is modelled as the output of autonomous dynamic system (exosystem) excited by nonzero initial condition of its state. Then the reference can be asymptotically tracked or the disturbance influence can be completely eliminated if the exosystem model is suitably reduplicated in the structure of the closed-loop system.

Initially the IMP was developed for LTI plants and exosystems with known parameters [26–28, 44–46, 68, 76, 147]. There are two main contributions of this stage of investigations: necessary and sufficient conditions for existence of a robust controller for LTI MIMO systems providing asymptotic tracking/rejection of exosignals generated by a linear exosystem with known parameters [28]; simple algebraic solution of the regulation problem [44]. In this case, the robust controller admits any perturbations of the plant parameters and the closed-loop control system remains stable. It was demonstrated that such a controller (*servocontroller*, in the terminology of those works) must contain stabilizing and compensating (*servocompensating*) components (two control loops). Design of the servocompensating component is based on a solution of the linear matrix equation [44] called *regulator equation* or the *Francis equation*. The properties of the Francis equation were further investigated in [50].

By now, nonadaptive implementation of the IMP with application to LTI MIMO plants is well-developed design tool and constitutes a recognized part of the linear control theory (see corresponding chapters and sections in [46, 64, 78, 142]).

In further works, the problem of output nonadaptive regulation was extended to nonlinear plants and exosystems (see the survey [19] and the monographs [25, 52]). The first solutions for some particular cases were proposed in [29, 66]. In [65], the problem of *local* output regulation was solved for the general class of multidimensional nonlinear plants and nonlinear exosystems. As it was demonstrated in [65], this problem is solvable if and only if a certain nonlinear partial differential equation is solvable that in the linear case is reduced to the linear matrix equation obtained by Francis. For this reason, this equation is called the *nonlinear regulator equation* [19] or the *Francis–Byrnes–Isidori equation* [1]. The solvability of the Francis–Byrnes–Isidori equation is investigated in [59]. To increase practical applicability of nonlinear regulation problem, several methods of approximate solutions of the nonlinear partial differential equation (PDE) yielding *approximate* or *practical* nonlinear output regulation were proposed [1, 54, 55, 73, 86].

In [65], the problem of nonlinear output regulation was solved locally, i.e., for all initial states of the plant and exosystem which are sufficiently close to the origin and, as a results, for small exosignals. The problem of nonlinear regulation for the case of possibly unbounded exosignals with small time derivatives was solved in [55].

The problems of robust regulation applied to nonlinear plants with uncertainties were investigated in [56–58]. Robust semiglobal or regional asymptotic regulation in nonlinear uncertain systems was obtained in [63, 70, 82, 129, 132, 141], while the papers [24, 30, 53, 131, 145] present solutions of global robust output regulation for different classes of nonlinear systems. The particular case of the global robust nonlinear output regulation that admits unbounded exogeneous signals is considered in [23], while the paper [140] investigates the case of MIMO nonlinear systems.

In all the works mentioned above, the exosystem is assumed to be known a priori, and all the uncertainties of the exosignal are caused by unknown initial conditions of the exosystem state. The latter means that we know exactly the exosignal waveform except its amplitude and phase. However, from the practical point of view, it is more realistic to assume that we know a priori only a class of possible disturbances (references) but not their exact waveforms. To cope with uncertain exosignals *adaptive implementation* of the IMP was proposed in [33, 116]. According to this approach, an uncertain exosignal is modelled as the output of a linear exosystem of known order but with unknown parameters, and an appropriate adaptive control is designed for the closed-loop system whose model is augmented by the uncertain exosystem.

During the last decades, the adaptive implementation of the IMP was successfully developed for solving the problems of uncertain reference tracking and uncertain disturbances rejection for linear plants with known [8, 15, 90, 101, 126, 130, 138] and unknown parameters [9–11, 94, 102–104, 108, 146]. A particular case of asymptotically stable but nonminimum phase known LTI SISO plant affected by a single biased sinusoid is considered in [87]. Extension of the adaptive IMP to the case of the exosystem of uncertain order is presented in [89]. A much more general case of uncertain minimum phase LTI SISO plants with known relative degree and the high frequency gain sign affected by an exosystem of unknown order and with unknown parameters is investigated in [92]. The main advantage of the proposed adaptive controller consists in the fact that the upper bound on the exosystem order is not required. If the internal model exactly models the excited frequencies of the actual exosystem, then exponential regulation is achieved. When the internal model undermodels the actual exosystem, the regulation error exponentially converges to a bounded residual set. However, some a priori known bounds are imposed on the plant uncertainties and exosystem frequencies. Extension of the latter approach to the case of nonminimum phase LTI SISO plants is presented in [93]. A particular case of the state derivative feedback (i.e., the case when the first derivative of the state is measured) is considered in [5, 6]. In [6], the problem of adaptive cancellation of unmatched unknown disturbance is solved for a more general class of strict-feedback LTI SISO plants with unknown parameters.

A comprehensive review of different methods (including adaptive IMP) used for rejection of periodic (not only harmonics) disturbances affecting linear plants is presented in [17].

In the works [31, 32, 49, 88], adaptive compensation of uncertain external disturbances was extended to the case of nonlinear plants with exactly known mathematical model. The problem of adaptive compensation of uncertain external disturbance affecting a nonlinear plant with unknown parameters is much more complex and challenging. In one of the pioneering works in this field [105], a nonlinear plant represented in the canonical parametric strict-feedback form with unknown parameters affected by a matched uncertain disturbance was considered. With the use of special disturbance observer, a suitable plant model parameterization was obtained enabling one to design an adaptive regulator for complete compensation of an uncertain disturbance and global stabilization of the closed-loop system. The case of uncertain nonlinear plants transformable into the normal canonical form was considered in [107, 133], while the papers [39, 40] consider the nonlinear plants in strict-feedback canonical form. Different classes of nonlinear systems with measurable output were investigated in [13, 21, 91, 144]. Extension to the case of uncertain nonlinear internal model was presented in [119]. A special class of nonlinear Euler–Lagrange systems was considered in [79, 81].

Recently, the problem of adaptive implementation of the IMP for systems with input delay draws attention of the control community. This is due to the fact that time delay is a common phenomenon observed in many real-world applications (for example, see the survey [128] or the monograph [47] and the references therein). By now, several solutions to the problem of adaptive disturbance rejection in systems with input delay are proposed and use the combination of appropriate prediction techniques with both indirect (identification based) adaptation [12, 14, 16, 120–125, 127] and with direct adaptation [7, 35, 38, 111, 112]. The problem of adaptive reference tracking in plants with input delay was considered in [36, 37].

Therefore, we can conclude that by now there exists a great variety of different nonadaptive as well as adaptive implementations of the IMP that solved many important theoretical as well as practical problems. However, often these solutions look unrelated and sometimes lead to similar results. The main purpose of this book is to give unified and systematic presentation of the adaptive implementation of the IMP applied to the problems of asymptotic rejection and/or tracking of a priori uncertain exogenous signals. The key idea behind the proposed approach is based on a special *canonical form* of the exosystem initially introduced in [101, 102, 105]. As it was demonstrated in subsequent works of different authors, this canonical form enables one to design, in a unified and systematic way, appropriate observers of exosignals and obtain appropriate parameterizations of external signals in the form of linear regressions with unknown constant parameters and measured regressors [109, 110].[9] It is worth noting that such parameterizations can be obtained using a unified and systematic way for different classes of linear and nonlinear plants, plants with measured state or output as well as for plants with unknown parameters. In own turn, such parameterizations enable one to employ well-known *basic* algorithms of adaptation and efficiently solve adaptive regulation problems for different classes of control systems.

In the next chapter, we introduce the *exosystem canonical form* and present a unified approach to design of observers for a class of regular uncertain exosignals considered in this book. In Chap. 3, on the basis of the plant parameterizations constructed in Chap. 2, we define *canonical error models* and derive for them *basic*

[9] Alternative approaches to design of the disturbance observers is presented in the survey [2, 3].

algorithms of adaptation. In subsequent chapters, with the use of these exosignal, observers and the basic algorithms of adaptation, we present solutions of adaptive and robust adaptive regulation problems for different classes of control systems.

1.5 Internal Model and Output Regulation: Practical Implementations

1.5.1 Vibration Control

Whether natural or artificial, vibrations are omnipresent and can be caused, for example, by seismic disturbances, acoustic signals, transport systems, different rotating mechanisms of pumps, mixers, engines, electrical drives, mills, conveyors, etc. As a rule, vibrations cause destructive effect on the environment, and, therefore, in such applications as industrial systems, transport or construction it is mandatory to provide vibroprotection. Usually, for vibroprotection passive elements (spring or fluid dampers, elastic materials) are used. However, for more efficient compensation (if possible by construction), it is necessary to provide active vibration control that, by applying a force, can counteract the vibration signal.

A system of mechanic vibrocompensation can be presented by the general scheme depicted in Fig. 1.4[10] and consists of a mechanical system affected by the vibration signal and the loop of active vibration control producing the counteracting force. Usually, in practice, the loop includes sensors detecting local displacements of the mechanical system (or/and velocities, or/and accelerations), a microcontroller and actuators generating local forces to reduce the vibrations over the whole system.

For more efficient compensation, the vibration force $\upsilon_\delta(t)$ can be represented as

$$\upsilon_\delta = \upsilon + \delta,$$

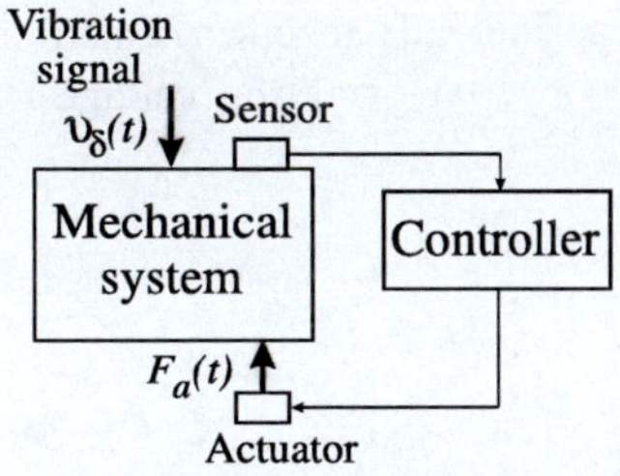

Fig. 1.4 General scheme of vibration control system

[10] For more details about numerous constructions of vibration control systems and their models, see the books [32, 77, 148].

where $\upsilon(t)=\sum_{i=1}^{N}\beta_i\sin(w_i t+\phi_i)$ is the regular component with unknown parameters β_i, w_i, ϕ_i and a number of harmonics N, δ is the irregular component. The regular component can be a result of spectral analysis of the signal υ_δ in which a number of dominating harmonics N is determined. Both components are assumed inaccessible for measurement.

For SISO control, the mechanical system response can be represented (under some assumptions) as the output of the linear plant model [32, 77]

$$y = W(s)\,[-u + \upsilon_\delta]\,, \tag{1.74}$$

where y is the mechanical system response, $u = F_a$ is the actuator force, $W(s)$ is the transfer function of the mechanical system.

The problem of vibrocompensation can be formulated as follows: to design a control that provides the boundedness of all the signals in the compensation system and drives the output y to a residual set with a small radius for small δ. If $\delta(t)\equiv 0$, the control must ensure the convergence of $y(t)$ to zero.

Some solutions to particular problems of adaptive vibrocompensation are proposed for mechanical benchmarks [77] (see also references to the preceding authors' papers therein), [72], hard disk and Blu-ray drives [80, 95, 142], seismic protection systems [18], engines [113, 114], rolling mills [75, 134].

1.5.2 *Ripple Reduction in AC–DC Converters*

Another problem of disturbance compensation that in the scope of this book is in ripple reduction in pulse width modulation (PWM) AC–DC converters widely used in power systems.

To formulate the problem of control, we consider the typical scheme of an AC–DC converter presented in Fig. 1.5. The scheme consists of a diode rectifier and a *boost converter* with a PWM device. The rectifier together with the filtering capacity C_1 periodically reverses direction to get the voltage $u_g(t)$ of direct current. The boost converter using the voltage $u_g(t)$ and the modulating signal from a PWM device

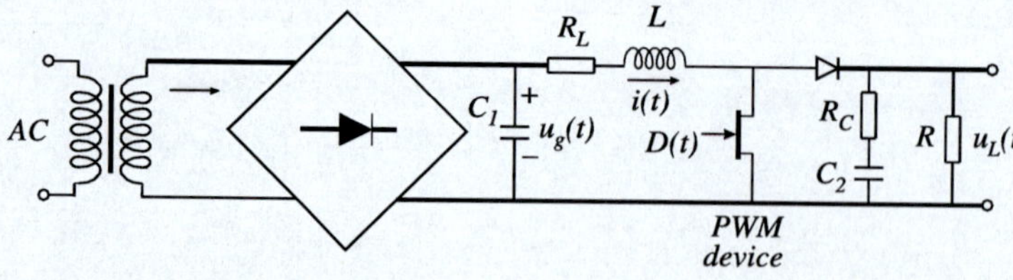

Fig. 1.5 Scheme of AC–DC converter

(e.g., a transistor) alternates between the values u_g and 0 at a switching frequency. This pulse waveform is low-pass-filtered with the capacitors C_1, C_2 and the inductor L. Controlling the duty ratio of the switch, i.e., providing a suitable modulation, we can control the DC component of this waveform. Note that the boost converter provides the output voltage $u_L = \frac{1}{1-D}u_g$, where D is the control signal – duty ratio (see [4, Chap. 2], [98]). Since $0 < D < 1$, the output voltage u_L is higher than u_g.

One of the typical obstacles in this control problem is ripple caused by incomplete compensation of the alternating voltage waveform due to nonlinear devices (transformer, diodes and a PWM device) and unmodelled dynamics of the rectifier circuits. As a result, the output DC voltage can contain undesirable residual variations that additionally heat DC circuit components and make distortions in such DC consumers as systems of audio and video signals processing. These variations contain the fundamental frequency of the AC source and additional harmonics — the result of nonlinear transformations in the circuit including PWM. In other words, the disturbance in the boost converter can be considered as a periodic signal that is additive to some averaged value of the voltage u_g and can be represented via a Fourier series with a truncated number of harmonics, that is,

$$u_g(t) = \bar{u}_g + \sum_{i=1}^{N} \beta_i \sin(2iwt + \phi_i), \tag{1.75}$$

where $\bar{u}_g$ is the constant value of the averaged input voltage $u_g(t)$, w is the fundamental frequency, β_i, ϕ_i are the amplitudes and phases of harmonics, $N \in \mathbb{N}$ is a number of harmonics. The parameters β_i, ϕ_i depend on the duty cycle of PWM device and characteristics of nonlinear components of the rectifier circuits. Since this dependence is very complex, these parameters can be assumed uncertain.

In order to design a ripple reduction controller, following to [98], we use the variables averaged over PWM cycles and introduce the control variation $\tilde{u} = u - u_{ss}$ and the perturbations notations $\tilde{x} = x - x_{ss}$, $\tilde{y} = y - y_{ss}$, where $u = D$, $x = [i, u_L]^\top$, $y = u_L$, u_{ss} is the steady-state constant duty ratio, x_{ss} and y_{ss} are the steady-state constant values of x and the output voltage u_L produced by u_{ss} when $u_g(t) \equiv \bar{u}_g$.

Perturbation nonlinear state-space model oriented to the ripple reduction can be represented in the form [98, 118]

$$\dot{\tilde{x}} = A\tilde{x} + A_u\tilde{x}\tilde{u} + b\tilde{u} + dv, \tag{1.76}$$

$$\tilde{y} = c^\top \tilde{x}, \tag{1.77}$$

where the matrices A, A_u and the vectors b and d have appropriate dimensions and depend on the constant circuit parameters R_L, R_C, L, C_2 and R, the vector c is given by $c = [0, 1]^\top$, while $v(t) = u_g(t) - \bar{u}_g$ is considered as a multisinusoidal disturbance to be compensated.

Note that the steady-state model inputs and the state can be calculated based on the desired output and the straightforward model analysis [98].

Thus, the problem is to design a control $\tilde{u}$ ensuring the boundedness of all the signals in the system and driving the perturbation $\tilde{y}$ to zero despite the influence of the disturbance υ

1.5.3 Ripple Reduction in Permanent Magnet Synchronous Motors

Permanent magnet synchronous motors (PMSM) are among the most common motors in industry, especially in high performance applications, due to low friction looses and low cost achieved by construction without rotor winding, brushes and slip rings on the rotor. New (however, rare-earth) magnetic materials with high flux densities enable to increase power characteristics of motors. At the same time, PMSM suffer from a drawback related to the ripple of the torque caused by the stator nonideally distributed rotor magnet flux and current magnetomotive forces as well as the deficiencies of the stator winding geometry and nonuniformity of the airgap [51, 60, 137].

Besides the development of motors construction (see [60, 137]), the ripple can be reduced by appropriate design of compensating/suppressing control based on physical models [34, 83, 84, 117, 139]. One of typical control-oriented $d-q$ models is given by the differential equations [34, 83, 84, 137]

$$L_d\dot{i}_d = -Ri_d + \omega_e L_q i_q + u_d, \tag{1.78}$$

$$L_q\dot{i}_q = -Ri_q - \omega_e\left(L_d i_d + \Psi_m\right) + u_q, \tag{1.79}$$

$$J\dot{\omega} = \tau_m - c\omega - \tau_r(q_m) - \tau_l, \tag{1.80}$$

$$\tau_m = \frac{3}{2}p\Psi_m i_q + (L_d - L_q)i_d i_q, \tag{1.81}$$

$$\dot{q}_m = \omega_e, \tag{1.82}$$

where L_d and L_q are the $d-q$ frame stator inductances, R is the stator winding resistance, i_d and i_q are the $d-q$ frame stator currents, u_d and u_q are the $d-q$ frame stator voltages, p is the number of pole pairs, ω_e and ω are the electrical and mechanical rotor speeds, respectively ($\omega_e = \frac{p}{2}\omega$), Ψ_m is the magnetic flux, c is the friction coefficient, τ_m is the electro-mechanical torque, τ_r is the ripple torque, q_m is the electrical angle, and τ_l is the torque of external loads.

The ripple torque represents the disturbance that can be given by the decomposition (see [84])

$$\tau_r(q_m) = \sum_{i=1}^{\infty} \beta_i \cos(iq_m + \phi_i),$$

where β_i are ϕ_i the amplitudes and phases, respectively. Since the dependence of τ_r from q_m is complex, the parameters β_i are ϕ_i are assumed unknown. If ω_e is

Fig. 1.6 Scheme of a vessel crane and the notations. (Adapted from [97], with permission from Elsevier)

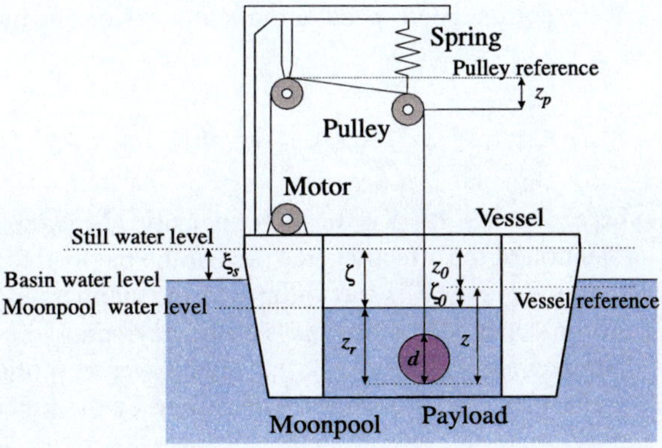

assumed to be a slowly time-varying function, the disturbance can be considered as an uncertain periodic signal of the form (1.37) .

The problem of ripple compensation can be formulated as follows: to design a control $u = [u_d, u_q]^\top$ that will stabilize the speed ω at some constant reference value with all closed-loop signals bounded.

1.5.4 Crane Vessel Control

Nowadays, the high-precision and safe underwater positioning of marine equipment is becoming more and more attractive due to the developing oil and natural gas industry, growing interest in ocean science, marine rescue and research works. The problem of positioning is carried out using cranes installed on vessels and equipped by controlled mechanisms. One of the widespread control problems that is in the focus of control engineers is the crane payload stabilization in the presence of waves.

As it was established by the experiments published in [42, 67] (see also [96]), a wave disturbance can be represented as a multisinusoidal function (1.37) with some additive irregular component.

Consider the typical scheme of a crane vessel depicted in Fig. 1.6. According to this scheme, the motion of the vessel is described by the following equations [67, 97]:

$$\ddot{z} = -\frac{k_p}{m}(z_m - z) - \frac{m_p + m}{m}\ddot{z}_0 + \frac{1}{m}f_z + g, \tag{1.83}$$

$$\dot{z}_m = u, \tag{1.84}$$

where m is the payload mass, m_p is the pulley mass, g is the gravity acceleration, f_z is the hydrodynamic force on the payload in the moonpool, k_p is the spring stiffness,

u is the commanded speed of the motor drive. The hydrodynamic force is calculated as

$$f_z = -\frac{1}{2}\rho C_D A_{pz}(z_r)\dot{z}_r|\dot{z}_r| - d_l(z_r)\dot{z}_r - \rho g \nabla(z_r, d) - \frac{\partial Z_{\ddot{z}_r}}{\partial z_r}(z_r)\dot{z}_r^2, \tag{1.85}$$

where $z_r = z - \zeta_0$, ρ is the water density, C_D is the drag coefficient, $A_{pz}(z_r)$ is the projection of the effective drag area of the payload to the vertical axis, d_l is the linear drag coefficient, ∇ is the volume of the submerged part of the payload, $Z_{\ddot{z}_r}$ is the position dependent added mass of the payload.

In the model (1.83)–(1.85), the heave vessel motion z_0 and the wave elevation ζ_0 are considered as disturbances described by the functions (see [96, 97])

$$z_0 = \sum_{i=1}^{N} \beta_{zi} \sin(w_{zi}t + \phi_{zi}),$$
$$\zeta_0 = \sum_{i=1}^{N} \beta_{\zeta i} \sin(w_{\zeta i}t + \phi_{\zeta i}),$$

where β_{zi}, $\beta_{\zeta i}$, w_{zi}, $w_{\zeta i}$, ϕ_{zi}, $\phi_{\zeta i}$ are some a priori unknown constants.

The problem of compensation is to design a control u that, despite the influence of disturbances, provides:

- if the payload is in the air and far enough from the moonpool, stabilization of the velocity $\dot{z} + \dot{z}_0$ at some desired level, say v_{des} (heave compensation);
- if the payload approaches the moonpool, stabilization of the velocity $\dot{z}_r$ at the level v_{des} (wave synchronization).

References

1. Aguilar, C.O., Krener, A.J.: Patchy solution of a Francis-Byrnes-Isidori partial differential equation. Int. J. of Robust Nonlinear Control **23**(9), 1046–1061 (2013)
2. Andrievsky, B.R., Furtat, I.B.: Disturbance observers: methods and applications. I. Methods. Autom. Remote Control **81**(9), 156–1610 (2020)
3. Andrievsky, B.R., Furtat, I.B.: Disturbance observers: methods and applications. II. Applications. Autom. Remote Control **81**(10), 177–1818 (2020)
4. Banerjee, S., Verghese, G.C.: Nonlinear Phenomena in Power Electronics: Bifurcations, Attractors, Chaos, and Nonlinear Control. Wiley-IEEE Press (2001)
5. Basturk, H.I., Krstić, M.: Adaptive cancelation of matched unknown sinusoidal disturbances for LTI systems by state derivative feedback. J. Dyn. Syst. Meas. Control **135**(1) (2013)
6. Basturk, H.I., Krstić, M.: State derivative feedback for adaptive cancellation of unmatched disturbances in unknown strict-feedback LTI systems. Automatica **50**, 2539–2545 (2014)
7. Basturk, H.I., Krstić, M.: Adaptive sinusoidal disturbance cancellation for unknown LTI systems despite input delay. Automatica **58**, 131–138 (2015)
8. Bobtsov, A.A.: An algorithm of robust output-based control of linear object with compensation of unknown deterministic disturbance. J. Comput. Syst. Sci. Int. **42**(2), 251–255 (2003)

9. Bobtsov, A.A.: Output control algorithm with the compensation of biased harmonic disturbances. Autom. Remote. Control. **69**(8), 1289–1296 (2008)
10. Bobtsov, A.A.: Adaptive output control with compensation of Biased harmonic disturbance. J. Comput. Syst. Sci. Int. **48**(1), 41–44 (2009)
11. Bobtsov, A.A., Kolubin, S.A., Kremlev, A.S., Pyrkin, A.A.: An iterative algorithm of adaptive output control with complete compensation for unknown sinusoidal disturbance. Autom. Remote Control 1327–1363 (2012)
12. Bobtsov, A.A., Kolyubin, S.A., Pyrkin, A.A.: Stabilization of a nonlinear plant with input delay and sinusoidal perturbation. Autom. Remote. Control. **76**(1), 16–23 (2015)
13. Bobtsov, A.A., Kremlev, A.S., Pyrkin, A.A.: Compensation of harmonic disturbances in nonlinear plants with parametric and functional uncertainty. Autom. Remote. Control. **72**(1), 111–118 (2011)
14. Bobtsov, A.A., Pyrkin, A.A.: The compensation of a harmonic perturbation under conditions of a delay in control. J. Comput. Syst. Sci. Int. **47**(4), 513–517 (2008)
15. Bobtsov, A.A., Pyrkin, A.A.: Compensation of unknown sinusoidal disturbances in linear plants of arbitrary relative degree. Autom. Remote. Control. **70**, 449–456 (2009)
16. Bobtsov, A.A., Pyrkin, A.A.: Cancelation of unknown multiharmonic disturbance for nonlinear plant with input delay. Int. J. Adapt. Control Signal Process. **26**, 302–315 (2012)
17. Bodson, M.: Rejection of periodic disturbances of unknown and time-varying frequency. Int. J. Adapt. Control Signal Process. **19**, 67–88 (2005)
18. Burdisso, R.A., Fuller, C.R., Suarez, L.E.: Adaptive feedforward control of structures subjected to seismic excitations. In: Proceedings of the American Control Conference, San Francisco, California, pp. 2104–2108 (1993)
19. Byrnes, C.I., Isidori, A.: Output regulation for nonlinear systems: an overview. Int. J. Robust and Nonlinear Control **10**, 323–337 (2000)
20. Chen, C.-T.: Linear Systems. Theory and Design, 3rd edn. Oxford University Press, NY, Oxford (1999)
21. Chen, J., Ding, Z.: Adaptive output regulation of output feedback systems with unknown nonlinear exosystems and unknown high-frequency gain sign. In: Proceedings of American Control Conference, Milwaukee, USA, pp. 4496–4501 (2018)
22. Chen, X., Tomizuka, M.: A Minimum parameter adaptive approach for rejecting multiple narrow-band disturbances with application to hard disk drives. IEEE Trans. Control Syst. Technol. **20**(2) (2012)
23. Chen, Z., Huang, J.: A general formulation and solvability of the global robust output regulation problem. IEEE Trans. Automatic Control **50**(4), 448–462
24. Chen, Z., Huang, J.: Uniform practical nonlinear output regulation. Asian J. Control **9**(3), 292–305
25. Chen, Z., Huang, J.: Stabilization and Regulation of Nonlinear Systems. Springer (2015)
26. Davison, E.D.: The output control of linear time-invariant multivariable systems with unmeasurable arbitrary disturbances. IEEE Trans. Autom. Control **17**, 621–630 (1972)
27. Davison, E.D.: A generalization of the output control of linear time-invariant multivariable systems with unmeasurable arbitrary disturbances. IEEE Trans. Autom. Control **20**, 788–792 (1972)
28. Davison, E.D.: The robust control of a servomechanism problem for linear time-invariant multivariable systems. IEEE Trans. Autom. Control **21**(1) (1976)
29. Di Benedetto, M.D.: Synthesis of an internal model for nonlinear output regulation. Int. J. Control **45**, 1023–1034 (1987)
30. Ding, Z.: Global output regulation of uncertain nonlinear systems with exogeneous signals. Automatica **37**, 113–119 (2001)
31. Ding, Z.: Universal disturbance rejection for nonlinear systems in output feedback form. IEEE Trans. Autom. Control **48**(7), 1222–1227 (2003)
32. Ding, Z.: Global stabilization and disturbance suppression of a class of nonlinear systems with uncertain internal model. Automatica **39**, 471–479 (2003)

33. Eliott, H., Goodwin, G.C.: Adaptive implementation of the internal model principle. In: Proceedings of 23rd Conference on Decision and Control, Las Vegas, USA, pp. 1292–1297 (1984)
34. Gan, W., Qui, L.: Torque and velocity ripple elimination of AC permanent magnet motor control systems using the internal model principle. IEEE/ASME Trans. Mechatron. **9**, 436–447 (2004)
35. Gerasimov, D.N., Paramonov, A.V., Nikiforov, V.O.: Adaptive disturbance compensation in delayed linear systems with arbitrary delay: internal model approach. Sci. Tech. J. Inf. Technol. Mech. Opt. (in Russian), **16**(6), 1023–1030 (2016)
36. Gerasimov, D.N., Miliushin, A.S., Nikiforov, V.O.: Algorithms of adaptive tracking of unknown multisinusoidal signals in linear systems with arbitrary input delay. Int. J. Adapt. Control Signal Process. **33**(6), 900–912 (2019)
37. Gerasimov, D.N., Paramonov, A.V., Nikiforov, V.O.: Adaptive tracking of unknown multisinusoidal signal in linear systems with arbitrary input delays and unknown sign of high frequency gain. IFAC-PapersOnLine **50**(1), 7052–7057 (2017)
38. Gerasimov, D.N., Paramonov, A.V., Nikiforov, V.O.: Algorithms of fast adaptive compensation of disturbance in linear systems with arbitrary input delay. Int. J. Control **93**(7), 1596–1604 (2020)
39. Gerasimov, D.N., Pashenko, A.V., Nikiforov, V.O.: Improved adaptive compensation of unmatched multisinusoidal disturbances in uncertain nonlinear plants. In: Proceedings of American Control Conference, Denver, USA, pp. 626–632 (2020)
40. Gerasimov, D.N., Pashenko, A.V., Nikiforov, V.O.: Improved adaptive servotracking for a class of nonlinear plants with unmatched uncertainties. IFAC-PapersOnLine **53**(2), 3835–3840 (2020)
41. Fairman, F.W.: Linear Control theory. The State Space approach. Wiley, Chichester, N.Y., Weinheim, Brisbane, Singapore, Toronto (1998)
42. Fossen, T.I.: Marine Control Systems: Guidance, Navigation and Control of Ships, Rigs and Underwater Vehicles. Marine Cybernetics, Trondheim (2002)
43. Fradkov, A.L., Miroshnik, I.V., Nikiforov, V.O.: Nonlinear and Adaptive Control of Complex Systems. Kluwer Academic Publishers (1999)
44. Francis, B.A.: The linear multivariable regulator problem. SIAMJ Control Optim. **14**, 48–505 (1976)
45. Francis, B.A., Sebakhy, O.A., Wonham, W.M.: Synthesis of multivariable regulators: the internal model principle. Appl. Math. Optim. **1**(1), 64–86 (1974)
46. Francis, B.A., Wonham, W.M.: The internal model principle for linear multivariable regulators. Appl. Math. Optim. **2**, 170–194 (1975)
47. Fridman, E.: Introduction to Time-Delay Systems. Springer International Publishing Switzerland (2014)
48. Fuller, C., Elliott, S., Nelson, P.: Active Control of Vibration. Academic, Cambridge (1999)
49. Han, J., Zhang, H., Wang, Y., Liu, X.: Anti-disturbance control for nonlinear system via adaptive disturbance observer. Int. J. Robust Nonlinear Control **27**, 2121–2144 (2017)
50. Hautus, M.: Linear matrix equations with applications to the regulator problem. In: Landau, I.D. (ed.) Qutils and Modeles Mathematique pour l'Automatique, Paris: C.N.R.S., pp. 399–412 (1983)
51. Holz, J., Springob, L.: Identification and compensation of torque ripple in high-precision permanent magnet motor drives. IEEE Trans. Ind. Electron. **43**, 309–320 (1996)
52. Huang, J.: Nonlinear Output Regulation: Theory and Applications. SIAM (2004)
53. Huang, J., Chen, Z.: A general framework for tackling the output regulation problem. IEEE Trans. Autom. Control **49**(12), 2203–2218 (2004)
54. Huang, J., Rugh, W.J.: An approximation method for the nonlinear servomechanism problem. IEEE Trans. Autom. Control **37**(7), 1009–1013 (1992)
55. Huang, J., Rugh, W.J.: Stabilization on zero-error manifolds and the nonlinear servomechanism problem. IEEE Trans. Autom. Control **37**, 1395–1398 (1992)

56. Huang, J., Rugh, W.J.: On a robust nonlinear servomechanism problem. IEEE Trans. Autom. Control **39**(7), 1510–1513 (1994)
57. Huang, J., Rugh, W.J.: Asymptotic tracking and disturbance rejection in uncertain nonlinear systems. IEEE Trans. Autom. Control **40**(6), 1118–1122 (1995)
58. Huang, J.: Remarks on the output regulation problem for nonlinear systems. IEEE Trans. Autom. Control **46**(12), 2028–2031 (2001)
59. Huang, J.: On the solvability of the regulator equations for a class of nonlinear systems. IEEE Trans. Autom. Control **48**(5), 880–885 (2003)
60. Jahns, T.M., Soong, W.L.: Pulsating torque minimization techniques for permanent magnet AC motor drives - a review. IEEE Trans. Ind. Electron. **43**, 321–330 (1996)
61. Ioannou, P., Kokotović, P.: Instability analysis and improvement of robustness of adaptive control. Automatica **20**(5), 583–594 (1984)
62. Ioannou, P., Sun, J.: Robust Adaptive Control. Prentice-Hall, NJ (1996)
63. Isidori, A.: A remark on the problem of semiglobal nonlinear output regulation. IEEE Trans. Autom. Control **42**(12), 1734–1738
64. Isidori, A.: Lectures in Feedback Design for Multivariable Systems. Springer International Publishing (2017)
65. Isidori, A., Byrnes, C.I.: Output regulation of nonlinear systems. IEEE Trans. Autom. Control **35**(2), 131–140 (1990)
66. Jie, H., Rugh, W.J.: On a nonlinear multivariable servomechanism problem, Tech. Rep. JHU/ECE-88/04.2 (1988)
67. Johansen, T.A., Fossen, T.I., Sagatun, S.I., Nielsen, F.G.: Wave synchronizing crane control during water entry in offshore moonpool operations-experimental results. IEEE J. Ocean. Eng. **28**(4) (2003)
68. Johnson, C.D.: Accommodation of external disturbances in linear regulator and servomechanism problem. IEEE Trans. Autom. Control **16**, 635–644 (1971)
69. Kailath, T.: Linear Systems. Prentice-Hall, Englewood Cliffs, NJ (1980)
70. Khalil, H.K.: On the design of robust servomechanisms for minimum phase nonlinear systems. Int. J. Robust Nonlinear Control **10**, 339–361 (2000)
71. Khalil, H.K.: Nonlinear Control. Prentice-Hall, NJ (2015)
72. Kinney, C.E., Fang, H., de Callafon, R.A., Alma, M.: Robust estimation and automatic controller tuning in vibration control of time varying harmonic disturbances. In: Proceedings of 18th IFAC World Congress, Milano, Italy, pp. 5401–5406 (2011)
73. Krener, A.J.: The construction of optimal linear and nonlinear regulators. In: Isidori, A., Tarn, T.J. (eds.) Systems, Models and Feedback: Theory and Applications, pp. 301–322. Birkhauser, Boston (1992)
74. Krstić, M., Kanellakopoulos, I., Kokotović, P.: Nonlinear and Adaptive Control Design. Wiley, NY (1995)
75. Kugi, A., Hass, W., Schlacher, K., Aistleitner, K., Frank, H., Günter, W.R.: Active compensation of roll eccentricity in rolling mills. IEEE Trans. Ind. Appl. **36**, 625–632 (2000)
76. Kwatny, H.G., Kalnitsky, K.C.: On alternative methodologies for the design of robust linear multivariable regulators. IEEE Trans. Autom. Control **23**(5), 930–933 (1978)
77. Landau, I.D., Airimitoaie, T.-B., Castellanos-Silva, A., Constantinescu, A.: Adaptive and Robust Active Vibration Control: Methodology and Tests. Springer (2017)
78. Lewin, W.S. (ed.), Control Handbook. (In two volumes). Jaico Publishing House, CRC Press in Cooperation with IEEE Press (1999)
79. Li, J., Du, J.: Adaptive disturbance cancellation for a class of MIMO nonlinear Euler-Lagrange systems under input saturation. ISA Trans. **96**, 14–23 (2020)
80. Liu, J.-J., Yang, Y.-P.: Stability of the frequency adaptive control technique and its application to compact disk drives. Control. Eng. Pract. **13**(5), 629–639 (2005)
81. Lu, M., Liu, L., Feng, G.: Adaptive tracking of uncertain Euler-Lagrange systems subjected to external disturbances. Automatica **104**, 207–219 (2019)
82. Mahmoud, N., Khalil, H.K.: Robust control for a nonlinear servomechanism problem. Int. J. Control **66**(6), 779–802 (1997)

83. Mandel, Y., Weiss, G.: Adaptive internal model-based suppression of torque ripple in brushless DC motor drives. Syst. Sci. Control Eng.: An Open Access J. **3**(1), 162–176 (2015)
84. Maier, S., Bals, J., Bodson, M.: Periodic disturbance rejection of a PMSM with adaptive control algorithms. In: Proceedings of the IEEE International Electric Machines and Drives Conference, pp. 1070–1075 (2011)
85. Marino, R., Tomei, P.: Nonlinear Control Design - Geometric, Adaptive and Robust. Prentice-Hall, Hemel Hempstead (1995)
86. Marconi, L., Praly, L.: Uniform practical nonlinear output regulation. IEEE Trans. Autom. Control **53**(5), 1184–1202 (2008)
87. Marino, R., Santosuosso, G.L., Tomei, P.: Robust adaptive compensation of biased sinusoidal disturbances with unknown frequency. Automatica **39**, 1755–1761 (2003)
88. Marino, R., Santosuosso, G.L.: Global compensation of unknown sinusoidal disturbances for a class of nonlinear nonminimum phase systems. IEEE Trans. Autom. Control **50**(11), 1816–1822 (2005)
89. Marino, R., Santosuosso, G.L.: Regulation of linear systems with unknown exosystems of uncertain order. IEEE Trans. Autom. Control **52**(2), 352–359 (2007)
90. Marino, R., Tomei, P.: Output regulation for linear systems via adaptive internal model. IEEE Trans. Autom. Control **48**, 2199–2202 (2003)
91. Marino, R., Tomei, P.: Adaptive tracking and disturbance rejection for uncertain nonlinear systems. IEEE Trans. Autom. Control **50**(1), 90–95 (2005)
92. Marino, R., Tomei, P.: An adaptive learning regulator for uncertain minimum phase systems with undermodeled unknown exosystems. Automatica **47**, 739–747 (2011)
93. Marino, R., Tomei, P.: Disturbance cancellation for linear systems by adaptive internal models. Automatica **49**, 1494–1500 (2013)
94. Marino, R., Tomei, P.: Output regulation for unknown stable systems. IEEE Trans. Autom. Control **60**(8), 2213–2218 (2015)
95. Martinez, J.J., Alma, M.: Improving playability of Blu-ray disc drives by using adaptive suppression of repetitive disturbances. Automatica **48**, 638–644 (2012)
96. Messineo, S., Celani, F., Egeland, O.: Crane feedback control in offshore moonpool operations. Control. Eng. Pract. **16**, 356–364 (2008)
97. Messineo, S., Serrani, A.: Offshore crane control based on adaptive external models. Automatica **9**, 2546–2556 (2009)
98. Middlebrook, R.D., Ćuk, S.: A general unified approach to modelling switching converter power stages. Int. J. Electron. **42**(6), 521–550 (1977)
99. Narendra, K.S., Annaswamy, A.M.: Stable Adaptive Systems. Prentice Hall (1989)
100. Narendra, K.S., Tripathi, S.S., Luders, G., Kudva, P.: Adaptive Control Using Lyapunov' s Direct Method, Technical Report No. CT-43, Becton Center, Yale University, New Haven, CT (1971)
101. Nikiforov, V.O.: Adaptive servocompensation of input disturbances. In: Proceedings of 13th World Congress of IFAC, San Francisco, USA, vol. 29, no. 1, pp. 5114–5119 (1996)
102. Nikiforov, V.O.: Adaptive servomechanism controller with an implicit reference model. Int. J. Control **68**(2), 277–286 (1997)
103. Nikiforov, V.O.: Adaptive stabilization of a linear plant subjected to external deterministic disturbances. J. Comput. Syst. Sci. Int. **36**(2), 258–261 (1997)
104. Nikiforov, V.O.: Adaptive controller rejecting uncertain deterministic disturbances in SISO systems. In: Proceedings of 4th European Control Conference, Brussels, Belgium, pp. 1207–1212 (1997)
105. Nikiforov, V.O.: Adaptive nonlinear tracking with complete compensation of unknown disturbances. Eur. J. Control **4**, 132–139 (1998)
106. Nikiforov, V.O.: Robust high-order tuner of simplified structure. Automatica **35**(8), 1409–1415 (1999)
107. Nikiforov, V.O.: Nonlinear servocompensation of unknown external disturbances. Automatica **37**(10), 1647–1653 (2001)

108. Nikiforov, V.O.: Adaptive and Robust Control with Disturbance Compensation. Nauka, St. Petersburg (2003).(in Russian)
109. Nikiforov, V.O.: Observers of external deterministic disturbances. Part I. Objects with known parameters. Autom. Remote Control **65**(10), 1531–1541 (2004)
110. Nikiforov, V.O.: Observers of external deterministic disturbances. Part II. Objects with unknown parameters. Autom. Remote Control **65**(11), 1724–1732 (2004)
111. Nikiforov, V.O., Paramonov, A.V., Gerasimov, D.N.: Adaptive control algorithms in MIMO linear systems with control delay. Autom. Remote. Control. **81**(6), 1091–1106 (2020)
112. Nikiforov, V.O., Paramonov, A.V., Gerasimov, D.N.: Adaptive servocompensation of unmatched disturbances in MIMO LTI plants with input delay. In: Proceedings of American Control Conference, New Orleans, USA, pp. 2426–2431 (2021)
113. Njeh, M., Cauet, S., Coirault, P.: Persistent disturbances rejection on internal combustion engine torque in hybrid electric vehicles. In: Proceedings of the 49th Conference on Decision and Control, Atlanta, pp. 6421–6426 (2010)
114. Orivuori, J., Zazas, I., Daley, S.: Active control of frequency varying disturbances in a diesel engine. Control. Eng. Pract. **20**, 1206–1219 (2012)
115. Ortega, R.: On Morse's new adaptive controller: parameter convergence and transient performance. IEEE Trans. Autom. Control **38**(8), 1191–1202 (1993)
116. Palaniswami, M., Goodwin, G.C.: An adaptive implementation of the internal model principle. In: Proceedings of American Control Conference, Minneapolis, USA (1987)
117. Petrovic, V., Ortega, R., Stankovic, A.M., Tadmor, G.: Design and implementation of an adaptive controller for torque ripple minimization in PM synchronous motors. IEEE Trans. Power Electron. **15**, 871–880 (2000)
118. Pham, H., Jung, H., Hu, T.: State-space approach to modeling and ripple reduction in AC-DC converters. IEEE Trans. Control Syst. Technol. **21**(5), 1949–1955 (2013)
119. Priscoli, F.D., Marconi, L., Isidori, A.: A new approach to adaptive nonlinear regulation. SIAM J. Control. Optim. **45**(3), 829–855 (2006)
120. Pyrkin, A.A.: Adaptive algorithm to compensate parametrically uncertain biased disturbance of a linear plant with delay in the control channel. Autom. Remote. Control. **71**(8), 1562–1577 (2010)
121. Pyrkin, A.A., Bobtsov, A.A.: Adaptive controller for linear system with input delay and output disturbance. IEEE Trans. Autom. Control **61**(12), 4229–4234 (2016)
122. Pyrkin, A.A., Bobtsov, A.A., Aranovskiy, S.V., Kolyubin, S.A., Gromov, V.S.: Adaptive Controller for Linear Plant with Parametric Uncertainties, Input delay and unknown disturbance. In: Proceedings of 19th IFAC World Congress, Cape Town, South Africa (2014)
123. Pyrkin, A.A., Bobtsov, A.A., Kolyubin, S.A., Vedyakov, A.A., Borisov, O.I., Gromov, V.S.: Stabilization of nonlinear system with input delay and biased sinusoidal disturbance. In: Proceedings of 19th IFAC World Congress, Cape Town, South Africa (2014)
124. Pyrkin, A.A., Bobtsov, A.A., Nikiforov, V.O., Vedyakov, A.A., Kolyubin, S.A., Borisov, O.I.: Output control approach for delayed linear systems with adaptive rejection of multiharmonic disturbance. In: Proceedings of 19th IFAC World Congress, Cape Town, South Africa (2014)
125. Pyrkin, A.A., Bobtsov, A.A., Nikiforov, V.O., Kolyubin, S.A., Vedyakov, A.A., Borisov, O.I., Gromov, V.S.: Compensation of polyharmonic disturbance of state and output of a linear system with delay in the control channel. Autom. Remote. Control. **76**(12), 2124–2142 (2015)
126. Pyrkin, A., Isidori, A.: Adaptive output regulation of right-invertible MIMO LTI systems, with application to vessel motion control. Eur. J. Control **46**, 63–79 (2019)
127. Pyrkin, A., Smishlyaev, A., Bekiaris-Liberis, N., Krictić, M.: Rejection of sinusoidal disturbance of unknown frequency for linear system with input delay. In: American Control Conference, Baltimore, USA, pp. 5688–5693 (2010)
128. Richard, J.-P.: Time-delay systems: an overview of some recent advantages and open problems. Automatica **39**, 1667–1694 (2003)
129. Seshagiri, S., Khalil, H.K.: Robust output regulation of minimum phase nonlinear systems using conditional servocontroller. Int. J. Robust Nonlinear Control **15**, 83–102 (2005)

130. Serrani, A.: Rejection of harmonic disturbances at the controller input via hybrid adaptive external model. Automatica **42**, 1977–1985 (2006)
131. Serrani, A., Isidori, A.: Global robust output regulation for a class of nonlinear systems. Syst. Control Lett. **39**, 133–139 (2000)
132. Serrani, A., Isidori, A., Marconi, L.: Semiglobal robust output regulation of minimum-phase nonlinear system. Int. J. Robust Nonlinear Control **10**, 379–396 (2000)
133. Serrani, A., Isidori, A., Marconi, L.: Semiglobal nonlinear output regulation with adaptive internal model. IEEE Trans. Autom. Control **46**(8), 1178–1194 (2001)
134. Shin, K., Jang, J., Kang, J., Song, S.: Compensation method for tension disturbance due to an unknown roll shape in a web transport system. IEEE Trans. Ind. Appl. **39**, 1422–1428 (2003)
135. Qu, Z.: Robust Control of Nonlinear Uncertain Systems. Wiley, NY (1998)
136. Trentelman, H.L., Stoorvogel, A.A., Hautus, M.: Control Theory for Linear Systems (2001). https://doi.org/10.1115/1.1497472
137. Wach, P.: Dynamics and Control of Electrical Drives. Springer (2011)
138. Wang, J., Aranovskiy, S.V., Bobtsov, A.A., Pyrkin, A.A.: Compensating for a multisinusoidal disturbance based on Youla-Kucera parametrization. Autom. Remote. Control. **78**(9), 1559–1571 (2017)
139. Wang, J., Gan, W.C., Qiu, L.: A gain scheduled controller for sinusoidal ripple elimination of AC PM motor systems. In: Proceedings of the 15th Mediterranean Conference on Control and Automation, Athens, pp. 1–6 (2007)
140. Wang, L., Isidori, A., Liu, Z., Su, H.: Robust output regulation for invertible nonlinear MIMO systems. Automatica **82**, 278–286 (2017)
141. Wei, X., Lu, M., Wang, J., Tsang, K.M., Deng, B., Che, Y.: Semi-global robust output regulation of minimum-phase nonlinear systems based on high-gain nonlinear internal model. Int. J. Control **83**(5), 1009–1024 (2010)
142. Wonham, W.M.: Linear Multivariable Control. A Geometric Approach, 3rd edn. Springer, New York, Berlin, Heidelberg, Tokyo (1985)
143. Wu, Z., Li, Y., Wang, P., Liu, M.: Dynamic head-disk interface modeling and adaptive control of a hybrid actuator for optical data storage systems. Int. J. Optomechatron. **9**, 62–88 (2015)
144. Xu, D., Huang, J., Jiang, Z.-P.: Global adaptive output regulation for a class of nonlinear systems with iISS inverse dynamics using output feedback. Automatica **49**, 2184–2191 (2013)
145. Yang, J., Ding, Z.: Global output robust regulation for a class of lower triangular nonlinear systems: a feedback dominant approach. Automatica **76**, 65–69 (2017)
146. Yilmaz, C.T., Basturk, H.I.: Output feedback control for unknown LTI systems driven by unknown periodic disturbances. Automatica **99**, 112–119 (2019)
147. Young, P.C., Willems, J.C.: An approach to the multivariable servomechanism problem. Int. J. Control **15**(5), 961–979 (1972)
148. Zapateiro de la Hoz, M., Pozo, F. (eds): Advances on Analysis and Control of Vibrations - Theory and Applications, InTech (2012). https://doi.org/10.5772/2586

Chapter 2
Exosignals: Models and Observers

Abstract The chapter presents a unified approach to design of observers for a class of regular uncertain exosignals considered in this book. This approach is based on employment of the exosystem *canonical form* and results in representation of the regular uncertain exosignal in the form of linear regression with a constant vector of unknown parameters and measured regressor. In own turn, this representation allows one to derive a suitable parameterizations of the plants affected by uncertain external disturbances. In Sect. 2.1, the exosignal model in the form of an autonomous linear system is introduced and discussed. Section 2.2 presents the fundamental Lemma 2.1 introducing the exosystem canonical form, while several useful but rather technical corollaries of this lemma are given in Appendix A. Section 2.3 presents the observer of measured reference exosignal. Finally, Sects. 2.4 and 2.5 present the observers of unmeasured disturbances affecting different classes of linear and nonlinear plants with known or unknown parameters.

2.1 Exosignal Models

In this chapter, we consider the *regular*[1] *exosignals* $\upsilon \in \mathbb{R}$, i.e., *references* to be tracked or *disturbances* to be rejected that are modelled as the output of an autonomous linear system (*exosystem*) of the form

$$\dot{\xi} = \Gamma \xi, \tag{2.1}$$

$$\upsilon = h^\top \xi, \tag{2.2}$$

where $\xi \in \mathbb{R}^q$ is the state of the exosystem with initial condition $\xi(0)$, Γ and h are the constant matrix and the vector of appropriate dimensions. Without loss of generality, it is assumed that the pair (Γ, h) is observable.

---

[1] The notion of the *regular* component of exosignal was introduced in Sect. 1.2, see Eq. (1.37).

V. Nikiforov and D. Gerasimov, *Adaptive Regulation*, Lecture Notes in Control and Information Sciences 491, https://doi.org/10.1007/978-3-030-96091-9_2

Then the solution $\upsilon(t, \xi(0))$ of (2.1), (2.2) determined by the initial condition $\xi(0)$ is expressed as

$$\upsilon(t, \xi(0)) = h^\top e^{\Gamma t} \xi(0), \tag{2.3}$$

where $e^{\Gamma t}$ is the matrix exponential. It is known from the theory of ordinary differential equations (see [1, Sect. 4.2], [30, Chap. 4]) that each element of the solution (2.3) is a linear combination of *modes* (also known as *characteristic modes*, or *natural modes*—see [11], p. 5) each of them is a time function of a certain class defined by a corresponding *eigenvalue* of Γ and its *index*.[2] Namely, these modes take the following forms:

- for a real eigenvalue α with index ρ:

$$\sum_{i=1}^{\rho} \beta_i t^{i-1} e^{\alpha t};$$

- for a pair of complex conjugate eigenvalues $\alpha \pm jw$ with index ρ:

$$\sum_{i=1}^{\rho} \beta_i t^{i-1} e^{\alpha t} \sin(wt + \varphi_i),$$

where α and w are some real numbers, β_i and φ_i are the constants dependent on the initial condition $\xi(0)$, j is the imaginary unit.

Therefore, the exosystem (2.1), (2.2) can generate the exosignals of the following types (and their linear combinations):

- constants (when $\alpha = w = 0$, $\rho = 1$);
- polynomial functions of time (when $\alpha = w = 0$, $\rho > 1$);
- growing and decaying exponents (when $\alpha \neq 0$, $w = 0$);
- sinusoidal functions with constant amplitudes (when $\alpha = 0$, $w \neq 0$, $\rho = 1$);
- sinusoidal functions with growing or decaying amplitudes (when $\alpha \neq 0$, $w \neq 0$, $\rho \geq 1$ or $\alpha = 0$, $w \neq 0$, $\rho > 1$).

The general approach to design of the model (2.1), (2.2) for a given exosignal υ consists in *successive differentiation* of υ. Let us illustrate this technique by the following examples.

Example 2.1 Consider the simplest case when an exosignal is given by a constant function, i.e., $\upsilon(t) = \beta_1$, where β_1 is a prescribed constant. Let us assign the exosignal as the state of the exosystem: $\xi_1 = \upsilon$. Taking the time derivative of ξ_1, we obtain

---

[2] The *index* ρ_i of an eigenvalue λ_i of a matrix Γ is defined as the multiplicity of this eigenvalue when it is considered as a root of the minimal polynomial of Γ. The minimal polynomial of a matrix Γ is the monic polynomial $\psi(\lambda_i)$ of the least degree such that $\psi(\Gamma) = O$ (for more details, see [1], p. 62 or [4]).

$\dot{\xi}_1 = 0$. Therefore, it is easy to see that a constant exosignal is described by the exosystem (2.1), (2.2) of the first order with $\Gamma = 0$, $h = 1$ and the initial state $\xi(0) = \beta_1$. In this case, the exosystem has the single eigenvalue $\lambda\{\Gamma\} = \{0\}$. ■

Example 2.2 Consider an exosignal represented by the sinusoidal function

$$\upsilon(t) = \beta_2 \sin(wt + \varphi), \tag{2.4}$$

where β_2 is the amplitude, w is the angular frequency, and φ is the phase. We assign the exosignal as the first element of the state vector: $\xi_1 = \upsilon$. Calculating the time derivative of ξ_1, we have

$$\dot{\xi}_1 = w\beta_2 \cos(wt + \varphi). \tag{2.5}$$

Now, we assign the time derivative $\dot{\upsilon} = \dot{\xi}_1$ as the second element of the state vector: $\xi_2 = \dot{\upsilon}$. Then, proceeding with the successive differentiation, we obtain

$$\dot{\xi}_2 = \ddot{\upsilon} = -w^2\beta_2 \sin(wt + \varphi) = -w^2\xi_1. \tag{2.6}$$

Combining Eqs. (2.5) and (2.6), we finally obtain

$$\dot{\xi}_1 = \xi_2, \quad \dot{\xi}_2 = -w^2\xi_1, \quad \upsilon = \xi_1.$$

Thus, for the exosystem (2.1), (2.2), we have

$$\xi = \begin{bmatrix} \xi_1 \\ \xi_2 \end{bmatrix}, \Gamma = \begin{bmatrix} 0 & 1 \\ -w^2 & 0 \end{bmatrix}, h = \begin{bmatrix} 1 \\ 0 \end{bmatrix} \tag{2.7}$$

and

$$\xi_1(0) = \upsilon(0) = \beta_2 \sin\varphi, \quad \xi_2(0) = \dot{\upsilon}(0) = w\beta_2 \cos\varphi. \tag{2.8}$$

In this case, the exosystem has a pair of imaginary conjugate eigenvalues $\lambda\{\Gamma\} = \{jw, -jw\}$. ■

Example 2.3 Now, let us construct the model (2.1), (2.2) for a biased harmonic

$$\upsilon(t) = \beta_1 + \beta_2 \sin(wt + \varphi),$$

where β_1 is the bias, β_2 is the amplitude, w is the angular frequency, and φ is the phase. Taking into account the results presented by Examples 2.1 and 2.2, it can be shown that in this case the exosystem can be described by Eqs. (2.1), (2.2) of the third-order with

$$\Gamma = \begin{bmatrix} 0 & 0 & 0 \\ 0 & 0 & 1 \\ 0 & -w^2 & 0 \end{bmatrix}, \quad h = \begin{bmatrix} 1 \\ 1 \\ 0 \end{bmatrix} \tag{2.9}$$

and

$$\xi_1(0) = \beta_1, \quad \xi_2(0) = \beta_2 \sin\varphi, \quad \xi_3(0) = w\beta_2 \cos\varphi.$$

In this case, the exosystem has three distinct eigenvalues $\lambda\{\Gamma\} = \{0, jw, -jw\}$. ■

It is worth noting that for a given exosignal υ the parameters and even dimension of the exosystem (2.1), (2.2) are not unique, i.e., exosystems with different parameters and different dimensions can generate the same exosignal.

The ability of exosystems with different parameters to generate the same exosignals is due to *equivalent state transformation*. Let $M \in \mathbb{R}^{q\times q}$ be a nonsingular matrix called *state transformation matrix*, and let $\bar{\xi} = M\xi$ be a *transformed state*. Then, for $\bar{\xi}$, the exosystem (2.1), (2.2) is described as

$$\dot{\bar{\xi}} = \bar{\Gamma}\bar{\xi}, \tag{2.10}$$

$$\upsilon = \bar{h}^{\top}\bar{\xi}, \tag{2.11}$$

where (see [1], p. 95)

$$\bar{\Gamma} = M\Gamma M^{-1}, \quad \bar{h}^{\top} = h^{\top}M^{-1}, \quad \bar{\xi}(0) = M\xi(0). \tag{2.12}$$

The state equations (2.10), (2.11) are called *equivalent* to (2.1), (2.2). As it will be shown in the next section, the equivalent state transformation plays a crucial role in parameterization of uncertain exosignals with the use of a special *canonical form* of exosystems.

Example 2.4 It is easy to check that the matrix

$$\bar{\Gamma} = \begin{bmatrix} 0 & w \\ -w & 0 \end{bmatrix}$$

has the same eigenvalues as Γ from Example 2.2, i.e., $\lambda\{\Gamma\} = \lambda\{\bar{\Gamma}\} = \{jw, -jw\}$. Therefore, it can be used to generate the sinusoidal exosignal (2.4). Let us find the state transformation matrix M and the vectors $\bar{h}$, $\bar{\xi}(0)$ corresponding to the signal (2.4). Solving the matrix equation

$$\bar{\Gamma}M = M\Gamma, \tag{2.13}$$

we obtain

$$M = \begin{bmatrix} wm_1 & m_2 \\ -wm_2 & m_1 \end{bmatrix},$$

and, therefore,

$$\bar{h} = \frac{1}{w(m_1^2 + m_2^2)} \begin{bmatrix} m_1 \\ -m_2 \end{bmatrix}, \quad \bar{\xi}(0) = \begin{bmatrix} wm_1\xi_1(0) + m_2\xi_2(0) \\ -wm_2\xi_1(0) + m_1\xi_2(0) \end{bmatrix},$$

where m_1 and m_2 are arbitrary nonzero real numbers. ■

The ability of exosystems of different dimensions to generate the same exosignal is determined by *initial excitation of the modes.*

Example 2.5 If in Example 2.3 we set

$$\xi_1(0) = \beta_1, \quad \xi_2(0) = 0, \quad \xi_3(0) = 0,$$

then the third-order exosystem (2.1), (2.2), (2.9) will generate a constant $\upsilon = \beta_1$ like the first order exosystem from Example 2.1. If we set

$$\xi_1(0) = 0, \quad \xi_2(0) = \beta_2 \sin\varphi, \quad \xi_3(0) = w\beta_2 \cos\varphi,$$

we obtain $\upsilon(t) = \beta_2 \sin(wt + \varphi)$ as in Example 2.2. ■

If the initial condition $\xi(0)$ does not excite all the modes, then we say that the exosystem is *redundant with respect to the exosignal* $\upsilon(t, \xi(0))$. To solve the output regulation problem, some design techniques require that all the modes of the exosystem should be excited [13, 14, 29]. The technique presented in this book does not require this assumption that increases flexibility of the approach.

In nonadaptive implementations of the output regulation problem, the parameters of Γ and h of the exosystem (2.1), (2.2) are assumed to be known a priori. In this case, a class of exosignals, for which the regulation problem is solved, is defined by all solutions of (2.1), (2.2) obtained for known Γ and h and for all possible initial conditions $\xi(0)$. This assumption implies that all the exosignal uncertainties are only represented by the initial condition $\xi(0)$. In the case of the sinusoidal exosignal, it means that the amplitude and the phase of this signal are unknown, while the frequency is exactly known. The latter is an essential restriction from the practical point of view.

The class of exosignals can be significantly extended if the parameters of Γ and h are assumed to be unknown. In this case, we will speak about *uncertain* exosignal. In this book, we consider a class of uncertain exosignals including all the solutions of (2.1), (2.2) obtained for unknown Γ and h of known dimensions and for all possible initial conditions $\xi(0)$. The known dimension of the matrix Γ determines the upper bound on the number of modes, however not all of them must be excited by a specific initial condition $\xi(0)$. Under this assumption, an uncertain signal generated, for example, by the second-order exosystem can be represented by an unknown constant as well as by a sinusoidal function with unknown frequency, amplitude and phase.

Following the arguments presented in [8, Sect. 4.1], we restrict a class of the exosystems for which the considered regulation problem has practical significance. First, note that the modes corresponding to the eigenvalues with negative real part are exponentially decaying. As a result, after stabilization of the closed-loop system the influence of such modes on the control error decreases exponentially. Therefore, the modes caused by eigenvalues with nonnegative real part are only of practical interest. Second, if some eigenvalues of the exosystem have positive real part, then the exosystem can generate unbounded exosignals. Such a case is of questionable practical importance because rejection or tracking of unbounded signals requires unbounded energy. Therefore, we will consider only bounded exosignals and will use assumption that all the eigenvalues of the exosystem have zero real parts and either they are simple, or the unbounded modes corresponding to multiple roots of the minimal polynomial of Γ are not excited by initial conditions.

Summarizing the above arguments, we define a class of uncertain exosignals for which different adaptive regulation problems will be solved in this book.

Assumption 2.1 The bounded exosignal υ is modelled as the output of the exosystem (2.1), (2.2) such that:

A.2.1.1. parameters of Γ and h are unknown;

A.2.1.2. dimension q of the exosystem is known;

A.2.1.3. the pair (Γ, h) is observable;

A.2.1.4. all the eigenvalues of the matrix Γ have zero real parts, and either they are simple, or the unbounded modes corresponding to multiple roots of the minimal polynomial of Γ are not excited by the initial conditions;

A.2.1.5. the state ξ is not accessible for measurements.

The second part of Assumption A.2.1.4 concerning unexcited modes allows one to correctly describe the case when an exosystem of a fixed order is used for modelling of different types of exosignals, with respect to some of which this exosystem can be redundant. Let us illustrate this statement by the following example.

Example 2.6 Consider the second-order exosystem (2.1), (2.2) with unknown parameters. Under Assumption 2.1, it can generate:

- Sinusoidal signal (2.4). For such a signal, Γ and h can be represented in the form (2.7), while the initial conditions are given by (2.8). In this case, the pair (G, h) is observable, and the eigenvalues of Γ have zero real parts and are simple.
- Constant signal $\upsilon = \beta$. For such a signal, Γ and h can be represented in the form

$$\Gamma = \begin{bmatrix} 0 & 1 \\ 0 & 0 \end{bmatrix}, \quad h = \begin{bmatrix} 1 \\ 0 \end{bmatrix},$$

while appropriate initial condition can be chosen as $\xi(0) = [\beta, 0]^\top$. In this case, the pair (G, h) is observable, the eigenvalue of Γ is zero with the index $\rho = 2 > 1$. However, the appropriately chosen initial condition $\xi(0)$ does not excite the unbounded mode. ■

Assumption 2.1 says nothing about accessibility of the exosignal υ because it depends on conditions of a certain regulation problem. In the tracking problem, υ

represents the reference that is measured, as usual. In the problem of disturbance compensation, the exosignal υ represents the disturbance that is unmeasured in the majority of practical applications.

2.2 Exosystem Canonical Form

The first step towards resolving the adaptive regulation problems is a special parameterization of the exosignal υ based on a *canonical form* of the exosystem (2.1), (2.2) [16, 17, 21]. As it will be shown later, this canonical form plays a crucial role in design of exosignal observers as well as adaptive controllers.

Lemma 2.1 *The exosignal $\upsilon \in \mathbb{R}$ satisfying Assumption 2.1 can be represented in the form of linear regression*

$$\upsilon = \theta^\top \bar{\xi}, \tag{2.14}$$

where $\theta \in \mathbb{R}^q$ is a vector of unknown parameters, while the bounded regressor $\bar{\xi} \in \mathbb{R}^q$ is the state of the model

$$\dot{\bar{\xi}} = G\bar{\xi} + l\upsilon \tag{2.15}$$

or

$$\dot{\bar{\xi}} = \left(G + l\theta^\top\right)\bar{\xi} \tag{2.16}$$

with a certain initial condition $\bar{\xi}(0)$, an arbitrary Hurwitz matrix $G \in \mathbb{R}^{q\times q}$ and a constant vector $l \in \mathbb{R}^q$ such that the pair (G, l) is controllable.

Proof Define

$$\bar{\xi} = M\xi, \tag{2.17}$$

where $M \in \mathbb{R}^{q\times q}$ is a solution of the Sylvester equation

$$M\Gamma - GM = lh^\top. \tag{2.18}$$

According to Assumption 2.1 and the conditions of Lemma 2.1, the matrices Γ and G have no common eigenvalues, the pair (Γ, h) is observable, and the pair (G, l) is controllable. Therefore, the Eq. (2.18) has a unique nonsingular solution M [1, Theorem 8.4]. By taking the time derivative of $\bar{\xi} = M\xi$, in view of (2.1), (2.2) and (2.18), we obtain

$$\dot{\bar{\xi}} = M\Gamma\xi = GM\xi + lh^\top\xi = G\bar{\xi} + l\upsilon.$$

This proves (2.15) for the initial state $\bar{\xi}(0) = M\xi(0)$. Finally, substituting $\xi = M^{-1}\bar{\xi}$ into (2.2), we obtain (2.14), where $\theta^\top = h^\top M^{-1}$. The latter proves (2.16). □

We make some remarks concerning the obtained results.

Remark 2.1 Two canonical forms (2.15) and (2.16) have been introduced. In spite of the fact that each of them can be derived from other by substitution $\upsilon = \theta^\top\bar{\xi}$, they will play different roles in the design of adaptive controllers addressed in this book. The model (2.15) will be used in design of suitable observers of the regressor $\bar{\xi}$ which will be called *exosignal (regressor or disturbance) observers.* If we obtain an asymptotic estimate $\hat{\bar{\xi}}$ of the regressor $\bar{\xi}$, then the linear regression (2.14) gives a wide opportunity to design adaptive controllers. The model (2.16) will be used in the Francis equation to match the disturbance with the control. Thereafter, the model (2.15) will be referred to as *exosystem canonical form with input*, while the model (2.16) will be referred to as *exosystem autonomous canonical form.* Since the state $\bar{\xi}$ is the same for the both models, we will call it just as *the state of the exosystem canonical form.* ■

Remark 2.2 It is easy to show that the exosystems (2.1), (2.2) and (2.16), (2.14) are equivalent and the matrix M defines equivalent state transformation. Indeed, in view of (2.18), we can write

$$M\Gamma = (G + l\theta^\top)M, \quad \theta^\top = h^\top M^{-1}.$$

The latter equations corresponds to (2.12) in which

$$\bar{\Gamma} = G + l\theta^\top, \quad \bar{h}^\top = \theta^\top.$$

■

Remark 2.3 Under Assumption A.2.1.1, the state transformation matrix M is unknown since it depends on an unknown matrix Γ and an unknown vector h (see Eq. (2.18)). However, to derive a linear regression (2.14) with an unknown vector θ it is only necessary to establish the existence of a nonsingular matrix M. As it is shown by Lemma 2.1 and its proof, the existence of a nonsingular M is ensured by the properties of the matrices Γ, G and the pairs (Γ, h), (G, l). ■

Example 2.7 Consider the second-order exosystem (2.1), (2.2) with

$$\Gamma = \begin{bmatrix} 0 & 1 \\ -1 & 0 \end{bmatrix}, \quad h = \begin{bmatrix} 1 \\ 0 \end{bmatrix}. \tag{2.19}$$

It is easy to check that $\lambda\{\Gamma\} = \{-j, j\}$, and the exosystem generates sinusoidal exosignal $\upsilon(t) = \beta \sin(t + \varphi)$, where the amplitude β and the phase φ are determined by the initial condition $\xi(0)$ (see Example 2.2). Let us select the following matrices of the exosystem canonical form (2.15) (or (2.16)):

$$G = \begin{bmatrix} 0 & 1 \\ -2 & -3 \end{bmatrix}, \quad l = \begin{bmatrix} 0 \\ 1 \end{bmatrix}.$$

In this case, the solution of the matrix equation (2.18) has the form

$$M = \begin{bmatrix} 0.1 & -0.3 \\ 0.3 & 0.1 \end{bmatrix},$$

and, therefore,

$$\theta^\top = h^\top M^{-1} = \begin{bmatrix} 1, & 3 \end{bmatrix}, \quad \bar{\Gamma} = (G + l\theta^\top) = \begin{bmatrix} 0 & 1 \\ -1 & 0 \end{bmatrix}, \quad \bar{\xi}(0) = M\xi(0).$$

Note that in this particular case, $\Gamma = \bar{\Gamma}$. ■

Example 2.8 Consider the third-order exosystem (2.1), (2.2) with

$$\Gamma = \begin{bmatrix} 0 & 1 & 0 \\ -1 & 0 & 0 \\ 0 & 0 & 0 \end{bmatrix}, \quad h = \begin{bmatrix} 1 \\ 0 \\ 1 \end{bmatrix}. \tag{2.20}$$

In this case, $\lambda\{\Gamma\} = \{0, -j, j\}$, and the exosystem generates the biased sinusoidal exosignal $\upsilon(t) = \beta_1 + \beta_2 \sin(t + \varphi)$ with the bias β_1, the amplitude β_2 and the phase φ determined by the initial condition $\xi(0)$ (see Example 2.3). Let us select the following matrices of the exosystem canonical form (2.15) (or (2.16)):

$$G = \begin{bmatrix} 0 & 1 & 0 \\ 0 & 0 & 1 \\ -6 & -11 & -6 \end{bmatrix}, \quad l = \begin{bmatrix} 0 \\ 0 \\ 1 \end{bmatrix}.$$

Then the solution of the matrix equation (2.18) has the form

$$M = \begin{bmatrix} 0 & -0.1 & 0.16(6) \\ 0.1 & 0 & 0 \\ 0 & 0.1 & 0 \end{bmatrix},$$

and therefore,

$$\theta^\top = h^\top M^{-1} = \begin{bmatrix} 6 & 10 & 6 \end{bmatrix}, \bar{\Gamma} = (G + l\theta^\top) = \begin{bmatrix} 0 & 1 & 0 \\ 0 & 0 & 1 \\ 0 & -1 & 0 \end{bmatrix}, \bar{\xi}(0) = M\xi(0).$$

■

Now, we consider a multidimensional exosignal $\upsilon \in \mathbb{R}^\rho$ generated by the exosystem

$$\dot{\xi} = \Gamma\xi,\ \upsilon = H\xi, \tag{2.21}$$

and accept the following assumption.

Assumption 2.2 The bounded multidimensional exosignal $\upsilon \in \mathbb{R}^\rho$ is modelled as the output of the exosystem (2.21) such that:

A.2.2.1. parameters of Γ are unknown;

A.2.2.2. dimension q of the exosystem is known and $\rho \leq q$;

A.2.2.3. $H \in \mathbb{R}^{\rho\times q}$ is such that the pair $(\Gamma, \bar{h})$ is observable, where $\bar{h}^\top = \sum_{i=1}^{\rho} H_i$, and H_i is the ith row of H;

A.2.2.4. all the eigenvalues of the matrix Γ have zero real parts, and either they are simple or the unbounded modes corresponding to multiple roots of the minimal polynomial of Γ are not excited by the initial conditions;

A.2.2.5. the state ξ is not accessible for measurements.

We will construct canonical forms of the multidimensional exosystem suitable for design of disturbance observers and adaptive controllers. It is worth noting that in the multidimensional case controllability and observability of certain matrix pairs is only necessary but not sufficient condition for existence of a nonsingular solution of the Sylvester matrix equation [1, Theorem 8.M4]. However, we can obtain a nonsingular solution if we choose special form for matrices of the exosystem canonical form.

Lemma 2.2 *The exosignal $\upsilon \in \mathbb{R}^\rho$ satisfying Assumption 2.2 can be represented in the form of linear regression*

$$\upsilon = \Theta\bar{\xi} \tag{2.22}$$

with a matrix of unknown parameters $\Theta \in \mathbb{R}^{\rho\times q}$ and the bounded regressor $\bar{\xi} \in \mathbb{R}^q$ generated by the model

$$\dot{\bar{\xi}} = G\bar{\xi} + L\upsilon \tag{2.23}$$

or

$$\dot{\bar{\xi}} = (G + L\Theta)\,\bar{\xi} \tag{2.24}$$

with a certain initial condition $\bar{\xi}(0)$, where $G \in \mathbb{R}^{q\times q}$ is a Hurwitz matrix given by

$$G = \begin{bmatrix} 0 & 1 & 0 & \cdots & 0 \\ 0 & 0 & 1 & \cdots & 0 \\ \vdots & \vdots & \vdots & \ddots & \vdots \\ 0 & 0 & 0 & \cdots & 1 \\ -\gamma_0 & -\gamma_1 & -\gamma_2 & \cdots & -\gamma_{q-1} \end{bmatrix}, \tag{2.25}$$

$L \in \mathbb{R}^{q\times\rho}$ is a matrix defined as

$$L = \begin{bmatrix} 0 & 0 & \cdots & 0 \\ 0 & 0 & \cdots & 0 \\ \vdots & \vdots & \ddots & \vdots \\ 0 & 0 & \cdots & 0 \\ \bar{\lambda} & \bar{\lambda} & \cdots & \bar{\lambda} \end{bmatrix}, \tag{2.26}$$

γ_i ($i = 0, 1, \ldots, q-1$) and $\bar{\lambda} > 0$ are constants.

Proof Define

$$\bar{\xi} = M\xi, \tag{2.27}$$

where $M \in \mathbb{R}^{q\times q}$ is a solution of the Sylvester equation

$$M\Gamma - GM = LH. \tag{2.28}$$

Taking the structure of the matrix L into account, we can write $LH = \bar{\lambda}\varsigma_q\bar{h}^\top$, where

$$\varsigma_q = \begin{bmatrix} 0 \\ 0 \\ \vdots \\ 1 \end{bmatrix}, \bar{h}^\top = \sum_{i=1}^{\rho} H_i.$$

Therefore, Eq. (2.28) can be rewritten in the form

$$M\Gamma - GM = \bar{\lambda}\varsigma_q\bar{h}^\top. \tag{2.29}$$

It is obvious that the pair (G, ς_q) is controllable, while the pair $(\Gamma, \bar{h})$ is observable by Assumption A.2.2.3. Therefore, in accordance with [1, Theorem 8.4], [12] Eq. (2.29) (and the equivalent Eq. (2.28)) has a unique nonsingular solution M. By taking the time derivative of $\bar{\xi} = M\xi$, in view of (2.21) and (2.28), we obtain

$$\dot{\bar{\xi}} = M\Gamma\xi = GM\xi + LH\xi = G\bar{\xi} + L\upsilon.$$

This proves (2.23) for the initial condition $\bar{\xi}(0) = M\xi(0)$. Finally, substituting $\xi = M^{-1}\bar{\xi}$ into $\upsilon = H\xi$, we obtain (2.22), where $\Theta = HM^{-1}$. The latter proves (2.24). □

As it was mentioned above, in the case of multidimensional uncertain exosignals, the observability of the pair (Γ, H) is not sufficient for design of a disturbance observer with the use of the Sylvester matrix equation. To overcome this problem, we use Assumption A.2.2.3 requiring that no modes of the exosignal are mutually cancelled when the elements υ_i are summed. In other words, the sum $\upsilon_s = \sum_{i=1}^{\rho} \upsilon_i$ must be composed of all modes determined by the matrix Γ (but not necessarily excited by a specific initial condition $\xi(0)$).

To verify Assumption A.2.2.3, one needs to use additional prior information about the presence of the exosignal modes in each υ_i. In this case, we can say that the exosystem (2.21) is *partially known*: its matrix Γ determining a set of exosignal modes is unknown, but we know the number and the waveforms of these modes and how these modes are mixed by the matrix H. In particular, Assumption A.2.2.3 holds if different elements υ_i contain the modes with different frequencies only.

Example 2.9 Consider two-dimensional exosignal $\upsilon = [\upsilon_1, \upsilon_2]^\top$ generated by the third-order exosystem (2.21) with

$$\Gamma = \begin{bmatrix} 0 & 1 & 0 \\ -w^2 & 0 & 0 \\ 0 & 0 & 0 \end{bmatrix}, \; H = \begin{bmatrix} 1 & 0 & 1 \\ 0 & 0 & 1 \end{bmatrix},$$

where $w > 0$. In this case, $\lambda\{\Gamma\} = \{0, -jw, jw\}$, and

$$\upsilon(t) = \begin{bmatrix} \upsilon_1(t) \\ \upsilon_2(t) \end{bmatrix} = \begin{bmatrix} \beta_1 + \beta_2 \sin(wt + \varphi) \\ \beta_1 \end{bmatrix},$$

where w, β_1, β_2 and φ are some constants. For the given Γ and H, we have $\bar{h}^\top = \left[1,\ 0,\ 2\right]$, and the pair $(\Gamma, \bar{h})$ is observable.

Let us select the following matrices of the exosystem canonical form:

$$G = \begin{bmatrix} 0 & 1 & 0 \\ 0 & 0 & 1 \\ -6 & -11 & -6 \end{bmatrix}, \; L = \begin{bmatrix} 0 & 0 \\ 0 & 0 \\ 1 & 1 \end{bmatrix}.$$

Then the solution of the matrix equation (2.28) has the form

$$M = \begin{bmatrix} 0 & -0.1 & 0.3(3) \\ 0.1 & 0 & 0 \\ 0 & 0.1 & 0 \end{bmatrix},$$

and, hence,

$$\Theta = HM^{-1} = \begin{bmatrix} 3 & 10 & 3 \\ 3 & 0 & 2 \end{bmatrix}, \; G + L\Theta = \begin{bmatrix} 0 & 1 & 0 \\ 0 & 0 & 1 \\ 0 & -1 & 0 \end{bmatrix}, \quad \bar{\xi}(0) = M\xi(0) = \begin{bmatrix} 0.6(6) \\ 0.1 \\ 0 \end{bmatrix}.$$

However, for

$$H = \begin{bmatrix} 1 & 0 & 1 \\ 0 & 0 & -1 \end{bmatrix}$$

when

$$v(t) = \begin{bmatrix} v_1(t) \\ v_2(t) \end{bmatrix} = \begin{bmatrix} \beta_1 + \beta_2 \sin(wt + \varphi) \\ -\beta_1 \end{bmatrix}, \tag{2.30}$$

we have $\bar{h} = \begin{bmatrix} 1, & 0, & 0 \end{bmatrix}^\top$, and the pair $(\Gamma, \bar{h})$ is not observable. In this case, the matrix equation (2.28) with the given matrices Γ, H, G and L has a singular solution. In practice, we can check this situation noting that the mode β_1 is cancelled in this sum: $v_s = \beta_2 \sin(wt + \varphi)$. ■

If we know a priori how the modes enter in v_i (as in the above example), we can provide existence of the parameterization (2.22) by a special choice of the matrix L. Consider

$$L = \begin{bmatrix} 0 & 0 & \cdots & 0 \\ \vdots & \vdots & \ddots & \vdots \\ 0 & 0 & \cdots & 0 \\ \bar{\lambda}_1 & \bar{\lambda}_2 & \cdots & \bar{\lambda}_\rho \end{bmatrix}, \tag{2.31}$$

where the constants $\bar{\lambda}_i$ are such that the signal $v_l = \sum_{i=1}^{\rho} \bar{\lambda}_i v_i$ contains all modes of v. In this case, the pair $(\Gamma, \bar{h}_l)$, with $\bar{h}_l^\top = \sum_{i=1}^{q} \bar{\lambda}_i H_i$, will be observable and the Sylvester equation (2.28) will have a nonsingular solution. For example, for the exosignal (2.30) we can choose $\bar{\lambda}_1 = 1, \bar{\lambda}_2 = 0$. In this scenario, $v_l = \beta_1 + \beta_2 \sin(t + \phi)$, $\bar{h}_l = \begin{bmatrix} 1, & 0, & 1 \end{bmatrix}^\top$, and the pair $(\Gamma, \bar{h}_l)$ is observable.

The above reasons lead us to the following assumption which is more relevant from the practical point of view than Assumption A.2.2.3.

Assumption A.2.2.3.a Based on prior knowledge about the exosignal modes, it is possible to construct a matrix L (2.31), such that the signal $v_l = \sum_{i=1}^{\rho} \bar{\lambda}_i v_i$ contains all modes of v.[3]

Remark 2.4 Usually in practical problems, the upper bounds on the mode amplitudes are known a priori. These upper bounds can be taken into account in order to construct an appropriate matrix L satisfying Assumption A.2.2.3.a. ■

An alternative approach to representation of the multidimensional exosignals assumes that each element v_i of the exosignal $v = [v_1, v_2, \ldots, v_\rho]^\top$ is modelled as a scalar output of the individual exosystem

$$\dot{\xi}_i = \Gamma_i \xi_i, \; v_i = h_i^\top \xi_i, \tag{2.32}$$

where $\xi_i \in \mathbb{R}^{q_i}$ is the state with an initial condition $\xi_i(0)$, Γ_i and h_i are the constant matrix and vector of appropriate dimensions, and each pair (Γ_i, h_i) is observable. The class of such uncertain multidimensional exosignals is defined by the following assumption.

---

[3] When we say that the signal v_l contains all modes, it means that these modes are observable from v_l considered as the output. However, not all modes can be excited by a specific initial condition $\xi(0)$.

Assumption 2.3 Each element υ_i of $\upsilon \in \mathbb{R}^\rho$ and each corresponding exosystem (2.32) satisfies Assumption 2.1.

It is easy to see that with the use of notations

$$\Gamma_d = \text{diag}\{\Gamma_i\}, \quad H_d = \text{diag}\left\{h_i^\top\right\}, \quad \xi_d = \left[\xi_1^\top, \xi_2^\top, \ldots, \xi_\rho^\top\right]^\top,$$

and $q = \sum_{i=1}^{\rho} q_i$ the set of individual exosystems (2.32) can be rewritten in the form of a single exosystem (2.21) satisfying Assumption 2.2 under which Lemma 2.2 defines canonical forms (2.23) and (2.24). However, taking into account the block-diagonal structures of Γ_d and H_d, we can propose *block-diagonal* canonical form of the exosystem defined by the following corollary of Lemma 2.1.

Corollary 2.1 *The exosignal $\upsilon \in \mathbb{R}^\rho$ satisfying Assumption 2.3 can be represented in the form of linear regression*

$$\upsilon = \Theta_d \bar{\xi}_d, \tag{2.33}$$

where $\Theta_d = \text{diag}\{\theta_i^\top\}$ is a block-diagonal matrix of unknown parameters, the regressor $\bar{\xi}_d \in \mathbb{R}^q$ is the state of the model

$$\dot{\bar{\xi}}_d = G_d \bar{\xi}_d + L_d \upsilon \tag{2.34}$$

or

$$\dot{\bar{\xi}}_d = (G_d + L_d \Theta_d)\, \bar{\xi}_d \tag{2.35}$$

with a certain initial condition $\bar{\xi}_d(0)$ and the block-diagonal matrices

$$G_d = \text{diag}\{G_i\}, \quad L_d = \text{diag}\{l_i\}, \tag{2.36}$$

while $G_i \in \mathbb{R}^{q_i \times q_i}$ is arbitrary Hurwitz matrix and constant vector $l_i \in \mathbb{R}^{q_i}$ is such that the pair (G_i, l_i) is controllable ($i = 1, 2, \ldots, \rho$).

To prove this corollary, we note that the models (2.34) and (2.35) can be splitted into the sets of individual ones

$$\dot{\bar{\xi}}_i = G_i \bar{\xi}_i + l_i \upsilon_i, \quad i = 1, 2, \ldots, \rho \tag{2.37}$$

or

$$\dot{\bar{\xi}}_i = \left(G_i + l_i \theta_i^\top\right) \bar{\xi}_i, \quad i = 1, 2, \ldots, \rho, \tag{2.38}$$

for which, in view of Lemma 2.1, there exist solution pairs (M_i, θ_i) of the Sylvester equations

$$M_i \Gamma_i - G_i M_i = l_i h_i^\top.$$

Remark 2.5 When the elements υ_i are modelled by individual exosystems, they can be assigned as the exosignal modes. In this case, the matrix H has block-diagonal form, and a multimode disturbance can be formed with the use of an input matrix gain of the plant. For example, for the linear plant

$$\dot{x} = Ax + Bu + D\upsilon,$$

a relevant mixing of the elements (modes) υ_i can be provided by appropriate choice of the input matrix gain D. ■

Several technical corollaries of Lemmas 2.1 and 2.2 used in this book for design of exosignal observers are presented in Appendix A.

In the next sections of this chapter, based on the exosystem canonical form and its properties, we will derive observers generating an exponential estimate $\hat{\xi}$ of the regressor $\bar{\xi}$ when parameters of the exosystem (and possibly parameters of the plant) are unknown. We will also use the Francis equation to match the disturbance with the control. In other words, we will derive suitable parameterizations of the plants affected by uncertain disturbances.

In general, we will distinguish between two cases regarding the exosignal availability. For the problem of tracking, the exosignal υ will represent the measured reference. Hence, the canonical form (2.14), (2.15) can be directly used for design of a reference observer. For the problem of disturbance rejection, the exosignal υ will represent the unmeasured disturbance. In this case, we will design more complex observers depending on the structure and the parameters of the plant and employing the measured variables only, i.e., the control, state or the output of the plant.

In order to simplify exposition of the key ideas, we will consider mostly the case of scalar exosignals. Extension to the case of multidimensional exosignals will be considered in Sect. 4.2.

2.3 Reference Observer

In the problem of reference tracking, the exosignal υ is the measured reference and its observer can be defined as follows [17, 25, 26].

Lemma 2.3 *If Assumption 2.1 holds, then:*

L.2.3.1. *the reference signal υ can be represented in the form*

$$\upsilon = \theta^\top \hat{\xi} + \epsilon, \tag{2.39}$$

where ϵ exponentially decays, $\theta \in \mathbb{R}^q$ is a vector of unknown parameters, and the bounded regressor $\hat{\xi} \in \mathbb{R}^q$ is generated by the filter

$$\dot{\hat{\xi}} = G\hat{\xi} + l\upsilon \tag{2.40}$$

with arbitrary initial condition $\hat{\xi}(0)$ *and the pair* (G, l) *from the exosystem canonical form with input* (2.15)*;*

L.2.3.2. *for the state* $\bar{\xi}$ *of the exosystem canonical form, the following equality holds:* $\bar{\xi} = \hat{\xi} + \epsilon_{\xi}$*, where the regressor estimate* $\hat{\xi}$ *is generated by the filter* (2.40)*, and* ϵ_{ξ} *exponentially decays.*

Proof Define $\epsilon_{\xi} = \bar{\xi} - \hat{\xi}$, where $\bar{\xi}$ is the state of the model (2.15) with the "true" initial condition $\bar{\xi}(0) = M\xi(0)$. Calculating the time derivative of ϵ_{ξ}, in view of (2.15) and (2.40), we obtain $\dot{\epsilon}_{\xi} = G\epsilon_{\xi}$. Substituting $\bar{\xi} = \hat{\xi} + \epsilon_{\xi}$ into (2.14), we obtain (2.39), where $\epsilon = \theta^{\top}\epsilon_{\xi}$. □

Corollary 2.2 *The* i*th time derivative of the reference* υ *can be represented in the form*

$$\upsilon^{(i)} = \theta^{\top}\left(G + l\theta^{\top}\right)^{i}\hat{\xi} + \epsilon_i,$$

where $\hat{\xi}$ *is generated by* (2.40)*,* $\epsilon_i = \theta^{\top}G^{i}\epsilon_{\xi}$ *exponentially decays, while* ϵ_{ξ} *satisfies the equation* $\dot{\epsilon}_{\xi} = G\epsilon_{\xi}$*.*

2.4 Disturbance Observers: Plants with Known Parameters

When the exosignal υ represents an unmeasured disturbance, the observer (2.40) is not implementable, and more complex observers involving the plant structure and parameters must be designed.

In this section, we assume that the plant parameters are known and demonstrate how an inaccessible for measurements uncertain disturbance can be represented in a form of linear regression with unknown parameters and regressor generated by a special observer (*disturbance observer*). We start with LTI plants and then extend the approach to different classes of nonlinear ones.

2.4.1 *Disturbance Observers for LTI Plants*

We will first consider LTI plants with measured state and matched disturbance and then the plants with unmatched disturbance and measured output.

2.4.1.1 Plants with Matched Disturbance and Measured State

Consider the LTI plant

$$\dot{x} = Ax + b(u + \upsilon), \tag{2.41}$$

where $x \in \mathbb{R}^n$ is the measured state, $u \in \mathbb{R}$ is the control, A and b are known and have appropriate dimensions, and the disturbance $\upsilon \in \mathbb{R}$ generated by the exosystem (2.1), (2.2) is not accessible for measurements.

Taking the exosystem canonical form with input (2.15) into account, we define the regressor estimate

$$\hat{\xi} = \eta + Nx, \tag{2.42}$$

where $\eta \in \mathbb{R}^q$ is generated by the filter

$$\dot{\eta} = G\eta + (GN - NA)\,x - Nbu, \tag{2.43}$$

a matrix $N \in \mathbb{R}^{q\times n}$ satisfies the equality

$$Nb = l, \tag{2.44}$$

and the pair (G, l) is from the exosystem canonical form (2.15). Then the properties of the disturbance observer (2.42)–(2.44) are established as follows [16, 18, 25, 26].

Lemma 2.4 *If Assumption 2.1 holds and there exists a matrix $N \in \mathbb{R}^{q\times n}$ satisfying (2.44), then:*

L.2.4.1. *for the state $\bar{\xi}$ of the exosystem canonical form, the following equality is valid: $\bar{\xi} = \hat{\xi} + \epsilon_\xi$, where the bounded regressor estimate $\hat{\xi}$ is generated by the observer (2.42)–(2.44), and ϵ_ξ exponentially decays;*

L.2.4.2. *the disturbance υ affecting the plant (2.41) can be represented in the form*

$$\upsilon = \theta^\top \hat{\xi} + \epsilon, \tag{2.45}$$

where $\hat{\xi}$ is generated by the observer (2.42)–(2.44), $\theta \in \mathbb{R}^q$ is a vector of unknown parameters, and ϵ exponentially decays.

Proof Define the estimation error $\epsilon_\xi = \bar{\xi} - \hat{\xi}$, where $\bar{\xi}$ is the state of (2.15). Then, taking its time derivative, in view of (2.15), (2.41)–(2.44), we obtain

$$\begin{aligned}\dot{\epsilon}_\xi &= G\bar{\xi} + l\upsilon - G\eta - GNx + NAx + Nbu - NAx - Nbu - Nb\upsilon\\ &= G\left(\bar{\xi} - \eta - Nx\right) = G\epsilon_\xi,\ \epsilon_\xi(0) = \bar{\xi}(0) - \eta(0) - Nx(0).\end{aligned} \tag{2.46}$$

The latter proves the exponential decaying of ϵ_ξ. Further, substituting $\bar{\xi} = \hat{\xi} + \epsilon_\xi$ into (2.14), we directly obtain (2.45) with $\epsilon = \theta^\top \epsilon_\xi$. □

The main significance of this lemma is that it represents the uncertain disturbance υ in the form of linear regression (2.45) with the measured regressor $\hat{\xi}$ generated by the implementable observer (2.42)–(2.44). Therefore, the plant (2.41) subjected to the external disturbance (2.1), (2.2) can be represented by the following model.

Proposition 2.1 *If Assumption 2.1 holds and there exists a matrix $N \in \mathbb{R}^{q\times n}$ satisfying (2.44), then the plant (2.41) subjected to the external disturbance (2.1), (2.2) can be represented in the form*

$$\dot{x} = Ax + b\left(u + \theta^{\top}\hat{\xi} + \epsilon\right), \tag{2.47}$$

where the regressor estimate $\hat{\xi}$ is generated by the observer (2.42)–(2.44), $\theta \in \mathbb{R}^q$ is a vector of unknown parameters, and ϵ exponentially decays.

This model is well known in adaptive control theory [7, 15] and can be used for design of different certainty equivalence adaptive controllers (see Sect. 4.1.2).

It is worth noting that under Assumption 2.1 the regressor estimate $\hat{\xi}$ is bounded regardless the behavior of x. This important property of the proposed observer allows one to design separately a disturbance compensation loop and a plant stabilization loop (see Chaps. 4–7).

2.4.1.2 Plants with Unmatched Disturbance and Measured State

Consider the LTI plant

$$\dot{x} = Ax + bu + d\upsilon, \quad y = c^{\top}x + d_n\upsilon, \tag{2.48}$$

where $x \in \mathbb{R}^n$ is the measured state, $y \in \mathbb{R}$ is the output (regulated) variable, $u \in \mathbb{R}$ is the control, A, b, d, c are known and have appropriate dimensions, the constant d_n is unknown, and $\upsilon \in \mathbb{R}$ is the inaccessible for measurements disturbance generated by (2.1), (2.2). We make the following assumption regarding the plant (2.48).

Assumption 2.4 The plant (2.48) is such that:

A.2.4.1. the triplet (A, b, c) is controllable and observable;

A.2.4.2. all zeros of the transfer function $W_u(s) = c^{\top}(sI - A)^{-1}b$ have negative real parts.

It is worth noting that if b and d are linearly independent and/or $d_n \neq 0$, then the disturbance υ is not matched with the control u. Therefore, to derive an appropriate parameterization of the plant subjected to the external disturbance, we will simultaneously solve the following two problems. The first one is to obtain an estimate of the state of the exosystem canonical form. This problem will be solved with the use of the exosystem canonical form with input (2.15). The second problem is to match the disturbance with the control. This problem will be solved with the use of the exosystem autonomous canonical form (2.16) and the Francis matrix equation.

Remark 2.6 The Francis matrix equation is solvable for stabilizable plants. However, we accept more restrictive Assumption A.2.4.1 that is sufficient (but not necessary) to solve the control problems addressed in the next chapters of this book. ■

Remark 2.7 In view of Lemma 1.1, to resolve the Francis matrix equation it is enough to require that none of the zeros of the transfer function $W_u(s)$ coincides with an eigenvalue of Γ. Despite the fact that the eigenvalues of Γ are unknown, they have zero real parts by virtue of Assumption A.2.1.4 . Therefore, to guarantee the existence of a solution to the Francis matrix equation, we strengthen the requirement to $W_u(s)$ and accept Assumption A.2.4.2. ■

Remark 2.8 If it is known that $d_n \neq 0$, we can calculate $d_n \upsilon = y - c^\top x$ (since x is measured) and use an observer like for the reference one (2.40) in order to estimate the state of the exosystem canonical form. However, d_n is unknown and possibly is equal to zero. Therefore, we will use a disturbance observer like (2.42)–(2.44) involving the knowledge of d. The case when the both d and d_n are unknown will be considered in Sect. 2.4.1.3. ■

To obtain an estimate of the state $\bar{\xi}$ of the exosystem canonical form, we use the observer (2.42), (2.43), where a $q \times n$ matrix N is such that

$$Nd = l. \tag{2.49}$$

In view of (2.15), using the same arguments as in the proof of Lemma 2.4, we can formulate the following statement.

Lemma 2.5 *If Assumption 2.1 holds and there exists a matrix $N \in \mathbb{R}^{q\times n}$ satisfying (2.49), then for the state $\bar{\xi}$ of the exosystem canonical form the following equality holds: $\bar{\xi} = \hat{\xi} + \epsilon_\xi$, where the bounded regressor estimate $\hat{\xi}$ is generated by the observer (2.42), (2.43), (2.49), and ϵ_ξ exponentially decays.*

Further, we match the disturbance υ with the control u. To this end, we consider υ as the output of the exosystem canonical form (2.14), (2.16) and rewrite the plant (2.48) in the form

$$\dot{x} = Ax + bu + d\theta^\top \bar{\xi}, \quad y = c^\top x + d_n \theta^\top \bar{\xi}. \tag{2.50}$$

It is easy to see that the plant (2.50) and the disturbance (2.14), (2.16) exactly duplicate the form of (1.48), (1.49) and (1.65) with corresponding substitutions: $\xi_\upsilon = \bar{\xi}$, $\Gamma_\upsilon = \Gamma + l\theta^\top$, $H_\upsilon = \theta^\top$, $B = b$, $D = d$, $C = c^\top$, $R = d_n$ and $Q = O$. Therefore, in accordance with Lemmas 1.1 and 1.2, there exists a solution pair $M_\upsilon \in \mathbb{R}^{n\times q}$ and $\Theta_\upsilon = \theta_\upsilon^\top \in \mathbb{R}^q$ of the matrix equation (1.67) that reduces the plant model to the form

$$\dot{e} = Ae + b(u + \bar{\upsilon}), \quad \varepsilon = c^\top e, \tag{2.51}$$

where $\bar{\upsilon} = \theta_\upsilon^\top \bar{\xi}$ is the *matched disturbance*, the state error $e = x - M_\upsilon \bar{\xi}$ is inaccessible for measurements (since M_υ is unknown), while the output

$$\varepsilon = c^\top e = c^\top x - c^\top M_\upsilon \bar{\xi} = c^\top x + d_n \theta_\upsilon^\top \bar{\xi} = y \tag{2.52}$$

is measured. In contrast to the initial plant model (2.48), the model (2.51) is affected by the disturbance $\bar{\upsilon}$ matched with the control u. However, inaccessibility of the new state vector e for measurements is the price paid for this matching.

Now, using Lemma 2.5, we substitute $\bar{\xi} = \hat{\bar{\xi}} + \epsilon_{\xi}$ into $\bar{\upsilon} = \theta_{\upsilon}^{\top}\bar{\xi}$ and (2.51), and prove the following statement.

Proposition 2.2 *If Assumptions 2.1 and 2.4 hold, then the plant (2.48), can be represented by the following model:*

$$\dot{e} = Ae + b\left(u + \theta_{\upsilon}^{\top}\hat{\bar{\xi}} + \epsilon\right), \quad y = c^{\top}e, \tag{2.53}$$

where $\theta_{\upsilon} \in \mathbb{R}^{q}$ is a vector of unknown parameters, the regressor $\hat{\bar{\xi}}$ is generated by the observer (2.42), (2.43), (2.49), the state e is not measured, and ϵ exponentially decays.

The error model (2.53) is well known in adaptive control theory [7, 15] and can be used for design of different certainty equivalence adaptive controllers.

2.4.1.3 Plants with Unmatched Disturbance and Measured Output

Now, we assume that the state x of the plant (2.48) is not accessible for measurements, while d and d_n are possibly unknown.

In this case, we are not able to use the observer (2.42), (2.43), (2.49) and are forced to construct a more sophisticated one involving measured signals y and u only. Below, we will present two disturbance observers, the first of which is suitable for design of certainty equivalence adaptive controller, while the second one can be used for the adaptive backstepping design.

Observer for certainty equivalence design. We rewrite the plant model in the form

$$\dot{x} = Ax + bu + dh^{\top}\xi, \quad y = c^{\top}x + d_n h^{\top}\xi, \tag{2.54}$$

which together with the exosystem (2.1), (2.2) exactly duplicates the form of (1.48), (1.49) and (1.65) with corresponding substitutions: $\xi_{\upsilon} = \xi$, $\Gamma_{\upsilon} = \Gamma$, $H_{\upsilon} = h^{\top}$, $B = b$, $D = d$, $C = c^{\top}$, $R = d_n$ and $Q = O$. Therefore, in accordance with Lemmas 1.1 and 1.2, there exists a solution pair $M_{\upsilon} \in \mathbb{R}^{n\times q}$ and $\Theta = h_{\upsilon} \in \mathbb{R}^{q}$ of the matrix equation (1.67) that, in view of (2.52), changes the plant model to the form

$$\dot{e} = Ae + b(u + \bar{\upsilon}), \qquad y = c^{\top}e, \tag{2.55}$$

where the state error $e = x - M_{\upsilon}\xi$ is inaccessible for measurements (since M_{υ} is unknown), and the matched disturbance denoted as $\bar{\upsilon} = h_{\upsilon}^{\top}\xi$ can be modelled as the output of the exosystem

$$\dot{\xi} = \Gamma\xi, \qquad \bar{\upsilon} = h_\upsilon^\top \xi. \tag{2.56}$$

Then, applying Lemma 2.1 to the model (2.56), we can rewrite it in the canonical form

$$\bar{\upsilon} = \theta^\top \bar{\xi}, \qquad \dot{\bar{\xi}} = G\bar{\xi} + l\bar{\upsilon}, \tag{2.57}$$

where $\bar{\xi}$ is the state vector of this canonical form, $\theta \in \mathbb{R}^q$ is a new vector of unknown parameters, and the pair (G, l) is from (2.15).

Remark 2.9 In a particular case when some zeroes of the transfer function $W_\upsilon(s) = d_n + c^\top(sI - A)^{-1}d$ coincide with some eigenvalues of Γ, the pair (Γ, h_υ) can be partially unobservable. However, for correct application of Lemma 2.1, we can assume that the matched disturbance can be represented in the form $\bar{\upsilon} = \bar{h}_\upsilon^\top \xi$, where $\bar{h}_\upsilon$ is such that the pair $(\Gamma, \bar{h}_\upsilon)$ is completely observable, but not all the modes of the exosystem are excited. ■

Now, we use the canonical form (2.57) in order to construct an observer of the matched disturbance $\bar{\upsilon}$. To this end, we introduce the state estimation filter

$$\dot{\hat{e}} = A\hat{e} + bu + l_e\left(y - c^\top\hat{e}\right), \tag{2.58}$$

where $\hat{e} \in \mathbb{R}^n$ has arbitrary initial condition $\hat{e}(0)$, and the vector of feedback gains $l_e \in \mathbb{R}^n$ is selected so that $A_e = A - l_e c^\top$ is Hurwitz [1, Sect. 8.4]. Therefore, it can be shown that $e = \hat{e} + e_\upsilon$, where e_υ is the solution of the equation

$$\dot{e}_\upsilon = A_e e_\upsilon + b\bar{\upsilon}, \quad e_\upsilon(0) = e(0) - \hat{e}(0). \tag{2.59}$$

Then, for the output variable, we have

$$y = c^\top\hat{e} + c^\top e_\upsilon = c^\top\left(\hat{e} + \epsilon_e\right) + \upsilon_f, \tag{2.60}$$

where $\upsilon_f = W_e(s)[\bar{\upsilon}]$ is the *filtered matched disturbance*, $W_e(s) = c^\top(sI - A_e)^{-1}b$ is the asymptotically stable transfer function with Hurwitz numerator (by virtue of Assumption A.2.4.2), and ϵ_y exponentially decays according to the equation $\dot{\epsilon}_e = A_e\epsilon_e$, $\epsilon_e(0) = e(0) - \hat{e}(0)$. Therefore,

$$\upsilon_f = y - c^\top\left(\hat{e} + \epsilon_e\right). \tag{2.61}$$

At the same time, applying Corollary A.3 to the filtered matched disturbance $\upsilon_f = W_e(s)[\bar{\upsilon}]$, where $\bar{\upsilon}$ is generated by (2.57), we can write $\upsilon_f = \theta^\top(\bar{\xi}_f + \epsilon_1)$, where ϵ_1 is the exponentially decaying term defined by (A.11), while the regressor $\bar{\xi}_f$ is the state of the filter

$$\dot{\bar{\xi}}_f = G\bar{\xi}_f + l\upsilon_f \tag{2.62}$$

with an arbitrary initial condition $\bar{\xi}_f(0)$ and the pair (G, l) from the exosystem canonical form. Then, replacing (2.61) in (2.62), we can conclude that

$$\bar{\xi}_f = \hat{\xi}_f + \epsilon_f, \tag{2.63}$$

where the regressor $\hat{\xi}_f$ is the state of the filter

$$\dot{\hat{\xi}}_f = G\hat{\xi}_f + l\left(y - c^\top \hat{e}\right) \tag{2.64}$$

with an arbitrary initial condition $\hat{\xi}_f(0)$ and the exponentially decaying term ϵ_f satisfying the equation $\dot{\epsilon}_f = G\epsilon_f - lc^\top \epsilon_e$, $\epsilon_f(0) = \bar{\xi}_f(0) - \hat{\xi}_f(0)$.

Finally, in view of Corollary A.5 (see Eq. (A.28)) and (2.63), we can write

$$\bar{\upsilon} = \bar{\theta}^\top \hat{\xi}_f + \bar{\epsilon},$$

where $\bar{\theta} \in \mathbb{R}^q$ is a new vector of unknown parameters, $\bar{\epsilon} = \bar{\theta}^\top(\epsilon_1 + \epsilon_f - \epsilon_3)$ exponentially decays, ϵ_3 is defined by (A.20).

Therefore, we have proved the following statement.

Proposition 2.3 *If Assumptions 2.1 and 2.4 hold, then the plant (2.48) with inaccessible state x can be represented by the following parameterized model:*

$$\dot{e} = Ae + b\left(u + \bar{\theta}^\top \hat{\xi}_f + \bar{\epsilon}\right), \quad y = c^\top e, \tag{2.65}$$

where $\bar{\theta} \in \mathbb{R}^q$ is a vector of unknown parameters, the regressor $\hat{\xi}_f$ is generated by the observer (2.58), (2.64), while $\bar{\epsilon}$ exponentially decays.

In the model (2.65), the disturbance represented by the linear regression $\bar{\theta}^\top \hat{\xi}_f$ with the measured regressor $\hat{\xi}_f$ is matched with the control u. This result allows one to design certainty equivalence adaptive controllers (see Sect. 4.1.3).

Observer for backstepping design. Now, we present an alternative approach to the unmatched disturbance parameterization that can be used in backstepping design. First, for design purposes we write the transfer functions

$$W_u(s) = c^\top(sI - A)^{-1}b = \frac{\beta(s)}{\alpha(s)}, \; W_\upsilon(s) = d_n + c^\top(sI - A)^{-1}d = \frac{\delta(s)}{\alpha(s)},$$

where

$$\alpha(s) = s^n + a_{n-1}s^{n-1} + \cdots + a_0, \; \beta(s) = b_m s^m + b_{m-1}s^{m-1} + \cdots + b_0,$$

$$\delta(s) = d_n s^n + d_{n-1}s^{n-1} + \cdots + d_0,$$

and define the known parameters $a_{n-1}, \ldots, a_0$ and $b_m, \ldots, b_0$.

Further, we represent the plant (2.48) in the observable canonical form

$$\dot{\bar{x}} = E\bar{x} - \bar{a}y + \bar{b}u + \bar{d}\upsilon, \tag{2.66}$$

$$y = \bar{c}^\top \bar{x} + d_n \upsilon \tag{2.67}$$

with the unmeasured state vector $\bar{x}$ and the parameters

$$E = \begin{bmatrix} 0 & \\ \vdots & I_{n-1} \\ 0 & \cdots 0 \end{bmatrix}, \bar{a} = \begin{bmatrix} a_{n-1} \\ \vdots \\ a_1 \\ a_0 \end{bmatrix}, \ \bar{b} = \begin{bmatrix} 0 \\ \vdots \\ 0 \\ b_m \\ \vdots \\ b_0 \end{bmatrix}, \ \bar{d} = \begin{bmatrix} d_{n-1} \\ \vdots \\ d_1 \\ d_0 \end{bmatrix}, \ \bar{c} = \begin{bmatrix} 1 \\ 0 \\ \vdots \\ 0 \end{bmatrix}.$$

We employ the following state estimation filter:

$$\dot{\hat{x}} = \bar{A}\hat{x} + \left(\bar{l} - \bar{a}\right) y + \bar{b}u, \tag{2.68}$$

where the vector of the feedback gains $\bar{l} = \left[\bar{l}_{n-1}, \bar{l}_{n-2}, \ldots, \bar{l}_0\right]^\top$ is chosen so that the matrix $\bar{A} = E - \bar{l}\bar{c}^\top$ is Hurwitz (see [1], Sect. 8.4). In view of (2.66), (2.67) and (2.68), we can show that

$$\bar{x} = \hat{x} + x_\upsilon,$$

where x_υ satisfies the equation

$$\dot{x}_\upsilon = \bar{A}x_\upsilon + \left(\bar{d} - d_n\bar{l}\right)\upsilon, \quad x_\upsilon(0) = x(0) - \hat{x}(0).$$

Then, for the output signal, we have

$$y = \bar{c}^\top\hat{x} + \bar{c}^\top x_\upsilon + d_n\upsilon = \bar{c}^\top\left(\hat{x}+\epsilon_x\right) + \upsilon_f, \tag{2.69}$$

where $\upsilon_f = W_x(s)[\upsilon]$ is the filtered disturbance, $W_x(s) = d_n + \bar{c}^\top(sI - \bar{A})^{-1}(\bar{d} - d_n\bar{l})$ is the asymptotically stable transfer function, and ϵ_x exponentially decays according to the equation $\dot{\epsilon}_x = \bar{A}\epsilon_x$, $\epsilon_x(0) = x(0) - \hat{x}(0)$. Therefore, $\upsilon_f = y - \bar{c}^\top(\hat{x}+\epsilon_x)$, and, in view of Corollary A.3, we obtain the following parameterization of the filtered disturbance:

$$\upsilon_f = \theta^\top\hat{\xi}_f + \epsilon_\upsilon, \tag{2.70}$$

where $\theta \in \mathbb{R}^q$ is a vector of unknown parameters, ϵ_υ exponentially decays, the regressor $\hat{\xi}_f$ is generated by the filter

$$\dot{\hat{\xi}}_f = G\hat{\xi}_f + l\left(y - \bar{c}^\top \hat{x}\right) \tag{2.71}$$

with an arbitrary initial condition $\hat{\xi}_f(0)$ and the pair (G, l) from the exosystem canonical form. For the signal ϵ_υ, we have $\epsilon_\upsilon = \theta^\top(\epsilon_1 + \epsilon_f)$, where ϵ_1 is defined in Appendix A by (A.11) with substitution $W(s) = W_x(s)$, while ϵ_f satisfies the equation $\dot{\epsilon}_f = G\epsilon_f - l\bar{c}^\top \epsilon_x$ with the initial condition $\epsilon_f(0) = \hat{\xi}_f(0)$.

As a result, for the plant output, we obtain

$$y = \bar{c}^\top \hat{x} + \theta^\top \hat{\xi}_f + \epsilon_y, \tag{2.72}$$

where the state estimate $\hat{x}$ is generated by the observer (2.68), the regressor $\hat{\xi}_f$ is generated by the filter (2.71), while $\epsilon_y = \epsilon_\upsilon + \bar{c}^\top \epsilon_x$ exponentially decays.

Then, calculating the first time derivative of the plant output y in view of (2.72) and (2.68) and adding the equations connecting y with u, we obtain the following model:

$$\dot{y} = -\bar{l}_{n-1}\hat{x}_1 + \hat{x}_2 + \left(\bar{l}_{n-1} - a_{n-1}\right) y + \theta^\top \dot{\hat{\xi}}_f + \dot{\epsilon}_y, \tag{2.73}$$

$$\dot{\hat{x}}_i = -\bar{l}_{n-i}\hat{x}_1 + \hat{x}_{i+1} + \left(\bar{l}_{n-i} - a_{n-i}\right) y, \quad i = 2, 3, \ldots, n-m-1, \tag{2.74}$$

$$\dot{\hat{x}}_{n-m} = -\bar{l}_m \hat{x}_1 + \hat{x}_{n-m+1} + \left(\bar{l}_m - a_m\right) y + b_m u. \tag{2.75}$$

The model (2.73)–(2.75) represents a chain of $n - m$ integrators connected by measured variables y and $\hat{x}_i$. The vector of unknown parameters θ appears in the first equation and, in the general case, is not matched with the control u. This model can be considered as a special case of linear cascade systems to which different types of adaptive backstepping design can be applied (see [9, Chap. 10]). We use the *adaptive modular backstepping with high-order tuners* [5, 6, 24, 28]. The detailed presentation of this design procedure is given in Appendix E.

At each step of the backstepping design we need the time derivative of $\hat{\xi}_f$ of an appropriate order. Such time derivatives can be calculated either analytically from Eq. (2.71) or based on Corollary A.1 from Appendix A. Both of these approaches lead to controllers with overparameterization and complicated analytical expressions. To avoid the overparameterization and simplify the analytical calculations, we form the *filtered regressor*[4]

$$\hat{\xi}_\kappa = \frac{\kappa_0}{\kappa(s)}\left[\hat{\xi}_f\right], \tag{2.76}$$

---

[4] An alternative method, that allows one to simplify adaptive backstepping design by complete removing analytical calculations of partial derivatives, is developed in [2, 3, 32]. This method is called *command filtered adaptive backstepping* and is based on employment of low-pass filters (called *command filters*) generating filtered state errors.

where $\kappa(s)$ is an arbitrary Hurwitz polynomial of the degree $r \geq (n-m)$, $\kappa_0 > 0$ is a constant. Therefore, the ith time derivative of the filtered regressor $\hat{\xi}_\kappa$ can be calculated as

$$\hat{\xi}_\kappa^{(i)} = \frac{\kappa_0 s^i}{\kappa(s)}\left[\hat{\xi}_f\right], \quad i = 0, 1, \ldots, r. \tag{2.77}$$

Using Corollary A.5, we have

$$\upsilon_f = \bar{\theta}^\top \hat{\xi}_\kappa + \epsilon_\kappa, \tag{2.78}$$

where $\bar{\theta} \in \mathbb{R}^q$ is a new vector of unknown parameters, ϵ_κ is the exponentially decaying term. Substituting the latter equation into (2.69), we obtain

$$y = \bar{c}^\top \hat{x} + \bar{\theta}^\top \hat{\xi}_\kappa + \bar{c}^\top \epsilon_x + \epsilon_\kappa. \tag{2.79}$$

Then, instead of (2.73)–(2.75), we can write:

$$\dot{y} = -\bar{l}_{n-1}\hat{x}_1 + \hat{x}_2 + \left(\bar{l}_{n-1} - a_{n-1}\right) y + \bar{\theta}^\top \dot{\hat{\xi}}_\kappa + \bar{\epsilon}_\kappa, \tag{2.80}$$

$$\dot{\hat{x}}_i = -\bar{l}_{n-i}\hat{x}_1 + \hat{x}_{i+1} + \left(\bar{l}_{n-i} - a_{n-i}\right) y, \quad i = 2, 3, \ldots, n-m-1, \tag{2.81}$$

$$\dot{\hat{x}}_{n-m} = -\bar{l}_m \hat{x}_1 + \hat{x}_{n-m+1} + \left(\bar{l}_m - a_m\right) y + b_m u, \tag{2.82}$$

where $\dot{\hat{\xi}}_\kappa$ is defined by (2.77), and $\bar{\epsilon}_\kappa = \bar{c}^\top \dot{\epsilon}_x + \dot{\epsilon}_\kappa$ exponentially decays.

It is worth noting that the model (2.80)–(2.82) does not contain additional m equations completing the plant model up to the full order of n. These omitted equations describe the *internal dynamics* of the plant (in terminology of [10, Sect. 8.1]). For the considered linear plant, these internal dynamics are linear and their stability properties are determined by the zeros of the transfer function $W_u(s)$ [8, Sect. 2.1], [10, Example 8.3]. Therefore, under Assumption A.2.4.2 the origin of the internal dynamics is asymptotically stable. Moreover, with a special choice of a control law, the internal dynamics can be made unobservable from the output y.[5] The latter two arguments allow one to use the reduced order model (2.80)–(2.82) for a control law design, while the behavior of the internal dynamics must be analyzed in the proof of stability of the closed-loop system.

Therefore, we have proved the following statement determining the parameterization which is suitable for backstepping design (see Sect. 4.1.3.2).

Proposition 2.4 *If Assumptions 2.1 and 2.4 hold, then the plant (2.48) with inaccessible state x admits the parameterization (2.80)–(2.82), where $\bar{\theta} \in \mathbb{R}^q$ is a vector of unknown parameters, $\hat{x}$ is the state of (2.68), $\hat{\xi}_\kappa$ is the regressor generated by the filters (2.71), (2.76), and $\bar{\epsilon}_\kappa$ is the exponentially decaying term.*

[5] The internal dynamics corresponding to zero output are called the *zero dynamics* of the system. For more details on zero dynamics, see [8, Sect. 6.3], [9, Sect. 2.2.3], [10, Sect. 8.1].

2.4.2 *Disturbance Observers for Nonlinear Plants*

In this section, we demonstrate how the above method of the disturbance parameterization can be extended to certain classes of nonlinear plants. Starting with consideration of a general form of nonlinear plants, we apply this method to globally defined *canonical normal form*, *strict-feedback canonical form* and *output-feedback canonical form*.

2.4.2.1 Plant in General form with Measured State

Consider the nonlinear plant of the form

$$\dot{x} = f(x) + \phi(x)u + d(x)\upsilon, \quad y = h(x), \tag{2.83}$$

where $x \in \mathbb{R}^n$ is the state available for measurement with the initial condition $x(0)$, $f(x)$, $\phi(x)$, $d(x)$ are the smooth (that is, infinitely differentiable) vector functions, $h(x)$ is the smooth function, υ is the external disturbance satisfying Assumption 2.1. We also assume that for any $x(0)$ and admissible u and υ there exists a unique solution $x(t, x(0))$ defined for all $t \geq 0$. The considered problem is to obtain a disturbance parameterization in the form of linear regression with implementable regressor estimate $\hat{\xi}$ like (2.45).

To this end, we define the regressor estimate as [19, 22, 25, 26]

$$\hat{\xi} = \eta + \pi(x), \tag{2.84}$$

where the auxiliary vector $\eta \in \mathbb{R}^q$ is generated by the nonlinear filter

$$\dot{\eta} = G\eta + G\pi(x) - \frac{\partial \pi(x)}{\partial x}\left(f(x) + \phi(x)u\right), \tag{2.85}$$

the vector function $\pi(x) \in \mathbb{R}^q$ satisfies the PDE

$$\frac{\partial \pi(x)}{\partial x} d(x) = l, \tag{2.86}$$

while the pair (G, l) is from the exosystem canonical form (2.15).

Remark 2.10 Compared to the case of linear systems, for nonlinear plant (2.83) one needs to solve the PDE (2.86) instead of the linear algebraic equation (2.49). This problem is simplified if the vector l is selected as $l = \left[0, \ldots, 0, l_q\right]^\top$ and, therefore, $\pi(x)$ can be defined as $\pi(x) = \left[0, \ldots, 0, \pi_q(x)\right]^\top$, where the function $\pi_q(x)$ is the solution of the scalar PDE

$$\frac{\partial \pi_q(x)}{\partial x} d(x) = l_q. \tag{2.87}$$

■

Proposition 2.5 *If Assumption 2.1 holds and there exists a vector function $\pi(x)$ satisfying (2.86), then the disturbance υ affecting the plant (2.83) can be represented in the form*

$$\upsilon = \theta^{\top}\hat{\xi}+\epsilon, \tag{2.88}$$

where the bounded regressor $\hat{\xi}$ is generated by the observer (2.84), (2.85), $\theta \in \mathbb{R}^q$ is a vector of unknown parameters, and ϵ exponentially decays.

Proof Define the estimation error $\epsilon_{\xi} = \bar{\xi} - \hat{\xi} = \bar{\xi} - \eta - \pi(x)$, where $\bar{\xi}$ is the state of the exosystem canonical form (2.15). Calculating the time derivative of ϵ_{ξ}, in view of (2.15), (2.83)–(2.86), we obtain

$$\dot{\epsilon}_{\xi} = G\bar{\xi} + l\upsilon - G\eta - G\pi(x) + \frac{\partial \pi(x)}{\partial x}(f(x) + \phi(x)u)$$
$$- \frac{\partial \pi(x)}{\partial x}(f(x) + \phi(x)u + d(x)\upsilon) = G\left(\bar{\xi} - \eta - \pi(x)\right) = G\epsilon_{\xi}.$$

The latter proves the exponential decaying of ϵ_{ξ}. Further, by substituting $\bar{\xi} = \hat{\xi}+\epsilon_{\xi}$ into (2.14) we directly obtain (2.88) with $\epsilon = \theta^{\top}\epsilon_{\xi}$. □

Therefore, the plant (2.83) subjected to the disturbance (2.1), (2.2) can be represented in the following parameterized form:

$$\dot{x} = f(x) + \phi(x)u + d(x)\left(\theta^{\top}\hat{\xi}+\epsilon\right), \tag{2.89}$$

where the regressor $\hat{\xi}$ is physically implementable.

In the next sections, the general approach presented by Proposition 2.5 will be applied to particular canonical forms of nonlinear plants.

2.4.2.2 Plant in the Normal Form

Assumption 2.5 The plant (2.83) with $f(0) = 0$, $\phi(0) = 0$ and $d(x) = \phi(x)$ has *a uniform relative degree* $\rho \leq n$, is *strongly minimum phase* and possesses globally defined normal form.[6]

This assumption means that for all $x \in \mathbb{R}^n$:

(i) $L_g h(x) = L_g L_f h(x) = \cdots = L_g L_f^{\rho-2} h(x) = 0$ and $L_g L_f^{\rho-1} h(x) \neq 0$;
(ii) it is possible to find $n - \rho$ functions $\bar{\pi}_1(x), \bar{\pi}_2(x), \ldots, \bar{\pi}_{n-\rho}(x)$ such that $L_g \bar{\pi}_i(x) = 0$ for $i = 1, 2, \ldots, \rho$ and the mapping

[6] For more details on the normal form, see [8, Sect. 6.2].

$$\Pi(x) = \begin{bmatrix} \bar{\pi}_1(x) \\ \vdots \\ \bar{\pi}_{n-\rho}(x) \\ h(x) \\ L_f h(x) \\ \vdots \\ L_f^{\rho-1} h(x) \end{bmatrix}$$

is *global diffeomorphism* defining the nonlinear change of coordinates

$$\begin{bmatrix} z(x) \\ \chi(x) \end{bmatrix} = \Pi(x),$$

where $z(x) = \left[\bar{\pi}_1(x), \bar{\pi}_2(x), \ldots, \bar{\pi}_{n-\rho}(x)\right]^\top$, $\chi(x) = \left[h(x), L_f h(x), \ldots, L_f^{\rho-1} h(x)\right]^\top$ are the $(n-\rho)$ and ρ-dimensional vector functions, respectively;

(iii) in the new coordinates z and χ the plant (2.83) takes the *normal* form

$$\dot{z} = f_0(z, \chi), \tag{2.90}$$
$$\dot{\chi} = E\chi + \varsigma_\rho \left(\phi_\chi(z, \chi) + \phi_u(z, \chi)\,(u + \upsilon)\right), \; y = \chi_1, \tag{2.91}$$

where

$$f_0(z, \chi) = \begin{bmatrix} L_f \bar{\pi}_1\left(\Pi^{-1}(z, \chi)\right) \\ \vdots \\ L_f \bar{\pi}_{n-\rho}\left(\Pi^{-1}(z, \chi)\right) \end{bmatrix}, \; E = \begin{bmatrix} 0 & I_{\rho-1} \\ \vdots & \\ 0 & \cdots 0 \end{bmatrix}, \; \varsigma_\rho = \begin{bmatrix} 0 \\ \vdots \\ 0 \\ 1 \end{bmatrix},$$

$$\phi_\chi(z, \chi) = L_f^\rho h\left(\Pi^{-1}(z, \chi)\right), \; \phi_u(z, \chi) = L_g L_f^{\rho-1} h\left(\Pi^{-1}(z, \chi)\right) \neq 0;$$

(iv) the dynamic subsystem (2.90), considered as a system with the input χ and the state z, is *input-to-state stable* (ISS).[7]

Now, to obtain the disturbance parameterization in the form of linear regression like (2.45), we introduce the regressor estimate [23, 25, 26]

$$\hat{\xi} = \eta + \pi(z, \chi), \tag{2.92}$$

where η is generated by the filter

[7] See Definition 6.1 in [8] and Definition B.9.

$$\dot{\eta} = G\eta + G\pi(z, \chi) - \frac{\partial \pi(z, \chi)}{\partial \chi}\left(E\chi + \varsigma_\rho \phi_\chi(z, \chi)\right)$$
$$- lu - \frac{\partial \pi(z, \chi)}{\partial z} f_0(z, \chi), \tag{2.93}$$

the vector function $\pi(z, \chi)$ satisfies the following PDE:

$$\frac{\partial \pi(z, \chi)}{\partial \chi} \varsigma_\rho \phi_u(z, \chi) = l, \tag{2.94}$$

while the pair (G, l) is from the exosystem canonical form (2.15).

If $l = [0, \ldots, 0, l_q]^\top$, then $\pi(z, \chi)$ can be defined as $\pi(z, \chi) = \left[0, \ldots, 0, \pi_q(z, \chi)\right]^\top$, where the function $\pi_q(z, \chi)$ is the solution of the scalar PDE

$$\frac{\partial \pi_q(z, \chi)}{\partial \chi} \phi_u(z, \chi) = l_q.$$

The disturbance parameterized model is defined by the following statement.

Proposition 2.6 *If Assumption 2.1 holds and there exists a vector function $\pi(z, \chi)$ satisfying (2.94), then the disturbance υ affecting the plant (2.90), (2.91) can be represented in the form*

$$\upsilon = \theta^\top \hat{\xi} + \epsilon, \tag{2.95}$$

where the regressor $\hat{\xi}$ is generated by the observer (2.92), (2.93), $\theta \in \mathbb{R}^q$ is a vector of unknown parameters, and the term ϵ exponentially decays.

The proposition is proved by taking the time derivative of the estimation error $\epsilon_\xi = \bar{\xi} - \hat{\xi} = \bar{\xi} - \eta - \pi(z, \chi)$ in view of (2.15), (2.90)–(2.94) that yields $\dot{\epsilon}_\xi = G\epsilon_\xi$ and (2.95) with $\epsilon = \theta^\top \epsilon_\xi$.

2.4.2.3 Plant with Unmatched Disturbance and Measured State

Now, let us assume that the nonlinear plant can be represented in (or transformed into) the *strict-feedback*[8] canonical form with external disturbance

$$\dot{x}_i = x_{i+1} + \phi_i(x_1, x_2, \ldots, x_i) + \bar{d}_{n-i}^\top \upsilon, \quad i = 1, 2, \ldots, n-1, \tag{2.96}$$
$$\dot{x}_n = \phi_n(x) + \phi_u(x)u + \bar{d}_0^\top \upsilon, \tag{2.97}$$
$$y = x_1, \tag{2.98}$$

[8] More general strict-feedback form is considered in [10, Sect. 9.5] and [9, Sect. 2.3.1].

where $\phi_i(x_1, x_2, \ldots, x_i)$ and $\phi_u(x)$ are known smooth scalar functions, $\bar{d}_i = [d_{i,1}, d_{i,2}, \ldots, d_{i,\rho}]^\top$ is the known constant vector ($i = 0, 1, \ldots, n-1$), and the multidimensional disturbance $\upsilon \in \mathbb{R}^\rho$ satisfies Assumption 2.2.

The strict-feedback canonical form plays an important role in the modern control theory. It became a startpoint for development of iterative design techniques such as backstepping to control cascade systems as well as systems with unmatched uncertainties [9, 10, 31]. Here we consider this form as a typical example of nonlinear system with unmatched disturbance.

For compactness of notation, let us rewrite Eqs. (2.96), (2.97) in the matrix form

$$\dot{x} = Ex + \phi(x) + \varsigma_n \phi_u(x)u + D\upsilon, \tag{2.99}$$

where

$$E = \begin{bmatrix} 0 & \\ \vdots & I_{n-1} \\ 0 \cdots 0 & \end{bmatrix}, \phi(x) = \begin{bmatrix} \phi_1(x_1) \\ \phi_2(x_1, x_2) \\ \vdots \\ \phi_n(x_1, x_2, \ldots, x_n) \end{bmatrix}, \varsigma_n = \begin{bmatrix} 0 \\ 0 \\ \vdots \\ 1 \end{bmatrix}, D = \begin{bmatrix} \bar{d}_{n-1}^\top \\ \vdots \\ \bar{d}_1^\top \\ \bar{d}_0^\top \end{bmatrix}.$$

Then the disturbance observer and the disturbance parameterization are defined as follows.

Lemma 2.6 *If Assumption 2.2 holds and there exists a matrix $N \in \mathbb{R}^{q \times n}$ such that*

$$ND = L, \tag{2.100}$$

then the disturbance υ can be represented in the form

$$\upsilon = \Theta\hat{\xi} + \epsilon, \tag{2.101}$$

where $\Theta \in \mathbb{R}^{\rho \times q}$ is a matrix of unknown parameters, $\hat{\xi}$ is the bounded regressor given by

$$\hat{\xi} = \eta + Nx, \tag{2.102}$$

$\eta \in \mathbb{R}^q$ is the vector generated by the filter

$$\dot{\eta} = G\eta + GNx - N\left(Ex + \phi(x) + \varsigma_n \phi_u(x)u\right), \tag{2.103}$$

and G and L are the matrices defined by (2.25), (2.26).

Proof Define the estimation error $\epsilon_\xi = \bar{\xi} - \hat{\xi}$, where $\bar{\xi}$ is the state of the canonical form (2.23). Calculating the time derivative of ϵ_ξ, in view of (2.23), (2.99), (2.100), (2.102), and (2.103), we obtain

$$\dot{\epsilon}_\xi = G\bar{\xi} + Lv - G\eta - GNx + N\left(Ex + \phi(x) + \varsigma_n\phi_u(x)u\right)$$
$$- NEx - N(\phi(x) + \varsigma_n\phi_u(x)u) - NDv = G(\bar{\xi} - G\eta - GNx) = G\epsilon_\xi.$$

The latter means that ϵ_ξ exponentially decays. Replacing $\bar{\xi} = \hat{\xi} + \epsilon_\xi$ into (2.22), we obtain (2.101), where $\epsilon = \Theta\epsilon_\xi$. □

Remark 2.11 If $\rho > n$, then Eq. (2.100) may not have a solution for a fixed prescribed L. However, we can try to resolve this equation using L of the form (2.31), where parameters $\bar{\lambda}_i$ are chosen together with N to satisfy Eq. (2.100) and Assumption A.2.2.3.a simultaneously (if possible). ■

In contrast to linear plants that allow us to match the disturbance with the control via a special state transformation (see Sect. 2.4.1.2), in nonlinear plant (2.96)–(2.98) the parameterized disturbance remains unmatched. To overcome this structural obstacle, iterative design techniques such as backstepping can be applied. For the plant (2.96)–(2.98), we use adaptive modular backstepping with high-order tuners (see Appendix E).

At each step of the iterative design technique, the time derivative of the regressor $\hat{\xi}$ of a corresponding order must be calculated and used in control. Following the approach from Sect. 2.4.1.3, we avoid the overparameterization of the controller and simplify analytical expressions introducing the filtered regressor and directly calculating its time derivative at each step of the design procedure. To this end, we define the filtered regressor

$$\hat{\xi}_\kappa = \frac{\kappa_0}{\kappa(s)}\left[\hat{\xi}\right], \tag{2.104}$$

where $\kappa(s)$ is a Hurwitz polynomial of degree no less than $n-1$, $\kappa_0 > 0$ is a constant. Then, introducing the diagonal transfer function matrix $W(s) = \frac{\kappa_0}{\kappa(s)} I_\rho$, we can extend the results of Corollaries A.2 and A.5 to the case of multidimensional disturbance $v \subset \mathbb{R}^\rho$. Therefore, we conclude that the disturbance can be represented in the form

$$v = \bar{\Theta}\hat{\xi}_\kappa + \epsilon_\kappa,$$

where $\bar{\Theta} \in \mathbb{R}^{\rho\times q}$ is a new matrix of unknown parameters, and ϵ_κ exponentially decays. It is obvious from (2.104) that we can directly derive the time derivatives of $\hat{\xi}_\kappa$ up to the order $n-1$ required for the backstepping procedure.

As a useful technical result, we represent the term $d_i^\top \bar{\Theta}\hat{\xi}_\kappa$ in the form

$$d_i^\top \bar{\Theta}\hat{\xi}_\kappa = \theta^\top \bar{D}_i\hat{\xi}_\kappa,$$

where $\theta = \left[\bar{\Theta}_1, \bar{\Theta}_2, \ldots, \bar{\Theta}_\rho\right]^\top \in \mathbb{R}^{q\rho}$ is a vector of unknown parameters aggregated from the rows $\bar{\Theta}_i$ of the matrix $\bar{\Theta}$,

$$\bar{D}_i = \begin{bmatrix} d_{i,1} I_q \\ d_{i,2} I_q \\ \vdots \\ d_{i,\rho} I_q \end{bmatrix} \in \mathbb{R}^{q\rho \times q},$$

$d_{i,j}$ are the elements of the vector d_i $(i = 0, 1, \ldots, n-1,\ j = 1, 2, \ldots, \rho)$.

Then the parameterized plant model is defined by the following statement.

Proposition 2.7 *If Assumption 2.2 holds and there exists a matrix* $N \in \mathbb{R}^{q \times n}$ *satisfying* (2.100), *then the plant* (2.96)–(2.98) *can be represented in the form*

$$\dot{x}_i = x_{i+1} + \phi_i(x_1, x_2, \ldots, x_i) + \theta^\top \bar{D}_{n-i} \hat{\xi}_\kappa + \bar{d}_{n-i}^\top \epsilon_\kappa, \quad i = 1, \ldots, n-1, \tag{2.105}$$

$$\dot{x}_n = \phi_n(x) + \phi_u(x) u + \theta^\top \bar{D}_0 \hat{\xi}_\kappa + \bar{d}_0^\top \epsilon_\kappa, \tag{2.106}$$

$$y = x_1, \tag{2.107}$$

where the filtered regressor $\hat{\xi}_\kappa$ *is generated by the observer* (2.102)–(2.104), *and* ϵ_κ *exponentially decays.*

Remark 2.12 The above proposition can be directly extended to the case of multidimensional exosignal υ satisfying Assumption 2.3. In this case, for construction of the observer (2.102), (2.103), one can use either matrices G and L of the form (2.25), (2.26) or the block-diagonal matrices (2.36). ■

2.4.2.4 Plant with Unmatched Disturbance and Measured Output

Now, we assume that the nonlinear plant can be represented in (or transformed into) the *output-feedback* canonical form with external disturbance

$$\dot{x} = Ex + \phi(y) + \bar{b}\phi_u(y)u + \bar{d}\upsilon, \tag{2.108}$$

$$y = \bar{c}^\top x + d_n \upsilon, \tag{2.109}$$

where the state $x \in \mathbb{R}^n$ is inaccessible for measurements,

$$E = \begin{bmatrix} 0 & \\ \vdots & I_{n-1} \\ 0 & \cdots\ 0 \end{bmatrix}, \quad \phi = \begin{bmatrix} \phi_1 \\ \phi_2 \\ \vdots \\ \phi_n \end{bmatrix}, \quad \bar{b} = \begin{bmatrix} 0 \\ \vdots \\ 0 \\ b_m \\ \vdots \\ b_0 \end{bmatrix}, \quad \bar{d} = \begin{bmatrix} d_{n-1} \\ \vdots \\ d_1 \\ d_0 \end{bmatrix}, \quad \bar{c} = \begin{bmatrix} 1 \\ 0 \\ \vdots \\ 0 \end{bmatrix},$$

$\phi_{i+1}(y)$ and $\phi_u(y)$ are the known smooth scalar functions, b_j are the known constant parameters, while d_i are possibly unknown constant parameters ($i = 0, 1, \ldots, n-1$, $j = 0, 1, \ldots, m$).

Additionally, we accept the following assumption.

Assumption 2.6 The polynomial $b_m s^m + \cdots + b_1 s + b_0$ is Hurwitz.

Under Assumption 2.6, the plant (2.108), (2.109) has the relative degree $n - m$ and linear asymptotically stable zero dynamics [9, Sect. 7.1.2].

The plant (2.108), (2.109) can be considered as a nonlinear generalization of the plant (2.66), (2.67). Therefore, the approach introduced in Sect. 2.4.1.3 can be used here with the main difference that the following nonlinear state observer is employed (instead of the linear one):

$$\dot{\hat{x}} = \bar{A}\hat{x} + \bar{l}y + \phi(y) + \bar{b}\phi_u(y)u, \tag{2.110}$$

where the vector of the feedback gains $\bar{l} = \left[\bar{l}_{n-1}, \bar{l}_{n-2}, \ldots, \bar{l}_0\right]^\top$ is chosen so that the matrix $\bar{A} = E - \bar{l}\bar{c}^\top$ is Hurwitz. We can shown that $x = \hat{x} + x_\upsilon$, and

$$y = \bar{c}^\top \hat{x} + \bar{c}^\top x_\upsilon + d_n \upsilon = \bar{c}^\top (\hat{x} + \epsilon_x) + \upsilon_f,$$

where x_υ satisfies the equation

$$\dot{x}_\upsilon = \bar{A}x_\upsilon + (\bar{d} - d_n\bar{l})\upsilon, \quad x_\upsilon(0) = x(0) - \hat{x}(0),$$

$\upsilon_f = W_x(s)[\upsilon]$ is the filtered disturbance, $W_x(s) = d_n + \bar{c}^\top (sI - \bar{A})^{-1}(\bar{d} - d_n\bar{l})$ is the asymptotically stable transfer function, and ϵ_x exponentially decays according to the equation $\dot{\epsilon}_x = \bar{A}\epsilon_x$, $\epsilon_x(0) = x(0) - \hat{x}(0)$.

Following the approach introduced in Sect. 2.4.1.3, we use the filters

$$\dot{\hat{\xi}}_f = G\hat{\xi}_f + l(y - \bar{c}^\top \hat{x}), \tag{2.111}$$

$$\hat{\xi}_\kappa = \frac{\kappa_0}{\kappa(s)}\left[\hat{\xi}_f\right], \tag{2.112}$$

where the pair (G, l) is from the exosystem canonical form (2.15), $\kappa(s)$ is an arbitrary Hurwitz polynomial of degree no less than $n - m$, and $\kappa_0 > 0$ is a constant. Then, in view of Corollaries A.3 and A.5 from Appendix A, the filtered disturbance υ_f can be represented in the form

$$\upsilon_f = \theta^\top \hat{\xi}_\kappa + \epsilon_\kappa,$$

where $\theta \in \mathbb{R}^q$ is a vector of unknown parameters, ϵ_κ is the exponentially decaying term. Therefore,

$$y = \bar{c}^\top \hat{x} + \theta^\top \hat{\xi}_\kappa + \bar{c}^\top \epsilon_x + \epsilon_\kappa. \tag{2.113}$$

Finally, calculating the first time derivative of the plant output y in view of (2.113), (2.110) and (2.112) and adding the equations connecting the output y with the control u, we obtain the parameterized plant model established by the following statement.

Proposition 2.8 *If Assumptions 2.1 and 2.6 hold, then the plant (2.108), (2.109) admits the following parameterization:*

$$\dot{y} = -\bar{l}_{n-1}\hat{x}_1 + \hat{x}_2 + \bar{l}_{n-1}y + \phi_{n-1}(y) + \theta^\top \dot{\hat{\xi}}_\kappa + \bar{\epsilon}_\kappa, \tag{2.114}$$

$$\dot{\hat{x}}_i = -\bar{l}_{n-i}\hat{x}_1 + \hat{x}_{j+1} + \bar{l}_{n-i}y + \phi_i(y), \quad i = 2, 3, \ldots, n-m-1, \tag{2.115}$$

$$\dot{\hat{x}}_{n-m} = -\bar{l}_m\hat{x}_1 + \hat{x}_{n-m+1} + \bar{l}_m y + \phi_m(y) + b_m\phi_u(y)u, \tag{2.116}$$

where $\theta \in \mathbb{R}^q$ is a vector of unknown parameters, the state estimate $\hat{x}$ is generated by the observer (2.110), the regressor $\hat{\xi}_\kappa$ is generated by the filters (2.111) and (2.112), and $\bar{\epsilon}_\kappa = \bar{c}^\top \dot{\epsilon}_x + \dot{\epsilon}_\kappa$ is the exponentially decaying term.

2.5 Disturbance Observers: Plants with Unknown Parameters

In this section, we assume that the plant parameters are unknown and demonstrate how in this case uncertain disturbance inaccessible for measurements can be represented in a form of linear regression with unknown constant parameters and measured regressor. The problem of unknown parameters of the plant will be overcome with the use of special more complicated disturbance observers and an *extended* vector of unknown parameters. As in Sect. 2.4, we start with LTI plants and then extend the approach to different classes of nonlinear ones.

2.5.1 *Disturbance Observers for LTI Plants*

2.5.1.1 Plants with Matched Disturbance and Measured State

Let us consider the LTI plant

$$\dot{x} = Ax + b_M(\psi_u u + \upsilon) \tag{2.117}$$

with the measured state $x \in \mathbb{R}^n$, the control $u \in \mathbb{R}$, unknown matrix $A \in \mathbb{R}^{n\times n}$ and unknown constant ψ_u. We assume that the uncertainties of the matrix A are matched with the control, and the matrix A satisfies the following assumption.

Assumption 2.7 The matrix A can be represented in the form

$$A = A_M + b_M \psi^\top, \tag{2.118}$$

where $A_M \in \mathbb{R}^{n\times n}$ and $b_M \in \mathbb{R}^n$ are known, while the vector of constant parameters $\psi \in \mathbb{R}^n$ is unknown.

In this case, we use the following filters:

$$\dot{\eta}_0 = G\eta_0 + (GN - NA_M)x, \tag{2.119}$$
$$\dot{\eta}_i = G\eta_i - lx_i, \quad i = 1, 2, \dots, n, \tag{2.120}$$
$$\dot{\eta}_u = G\eta_u - lu, \tag{2.121}$$

where $\eta_0, \eta_i, \eta_u \in \mathbb{R}^q$ are the states of the corresponding filters with arbitrary initial conditions, x_i is the ith element of x, the pair (G, l) is from the exosystem canonical form (2.15), and the matrix $N \in \mathbb{R}^{q\times n}$ satisfies the equality

$$Nb_M = l. \tag{2.122}$$

The properties of the disturbance observer (2.119)–(2.122) are established as follows [20, 25, 27].

Lemma 2.7 *If Assumptions 2.1 and 2.7 hold and there exists a matrix $N \in \mathbb{R}^{q\times n}$ satisfying (2.122), then the disturbance υ affecting the plant (2.117), (2.118) can be represented in the form*

$$\upsilon = \theta^\top \hat{\xi} + \epsilon, \tag{2.123}$$

where $\theta \in \mathbb{R}^q$ is a vector of unknown parameters, ϵ is the exponentially decaying term, $\hat{\xi}$ is the bounded regressor defined as

$$\hat{\xi} = \eta_0 + Nx + \sum_{i=1}^{n} \psi_i \eta_i + \psi_u \eta_u, \tag{2.124}$$

ψ_i is the ith element of the vector ψ, and η_0, η_i, η_u are the states of the filters (2.119)–(2.122).

Proof Define the regressor estimation error as

$$\epsilon_\xi = \bar{\xi} - \hat{\xi} = \bar{\xi} - \eta_0 - Nx - \sum_{i=1}^{n} \psi_i \eta_i - \psi_u \eta_u,$$

where $\bar{\xi}$ is the state of the exosystem canonical form (2.15) and calculate its time derivative in view of (2.15) and (2.117)–(2.122):

$$
\begin{aligned}
\dot{\epsilon}_\xi &= G\xi + lv - G\eta_0 - GNx + NA_M x - NA_M x - Nb_M\psi^\top x - Nb_M\psi_u u - Nb_M v \\
&\quad - G\sum_{i=1}^{n}\psi_i\eta_i + l\sum_{i=1}^{n}\psi_i x_i - G\psi_u\eta_u + \psi_u l u \\
&= G\left(\xi - \eta_0 - Nx - \sum_{i=1}^{n}\psi_i\eta_i - \psi_u\eta_u\right) = G\epsilon_\xi .
\end{aligned}
$$

The latter proves the exponential decaying of ϵ_ξ. Further, substituting $\xi = \hat{\xi} + \epsilon_\xi$ into (2.14), we directly obtain (2.123) with $\epsilon = \theta^\top\epsilon_\xi$. □

However, the regressor (2.124) is not implementable since it contains unknown parameters ψ_i and ψ_u. To derive an appropriate parameterization of the plant (2.117), (2.118), we define the extended vector of unknown parameters

$$
\vartheta = \left[\psi^\top, \theta^\top, \psi_1\theta^\top, \psi_2\theta^\top, \ldots, \psi_n\theta^\top, \psi_u\theta^\top\right]^\top \tag{2.125}
$$

and the extended regressor

$$
\omega = \left[x^\top, (\eta_0 + Nx)^\top, \eta_1^\top, \eta_2^\top, \ldots, \eta_n^\top, \eta_u^\top\right]^\top . \tag{2.126}
$$

Proposition 2.9 *If Assumptions 2.1 and 2.7 hold and there exists a matrix $N \in \mathbb{R}^{q\times n}$ satisfying (2.122), then the plant (2.117), (2.118) can be represented in the form*

$$
\dot{x} = A_M x + b_M(\vartheta^\top\omega + \psi_u u + \epsilon), \tag{2.127}
$$

where $\vartheta \in \mathbb{R}^{n+(n+2)q}$ is a vector of unknown parameters, the regressor ω is given by (2.126), and η_0, η_i, η_u are generated by the filters (2.119)–(2.122).

The proposition is proved by direct substitution of (2.123), (2.124) into (2.117) in view of Assumption 2.7 and by combining the corresponding terms in the extended vectors (2.125), (2.126).

Remark 2.13 In contrast to the plants with known parameters, in the parameterization (2.127), the boundedness of the regressor ω is not established since ω depends on the state x, which is not known to be bounded a priori. This fact should be taken into account in design of adaptive control and in the proof of the closed-loop system stability as well. ■

2.5.1.2 Plants with Unmatched Disturbance and Measured State

Consider the LTI plant

$$
\dot{x} = Ax + bu + dv, \quad y = c^\top x + d_n v, \tag{2.128}
$$

where $x \in \mathbb{R}^n$ is the measured state, $u \in \mathbb{R}$ is the control, $y \in \mathbb{R}$ is the output, the matrix $A \in \mathbb{R}^{n \times n}$, the vectors $c,\ b,\ \ \in \mathbb{R}^n$ and the constant d_n are unknown, while the vector $d \in \mathbb{R}^n$ is known.

Additionally, we make the following assumptions.

Assumption 2.8 The plant (2.128) is such that:

A.2.4.1. the matrix A and the vector b can be represented in the forms

$$A = A_M + b_M \psi^\top, \quad b = b_M \psi_u, \tag{2.129}$$

where $A_M \in \mathbb{R}^{n \times n}$, $b_M \in \mathbb{R}^n$ are known, while the vector of constant parameters $\psi \in \mathbb{R}^n$ and the constant ψ_u are unknown.

A.2.4.2. the zeros of the transfer function $W_u(s) = c^\top (sI - A_M)^{-1} b_M$ have negative real parts.

We derive an appropriate parameterization of the plant (2.128) in the following three steps.

First, we obtain a plant model with a Hurwitz system matrix. Such a model will be suitable for design of certainty equivalence adaptive controllers. To this end, we introduce the vector of constant gains $k \in \mathbb{R}^n$ such that the matrix $A_\Sigma = A_M - b_M k^\top$ is Hurwitz and after routine calculations we get

$$\dot{x} = A_\Sigma x + b_M (\bar{\psi}^\top x + \psi_u u) + d\upsilon,$$

where $\bar{\psi} = \psi + k$.

Second, we match the disturbance with the control. For this purpose, we rewrite the plant model in the form

$$\dot{x} = A_\Sigma x + b_M (\bar{\psi}^\top x + \psi_u u) + d h^\top \xi, \quad y = c^\top x + d_n h^\top \xi$$

which together with the exosystem (2.1), (2.2) exactly duplicates the form of (1.48), (1.49) and (1.65) (with corresponding change of notations). Then, according to Lemmas 1.1 and 1.2, there exists a solution pair $M_\upsilon \in \mathbb{R}^{n \times q}$ and $\Theta_\upsilon = h_\upsilon \in \mathbb{R}^q$ of the matrix equation (1.67) that transforms the plant model into the form

$$\dot{e} = A_\Sigma e + b_M (\bar{\psi}^\top x + \psi_u u + \bar{\upsilon}),\ y = c^\top e, \tag{2.130}$$

where $e = x - M_\upsilon \xi$ is the new inaccessible for measurement state, $\bar{\upsilon}$ is the matched disturbance modelled as the output of the exosystem[9]

$$\dot{\xi} = \Gamma \xi,\ \bar{\upsilon} = h_\upsilon^\top \xi.$$

[9] See Remark 2.9 concerning zeros of the transfer function $W_\upsilon(s) = d_n + c^\top (sI - A_\Sigma)^{-1} d$.

Third, we obtain a suitable parameterization of the matched disturbance $\bar{\upsilon}$. Applying the approach from Sect. 2.5.1.1, we represent the disturbance $\bar{\upsilon}$ as (see Lemma 2.7)

$$\bar{\upsilon} = \theta^\top \hat{\xi} + \epsilon, \tag{2.131}$$

where $\theta \in \mathbb{R}^q$ is a vector of unknown parameters, ϵ is the exponentially decaying term, $\hat{\xi}$ is the regressor given by the equality

$$\hat{\xi} = \eta_0 + Nx + \sum_{i=1}^{n} \bar{\psi}_i \eta_i + \psi_u \eta_u, \tag{2.132}$$

in which $\bar{\psi}_i$ is the ith element of the vector $\bar{\psi}$, and η_0, η_i, η_u are the vectors generated by the filters

$$\dot{\eta}_0 = G\eta_0 + (GN - NA_\Sigma)x, \tag{2.133}$$

$$\dot{\eta}_i = G\eta_i - Nb_M x_i, \qquad i = 1, 2, \ldots, n, \tag{2.134}$$

$$\dot{\eta}_u = G\eta_u - Nb_M u \tag{2.135}$$

with the ith element of x denoted by x_i. The pair (G, l) is from the exosystem canonical form (2.15), and the matrix $N \in \mathbb{R}^{q \times n}$ is such that

$$Nd = l. \tag{2.136}$$

Now, substituting (2.132) and (2.131) into (2.130), we prove the following proposition.

Proposition 2.10 *If Assumptions 2.1 and 2.8 hold and there exists a matrix $N \in \mathbb{R}^{q \times n}$ satisfying (2.136), then the plant (2.128) can be represented in the form*

$$\dot{e} = A_\Sigma e + b_M(\vartheta^\top \omega + \psi_u u + \epsilon), \; y = c^\top e, \tag{2.137}$$

where $\vartheta \in \mathbb{R}^{n+(n+2)q}$ is an extended vector of unknown parameters given by

$$\vartheta = \left[\bar{\psi}^\top, \theta^\top, \bar{\psi}_1\theta^\top, \bar{\psi}_2\theta^\top, \ldots, \bar{\psi}_n\theta^\top, \psi_u\theta^\top\right]^\top, \tag{2.138}$$

$\omega \in \mathbb{R}^{n+(n+2)q}$ is the extended regressor given by

$$\omega = \left[x^\top, (\eta_0 + Nx)^\top, \eta_1^\top, \eta_2^\top, \ldots, \eta_n^\top, \eta_u^\top\right]^\top, \tag{2.139}$$

η_0, η_i, η_u are generated by the filters (2.133)–(2.135), and ϵ is the exponentially decaying term.

2.5.1.3 Plants with Unmatched Disturbance and Measured Output

Let us consider a controllable and observable LTI plant of the form (2.128) with the inaccessible state x and corresponding transfer functions

$$W_u(s) = c^\top (sI - A)^{-1} b = \frac{\beta(s)}{\alpha(s)}, \quad W_v(s) = d_n + c^\top (sI - A)^{-1} d = \frac{\delta(s)}{\alpha(s)},$$

where the polynomials

$$\alpha(s) = s^n + a_{n-1}s^{n-1} + \cdots + a_0, \beta(s) = b_m s^m + b_{m-1}s^{m-1} + \cdots + b_0,$$
$$\delta(s) = d_n s^n + d_{n-1}s^{n-1} + \cdots + d_0$$

have unknown parameters a_i, b_i and d_i.

We make the following assumption regarding the plant (2.128).

Assumption 2.9 The zeroes of the transfer function $W_u(s)$ (i.e., the roots of the polynomial $\beta(s)$) have negative real parts.

We also represent the plant (2.128) in the observable canonical form

$$\dot{\bar{x}} = E\bar{x} - \bar{a}y + \bar{b}u + \bar{d}v, \tag{2.140}$$
$$y = \bar{c}^\top \bar{x} + d_n v, \tag{2.141}$$

where $\bar{x}$ is the unmeasured state,

$$E = \begin{bmatrix} 0 & \\ \vdots & I_{n-1} \\ 0 & \cdots \; 0 \end{bmatrix}, \quad \bar{a} = \begin{bmatrix} a_{n-1} \\ \vdots \\ a_1 \\ a_0 \end{bmatrix}, \quad \bar{b} = \begin{bmatrix} 0 \\ \vdots \\ 0 \\ b_m \\ \vdots \\ b_0 \end{bmatrix}, \quad \bar{d} = \begin{bmatrix} d_{n-1} \\ \vdots \\ d_1 \\ d_0 \end{bmatrix}, \quad \bar{c} = \begin{bmatrix} 1 \\ 0 \\ \vdots \\ 0 \end{bmatrix}.$$

The plant parameterization employing the plant unknown parameters is defined by the following lemma [9, Sect. 10.1].

Lemma 2.8 *Consider the state estimation filters*

$$\dot{\zeta}_n = \bar{A}\zeta_n + \bar{l}y, \tag{2.142}$$
$$\dot{\zeta}_i = \bar{A}\zeta_i + \varsigma_{n-i}y, \quad i = 0, 1, \ldots, n-1, \tag{2.143}$$
$$\dot{v}_j = \bar{A}v_j + \varsigma_{n-j}u, \quad j = 0, 1, \ldots, m, \tag{2.144}$$

where $\zeta_i, \zeta_n, v_j \in \mathbb{R}^n$ are the states of the corresponding filters with arbitrary initial conditions, the vector of feedback gains $\bar{l} = \left[\bar{l}_{n-1}, \bar{l}_{n-2}, \ldots, \bar{l}_0\right]^\top$ is chosen so that *the matrix* $\bar{A} = E - \bar{l}\bar{c}^\top$ *is Hurwitz,* ς_i *is the ith coordinate vector in* $\mathbb{R}^n$. *Then the output of the plant (2.128) can be represented in the form*

$$y = \zeta_{n,1} + \psi^\top \varphi + \upsilon_f + \epsilon_y, \tag{2.145}$$

where the vector of unknown parameters ψ *and the regressor* φ *are given by*

$$\psi = \left[b_m, b_{m-1}, \ldots, b_0, -a_{n-1}, -a_{n-2}, \ldots, -a_0\right]^\top, \tag{2.146}$$

$$\varphi = \left[\nu_{m,1}, \nu_{m-1,1}, \ldots, \nu_{0,1}, \zeta_{n-1,1}, \zeta_{n-2,1}, \ldots, \zeta_{0,1}\right]^\top, \tag{2.147}$$

υ_f *is the filtered disturbance defined as*

$$\upsilon_f = W_f(s)[\upsilon], \;\; W_f(s) = \frac{d_n s^n + d_{n-1}s^{n-1} + \cdots + d_0}{s^n + \bar{l}_{n-1}s^{n-1} + \cdots + \bar{l}_0}, \tag{2.148}$$

and ϵ_y *is the exponentially decaying term.*

Proof Introduce the state estimate

$$\hat{x} = \zeta_n - \sum_{i=0}^{n-1} a_i \zeta_i + \sum_{j=0}^{m} b_j \nu_j$$

and the state estimation error $x_\upsilon = \bar{x} - \hat{x}$. Then we calculate the time derivative of x_υ in view of (2.140) and (2.142)–(2.144):

$$\begin{aligned}\dot{x}_\upsilon &= E\bar{x} - ay + \bar{b}u + \bar{d}\upsilon - \bar{A}\zeta_n - \bar{l}y + \bar{A}\sum_{i=0}^{n-1} a_i\zeta_i + \sum_{i=0}^{n-1} a_i\varsigma_{n-i}y \\ &- \bar{A}\sum_{j=0}^{m} b_j\nu_j - \sum_{j=0}^{m} b_j\varsigma_{n-j}u = \bar{A}\left(\bar{x} - \zeta_n + \sum_{i=0}^{n-1} a_i\zeta_i - \sum_{j=0}^{m} b_j\nu_j\right) + (\bar{d} - d_n\bar{l})\upsilon.\end{aligned}$$

Therefore,

$$\dot{x}_\upsilon = \bar{A}x_\upsilon + (\bar{d} - d_n\bar{l})\upsilon, \;\; x_\upsilon(0) = \bar{x}(0) - \hat{x}(0). \tag{2.149}$$

Substituting $\bar{x} = \hat{x} + x_\upsilon$ into (2.141) and collecting the corresponding terms into the vectors ψ and $\bar{\varphi}$, we directly obtain (2.145) with

$$\upsilon_f = \left(d_n + \bar{c}^\top(sI - \bar{A})^{-1}(\bar{d} - d_n\bar{l})\right)[\upsilon] = W_f(s)[\upsilon]$$

and $\epsilon_y = \bar{c}^\top \epsilon_x$, where ϵ_x satisfies the equation $\dot{\epsilon}_x = \bar{A}\epsilon_x$, $\epsilon_x(0) = \bar{x}(0) - \hat{x}(0)$. □

Corollary 2.3 *The time derivative of the plant output can be represented in the form*

$$\dot{y} = \beta_1 + \psi^{\top}\bar{\varphi} + \bar{\upsilon}_f + \epsilon_d, \tag{2.150}$$

where

$$\beta_1 = -\bar{l}_{n-1}\zeta_{n,1} + \zeta_{n,2} + \bar{l}_{n-1}y, \tag{2.151}$$

$$\begin{aligned}\bar{\varphi} = \big[&-\bar{l}_{n-1}\nu_{m,1} + \nu_{m,2}, -\bar{l}_{n-1}\nu_{m-1,1} + \nu_{m-1,2}, \ldots,\\ &-\bar{l}_{n-1}\nu_{0,1} + \nu_{0,2}, -\bar{l}_{n-1}\zeta_{n-1,1} + \zeta_{n-1,2} + y,\\ &-\bar{l}_{n-1}\zeta_{n-2,1} + \zeta_{n-2,2}, \ldots, -\bar{l}_{n-1}\zeta_{0,1} + \zeta_{0,2}\big]^{\top}\end{aligned} \tag{2.152}$$

are the known functions, ϵ_d is the exponentially decaying term, $\bar{\upsilon}_f$ is the new filtered disturbance defined as $\bar{\upsilon}_f = sW_f(s)[\upsilon] = W_f(s)[\dot{\upsilon}]$.

Corollary 2.3 is proved by taking the time derivative of $\zeta_{n,1}$ and φ in view of (2.142)–(2.144).

The model (2.150) contains the vector of the plant unknown parameters ψ and the filtered uncertain disturbance $\bar{\upsilon}_f$. However, in the general case when $m < n - 1$, it does not contain the control u explicitly. It is easy to check that there are $n - m - 1$ integrators between $\psi^{\top}\bar{\varphi}$, $\bar{\upsilon}_f$ and u. Therefore, in this model, the uncertainties are not matched with the control. Hence, iterative design tools like adaptive backstepping must be applied. To derive a parameterized model of the plant suitable for adaptive backstepping design, we need to obtain a parameterization of the derivative $\dot{y}$ and augment this model by $n - m - 1$ equations connecting the control u with $\dot{y}$. To this end, we use the vector ν_m, because its first element $\nu_{m,1}$ is connected with the control u via the minimum number of integrators equal to $n - m$.

A model which will be used in adaptive backstepping design is defined by the following proposition [20, 25, 27].

Proposition 2.11 *Together with the plant* (2.128) *and the state estimation filters* (2.142)–(2.144), *consider the disturbance estimation filters*

$$\dot{\eta}_0 = G\eta_0 + Gly - l\beta_1, \tag{2.153}$$

$$\dot{\eta}_i = G\eta_i - l\bar{\varphi}_i, \quad i = 1, 2, \ldots, n + m + 1 \tag{2.154}$$

with arbitrary initial conditions, where the pair (G, l) is from the exosystem canonical form (2.15), *$\bar{\varphi}_i$ is the ith element of the regressor $\bar{\varphi}$* (2.152). *Then, under Assumptions 2.1 and 2.9, the plant* (2.128) *with inaccessible state x admits the following parameterization:*

$$\dot{y} = \beta_1 + \vartheta^{\top}\omega + \bar{\epsilon}_d, \tag{2.155}$$

$$\dot{\nu}_{m,i} = -\bar{l}_{n-i}\nu_{m,1} + \nu_{n,i+1}, \quad i = 2, 3, \ldots, n - m - 1, \tag{2.156}$$

$$\dot{\nu}_{m,n-m} = -\bar{l}_{n-m}\nu_{m,1} + \nu_{m,n-m+1} + u, \tag{2.157}$$

where β_1 is defined by (2.151), the extended $((n+m+1)(q+1)+q)$-dimensional vector of unknown parameters ϑ and the extended regressor ω are given by

$$\vartheta = \left[\psi^\top, \theta^\top, \psi_1\theta^\top, \ldots, \psi_{n+m+1}\theta^\top\right]^\top, \tag{2.158}$$

$$\omega = \left[\bar{\varphi}^\top, (\eta_0 + ly)^\top, \eta_1^\top, \ldots, \eta_{n+m+1}^\top\right]^\top, \tag{2.159}$$

and $\bar{\epsilon}_d$ is the exponentially decaying term.

Proof First, we derive the parameterization of $\bar{\upsilon}_f$. In view of Remark A.1, we can write

$$\dot{\upsilon} = h^\top \Gamma \xi + \upsilon(0) D_\delta(t),$$

where $D_\delta(t) = \frac{d}{dt}1(t)$ is the Dirac delta function. Let us define $\upsilon_d = h^\top \Gamma \xi$ and represent υ_d as the output of the exosystem

$$\upsilon_d = \bar{h}^\top \xi, \quad \dot{\xi} = \Gamma \xi,$$

where $\bar{h}^\top = h^\top \Gamma$. Then we can write

$$\bar{\upsilon}_f = W_f(s)\,[\dot{\upsilon}] = W_f(s)\,[\upsilon_d + \upsilon(0) D_\delta(t)] = W_f(s)\,[\upsilon_d] + \epsilon_\delta,$$

where $\epsilon_\delta = W_f(s)[\upsilon(0) D_\delta(t)]$ exponentially decays. In view of Corollary A.3, the filtered disturbance $\bar{\upsilon}_f = W_f(s)[\upsilon_d]$ can be represented in the form

$$\bar{\upsilon}_f = \theta^\top \bar{\xi}_f + \bar{\epsilon}_\delta,$$

where $\theta \in \mathbb{R}^q$ is a vector of unknown parameters, $\bar{\xi}_f$ is the regressor generated by the virtual model

$$\dot{\bar{\xi}}_f = G\bar{\xi}_f + l\bar{\upsilon}_f, \tag{2.160}$$

the pair (G, l) is from the exosystem canonical form (2.15), while $\bar{\epsilon}_\delta$ exponentially decays.

Let us introduce the estimate of $\bar{\xi}_f$ as

$$\hat{\xi}_f = \eta_0 + \sum_{i=0}^{n+m+1} \psi_i \eta_i + ly$$

and define the estimation error $\bar{\epsilon}_f = \bar{\xi}_f - \hat{\xi}_f$. Then, in view of (2.150), (2.160) and (2.153), (2.154), we calculate the time derivative of $\bar{\epsilon}_f$:

$$\begin{aligned}\dot{\bar{\epsilon}}_f &= G\bar{\xi}_f + l\bar{\upsilon}_f - G\eta_0 - Gly + l\beta_1 - G\sum_{i=0}^{n+m+1}\psi_i\eta_i + l\sum_{i=0}^{n+m+1}\psi_i\bar{\varphi}_i \\ &- l\beta_1 - l\psi^\top\bar{\varphi} - l\bar{\upsilon}_f - l\epsilon_d = G\left(\bar{\xi}_f - \eta_0 - ly - \sum_{i=0}^{n+m+1}\psi_i\eta_i\right) - l\epsilon_d \\ &= G\bar{\epsilon}_f - l\epsilon_d.\end{aligned}$$

Hence, $\bar{\epsilon}_f\ \bar{\upsilon}_f = \theta^\top(\hat{\xi}_f + \bar{\epsilon}_f) + \bar{\epsilon}_\delta$ into (2.150) and collecting appropriate variables into the vectors ϑ and ω, we obtain (2.155). □

Again, as in Sect. 2.4.1.3, m equations describing the internal dynamics of the plant are omitted in the model (2.155)–(2.157). In the considered case, these dynamics are linear and, under Assumption 2.9, asymptotically stable. Therefore, Assumption 2.9 allows one to reduce the parameterized model to the $n - m$ equations (2.155)–(2.157) and to simplify design of a control law.

Finally, to explicitly represent the dependence of $\dot{y}$ on the variable $\nu_{m,2}$, which will be used in backstepping design as the first *virtual control* (see Appendix E), we rewrite Eq. (2.155) in the form

$$\dot{y} = \beta_1 + \vartheta^\top\bar{\omega} + b_m(-\bar{l}_{n-1}\nu_{m,1} + \nu_{m,2}) + \bar{\epsilon}_d, \tag{2.161}$$

where $\bar{\omega} = \left[\bar{\varphi}_0^\top, (\eta_0 + ly)^\top, \eta_1^\top, \ldots, \eta_{n+m+1}^\top\right]^\top$, and $\bar{\varphi}_0$ is obtained from $\bar{\varphi}$ (2.152) by zeroing of the first element, i.e.,

$$\begin{aligned}\bar{\varphi}_0 = \big[&0, -\bar{l}_{n-1}\nu_{m-1,1} + \nu_{m-1,2}, \ldots, -\bar{l}_{n-1}\nu_{0,1} + \nu_{0,2}, \\ &- \bar{l}_{n-1}\zeta_{n-1,1} + \zeta_{n-1,2} + y, -\bar{l}_{n-1}\zeta_{n-2,1} + \zeta_{n-2,2}, \ldots, -\bar{l}_{n-1}\zeta_{0,1} + \zeta_{0,2}\big]^\top.\end{aligned} \tag{2.162}$$

2.5.2 *Disturbance Observers for Nonlinear Plants*

2.5.2.1 Case Study

To present the underlying ideas in design of disturbance observers for nonlinear plants with unknown parameters, we first consider the nonlinear plant

$$\dot{x} = f(x) + \phi(x)u + d(x)\upsilon, \tag{2.163}$$

in which the control u, the plant uncertainties and the disturbance υ are matched due to the following parameterization:

$$f(x) = f_0(x) + d(x)\psi^\top\varphi(x), \phi(x) = d(x)\psi_u u, \tag{2.164}$$

where the vector $\psi \in \mathbb{R}^r$ and the constant ψ_u are unknown, and $f(x)$, $f_0(x)$, $\phi(x)$, $d(x)$, $\varphi(x)$ are the known smooth vector functions of appropriate dimensions.

In this case, we introduce the following filters of the disturbance observer:

$$\dot{\eta}_0 = G\eta_0 + G\pi(x) - \frac{\partial \pi(x)}{\partial x} f_0(x), \tag{2.165}$$

$$\dot{\eta}_i = G\eta_i - l\varphi_i(x), i = 1, 2, \ldots, r, \tag{2.166}$$

$$\dot{\eta}_u = G\eta_u - lu, \tag{2.167}$$

where $\eta_0,\ \eta_i,\ \eta_u \in \mathbb{R}^q$ are the states of the filters with arbitrary initial conditions, the pair (G, l) is from the exosystem canonical form (2.15), φ_i is the ith element of the vector φ, and the function $\pi(x)$ is a solution of the PDE[10]

$$\frac{\partial \pi(x)}{\partial x} d(x) = l. \tag{2.168}$$

The disturbance parameterization obtained with the use of the filters (2.165)–(2.168) is defined by the following lemma.

Lemma 2.9 *If Assumption 2.1 holds and there exists a function $\pi(x)$ satisfying (2.168), then the disturbance υ affecting the plant (2.163), (2.164) can be represented in the form*

$$\upsilon = \theta^\top \hat{\xi} + \epsilon, \tag{2.169}$$

where $\theta \in \mathbb{R}^q$ is a vector of unknown parameters, ϵ is the exponentially decaying term, $\hat{\xi}$ is the bounded regressor estimate given by

$$\hat{\xi} = \eta_0 + \pi(x) + \sum_{i=1}^{r} \psi_i \eta_i + \psi_u \eta_u, \tag{2.170}$$

and $\eta_0,\ \eta_i,\ \eta_u$ are generated by the filters (2.165)–(2.167).

Proof Define the disturbance estimation error:

$$\epsilon_\xi = \bar{\xi} - \hat{\xi} = \bar{\xi} - \eta_0 - \pi(x) - \sum_{i=1}^{r} \psi_i \eta_i - \psi_u \eta_u,$$

where $\bar{\xi}$ is the state of the exosystem canonical form (2.15). Calculating the time derivative of ϵ_ξ, in view of (2.15), (2.163), (2.164) and (2.163)–(2.168), we obtain

[10] See Remark 2.10 concerning possible simplification of this PDE.

$$
\begin{aligned}
\dot{\epsilon}_\xi &= G\bar{\xi} + lv - G\eta_0 - G\pi(x) + \frac{\partial \pi(x)}{\partial x} f_0(x) \\
&- \frac{\partial \pi(x)}{\partial x}\left(f_0(x) + d(x)\psi^\top \varphi(x) + \phi(x)u + d(x)v\right) \\
&- \sum_{i=1}^{r} \psi_i \left(G\eta_i - l\varphi_i\right) - \psi_u(G\eta_u - lu) \\
&= G\left(\xi - \eta_0 - \pi(x) - \sum_{i=1}^{r} \psi_i \eta_i - \psi_u u\right) = G\epsilon_\xi .
\end{aligned}
$$

The latter proves the exponential decaying of ϵ_ξ. Further, substituting $\xi = \hat{\xi} + \epsilon_\xi$ into (2.14) we directly obtain (2.169) with $\epsilon = \theta^\top \epsilon_\xi$. □

However, the regressor estimate (2.170) is not implementable since it contains unknown parameters ψ and ψ_u. To derive an appropriate parameterization of the plant (2.163), (2.164), we introduce the extended vector of unknown parameters

$$
\vartheta = \left[\psi^\top, \theta^\top, \psi_1\theta^\top, \psi_2\theta^\top, \ldots, \psi_r\theta^\top, \psi_u\theta^\top\right]^\top \tag{2.171}
$$

and the extended regressor

$$
\omega = \left[\varphi^\top, (\eta_0 + \pi(x))^\top, \eta_1^\top, \eta_2^\top, \ldots, \eta_r^\top, \eta_u^\top\right]^\top . \tag{2.172}
$$

Proposition 2.12 *If Assumption 2.1 holds and there exists a function $\pi(x)$ satisfying (2.168), then the plant (2.163), (2.164) can be represented in the form*

$$
\dot{x} = f_0(x) + d(x)\left(\vartheta^\top \omega + \psi_u u + \epsilon\right), \tag{2.173}
$$

where $\vartheta \in \mathbb{R}^{r+(r+2)q}$ is a vector of unknown parameters, ω is the vector defined by (2.172), and η_0, η_i, η_u are generated by the filters (2.165)–(2.167).

The proposition is proved by direct substitution of (2.169), (2.170) into (2.163), (2.164) in view of (2.171), (2.172).

In the next sections, the general approach presented by Proposition 2.12 will be applied to particular canonical forms of nonlinear plants.

2.5.2.2 Plant with Unmatched Uncertainties and Measured State

Consider an uncertain nonlinear plant represented in the parametric strict-feedback canonical form with external disturbances

$$\dot{x}_i = x_{i+1} + \psi^\top \phi_i(x_1, x_2, \cdots, x_i) + \bar{d}_{n-i}^\top \upsilon, \quad i = 1, \ldots, n-1, \tag{2.174}$$
$$\dot{x}_n = \psi^\top \phi_n(x) + \phi_x(x) + \phi_u(x)u + \bar{d}_0^\top \upsilon, \tag{2.175}$$
$$y = x_1, \tag{2.176}$$

where x_i is the ith element of the measured state vector $x \in \mathbb{R}^n$, $u \in \mathbb{R}$ is the control, $y \in \mathbb{R}$ is the regulated variable, $\psi \in \mathbb{R}^k$ is a constant vector of unknown parameters, $\phi_x(x), \phi_u(x), \phi_i(\cdot)$ are the known smooth scalar functions, $\bar{d}_i = \left[d_{i,1}, d_{i,2}, \ldots, d_{i,\rho}\right]^\top$ is the known constant vector ($i = 0, 1, \ldots, n-1$), $\upsilon \in \mathbb{R}^\rho$ is the multidimensional disturbance satisfying Assumption 2.2.

We also represent the plant (2.174)–(2.176) in the matrix form

$$\dot{x} = Ex + \varsigma_n \left(\phi_x(x) + \phi_u(x)u\right) + \Phi(x)\psi + D\upsilon, \quad y = \varsigma_1^\top x, \tag{2.177}$$

where

$$E = \begin{bmatrix} 0 & \\ \vdots & I_{n-1} \\ 0 \cdots 0 & \end{bmatrix}, \Phi = \begin{bmatrix} \phi_1^\top \\ \phi_2^\top \\ \vdots \\ \phi_n^\top \end{bmatrix}, D = \begin{bmatrix} \bar{d}_{n-1}^\top \\ \vdots \\ \bar{d}_1^\top \\ \bar{d}_0^\top \end{bmatrix}.$$

In the considered case, the disturbance observer as well as the corresponding disturbance parameterization are defined by the following lemma.[11]

Lemma 2.10 *If Assumption 2.2 holds and there exists a matrix $N \in \mathbb{R}^{q \times n}$ such that*

$$ND = L, \tag{2.178}$$

then the disturbance υ affecting the plant (2.174)–(2.176) can be represented in the form

$$\upsilon = T_0\omega + \epsilon_\upsilon, \tag{2.179}$$

where $\rho \times q(k+1)$-dimensional extended matrix of unknown parameters T_0 and $q(k+1)$-dimensional extended regressor ω are given by

$$T_0 = \left[\Theta, \psi_1\Theta, \psi_2\Theta, \ldots, \psi_k\Theta\right], \tag{2.180}$$
$$\omega = \left[(\eta + Nx)^\top, \Lambda_1^\top, \Lambda_2^\top, \ldots, \Lambda_k^\top\right]^\top, \tag{2.181}$$

ψ_i is the ith element of the vector ψ, Λ_i is the ith column of the matrix Λ, ϵ_υ is the exponentially decaying term, the pair (G, L) is defined by (2.25), (2.26), $\Theta \in \mathbb{R}^{\rho \times q}$ is a matrix of unknown parameters form the exosystem canonical form (2.22), (2.23), while $\eta \in \mathbb{R}^q$ and $\Lambda \in \mathbb{R}^{q \times k}$ are generated by the filters

[11] This lemma extends the results of [21] to the case of multidimensional disturbance.

$$\dot{\eta} = G\eta + GNx - NEx - N\varsigma_n\left(\phi_x(x) + \phi_u(x)u\right), \tag{2.182}$$
$$\dot{\Lambda} = G\Lambda - N\Phi(x) \tag{2.183}$$

with arbitrary initial conditions.

Proof Consider the exosystem canonical form (2.23) with the state $\bar{\xi}$ and define the state estimation error

$$\epsilon_{\xi} = \bar{\xi} - \eta - Nx - \Lambda\psi. \tag{2.184}$$

Calculating its time derivative, in view of (2.23), (2.177), (2.182), (2.183) and (2.178), we obtain

$$\begin{aligned}\dot{\epsilon}_{\xi} &= G\bar{\xi} + L\upsilon - G\eta - GNx + NEx + N\varsigma_n\left(\phi_x(x) + \phi_u(x)u\right)\\ &- NEx - N\varsigma_n(\phi_x(x) + \phi_u(x)u) - N\Phi(x)\psi - ND\upsilon\\ &- G\Lambda\psi + N\Phi(x)\psi = G(\bar{\xi} - \eta - Nx - \Lambda\psi) = G\epsilon_{\xi}.\end{aligned}$$

The latter means that ϵ_{ξ} exponentially decays, and, hence,

$$\bar{\xi} = \eta + Nx + \Lambda\psi + \epsilon_{\xi}. \tag{2.185}$$

Replacing (2.185) in (2.22), we directly obtain (2.179) with $\epsilon = \Theta\epsilon_{\xi}$ and T_0, ω defined by (2.180) and (2.181), respectively. □

Remark 2.14 The above lemma can be directly extended to the case of multidimensional exosignal υ satisfying Assumption 2.3. In this case, for construction of the observer (2.178), (2.182), and (2.183), one can use either matrices G and L of the form (2.25), (2.26) or the block-diagonal matrices (2.36). ■

Since in the plant (2.174)–(2.176) neither the unknown parameters ψ, nor the uncertain disturbances υ_i are matched with the control u, we will use the backstepping to design an adaptive controller (see Appendix E). At each design step of this procedure the corresponding time derivative of υ and, as a results, of ω must be calculated. The HOTD of ω can be calculated analytically based on the expressions (2.182), (2.183). However, to simplify analytical calculations, we will use the filtered regressor. To this end, instead of parameterization (2.22), consider expression

$$\upsilon = \bar{\Theta}\bar{\xi}_{\kappa} + \epsilon_{\kappa}, \tag{2.186}$$

where $\bar{\Theta}$ is a new matrix of unknown parameters, $\bar{\xi}_{\kappa} = \frac{\kappa_o}{\kappa(s)}[\xi]$ is the filtered regressor, $\kappa(s)$ is an arbitrary Hurwitz polynomial of degree no less than $n-1$, $\kappa_0 > 0$ is some constant, and ϵ_{κ} is the exponentially decaying term. Validity of the latter equality follows from Corollaries A.2 and A.5 extended to the case of multidimensional disturbance $\upsilon \in \mathbb{R}^{\rho}$ and the diagonal transfer function matrix $W(s) = \frac{\kappa_0}{\kappa(s)} I_{\rho}$.

Then, in view of (2.185), we can write

$$\bar{\xi}_\kappa = \frac{\kappa_0}{\kappa(s)}[\eta + Nx + \Lambda\psi + \epsilon_\xi].$$

Substituting the latter equation into (2.186) and combining the corresponding terms in the extended regressor

$$\omega_\kappa = \frac{\kappa_0}{\kappa(s)}[\omega] \tag{2.187}$$

and in a new matrix of unknown parameters $\bar{T}_0 \in \mathbb{R}^{\rho \times q(k+1)}$, we finally obtain

$$\upsilon = \bar{T}_0 \omega_\kappa + \bar{\epsilon}_\kappa, \tag{2.188}$$

where $\bar{\epsilon}_\kappa = \bar{\Theta}\epsilon_\xi + \epsilon_\kappa$ exponentially decays.

Therefore, we have proved the following statement defining the appropriate plant model parameterization.

Proposition 2.13 *Consider the plant (2.174)–(2.176) affected by the disturbance υ satisfying Assumption 2.2. Then the parameterized plant model can be represented in the form*

$$\dot{x} = Ex + \varsigma_n\left(\phi_x(x) + \phi_u(x)u\right) + \Phi(x)\psi + T\omega_\kappa + \epsilon, \tag{2.189}$$

where $T = D\bar{T}_0$ is the $n \times q(k+1)$-dimensional matrix of unknown parameters, the filtered regressor ω_κ is defined by (2.187), (2.181) with η and Λ_i generated by the filters (2.182), (2.183), and $\epsilon = D(\bar{\Theta}\epsilon_\xi + \epsilon_\kappa)$ is the exponentially decaying n-dimensional vector.

2.5.2.3 Plant with Unmatched Uncertainties and Measured Output

Consider a nonlinear plant represented in the *parametric output-feedback* canonical form with the disturbance

$$\dot{x} = Ex + \phi_x(y) + \sum_{i=1}^{k} \phi_i(y)\psi_i + \bar{b}\phi_u(y)u + \bar{d}\upsilon, \tag{2.190}$$

$$y = \bar{c}^\top x + d_n\upsilon, \tag{2.191}$$

where $x \in \mathbb{R}^n$ is the inaccessible state,

$$E = \begin{bmatrix} 0 & \\ \vdots & I_{n-1} \\ 0 \cdots 0 & \end{bmatrix}, \quad \bar{b} = \begin{bmatrix} 0 \\ \vdots \\ 0 \\ b_m \\ \vdots \\ b_0 \end{bmatrix}, \quad \bar{d} = \begin{bmatrix} d_{n-1} \\ \vdots \\ d_1 \\ d_0 \end{bmatrix}, \quad \bar{c} = \begin{bmatrix} 1 \\ 0 \\ \vdots \\ 0 \end{bmatrix},$$

$\phi_i(y) \in \mathbb{R}^n$, $\phi_x(y) \in \mathbb{R}^n$ are the known smooth vector functions, $\phi_u(y)$ is the known smooth function, b_i, d_i, d_n, ψ_i are unknown parameters.

We make the following assumption regarding the plant (2.190), (2.191).

Assumption 2.10 The polynomial $\beta(s) = b_m s^m + \cdots + b_1 s + b_0$ is Hurwitz.

The plant output parameterization employing the plant unknown parameters is defined by the following lemma (see [9], p. 308).

Lemma 2.11 *Consider the state estimation filters*

$$\dot{\zeta}_n = \bar{A}\zeta_n + \bar{l}y + \phi_x(y), \tag{2.192}$$
$$\dot{\zeta}_i = \bar{A}\zeta_i + \phi_i(y), i = 1, 2, \ldots, k, \tag{2.193}$$
$$\dot{\nu}_j = \bar{A}\nu_j + \varsigma_{n-j}\phi_u(y)u, \ j = 0, 1, \ldots, m, \tag{2.194}$$

where ζ_i, ζ_n, $\nu_j \in \mathbb{R}^n$ are the states of the corresponding filters with arbitrary initial conditions, the vector of feedback gains $\bar{l} = \left[\bar{l}_{n-1}, \bar{l}_{n-2}, \ldots, \bar{l}_0\right]^\top$ is chosen so that the matrix $\bar{A} = E - \bar{l}\bar{c}^\top$ is Hurwitz, ς_i is the ith coordinate vector in $\mathbb{R}^n$. Then the plant output can be represented in the form

$$y = \zeta_{n,1} + \bar{\psi}^\top \varphi + \upsilon_f + \epsilon_y \tag{2.195}$$

with the vector of the plant unknown parameters $\bar{\psi}$ and the regressor φ given by

$$\bar{\psi} = [b_m, b_{m-1}, \ldots, b_0, \psi_k, \psi_{k-1}, \ldots, \psi_1]^\top, \tag{2.196}$$
$$\varphi = [\nu_{m,1}, \nu_{m-1}, \ldots, \nu_{0,1}, \zeta_{k,1}, \zeta_{k-1,1}, \ldots, \zeta_{1,1}]^\top, \tag{2.197}$$

the filtered disturbance υ_f defined as

$$\upsilon_f = W_f(s)[\upsilon], \quad W_f(s) = \frac{d_n s^n + d_{n-1}s^{n-1} + \cdots + d_0}{s^n + \bar{l}_{n-1}s^{n-1} + \cdots + \bar{l}_0},$$

and the exponentially decaying term ϵ_y.

Proof Introduce the state estimate

$$\hat{x} = \zeta_n - \sum_{i=1}^{k} \psi_i \zeta_i + \sum_{j=0}^{m} b_j \nu_j$$

and the state estimation error $x_\upsilon = x - \hat{x}$. Then, calculating the time derivative of x_υ, in view of (2.192)–(2.194) and the plant model (2.190), (2.191), we obtain

$$\dot{x}_\upsilon = \bar{A}x_\upsilon + \left(\bar{d} - \bar{l}d_n\right)\upsilon.$$

Parameterization (2.195)–(2.197) is derived by substitution of $x = \hat{x} + x_\upsilon$ into (2.191). □

Corollary 2.4 *The first time derivative of the plant output can be represented in the form*

$$\dot{y} = \bar{\zeta}_1 + \bar{\psi}^\top \bar{\varphi} + \bar{\upsilon}_f + \epsilon_d, \tag{2.198}$$

where

$$\bar{\zeta}_1 = -\bar{l}_{n-1}\zeta_{n,1} + \zeta_{n,2} + \bar{l}_{n-1}y + \phi_{x1}(y), \tag{2.199}$$

$$\begin{aligned}\bar{\varphi} = \big[&-\bar{l}_{n-1}\nu_{m,1} + \nu_{m,2}, -\bar{l}_{n-1}\nu_{m-1,1} + \nu_{m-1,2}, \ldots, -\bar{l}_{n-1}\nu_{0,1} + \nu_{0,2}\,, \\ &- \bar{l}_{n-1}\zeta_{k,1} + \zeta_{k,2} + \phi_{k,1}(y), \ldots, -\bar{l}_{n-1}\zeta_{1,1} + \zeta_{1,2} + \phi_{1,1}(y)\big]^\top\end{aligned} \tag{2.200}$$

are the known functions, $\phi_{i,1}$ is the first element of the vector ϕ_i $(i = 1, 2, \ldots, k)$, ϕ_{x1} is the first element of the vector ϕ_x, $\bar{\upsilon}_f$ is the new filtered disturbance defined as $\bar{\upsilon}_f = W_f(s)[\dot{\upsilon}]$, and ϵ_d is the exponentially decaying term.

It is worth noting that the model (2.198) has the same structure as the model (2.150) derived for the linear plant. Therefore, using the same arguments as in Sect. 2.5.1.3, we can prove the following statement.

Proposition 2.14 *Together with the plant* (2.190), (2.191) *and the state estimation filters* (2.192)–(2.194), *consider the disturbance observer*

$$\dot{\eta}_0 = G\eta_0 + Gly - l\bar{\zeta}_1, \tag{2.201}$$

$$\dot{\eta}_i = G\eta_i - l\bar{\varphi}_i, \qquad i = 1, 2, \ldots, k + m + 1, \tag{2.202}$$

where the pair (G, l) is from the exosystem canonical form (2.15), *$\bar{\varphi}_i$ is the ith element of the regressor $\bar{\varphi}$* (2.200). *Then, under Assumptions 2.1 and 2.10, the plant* (2.190) *admits the following parameterization:*

$$\dot{y} = \bar{\zeta}_1 + \vartheta^\top \omega + \bar{\epsilon}_d, \tag{2.203}$$

$$\dot{\nu}_{m,i} = -\bar{l}_{n-i}\nu_{m,1} + \nu_{n,i+1}, \qquad i = 2, 3, \ldots, n - m - 1, \tag{2.204}$$

$$\dot{\nu}_{m,n-m} = -\bar{l}_{n-m}\nu_{m,1} + \nu_{m,n-m+1} + \phi_u(y)u, \tag{2.205}$$

where $\bar{\zeta}_1$ is defined by (2.199), *the extended $(q + (k + m + 1)(q + 1))$-dimensional vector of unknown parameters ϑ and the extended regressor ω are given by*

$$\vartheta = \left[\bar{\psi}^\top, \theta^\top, \bar{\psi}_1\theta^\top, \ldots, \bar{\psi}_{k+m+1}\theta^\top\right]^\top, \tag{2.206}$$

$$\omega = \left[\bar{\varphi}^\top, (\eta_0 + l y)^\top, \eta_1^\top, \ldots, \eta_{k+m+1}^\top\right]^\top, \tag{2.207}$$

and $\bar{\epsilon}_d$ is the exponentially decaying term.

Finally, as shown in Sect. 2.5.1.3, to explicitly represent the dependence of $\dot{y}$ on the variable $\nu_{m,2}$, which will be used in backstepping design as the first *virtual control* (see Appendix E), we rewrite Eq. (2.203) in the form

$$\dot{y} = \bar{\zeta}_1 + \vartheta^\top \bar{\omega} + b_m(-\bar{l}_{n-1}\nu_{m,1} + \nu_{m,2}) + \bar{\epsilon}_d, \tag{2.208}$$

where $\bar{\omega} = \left[\bar{\varphi}_0^\top, (\eta_0 + l y)^\top, \eta_1^\top, \ldots, \eta_{k+m+1}^\top\right]^\top$, and $\bar{\varphi}_0$ is obtained from $\bar{\varphi}$ (2.200) by zeroing of the first element, i.e.,

$$\begin{aligned} \bar{\varphi}_0 = \big[&0, -\bar{l}_{n-1}\nu_{m-1,1} + \nu_{m-1,2}, \ldots, -\bar{l}_{n-1}\nu_{0,1} + \nu_{0,2}\,, \\ &- \bar{l}_{n-1}\zeta_{k,1} + \zeta_{k,2} + \phi_{k,1}(y), \ldots, -\bar{l}_{n-1}\zeta_{1,1} + \zeta_{1,2} + \phi_{1,1}(y)\big]^\top . \end{aligned} \tag{2.209}$$

References

1. Chen, C.-T.: Linear Systems. Theory and Design, 3rd edn. Oxford University Press, New York (1999)
2. Dong, W., Farrell, J.A., Polycarpou, M.M., Djapic, V., Sharma, M.: Command filtered adaptive backstepping. IEEE Trans. Control Syst. Technol. **20**(3), 566–580 (2012)
3. Farrell, J.A., Polycarpou, M., Sharma, M., Dong, W.: Command filtered backstepping. IEEE Trans. Autom. Control **54**(6), 1391–1395 (2009)
4. Gantmacher, F.R.: The Theory of Matrices, vol. 1. Chelsea Publishing Company, New York (1960)
5. Gerasimov, D.N., Pashenko, A.V., Nikiforov, V.O.: Improved adaptive compensation of unmatched multisinusoidal disturbances in uncertain nonlinear plants. In: American Control Conference (ACC'20), pp. 626–632 (2020)
6. Gerasimov, D.N., Pashenko, A.V., Nikiforov, V.O.: Improved adaptive servotracking for a class of nonlinear plants with unmatched uncertainties. IFAC-PapersOnLine **53**(2), 3835–3840 (2020)
7. Ioannou, P., Sun, J.: Robust Adaptive Control. Prentice-Hall, Englewood Cliffs (1996)
8. Isidori, A.: Lectures in Feedback Design for Multivariable Systems. Springer International Publishing, Berlin (2017)
9. Krstić, M., Kanellakopoulos, I., Kokotović, P.: Nonlinear and Adaptive Control Design. Wiley, New York (1995)
10. Khalil, H.K.: Nonlinear Control. Prentice-Hall, Englewood Cliffs (2015)
11. Levine, W.S. (ed.): The Control Handbook. CRC Press, IEEE Press, Boca Raton (2000)
12. Luenberger, D.G.: Invertible solutions to the operator equation $TA - BT = C$. Proc. Amer. Math. Soc. **16**(6), 1226–1229 (1965)
13. Marino, R., Santosuosso, G.L.: Regulation of linear systems with unknown exosystems of uncertain order. IEEE Trans. Autom. Control **52**(2), 352–359 (2007)
14. Marino, R., Tomei, P.: Output regulation for linear systems via adaptive internal model. IEEE Trans. Autom. Control **48**, 2199–2202 (2003)

15. Narendra, K.S., Annaswamy, A.M.: Stable Adaptive Systems. Prentice Hall, Hoboken (1989)
16. Nikiforov, V.O.: Adaptive servocompensation of input disturbances. In: Proceedings of the 13th World Congress of IFAC, San Francisco USA, vol. 29, no. 1, pp. 5114–5119 (1996)
17. Nikiforov, V.O.: Adaptive servomechanism controller with an implicit reference model. Int. J. Control **68**(2), 277–286 (1997)
18. Nikiforov, V.O.: Adaptive stabilization of a linear plant subjected to external deterministic disturbances. J. Comput. Syst. Sci. Int. **36**(2), 258–261 (1997)
19. Nikiforov, V.O.: Nonlinear control system with compensation of external deterministic disturbances. J. Comput. Syst. Sci. Int. **36**(4), 564–568 (1997)
20. Nikiforov, V.O.: Adaptive controller rejecting uncertain deterministic disturbances in SISO systems. In: 4th European Control Conference Belgium, Brussels (1997)
21. Nikiforov, V.O.: Adaptive nonlinear tracking with complete compensation of unknown disturbances. Eur. J. Control **4**, 132–139 (1998)
22. Nikiforov, V.O. : Adaptive nonlinear servocompensation of unknown external disturbances. In: 14th World Congress of IFAC, vol. 1, Beijing, P.R. China, pp. 283–289 (1999)
23. Nikiforov, V.O.: Nonlinear Servocompensation of unknown external disturbances. Automatica **37**(10), 1647–1653 (2001)
24. Nikiforov, V.O., Voronov, K.V.: Adaptive backstepping with a high-order tuner. Automatica **37**(12), 1953–1960 (2001)
25. Nikiforov, V.O.: Adaptive and Robust Control with Disturbance Compensation. Nauka, St. Petersburg (2003). (in Russian)
26. Nikiforov, V.O.: Observers of external deterministic disturbances. Part I. Objects with known parameters. Autom. Remote Control **65**(10), 1531–1541 (2004)
27. Nikiforov, V.O.: Observers of external deterministic disturbances. Part II. Objects with unknown parameters. Autom. Remote Control **65**(11), 1724–1732 (2004)
28. Nikiforov, V.O., Gerasimov, D.N., Pashenko, A.V.: Modular adaptive backstepping design with a high-order tuner. IEEE Trans. Autom. Control. https://doi.org/10.1109/TAC.2021.3091442. (to appear)
29. Obregon-Pulido, G., Castillo-Toledo, B., Loukianov, A.G.: A structurally stable globally adaptive internal model regulator for MIMO linear systems. IEEE Trans. Autom. Control **56**, 160–165 (2011)
30. Polyanin, A.D., Zaitsev, V.F.: Handbook of Ordinary Differential Equations: Exact Solutions, Methods, and Problems. Taylor & Francis, Milton Park (2018)
31. Sepulchre, R., Jankovic, M., Kokotović, P.: Constructive Nonlinear Control. Springer, London (1997)
32. Zheng, X., Yang, X.: Command filter and universal approximator based backstepping control design for strict-feedback nonlinear systems with uncertainty. IEEE Trans. Autom. Control **65**(3), 1310–1317 (2020)

Chapter 3
Basic Algorithms of Adaptation

Abstract In this chapter, based on the plant parameterizations presented in Chap. 2 we define canonical *error models* and derive for them *basic* algorithms of adaptation. Constructing such algorithms and establishing their convergence properties, we provide theoretical background that will be used in the next chapters for design of different adaptive controllers addressed in this book. Section 3.1 introduces canonical error models and defines classes of the regressor to be considered. Section 3.2 presents basic algorithms of adaptation for the static error model, while Sects. 3.3 and 3.4 present basic algorithms for the dynamic error models with measured state and output, respectively. Robust modifications of the basic algorithms of adaptation are presented in Sect. 3.5. The design methods used in the chapter are restricted to the so-called *Lyapunov-based design*. The relevant Lyapunov stability notions and theorems are briefly reviewed in Appendix B, while Appendix C presents norms and some useful properties of functions used in the stability analysis.

3.1 Parameterizations and Error Models

In Chap. 2, we introduce the reference and disturbance observers enabling one to represent the regular uncertain exosignals in the form of linear regression with a constant vector of unknown parameters and measured regressor. In own turn, with the use of such representation of the exosignals suitable parameterizations of different classes of linear and nonlinear plants affected by uncertain external disturbances are derived. In this section, we will augment these parameterizations by adjustable feedback loops and obtain relevant *error models*, the concept of which was introduced in [25] and developed in [26, Chap. 7]. Then we will generalize the classes of these error models and introduce their canonical forms used for design of algorithms of adaptation. Since these algorithms of adaptation are designed for sufficiently wide classes of the error models independently of particular control problems, we will call them *basic* algorithms of adaptation.

V. Nikiforov and D. Gerasimov, *Adaptive Regulation*, Lecture Notes in Control and Information Sciences 491, https://doi.org/10.1007/978-3-030-96091-9_3

Static model of parameterization. In Sect. 2.3, with the use of the reference observer we obtained the static parameterization of the exosignal in the form of expression (2.39). If we define the estimation of the reference signal as $\hat{\upsilon} = \hat{\theta}^\top \hat{\xi}$, where $\hat{\theta}$ is the vector of adjustable parameters generated by an adaptation algorithm, then we can obtain the following static model of the estimation error:

$$\tilde{\upsilon} = \tilde{\theta}^\top \hat{\xi} + \epsilon, \tag{3.1}$$

where $\tilde{\upsilon} = \upsilon - \hat{\upsilon}$ is the estimation error, $\tilde{\theta} = \theta - \hat{\theta}$ is the vector of parametric errors, $\hat{\xi}$ is the measured regressor, and ϵ is some exponentially decaying term. Defining an adaptation algorithm as

$$\dot{\hat{\theta}} = \Psi_\upsilon(\tilde{\upsilon}, \hat{\xi}),$$

where the function $\Psi_\upsilon(\tilde{\upsilon}, \hat{\xi})$ is to be found, in view of equality $\dot{\tilde{\theta}} = -\dot{\hat{\theta}}$, we derive the following *parametric error model*:

$$\dot{\tilde{\theta}} = -\Psi_\upsilon(\tilde{\upsilon}, \hat{\xi}). \tag{3.2}$$

With a slight abuse of notion, we will refer to the Eqs. (3.1) and (3.2) as the *static error model* in spite of the fact that the parametric error model (3.2) is a dynamic subsystem.

Then the objective of adaptation law design can be formulated as follows: to determine a function $\Psi_\upsilon(\tilde{\upsilon}, \hat{\xi})$ such that the origin $\tilde{\theta} = 0$ of the parametric error model (3.2) is stable and $\tilde{\upsilon}(t) \to 0$ as $t \to \infty$.

Dynamic model of parameterization. As it was demonstrated in Chap. 2, if a plant is affected by an external regular disturbance, then with the use of special disturbance observer the plant can be represented by a parameterized model. One of the most general forms of such models is described by Eq. (2.173). In order to focus the main attention on the compensation of the plant uncertainties, we introduce a nominal (i.e., without uncertainties) plant model of the form

$$\dot{x} = f_0(x) + d(x)u \tag{3.3}$$

and assume that there exists a smooth control law $u = u_s(x)$ ensuring the global asymptotic stability of the closed-loop nominal plant model $\dot{x} = f_\Sigma(x)$ with $f_\Sigma(x) = f_0(x) + d(x)u_s(x)$.

If we choose the control law $u = \hat{\vartheta}_\psi u_s(x) - \hat{\vartheta}^\top \omega$, where $\hat{\vartheta}$ is the vector of adjustable parameters and $\hat{\vartheta}_\psi$ is the auxiliary adjustable parameter, then replacing this control law in (2.173) we obtain the following dynamic model of the state error:

$$\dot{x} = f_\Sigma(x) + \psi_u d(x)(\tilde{\vartheta}^\top \bar{\omega} + \bar{\epsilon}), \tag{3.4}$$

where $\tilde{\vartheta} = \bar{\vartheta} - \hat{\vartheta}_*$ is the vector of parametric errors, $\bar{\vartheta} = \left[\frac{\vartheta^\top}{\psi_u}, \frac{1}{\psi_u}\right]^\top$ is the extended vector of unknown parameters, $\hat{\vartheta}_* = \left[\hat{\vartheta}^\top, \hat{\vartheta}_\psi\right]^\top$ is the extended vector of adjustable parameters, $\bar{\omega} = \left[\omega^\top, -u_s\right]^\top$ is the extended regressor, and $\bar{\epsilon} = \frac{\epsilon}{\psi_u}$ is the exponentially decaying term.

We define an adaptation algorithm

$$\dot{\hat{\vartheta}}_* = \Psi_x(x, \bar{\omega}),$$

where the function $\Psi_x(x, \bar{\omega})$ is to be found, and obtain the following parametric error model:

$$\dot{\tilde{\vartheta}} = -\Psi_x(x, \bar{\omega}). \tag{3.5}$$

We will refer to the Eqs. (3.4) and (3.5) as the *dynamic error model.*

Then the objective of the adaptation algorithm design can be formulated as follows: to determine a function $\Psi_x(x, \bar{\omega})$ so that the origin $(x, \tilde{\vartheta}) = (0, 0)$ of the dynamic error model (3.4), (3.5) is stable and $|x(t)| \to 0$ as $t \to \infty$.

It is worth noting that, in contrast to the problems of identification or identification-based adaptive control, asymptotic convergence of the parametric errors to zero is not required in the problems of adaptive control addressed in this book. However, asymptotic parametric convergence, as usual, accelerates and improves transients in adaptive systems, while exponential parametric convergence additionally provides robustness of the closed-loop adaptive system.

Canonical error models. Now, starting from the particular error models represented above, we define generalized canonical error models that will be used in design of basic algorithms of adaptation. For the sake of notation unification, we will use the following common notations for different error models: $e \in \mathbb{R}^n$ is the state error, $\varepsilon \in \mathbb{R}$ is the output error, $\theta \in \mathbb{R}^q$ is a vector of unknown parameters, $\hat{\theta} \in \mathbb{R}^q$ is the vector of adjustable parameters generated by an adaptation algorithm, $\tilde{\theta} = \theta - \hat{\theta}$ is the vector of parametric errors, $\omega \in \mathbb{R}^q$ is a regressor, ϵ is an exponentially decaying term.

We will consider the following two main classes of the error models:

- **the static error model** consisting of the static *control (estimation) error model*

$$\varepsilon = \tilde{\theta}^\top \omega + \epsilon \tag{3.6}$$

and the *parametric error model*

$$\dot{\tilde{\theta}} = -\Psi_\varepsilon(\varepsilon, \omega); \tag{3.7}$$

- **the dynamic error model** consisting of the dynamic *control error model*

$$\dot{e} = f_\Sigma(e) + \phi(e)(\tilde{\theta}^\top \omega + \epsilon) \tag{3.8}$$

and the *parametric error model*

$$\dot{\tilde{\theta}} = -\Psi_e(e, \omega). \tag{3.9}$$

In the above equations, the functions Ψ_ε and Ψ_e define adaptation algorithms. The methods of design of appropriate adaptation algorithms are in the scope of the next sections of the chapter.

Before we start consideration of the basic algorithms of adaptation, we make the following remarks concerning the dynamic error model as well as the properties of the regressor.

The general form of the dynamic control error model (3.8) can be reduced to particular cases including the linear model with measured state error e

$$\dot{e} = A_\Sigma e + b(\tilde{\theta}^\top \omega + \epsilon) \tag{3.10}$$

and the linear model with measured scalar output $\varepsilon = c^\top e$. Therefore, the class of dynamic error models includes several subclasses given by particular models considered in the next sections of this chapter.

The regressor ω can have different nature depending on a specific control problem, and, as a result, it can have different properties. In the problems of adaptive regulation in the plants with known parameters, the regressor represents an estimate of the exosystem state and can be considered as a function of time $\omega(t)$ that is known to be bounded a priori (see Sect. 2.4). In the problems of adaptive regulation in the plants with unknown parameters, the regressor contains known bounded functions of time together with some functions of the state error e or output error ε (see Sect. 2.5). In this case, the regressor denoted by $\omega(e, t)$ or $\omega(\varepsilon, t)$ is not known to be bounded a priori (due to its dependence on e or ε). To simplify the notation, sometimes we will denote such a regressor by $\omega(t)$ keeping in mind that its boundedness is not established.

Following the idea proposed in [26] (see Chap. 7), we will analyze the properties of the basic adaptation algorithms regarding different classes of the regressors considered in this book. These classes are presented by the following definitions (see also Appendix C).

Definition 3.1 A regressor $\omega(e, t) : \mathbb{R}^n \times \mathbb{R}_+ \to \mathbb{R}^q$ is said to belong to the class $\mathcal{L}$ (i.e., $\omega \in \mathcal{L}$) if it is locally Lipschitz in e uniformly in t.

Definition 3.2 A regressor $\omega(t) : \mathbb{R}_+ \to \mathbb{R}^q$ is said to belong to the class $\mathcal{PC}_{[0,\infty)}$ (i.e., $\omega \in \mathcal{PC}_{[0,\infty)}$) if it is piecewise continuous with bounded discontinuities and is defined on the time interval $[0, \infty)$.

Definition 3.3 A regressor $\omega(t) : \mathbb{R}_+ \to \mathbb{R}^q$ is said to belong to the class $\mathcal{L}_\infty$ (i.e., $\omega \in \mathcal{L}_\infty$) if it is piecewise continuous and globally bounded.

Definition 3.4 A regressor $\omega(t) : \mathbb{R}_+ \to \mathbb{R}^q$ is said to belong to the class $\mathcal{PE}$ (i.e., $\omega \in \mathcal{PE}$) if it is persistently exciting, i.e., it is bounded piecewise continuous and there exist constants $T > 0$, $\alpha > 0$ and $t_0 > 0$ such that

$$\int_{t}^{t+T} \omega(\tau)\omega^{\top}(\tau)d\tau \succeq \alpha I \text{ for all } t \geq t_0. \tag{3.11}$$

The class $\mathcal{L}$ of the regressors is more general and reflects their possible dependence on the state error e as well as on external signals represented by known functions of time. Regressors that belong to $\mathcal{L}$ satisfy the conditions of the global existence and uniqueness of differential equations solutions [15] (see Lemmas 1.1–1.3 in Sect. 1.1) and can be used for analysis of the closed-loop stability of the dynamic error models. Note that the boundedness of e (and, therefore, ω) in the closed-loop control system must be established.

The class $\mathcal{PC}_{[0,\infty)}$ of the regressors is defined for all $t \geq 0$, however, they can grow unbounded as $t \to \infty$. The class $\mathcal{PC}_{[0,\infty)}$ can be used for analysis of the signal properties in the closed-loop adaptive systems with static and dynamic error models when the boundedness of the regressor is not established yet. This situation takes place in the problems of adaptive regulation in the plants with unknown parameters. Regressors $\omega(t) \in \mathcal{L}_\infty$ are bounded a priori and, for example, are conventional for the problems of adaptive regulation in the plants with known parameters.

As it will be shown later, the regressors of the class $\mathcal{PE}$ provide the strongest stability property, namely, the exponential stability of the parametric error model. The inequality (3.11), also known as the *persistent excitation condition* (the PE condition), means that, despite the singularity of the matrix $\omega(t)\omega^{\top}(t)$ for any t, the integral of the this matrix is uniformly positive definite over any time interval $[t, t+T]$ starting from $t = t_0$.

3.2 Algorithms for Static Error Model

In this section, we consider the static output error model

$$\varepsilon = \tilde{\theta}^{\top}\omega + \epsilon, \tag{3.12}$$

where the *output error* $\varepsilon \in \mathbb{R}$ and the *regressor* $\omega : \mathbb{R}_+ \to \mathbb{R}^q$ are measured, $\tilde{\theta} = \theta - \hat{\theta}$ is the *parametric error*, $\hat{\theta}$ is the vector of adjustable parameters, θ is a vector of unknown parameters, and the *unmodelled disturbance* ϵ exponentially decays. We also assume that the *output variable*

$$y = \theta^{\top}\omega + \epsilon \tag{3.13}$$

can be measured or can be calculated as

$$y = \varepsilon + \hat{\theta}^{\top}\omega. \tag{3.14}$$

The design problem considered is to define a function $\Psi_\varepsilon(\varepsilon, \omega)$ such that the adaptation algorithm

$$\dot{\hat{\theta}} = \Psi_\varepsilon(\varepsilon, \omega)$$

provides the boundedness of $\hat{\theta}$ and zeroing of ε.

3.2.1 Gradient Algorithm of Adaptation

So-called *gradient* algorithm of adaptation is defined as

$$\dot{\hat{\theta}} = \gamma\omega\varepsilon, \tag{3.15}$$

where $\gamma > 0$ is a designer-chosen coefficient called *adaptation gain.*

Remark 3.1 The term *gradient* algorithm of adaptation can be explained as follows. Let us introduce the cost function

$$J(\hat{\theta}) = \frac{1}{2}\varepsilon^2(\hat{\theta}), \tag{3.16}$$

where the output error ε is considered as a function of $\hat{\theta}$. Therefore, to reduce ε to zero one can adjust $\hat{\theta}$ to achieve the minimum of the cost function (3.16). According to a so-called *steepest descent* or *gradient method* [5, 8, 12, 20], $\hat{\theta}$ is adjusted in the direction, along which the function (3.16) decreases most rapidly, i.e., along its negative gradient. In continuous time domain, the gradient algorithm takes the form

$$\dot{\hat{\theta}} = -\gamma\nabla J(\hat{\theta}), \tag{3.17}$$

where

$$\nabla J = \left(\frac{\partial J(\hat{\theta})}{\partial \hat{\theta}}\right)^\top$$

is the gradients of $J(\hat{\theta})$ with respect to $\hat{\theta}$. In view of (3.12), it is easy to see that

$$\frac{\partial J(\hat{\theta})}{\partial \hat{\theta}} = \varepsilon\frac{\partial \varepsilon(\hat{\theta})}{\partial \hat{\theta}}, \quad \frac{\partial \varepsilon(\hat{\theta})}{\partial \hat{\theta}} = -\omega.$$

Therefore, the gradient algorithm (3.17) is equivalent to (3.15), and, hence, the algorithm (3.15) provides gradient minimization of the cost function (3.16). ■

Let us first consider the case when $\epsilon(t) \equiv 0$. Then the algorithm (3.15) yields the parametric error model

$$\dot{\tilde{\theta}} = -\gamma\omega\omega^\top\tilde{\theta} \tag{3.18}$$

and ensures the following convergence properties [12, 26, 30].

Proposition 3.1 *The gradient algorithm of adaptation (3.15) when applied to the output error model (3.12) with $\epsilon(t) \equiv 0$ guarantees that:*

P.3.1.1. *if $\omega \in \mathcal{PC}_{[0,\infty)}$, then $\hat\theta \in \mathcal{L}_\infty$, $\varepsilon \in \mathcal{L}_2$;*

P.3.1.2. *if $\omega, \dot\omega \in \mathcal{L}_\infty$, then $\varepsilon, \hat\theta, \dot{\hat\theta} \in \mathcal{L}_\infty$, $\dot{\hat\theta} \in \mathcal{L}_2$, and $\varepsilon(t), |\dot{\hat\theta}(t)| \to 0$ as $t \to \infty$;*

P.3.1.3. *if $\omega \in \mathcal{PE}$, then, in addition to P.3.1.2, $|\tilde\theta(t)|$ tends to zero exponentially fast and there exists an optimal value of γ for which the rate of convergence is maximum.*

The properties established by Proposition 3.1 are well known in theory of adaptive control [12, Sect. 4.2.1], [26, Sect. 7.2], [30, Sect. 2.4.1]. However, for the sake of methodological purposes, here we recall the proof of some properties.

Proof Introduce the Lyapunov function

$$V(\tilde\theta) = \frac{1}{2\gamma}\tilde\theta^\top\tilde\theta \tag{3.19}$$

the time derivative of which along the solutions of (3.18) is given by

$$\dot V = \tilde\theta^\top\dot{\tilde\theta} = -\tilde\theta^\top\omega\varepsilon = -\varepsilon^2 \le 0. \tag{3.20}$$

Therefore, according to Theorem B.1, the origin of (3.18) is uniformly stable, and $\hat\theta \in \mathcal{L}_\infty$. Since $V > 0$ and $\dot V \le 0$, we conclude that $\lim_{t\to\infty} V(\tilde\theta(t)) = V_\infty$ exists and is finite. In view (3.20), we have

$$\int_0^\infty \varepsilon^2(\tau)d\tau = -\int_0^\infty \dot V(\tau)d\tau = V(\tilde\theta(0)) - V_\infty < \infty.$$

Therefore, $\varepsilon \in \mathcal{L}_2$. This completes the proof of Property P.3.1.1.

If $\omega \in \mathcal{L}_\infty$, then $\varepsilon, \dot{\hat\theta} \in \mathcal{L}_\infty$, and from (3.15) we obtain that $\dot{\hat\theta} \in \mathcal{L}_2$. If $\omega, \dot\omega \in \mathcal{L}_\infty$, then $\dot\varepsilon = (\dot\omega^\top\theta - \omega^\top\dot{\hat\theta}) \in \mathcal{L}_\infty$. Since $\varepsilon \in \mathcal{L}_\infty$ and $\dot\varepsilon \in \mathcal{L}_2$, by the Barbalat's lemma (see Lemma C.1 in Appendix C), it follows that $\varepsilon(t) \to 0$ as $t \to \infty$, and from (3.15) we obtain that $|\dot{\hat\theta}(t)| \to 0$ as $t \to \infty$. Property P.3.1.2 is proved.

The proof of Property P.3.1.3 can be found in [12, Sect. 4.8], [26, Sect. 6.5.3] or [30, Sect. 2.5]. □

It is worth noting that, in general, zeroing of the output error ε does not mean that the vector of adjustable parameters $\hat\theta(t)$ converges to the "true" value θ. It is seen from Property P.3.1.2 that $\varepsilon(t) \to 0$ as $t \to \infty$ for any $\omega, \dot\omega \in \mathcal{L}_\infty$. The convergence of the vector of parametric errors $\tilde\theta(t)$ (i.e., stability properties of the parametric error model (3.18)) primary depends on the properties of the regressor $\omega(t)$. Necessary conditions for *global asymptotic* (but not exponential) stability of (3.18) as well as sufficient (but not necessary) condition were established in [3]. Unfortunately, these conditions are highly technical and hard to verify in practical examples.

Now, consider the case when the error model (3.12) contains the exponentially decaying term ϵ. Then the gradient algorithm of adaptation (3.15) yields the following parametric error model:

$$\dot{\tilde{\theta}} = -\gamma\omega\varepsilon = -\gamma\omega\omega^{\top}\tilde{\theta} - \gamma\omega\epsilon. \tag{3.21}$$

To analyze the properties of (3.21), we choose the Lyapunov function

$$V(\tilde{\theta}, \epsilon) = \frac{1}{2\gamma}\tilde{\theta}^{\top}\tilde{\theta} + \int_t^{\infty} \epsilon^2(\tau)d\tau. \tag{3.22}$$

Its time derivative along the solutions of (3.21) takes the form

$$\dot{V}(\tilde{\theta}, \epsilon) = -\tilde{\theta}^{\top}\omega\varepsilon - \epsilon^2 = -\varepsilon^2 + \varepsilon\epsilon - \epsilon^2.$$

Completing the squares, we obtain

$$\dot{V}(\tilde{\theta}, \epsilon) \leq -\frac{1}{2}\varepsilon^2 - \frac{1}{2}(\varepsilon - \epsilon)^2 - \frac{1}{2}\epsilon^2 \leq -\frac{1}{2}\varepsilon^2.$$

The latter proves Property P.3.1.1. Then, with the use of the same arguments as in the proof of Proposition 3.1, we can prove Properties P.3.1.2 and P.3.1.3 for the parametric error model (3.21). Therefore, we formulate the following statement.

Corollary 3.1 *The gradient algorithm of adaptation* (3.15) *provides the properties established by Proposition 3.1 for the error model* (3.12) *with exponentially decaying* ϵ.

If in a control problem the boundedness of the regressor ω cannot be established a priori (for example, in the problems of adaptive regulation of possibly unstable plants with unknown parameters), then instead of the adaptation algorithm (3.15) different types of adaptation algorithms with *normalization* are applied [12, Sect. 4.3], [26, Sect. 5.4], [30, Sect. 2.4.1]. One of such algorithms is given by

$$\dot{\hat{\theta}} = \gamma\frac{\omega\varepsilon}{1 + \omega^{\top}\omega}, \tag{3.23}$$

where $(1 + \omega^{\top}\omega)$ is the *normalizing term*. Normalization enables one to guarantee the following properties that are crucial for stability analysis of adaptive systems when the boundedness of the closed-loop signals must be proved [12, Theorem 4.3.2].

Proposition 3.2 *For the error model* (3.12), *the gradient algorithm of adaptation with normalization* (3.23) *provides Properties P.3.1.2 and P.3.1.3 established by Proposition 3.1 and, additionally:*

P.3.2.1. *if* $\omega \in \mathcal{PC}_{[0,\infty)}$, *then it ensures that*

$$\hat{\theta}, \dot{\hat{\theta}} \in \mathcal{L}_\infty \quad \text{and} \quad \frac{\varepsilon}{\sqrt{1+\omega^\top \omega}}, \ \dot{\hat{\theta}} \in \mathcal{L}_2.$$

Proof For the sake of brevity of the proof presentation, we focus on the case when $\epsilon(t) \equiv 0$[1] and the parametric error model of the form

$$\dot{\tilde{\theta}} = -\gamma \frac{\omega \omega^\top}{1+\omega^\top \omega} \tilde{\theta}. \tag{3.24}$$

Calculating the time derivative of the Lyapunov function (3.19) along the solutions of (3.24), we obtain

$$\dot{V} = -\frac{\varepsilon^2}{1+\omega^\top \omega} \leq 0 \tag{3.25}$$

and prove the boundedness of $\hat{\theta}$. Further, in view (3.25), we have

$$\int_0^\infty \frac{\varepsilon^2(\tau)}{1+\omega^\top(\tau)\omega(\tau)} d\tau = -\int_0^\infty \dot{V}(\tau) d\tau = V(\tilde{\theta}(0)) - V_\infty < \infty.$$

Therefore,

$$\frac{\varepsilon}{\sqrt{1+\omega^\top \omega}} \in \mathcal{L}_2,$$

and

$$\dot{\tilde{\theta}} = -\gamma \frac{\omega}{\sqrt{1+\omega^\top \omega}} \frac{\varepsilon}{\sqrt{1+\omega^\top \omega}} \in \mathcal{L}_2.$$

The boundedness of $\dot{\hat{\theta}}$ directly follows from (3.23). This completes the proof of Property P.3.2.1. Properties P.3.1.2 and P.3.1.3 are proved with the arguments similar to those in the proof of Proposition 3.1. Finally, with the use technique similar to that in the proof of Corollary 3.1, all the results of Proposition 3.2 are directly extended to the case of the exponentially vanishing ϵ. □

The main drawback of the gradient algorithm of adaptation consists in the fact that even under the PE condition there exists an optimal value of the adaptation gain γ for which the rate of parametric convergence is maximum. Therefore, the parametric convergence (and, as a result, the output error convergence) cannot be accelerated arbitrarily even in theory. Moreover, if the PE condition is not satisfied, a little can be said about the transient performance. In the next two sections, we will consider adaptation algorithms with improved parametric convergence.

---

[1] All the results are directly extended to the case of exponentially decaying ϵ [26, Sect. 5.5].

3.2.2 Adaptation Algorithm with Dynamic Regressor Extension

In [19], it was proposed to extend the dimension of the output error model (or of the output variable model) by applying multiple filters. According to this technique, $p \geq (q-1)$ distinct *proper asymptotically stable and minimum phase* (PASMP) transfer functions $H_i(s)$ are introduced. In particular case, $H_i(s)$ can be selected as the first order filters $H_i(s) = \frac{\alpha_i}{s+\beta_i}$, where $\beta_j > 0$, and $\beta_i \neq \beta_j$ for $i \neq j$.

Define the transfer functions vector

$$\bar{H}(s) = \left[1, H_1(s), \ldots, H_p(s)\right]^\top .$$

Then the extended output $y_H \in \mathbb{R}^{p+1}$ and the extended regressor $W^\top \in \mathbb{R}^{(p+1)\times q}$ are given by

$$y_H = \bar{H}(s)[y] = \left[y, H_1(s)[y], \ldots, H_p(s)[y]\right]^\top \tag{3.26}$$

$$W^\top = \bar{H}(s)[\omega^\top] = \begin{bmatrix} \omega^\top \\ H_1(s)[\omega^\top] \\ \vdots \\ H_p(s)[\omega^\top] \end{bmatrix}$$
$$= \begin{bmatrix} \omega_1 & \omega_2 & \cdots & \omega_q \\ H_1(s)[\omega_1] & H_1(s)[\omega_2] & \cdots & H_1(s)[\omega_q] \\ \vdots & \vdots & \ddots & \vdots \\ H_p(s)[\omega_1] & H_p(s)[\omega_2] & \cdots & H_p(s)[\omega_q] \end{bmatrix}. \tag{3.27}$$

Following the terminology proposed in [1, 29], we will call such an approach *dynamic regressor extension* (DRE).[2]

Let us first consider the case when $\epsilon(t) \equiv 0$. Then, in view of (3.13), it is obvious that

$$y_H = W^\top \theta. \tag{3.28}$$

Now, introduce the dynamically extended output error

$$\varepsilon_H = y_H - W^\top \hat{\theta} \tag{3.29}$$

which leads to the dynamically extended output error model

$$\varepsilon_H = W^\top \tilde{\theta}. \tag{3.30}$$

Then the gradient minimization of the cost function $J_H(\hat{\theta}) = \frac{1}{2}|\varepsilon_H(\hat{\theta})|^2$ gives the algorithm of adaptation

[2] In [26], this method is referred to as *multiple models* (see Sect. 4.4.6).

$$\dot{\hat{\theta}} = \gamma W \varepsilon_H \tag{3.31}$$

and the corresponding closed-loop parametric error model

$$\dot{\tilde{\theta}} = -\gamma W W^\top \tilde{\theta}. \tag{3.32}$$

We denote by $\lambda_W(t)$ the minimum eigenvalue of the matrix $WW^\top$ and formulate the following properties ensured by the algorithm of adaptation with DRE.

Proposition 3.3 *The adaptation algorithm with DRE* (3.26), (3.27), (3.29), (3.31) *when applied to the output error model* (3.12) *with* $\epsilon(t) \equiv 0$ *guarantees that:*

P.3.3.1. *if* $\omega \in \mathcal{PC}_{[0,\infty)}$, *then* $\hat{\theta} \in \mathcal{L}_\infty$, $W^\top \tilde{\theta} \in \mathcal{L}_2$;

P.3.3.2. *if* $\omega, \dot{\omega} \in \mathcal{L}_\infty$, *then* $\varepsilon, \varepsilon_H, \hat{\theta}, \dot{\hat{\theta}}, \in \mathcal{L}_\infty$, $\dot{\hat{\theta}} \in \mathcal{L}_2$, *and* $\varepsilon(t)$, $|\varepsilon_H(t)|$, $|\dot{\hat{\theta}}(t)| \to 0$ *as* $t \to \infty$;

P.3.3.3. *if* $\omega, \dot{\omega} \in \mathcal{L}_\infty$, *and* $\lambda_W(t) \notin \mathcal{L}_1$, *then, additionally to P.3.3.2,* $|\tilde{\theta}(t)| \to 0$ *as* $t \to \infty$;

P.3.3.4. *if* $\omega \in \mathcal{PE}$, *then, additionally to P.3.3.2,* $|\tilde{\theta}(t)|$ *tends to zero exponentially and the rate of parametric convergence can be made arbitrarily fast by increasing* γ.

Proof Properties P.3.3.1 and P.3.3.2 are almost direct extension of Properties P.3.1.1 and P.3.1.2 of Proposition 3.1 to the case of the matrix regressor W.

To prove Property P.3.3.3, we note that the parametric error model (3.32) admits the following differential inequality:

$$\frac{d}{dt}|\tilde{\theta}|^2 = -2\gamma \tilde{\theta}^\top W W^\top \tilde{\theta} \le -2\gamma \lambda_W(t) |\tilde{\theta}|^2. \tag{3.33}$$

The solution of latter gives the upper bound of the squared norm $|\tilde{\theta}|$:

$$|\tilde{\theta}(t)|^2 \le \mathrm{e}^{-2\gamma \int_0^t \lambda_W(\tau)d\tau} |\tilde{\theta}(0)|^2. \tag{3.34}$$

Therefore, if $\int_0^t \lambda_W(\tau)d\tau \to \infty$ as $t \to \infty$ (i.e., $\lambda_W \notin \mathcal{L}_1$), then $|\tilde{\theta}(t)| \to 0$ as $t \to \infty$.

To prove Property P.3.3.4, note that if $\omega \in \mathcal{PE}$, then $W_i = H_i(s)[\omega] \in \mathcal{PE}$ (see Lemma 6.8 and Corollary 6.1 in [26]). Therefore, there exist $T_i > 0$, $\alpha_i > 0$ and t_{0i} such that

$$\int_t^{t+T_i} W_i(\tau) W_i^\top(\tau) d\tau \succeq \alpha_i I \;\; \text{for } \forall t \ge t_{0i}.$$

Hence,

$$\int_t^{t+T} W(\tau) W^\top(\tau) d\tau = \int_t^{t+T} \sum_{i=0}^p W_i(\tau) W_i^\top(\tau) d\tau \succeq \sum_{i=0}^p \alpha_i I \succ 0 \;\; \text{for } \forall t \ge t_0,$$

where $W_0 = \omega$, $T \geq \max_{i=1,2,\ldots,p}\{T_i\}$, $t_0 \geq \max_{i=1,2,\ldots,p}\{t_{0i}\}$. The latter means that the origin of (3.32) is exponentially stable. If $\omega \in \mathcal{PE}$ and $H_i(s)$ are different, then the columns of W are linearly independent and the arbitrarily fast convergence can be provided in theory by increasing γ [26, Sect. 4.4.6]. □

Comparing the properties established by Propositions 3.1 and 3.3, we see that the main contribution of the DRE consists in Properties P.3.3.3 and P.3.3.4 that are not achievable for the standard gradient algorithm of adaptation.

Now, consider the case when the model (3.13) contains the exponentially decaying term ϵ. Then we have

$$\varepsilon_H = W^\top \tilde{\theta} + \epsilon_H, \tag{3.35}$$

where $\epsilon_H = \bar{H}(s)[\epsilon]$ exponentially decays, and the algorithm of adaptation (3.31) yields the following parametric error model:

$$\dot{\tilde{\theta}} = -\gamma W W^\top \tilde{\theta} - \gamma W \epsilon_H.$$

Using the same approach as in the proof of Proposition 3.2, we can show that all the properties established by Proposition 3.3 are preserved for the case of the exponentially decaying term ϵ with the minor remark: the *arbitrarily fast* convergence is achieved in theory after some transient period determined by the decreasing rate of ϵ.

Corollary 3.2 *For the output error model* (3.12) *with exponentially decaying term* ϵ*, the adaptation algorithm with DRE* (3.26), (3.27), (3.29), (3.31) *provides Properties P.3.3.1, P.3.3.2, P.3.3.3 and the following one:*

P.3.2.1. *if* $\omega \in \mathcal{PE}$*, then, additionally to P.3.3.2,* $|\tilde{\theta}(t)|$ *tends to zero exponentially fast and the rate of parametric convergence can be increased by increasing* γ*.*

The adaptation algorithm with DRE admits several forms of normalization including the following ones:

$$\dot{\hat{\theta}} = \gamma (I + W^\top W)^{-1} W \varepsilon_H$$

or

$$\dot{\hat{\theta}} = \frac{\gamma}{1 + |W|_F^2} W \varepsilon_H,$$

where $|W|_F$ denotes the Frobenius norm of W. As it was mentioned in [18, p. 240], the use of the Frobenius norm is more preferable since it allows us to avoid algebraic burden in stability analysis and eliminates the need for online matrix inversion. The use of normalization enables one to ensure the following additional properties for $\omega \in \mathcal{PC}_{[0,\infty)}$:

$$\hat{\theta}, \dot{\hat{\theta}} \in \mathcal{L}_\infty \quad \text{and} \quad \frac{\varepsilon_H}{\sqrt{1 + |W|_F^2}}, \; \dot{\hat{\theta}} \in \mathcal{L}_2.$$

3.2.3 Adaptation Algorithm with Memory Regressor Extension

An alternative approach for improvement of the parametric convergence was proposed by Kreisselmeier initially for adaptive observers [16] and later for model reference adaptive control (MRAC) [17]. According to this approach, a single PASMP transfer function $L(s)$ is introduced. In particular, $L(s)$ can be selected as the simplest first order filter $L(s) = \frac{1}{s+1}$.

Introduce the filtered product of y by ω defined as

$$y_L = L(s)[\omega y], \quad y_L \in \mathbb{R}^q \tag{3.36}$$

or

$$y_L = L(s)\left[\omega\varepsilon + \omega\omega^\top\hat{\theta}\right]. \tag{3.37}$$

Let us first consider the case, when $\epsilon(t) \equiv 0$. Then it can be shown that

$$y_L = \Omega\theta, \tag{3.38}$$

where the matrix regressor $\Omega \in \mathbb{R}^{q\times q}$ is given by

$$\Omega = L(s)\left[\omega\omega^\top\right] = \begin{bmatrix} L(s)\left[\omega_1^2\right] & L(s)\left[\omega_1\omega_2\right] & \cdots & L(s)[\omega_1\omega_q] \\ L(s)[\omega_2\omega_1] & L(s)\left[\omega_2^2\right] & \cdots & L(s)[\omega_2\omega_q] \\ \vdots & \vdots & \ddots & \vdots \\ L(s)[\omega_q\omega_1] & L(s)[\omega_q\omega_2] & \cdots & L(s)\left[\omega_q^2\right] \end{bmatrix}. \tag{3.39}$$

Motivated by the regression (3.38), we introduce the extended output error of the form

$$\varepsilon_L = y_L - \Omega\hat{\theta}. \tag{3.40}$$

Since, in view of (3.37),

$$y_L - \Omega\hat{\theta} = L(s)\left[\omega\omega^\top\tilde{\theta}\right] + L(s)\left[\omega\omega^\top\hat{\theta}\right] - \Omega\hat{\theta} = \Omega\tilde{\theta}, \tag{3.41}$$

the error ε_L can be represented as

$$\varepsilon_L = \Omega\tilde{\theta}. \tag{3.42}$$

Therefore, the gradient minimization of the corresponding cost function $J_L(\hat{\theta}) = \frac{1}{2}|\varepsilon_L(\hat{\theta})|^2$, in view of symmetric property of Ω (i.e., $\Omega^\top = \Omega$), gives the algorithm of adaptation

$$\dot{\hat{\theta}} = \gamma \Omega \left(y_{L} - \Omega \hat{\theta} \right) \tag{3.43}$$

that yields the closed-loop parametric error model of the form

$$\dot{\tilde{\theta}} = -\gamma \Omega^2 \tilde{\theta}.$$

It worth noting that we can simplify the adaptation algorithm by selecting the PASMP transfer functions $L(s)$ with specific properties. To this end, note that for the first order filter $L(s) = \frac{b_0}{s+a}$ with $a, b_0 > 0$ we have

$$\Omega = \frac{b_0}{s+a} \left[\omega \omega^\top \right] = b_0 \int_0^t \mathrm{e}^{-a(t-\tau)} \omega(\tau) \omega^\top(\tau) d\tau.$$

Since $b_0 \mathrm{e}^{-a(t-\tau)} \geq 0$ for all $\tau \in [0, t]$, from $\omega \omega^\top \succeq 0$ we have that $\Omega \succeq 0$ [16], [26, Sect. 4.4.7], [12, Sect. 4.3.5]. Therefore, if we choose $L(s)$ as a chain of the first order filters

$$L(s) = b_0 \prod_{i=1}^{p} \frac{1}{(s + a_i)}, \tag{3.44}$$

where $b_0 > 0, a_i > 0\,(i = 1, 2, \ldots, p)$, then $\Omega(t) = L(s) \left[\omega(t) \omega^\top(t) \right] \succeq 0$. Indeed, let (A, b, c) be a minimal realization of the transfer function $L(s)$ (3.44), i.e., $L(s) = c^\top (sI - A)^{-1} b$. In particular, the triple (A, b, c) can be chosen in the form

$$A = \begin{bmatrix} -a_1 & 1 & 0 & \cdots & 0 \\ 0 & -a_2 & 1 & \cdots & 0 \\ \vdots & \ddots & \ddots & \ddots & \vdots \\ 0 & 0 & 0 & \cdots & 1 \\ 0 & 0 & 0 & \cdots & -a_p \end{bmatrix}, \quad b = \begin{bmatrix} 0 \\ 0 \\ \vdots \\ b_0 \end{bmatrix}, \quad c = \begin{bmatrix} 1 \\ 0 \\ \vdots \\ 0 \end{bmatrix}.$$

In this case, the vectors $b, c \in \mathbb{R}_+$, and A is the Metzler matrix.[3] Therefore, by virtue of [13, Theorem B.5], the impulse response of the filter (3.44) given by $h(t) = c^\top \mathrm{e}^{At} b$ is positive for all $t \geq 0$, $\tau \in [0, t]$. Since

$$\Omega(t) = L(s) \left[\omega(t) \omega^\top(t) \right] = \int_0^t h(t-\tau) \omega(\tau) \omega^\top(\tau) d\tau,$$

where $h(t-\tau) \geq 0$, then we conclude that $\Omega(t) \succeq 0$. It allows one to remove the factor $\Omega \succeq 0$ in front of the brackets of (3.43) and get the simplified algorithm of adaptation

$$\dot{\hat{\theta}} = \gamma \left(y_{L} - \Omega \hat{\theta} \right) \tag{3.45}$$

[3] Metzler matrix is a square matrix with nonnegative off-diagonal elements.

that yields the closed-loop parametric error model

$$\dot{\tilde{\theta}} = -\gamma \Omega \tilde{\theta}. \tag{3.46}$$

In this book, we will only use the transfer function $L(s)$ of the form (3.44) and the adaptation algorithm (3.45).

In particular, if $L(s) = \frac{1}{s+1}$, then the auxiliary filters of the Kreisselmeier's algorithm (3.37) and (3.39) can be represented in the following state-space form [16]:

$$\dot{\Omega} = -\Omega + \omega\omega^\top, \quad \dot{y}_L = -y_L + \omega(\varepsilon - \omega^\top \hat{\theta}).$$

The main distinguishing feature of the Kreisselmeier's algorithm consists in the use of the past values of the matrix $\omega(t)\omega^\top(t)$ provided by the filter $L(s)$. For example, the first order filter $L(s) = \frac{1}{s+1}$ collects and suitably weights all past values of the input. In other words, it provides time memory with forgetting. In this context, following the terminology proposed in [7, 29], we can say that $L(s)$ represents an "operator with memory", and the Kreisselmeier's scheme provides *memory regressor extension* (MRE).

We denote by $\lambda_\Omega(t)$ the minimum eigenvalue of the matrix Ω and formulate the following properties guaranteed by the algorithm of adaptation with MRE (see also Theorem 4.3.3 and Sect. 4.8.4 of [12]).

Proposition 3.4 *The adaptation algorithm with MRE* (3.37), (3.39), (3.45) *when applied to the output error model* (3.12) *with* $\epsilon(t) \equiv 0$ *guarantees that:*

P.3.4.1. *if* $\omega \in \mathcal{PC}_{[0,\infty)}$, *then* $\hat{\theta} \in \mathcal{L}_\infty$, *and* $(\tilde{\theta}^\top \Omega \tilde{\theta})^{1/2} \in \mathcal{L}_2$*;*

P.3.4.2. *If* $\omega, \dot{\omega} \in \mathcal{L}_\infty$, *then* $\varepsilon, \hat{\theta}, \dot{\hat{\theta}} \in \mathcal{L}_\infty$, $\Omega\tilde{\theta}, \dot{\hat{\theta}} \in \mathcal{L}_2$, *and* $\varepsilon(t), |\dot{\hat{\theta}}(t)| \to 0$ *as* $t \to \infty$*;*

P.3.4.3. *if* $\omega, \dot{\omega} \in \mathcal{L}_\infty$ *and* $\lambda_\Omega(t) \notin \mathcal{L}_1$, *then, additionally to P.3.3.2,* $|\tilde{\theta}(t)| \to 0$ *as* $t \to \infty$*;*

P.3.4.4. *if* $\omega \in \mathcal{PE}$, *then, additionally to P.3.4.2,* $|\tilde{\theta}(t)|$ *tends to zero exponentially and the rate of parametric convergence can be made arbitrarily fast by increasing* γ.

Proof Consider the Lyapunov function $V(\tilde{\theta}) = \frac{1}{2}|\tilde{\theta}|^2$, the time derivative of which along the solutions of (3.46) takes the form $\dot{V} = -\gamma\tilde{\theta}^\top \Omega \tilde{\theta} \leq 0$. Therefore, $\tilde{\theta}, \hat{\theta} \in \mathcal{L}_\infty$. Further, integrating $\dot{V}$, we obtain

$$\int_0^\infty \gamma \tilde{\theta}^\top(\tau)\Omega(\tau)\tilde{\theta}(\tau)d\tau = -\int_0^\infty \dot{V}(\tau)d\tau \leq V(\tilde{\theta}(0)) - V(\infty) \leq \infty.$$

Therefore, $(\tilde{\theta}^\top \Omega \tilde{\theta})^{1/2} = |\Omega^{1/2}\tilde{\theta}| \in \mathcal{L}_2$. This completes the proof of P.3.4.1.

If $\omega \in \mathcal{L}_\infty$, then there exists a positive constant c such that $\Omega \leq cI$ and, hence,

$$\int_0^\infty \tilde{\theta}^\top \Omega^2 \tilde{\theta} d\tau \leq c \int_0^\infty |\Omega^{1/2}\tilde{\theta}|^2 d\tau \leq \infty.$$

Thus, $\Omega\tilde{\theta} \in \mathcal{L}_2$. Since $\frac{d}{dt}\left|\Omega\tilde{\theta}\right| \in \mathcal{L}_\infty$, we obtain that $|\Omega(t)\tilde{\theta}(t)| \to 0$ and $|\dot{\hat{\theta}}(t)| \to 0$ as $t \to \infty$. Since $|\dot{\hat{\theta}}(t)| \to 0$, then (for rigorous proof see Lemma 1 in [7])

$$\left|L(s)\left[\omega(t)\omega^\top(t)\hat{\theta}(t)\right] - \Omega(t)\hat{\theta}(t)\right| \to 0 \quad \text{as} \quad t \to \infty.$$

As a result, from (3.41) we have $|L(s)[\omega(t)\varepsilon(t)]| \to 0$ as $t \to \infty$. Since $\dot{\omega}, \dot{\varepsilon} \in \mathcal{L}_\infty$, then due to the properties of linear low pass filters the latter means that $\varepsilon(t) \to 0$ as $t \to \infty$. Property P.3.4.2 is proved.

To prove Property P.3.4.3, we note that, in view of (3.46), the following inequality holds:

$$\frac{d}{dt}|\tilde{\theta}|^2 = -2\gamma\tilde{\theta}^\top\Omega^\top\tilde{\theta} \le -2\gamma\lambda_\Omega(t)|\tilde{\theta}|^2. \tag{3.47}$$

Solving the latter, we obtain the upper bound of the squared norm $|\tilde{\theta}|$:

$$|\tilde{\theta}(t)|^2 \le \mathrm{e}^{-2\gamma\int_0^t \lambda_\Omega(\tau)d\tau}|\tilde{\theta}(0)|^2. \tag{3.48}$$

Therefore, if $\int_0^t \lambda_\Omega(\tau)d\tau \to \infty$ as $t \to \infty$ (i.e., $\lambda_\Omega \notin \mathcal{L}_1$), then $|\tilde{\theta}(t)| \to 0$ as $t \to \infty$. Property P.3.4.3 is proved.

The proof of Property P.3.4.4 can be found in [12, Sect. 4.8.4]. □

Now, we extend the properties of Proposition 3.4 to the output error model (3.12) with the exponentially decaying term ϵ. In this case, we have

$$\varepsilon_L = \Omega\tilde{\theta} + L(s)[\omega\epsilon], \tag{3.49}$$

and the parametric error model takes the form

$$\dot{\tilde{\theta}} = -\gamma\Omega\tilde{\theta} - \gamma L(s)[\omega\epsilon]. \tag{3.50}$$

Evaluating the time derivative of the Lyapunov function (3.22) along the solutions (3.50) yields

$$\begin{aligned}
\dot{V} &= -\tilde{\theta}^\top\Omega\tilde{\theta} - \tilde{\theta}^\top L(s)[\omega\epsilon] - \epsilon^2 = -\tilde{\theta}^\top\Omega\tilde{\theta} - \tilde{\theta}^\top(t)\int_0^t h(t-\tau)\omega(\tau)\epsilon(\tau)d\tau - \epsilon^2 \\
&\le -\tilde{\theta}^\top\Omega\tilde{\theta} + \int_0^t h(t-\tau)\left|\tilde{\theta}^\top(t)\omega(\tau)\epsilon(\tau)\right|d\tau - \epsilon^2 \\
&\le -\tilde{\theta}^\top\Omega\tilde{\theta} + \left(\int_0^t h(t-\tau)\left(\tilde{\theta}^\top(t)\omega(\tau)\right)^2 d\tau\right)^{\frac{1}{2}}\left(\int_0^t h(t-\tau)\epsilon^2(\tau)d\tau\right)^{\frac{1}{2}} - \epsilon^2,
\end{aligned}$$

where $h(t-\tau) \ge 0$ is the impulse response of the filter (3.44). The latter inequality is obtained by applying the Schwartz inequality (see [12, Eqs. (3.2.3) and (8.5.58)]).

In the sequel, for the sake of simplicity and without loss of generality, we assume that $b_0 = \prod_{i=1}^{p} a_i$ and $\| L(s) \|_\infty = 1$. Taking into account this fact and the relations

$$\left(\tilde{\theta}^\top(t)\omega(\tau)\right)^2 = \tilde{\theta}^\top(t)\omega(\tau)\omega^\top(\tau)\tilde{\theta}(t),$$

$$\int_0^t h(t-\tau)\left(\tilde{\theta}^\top(t)\omega(\tau)\right)^2 d\tau \le \tilde{\theta}^\top(t)\Omega(t)\tilde{\theta}(t), \quad \int_0^t h(t-\tau)\epsilon^2(\tau)d\tau \le \epsilon^2,$$

we have

$$\dot{V} \le -\tilde{\theta}^\top \Omega\tilde{\theta} + (\tilde{\theta}^\top \Omega\tilde{\theta})^{\frac{1}{2}}|\epsilon| - \epsilon^2.$$

Completing the squares, we obtain

$$\dot{V} \le -\frac{1}{2}\tilde{\theta}^\top \Omega\tilde{\theta} - \frac{1}{2}\left((\tilde{\theta}^\top \Omega\tilde{\theta})^{\frac{1}{2}} - |\epsilon|\right)^2 - \frac{1}{2}\epsilon^2 \le -\frac{1}{2}\tilde{\theta}^\top \Omega\tilde{\theta}.$$

The latter proves P.3.4.1. Remaining properties can be proved in a way similar to that described in the proof of Proposition 3.4. Therefore, we are in position to formulate the following statement.

Corollary 3.3 *For the output error model* (3.12) *with exponentially decaying term* ϵ, *the adaptation algorithm with MRE* (3.37), (3.39), (3.45) *provides Properties P.3.4.1, P.3.4.2, P.3.4.3 and the following one:*

P.3.3.1. *if* $\omega \in \mathcal{PE}$, *then, additionally to P.3.4.2,* $|\tilde{\theta}(t)|$ *tends to zero exponentially fast and the rate of parametric convergence can be increased by increasing* γ.

If the boundedness of the regressor ω cannot be established a priori, then a normalized version of the Kreisselmeier's algorithm can be used by applying instead of (3.37) and (3.39) the filters

$$y_{_L} = L(s)\left[\tfrac{\omega y}{1+\omega^\top\omega}\right], \quad \Omega = L(s)\left[\tfrac{\omega\omega^\top}{1+\omega^\top\omega}\right]. \tag{3.51}$$

It can be shown that adaptation algorithm with MRE with the normalized filters (3.51) ensures the following properties [12, Theorem 4.3.3]:

$$\hat{\theta}, \dot{\hat{\theta}} \in \mathcal{L}_\infty \quad \text{and} \quad \frac{\varepsilon}{\sqrt{1+\omega^\top\omega}}, \ \dot{\hat{\theta}} \in \mathcal{L}_2.$$

As it was mentioned above, the main advantages of the adaptation algorithms with DRE and MRE consist in Properties P.3.3.3, P.3.3.4 and P.3.4.3, P.3.4.4 that are not attainable for the standard gradient algorithm. We illustrate this statement by the following example.

Example 3.1 Let us consider the output error model (3.12) with $q = 2$ and $\epsilon(t) \equiv 0$. It is obvious that for the parametric error model (3.18) obtained from the gradient algorithm of adaptation (3.15) the matrix $\omega\omega^\top$ is singular for any $t \geq 0$ and for any ω, i.e., $\lambda_\omega(t) \equiv 0$, where λ_ω is the minimum eigenvalue of $\omega\omega^\top$. Therefore, the following question about distinct properties of adaptation algorithms with extended regressor arises: is it possible to point out such ω and select such operators $H(s)$ and $L(s)$ that $\lambda_W(t) \not\equiv 0$, $\lambda_\Omega(t) \not\equiv 0$?

To answer this question, we consider the regressor given by (see [1])

$$\omega(t) = \left[1,\ \upsilon(t) + \dot{\upsilon}(t)\right]^\top, \tag{3.52}$$

where υ is a continuous function such that $\upsilon, \dot{\upsilon} \in \mathcal{L}_\infty$, $\dot{\upsilon} \notin \mathcal{L}_2$, and

$$\lim_{t\to\infty} \upsilon(t) = \lim_{t\to\infty} \dot{\upsilon}(t) = 0. \tag{3.53}$$

It is obvious that $\omega \notin \mathcal{PE}$, and, moreover, the regressor elements vanish as $t \to \infty$. For example, the function $\upsilon(t) = \sin(t)(1+t)^{-\frac{1}{2}}$ satisfies these conditions and constitutes the following regressor [1]:

$$\omega(t) = \left[1,\ \frac{\sin(t) + \cos(t)}{(1+t)^{\frac{1}{2}}} - \frac{\sin(t)}{2(1+t)^{\frac{3}{2}}}\right]^\top. \tag{3.54}$$

In [7], it was proved that for $H_1(s) = \frac{1}{s+1}$ and $W^\top = [\omega,\ H_1(s)[\omega]]^\top$ we obtain $\lambda_W(t) \geq \dot{\upsilon}^2$, and, hence, $\lambda_W \notin \mathcal{L}_1$. For $L(s) = \frac{1}{s+1}$, we have $\lambda_\Omega \geq cL(s)\left[\dot{\upsilon}^2\right]$ (where c is some constant), and, therefore, $\lambda_\Omega \notin \mathcal{L}_1$. As a result, we conclude that the adaptation algorithms with DRE and MRE provide Properties P.3.3.3 and P.3.4.3 for the regressor (3.54), while the gradient algorithm of adaptation does not.

To illustrate this distinctive property, we compare the gradient adaptation algorithm (3.15) with the algorithms with DRE (3.31) and with MRE (3.45). We use the regression (3.12) with $\epsilon(t) \equiv 0$ and the regressor (3.54). In all the experiments, we set $\theta = [-3, 3]^\top$ and $\hat{\theta}(0) = [0, 0]^\top$.

Simulation results demonstrating the not-in-$\mathcal{L}_1$ properties of λ_W, λ_Ω and the parametric convergence properties are presented in Figs. 3.1 and 3.2, respectively.

As it is seen from the plots $\int_0^t \lambda_\Omega(\tau)d\tau$ and $\int_0^t \lambda_W(\tau)d\tau$ in Fig. 3.1, despite the fact that $\omega \notin \mathcal{PE}$ we have λ_Ω, $\lambda_W \notin \mathcal{L}_1$. Thus, we can assume that the algorithms of adaptation with DRE and MRE provide asymptotic parametric convergence even in the case of "bad" regressor (3.54).

Figure 3.2 demonstrates the transients of the Eucledian norms $|\tilde{\theta}(t)|$ calculated via the considered algorithms of adaptation for different gains γ. For comparison purposes, Fig. 3.2a demonstrates the transients provided by the gradient algorithm (3.15), Fig. 3.2b shows the norms $|\tilde{\theta}(t)|$ provided by the algorithm (3.45) with DRE and $W = \left[\omega,\ \frac{1}{s+1}[\omega]\right]$, while Fig. 3.2c shows $|\tilde{\theta}(t)|$ for the algorithm (3.31) with MRE and $L(s) = \frac{1}{s+1}$.

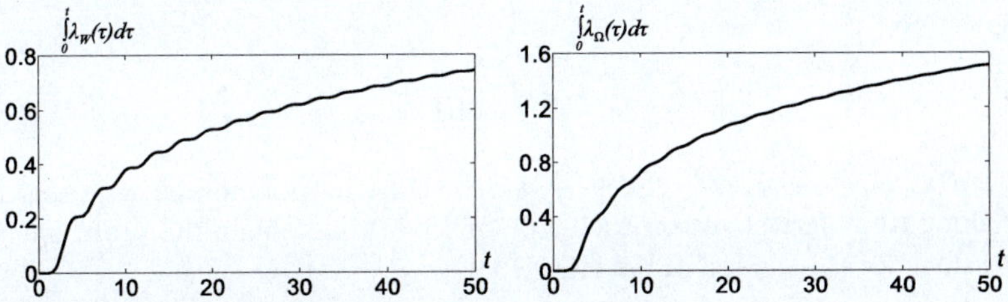

Fig. 3.1 Example 3.1: evolutions of the functions $\int_0^t \lambda_W(\tau)d\tau$ and $\int_0^t \lambda_\Omega(\tau)d\tau$

Fig. 3.2 Example 3.1: transients of $|\tilde{\theta}(t)|$ provided by: **a** the gradient adaptation algorithm; **b** the adaptation algorithm with DRE; **c** the adaptation algorithm with MRE

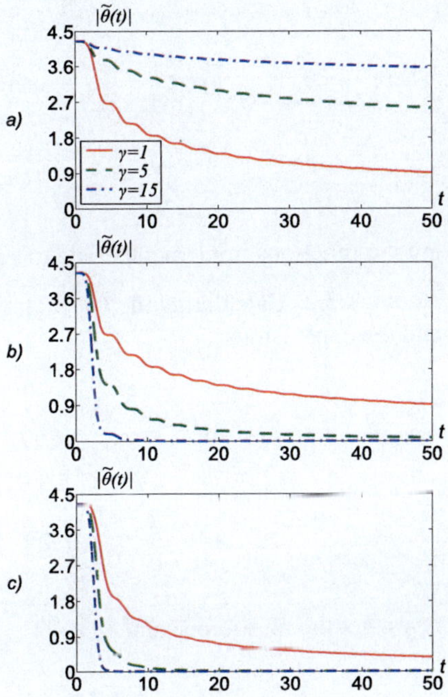

As seen from these plots, in spite of the fact that $\omega \notin \mathcal{PE}$, the algorithms with DRE and MRE asymptotically drive the parametric errors to zero and can be potentially used for acceleration of parameters tuning. ■

The MRE allows one to obtain one more useful property that can be important for design of certain adaptive controllers. As it was shown in [7], the MRE scheme is able to generate not only the adjustable parameters $\hat{\theta}$ but also their high-order time derivatives (HOTD) up to a prescribed order. In this regard, the algorithm was classified to a type of adaptation algorithms called *high-order tuners*. High-order tuners using alternative techniques for calculation of the HOTD of the adjustable parameters were proposed and developed in [23, 27, 28].

Selecting

$$L(s) = \frac{1}{d(s)},$$

where $d(s) = s^p + d_{p-1}s^{p-1} + \cdots + d_0$ is a Hurwitz polynomial with real roots, and taking time derivatives on each side of (3.45), we obtain the set of analytical expressions for calculation of the HOTD:

$$\hat{\theta}^{(i+1)} = \gamma \left(y_L^{(i)} - \sum_{j=0}^{i} C_j^i \Omega^{(i-j)} \hat{\theta}^{(j)} \right), \tag{3.55}$$

where $i = 0, 1, \ldots, p$, $C_j^i = \frac{i!}{j!(i-j)!}$ are the binomial coefficients, and

$$\Omega^{(i-j)} = \frac{s^{i-j}}{d(s)} \left[\omega\omega^\top\right], \quad y_L^{(i)} = \frac{s^i}{d(s)} [\omega y]$$

are the functions implemented due to sufficiently large order of $d(s)$.

Remark 3.2 Calculating the $(p+1)$st order time derivative of $\hat{\theta}$, in view of (3.14) and the expressions

$$y_L^{(p)} = \frac{s^p}{d(s)} [\omega y] = \omega y - \sum_{j=0}^{p-1} d_j y_L^{(j)},$$

$$\Omega^{(p)} = \frac{s^p}{d(s)} \left[\omega\omega^\top\right] = \omega\omega^\top - \sum_{j=0}^{p-1} d_j \Omega^{(j)},$$

we get the high-order tuner

$$\hat{\theta}^{(p+1)} = \gamma \left(\omega\varepsilon - \sum_{j=0}^{p-1} d_j \left(y_L^{(j)} - \Omega^{(j)}\hat{\theta} \right) - \sum_{j=1}^{p} C_j^p \Omega^{(p-j)} \hat{\theta}^{(j)} \right)$$

which can be potentially used for closed-loop calculation of the HOTD of the adjustable parameters. ■

3.3 Algorithms for Dynamic Error Model with Accessible State

In this section, we consider the dynamic state error model

$$\dot{e} = f(e) + \phi(e) \left(u + \theta^\top \omega(e, t) + \epsilon \right), \tag{3.56}$$

where $e \in \mathbb{R}^n$ is the state error, $\omega \in \mathbb{R}^q$ is the regressor, $u \in \mathbb{R}$ is the control, $\theta \in \mathbb{R}^q$ is a vector of unknown parameters, ϵ is the exponentially decaying term, $f(e)$ and $\phi(e)$ are smooth vector functions, $f(0) = 0$. The error e and the regressor ω are measured.

Remark 3.3 In the model (3.56), the control and the uncertainties are matched what allows us to design a *certainty equivalence* adaptive controller. If the matching is not possible, then to overcome the structural obstacle caused by unmatched uncertainties the special design tools called *adaptive backstepping with tuning functions* and *modular adaptive backstepping with high-order tuner* are used (see Appendix E). Adaptive backstepping involves an iterative design procedure with a special change of coordinates which in turn leads to linear parameterization allowing one to employ basic algorithms of adaptation. ■

To focus on the problem of counteraction to the influence of uncertainties, following [18, Sect. 2.2.1] and [9, Sect. 6.1], we introduce the *nominal* (i.e., without any uncertainties and disturbances) plant model

$$\dot{e} = f(e) + \phi(e)u, \tag{3.57}$$

and make the following assumption.

Assumption 3.1 There exist a smooth control law

$$u = u_s(e), \quad u_s(0) = 0 \tag{3.58}$$

and a smooth Lyapunov function $V_e(e)$ such that

$$c_1|e|^2 \le V_e(e) \le c_2|e|^2, \tag{3.59}$$

$$\left|\frac{\partial V_e(e)}{\partial e}\right| \le c_3|e|, \tag{3.60}$$

$$\dot{V}_e = \frac{\partial V_e(e)}{\partial e}(f(e) + \phi(e)u_s(e)) \le -c_4|e|^2, \tag{3.61}$$

where c_1, c_2, c_3, c_4 are some positive constants.

In view of Theorem B.3, Assumption 3.1 means that we know the control law (3.58) ensuring the global exponential stability of the nominal plant model (3.57) as well as the appropriate Lyapunov function $V_e(e)$ used to prove this property.

Now, to stabilize the plant with uncertainties we apply the adjustable controller

$$u = u_s(e) - \hat{\theta}^\top \omega(e, t), \tag{3.62}$$

where $\hat{\theta}$ is the vector of adjustable parameters generated by an adaptation algorithm of the form

$$\dot{\hat{\theta}} = \Psi_e(e,t). \tag{3.63}$$

Thus, the considered problem is to determine a function $\Psi_e(e,t)$ such that all the closed-loop signals are bounded and $|e(t)| \to 0$ as $t \to \infty$. Such a design problem will be called *adaptive stabilization*. We will start with the case when $\epsilon(t) \equiv 0$ and further extend the approach to the case of exponentially decaying term ϵ.

3.3.1 Adaptive Stabilization

Under condition that $\epsilon(t) \equiv 0$, a basic algorithm of adaptation used for adaptive stabilization of the plant (3.56) is defined by the following proposition [9, Theorem 6.1].

Proposition 3.5 *If Assumption 3.1 holds, $\omega \in \mathcal{L}$ and $\epsilon(t) \equiv 0$, then the control law (3.62) with the adaptation algorithm*

$$\dot{\hat{\theta}} = \gamma\omega(e,t)\frac{\partial V_e(e)}{\partial e}\phi(e) \tag{3.64}$$

when applied to the plant (3.56) guarantees for any $\gamma > 0$ and for all initial conditions $e(0)$, $\hat{\theta}(0)$ the following properties of the closed-loop system:

P.3.5.1. *e, $\hat{\theta} \in \mathcal{L}_\infty$ and $|e(t)|$, $|\dot{\hat{\theta}}(t)| \to 0$ as $t \to \infty$;*

P.3.5.2. *$\|e\|_2 \le \frac{1}{c_4}\sqrt{V(0)}$, $\|e\|_\infty \le \frac{1}{c_1}\sqrt{V(0)}$, where $V(0)$ is the initial value of the Lyapunov function*

$$V(e,\tilde{\theta}) = V_e(e) + \frac{1}{2\gamma}\tilde{\theta}^\top\tilde{\theta}, \tag{3.65}$$

$\tilde{\theta} = \theta - \hat{\theta}$ is the vector of parametric errors.

Proof Replacing (3.62) in (3.56) and taking into account that $\epsilon(t) \equiv 0$ and $\dot{\tilde{\theta}} = -\dot{\hat{\theta}}$, we obtain the closed-loop error model

$$\dot{e} = f_\Sigma(e) + \phi(e)\tilde{\theta}^\top\omega(e,t), \tag{3.66}$$

$$\dot{\tilde{\theta}} = -\gamma\omega(e,t)\frac{\partial V_e(e)}{\partial e}\phi(e), \tag{3.67}$$

where $f_\Sigma = f + \phi u_s$. Then the time derivative of the Lyapunov function (3.65) along the solutions of (3.66), (3.67), in view of (3.61), takes the form

$$\begin{aligned}\dot{V} &= \frac{\partial V_e}{\partial e}\left(f_\Sigma + \phi\tilde{\theta}^\top\omega\right) - \frac{1}{\gamma}\tilde{\theta}^\top\dot{\hat{\theta}} \\ &= \frac{\partial V_e}{\partial e} f_\Sigma + \frac{\partial V_e}{\partial e}\phi\tilde{\theta}^\top\omega - \tilde{\theta}^\top\omega\frac{\partial V_e}{\partial e}\phi = \frac{\partial V_e}{\partial e} f_\Sigma \\ &\le -c_4|e|^2. \end{aligned} \tag{3.68}$$

In view of Theorem B.3, the latter means that the origin $(e, \tilde{\theta}) = (0, 0)$ is uniformly stable, and, therefore, $e, \hat{\theta} \in \mathcal{L}_\infty$. Since $\omega \in \mathcal{L}$, by virtue of Theorem B.4, $|e(t)| \to 0$ as $t \to \infty$. Since e is bounded and $|e(t)| \to 0$, from (3.64) we obtain that $|\dot{\hat{\theta}}(t)| \to 0$ as $t \to \infty$. Property P.3.5.1 is proved.

Further, from the inequality $\dot{V} \le -c_4|e|^2$ we have

$$\|e(t)\|_2^2 = \int_0^\infty |e(\tau)|^2 d\tau \le \frac{1}{c_4}(V(0) - V(\infty)) \le \frac{1}{c_4}V(0).$$

Now, taking into account that $V(t)$ is the nonincreasing function of time, we can write

$$V(t) = V_e(e(t)) + \frac{1}{2\gamma}\tilde{\theta}(t)^\top\tilde{\theta}(t) \le V(0).$$

This, in view of (3.59), completes the proof of P.3.5.2. □

Remark 3.4 Because we use an appropriate Lyapunov function to derive an adaptation algorithm as well as to prove the properties of the closed-loop adaptive system, we will refer this procedure as *Lyapunov-based design*, while the corresponding adaptation algorithm will be called *Lyapunov-based algorithm of adaptation* [18, Sect. 1.1.4]. ■

Remark 3.5 All the properties established by Proposition 3.5 can be directly extended to the case of the state error model

$$\dot{e} = f(e) + \psi_u\phi(e)\left(\theta^\top\omega(e, t) + u\right),$$

where $\psi_u \in \mathbb{R}$ is an unknown *control gain* with known sign. Without loss of generality, we assume that $\psi_u > 0$. In this case, the control (3.62) results in the closed-loop state error model

$$\dot{e} = f_\Sigma(e) + \psi_u\phi(e)\tilde{\theta}^\top\omega(e, t). \tag{3.69}$$

Then, instead of the Lyapunov function (3.65), we use

$$V_\psi(e, \tilde{\theta}) = V_e(e) + \frac{\psi_u}{2\gamma}\tilde{\theta}^\top\tilde{\theta}$$

whose time derivative along the solutions of (3.69) and (3.67) takes the form

$$\dot{V}_\psi = \frac{\partial V_e}{\partial e} f_\Sigma + \psi_u \frac{\partial V_e}{\partial e} \phi \tilde{\theta}^\top \omega - \psi_u \tilde{\theta}^\top \omega \frac{\partial V_e}{\partial e} \phi = \frac{\partial V_e}{\partial e} f_\Sigma \le -c_4 |e|^2.$$

The latter inequality is identical to (3.68) that enables us to prove all the properties of Proposition 3.5 with the use of V_ψ instead of V. ■

Now, we consider a particular case of (3.56) given by the linear state error model

$$\dot{e} = Ae + b\left(u + \theta^\top \omega(e,t) + \epsilon\right), \tag{3.70}$$

where the pair (A, b) is known and controllable. The certainty equivalence adaptive controller can be chosen in the form

$$u = -k^\top e - \hat{\theta}^\top \omega, \tag{3.71}$$

where the vector of feedback gains k is such that the matrix $A_\Sigma = A - bk^\top$ is Hurwitz. Substituting (3.71) into (3.70), we obtain the closed-loop state error model

$$\dot{e} = A_\Sigma e + b\left(\tilde{\theta}^\top \omega(e,t) + \epsilon\right). \tag{3.72}$$

The corresponding Lyapunov function $V_e(e)$ can be selected as $V_e(e) = \frac{1}{2} e^\top P e$, where $P = P^\top \succ 0$ is the solution of the Lyapunov matrix equation $A_\Sigma^\top P + PA_\Sigma = -Q$ with arbitrary $Q = Q^\top \succ 0$. In this case,

$$\frac{\partial V_e}{\partial e} b = e^\top P b = b^\top P e,$$

and, hence, the adaptation algorithm (3.64) takes the form

$$\dot{\hat{\theta}} = \gamma \omega(e,t) e^\top P b = \gamma \omega(e,t) b^\top P e. \tag{3.73}$$

Then the time derivative of the Lyapunov function

$$V_\epsilon(e, \tilde{\theta}, \epsilon) = \frac{1}{2} e^\top P e + \frac{1}{2\gamma} \tilde{\theta}^\top \tilde{\theta} + \frac{|Pb|^2}{\lambda_Q} \int_t^\infty \epsilon^2(\tau) d\tau, \tag{3.74}$$

where λ_Q is the minimum eigenvalue of Q, along the solutions of the closed-loop error model is given by

$$\dot{V}_\epsilon = -\frac{1}{2} e^\top Q e + e^\top P b \epsilon - \frac{|Pb|^2}{\lambda_Q} \epsilon^2 \le -\frac{\lambda_Q}{2} |e|^2 + |e||Pb||\epsilon| - \frac{|Pb|^2}{\lambda_Q} |\epsilon|^2.$$

Completing the squares, we obtain

$$\dot{V}_\epsilon \leq -\frac{\lambda_Q}{4}|e|^2 - \left(\frac{\sqrt{\lambda_Q}}{2}|e| - \frac{|Pb|}{\sqrt{\lambda_Q}}|\epsilon|\right)^2 \leq -\frac{\lambda_Q}{4}|e|^2.$$

The latter inequality being a particular case of (3.68) enables us to prove the first property of the following proposition as a direct corollary of Proposition 3.5.

Proposition 3.6 *The control law* (3.71) *with the adaptation algorithm* (3.73) *when applied to the plant* (3.70) *guarantees for any* $\gamma > 0$ *and for all initial conditions* $e(0)$, $\hat{\theta}(0)$ *that:*

P.3.6.1. *if* $\omega \in \mathcal{L}$, *then* $e, \hat{\theta} \in \mathcal{L}_\infty$ *and* $|e(t)|, |\dot{\hat{\theta}}(t)| \to 0$ *as* $t \to \infty$;

P.3.6.2. *if* $\omega(e,t) = \omega(t) \in \mathcal{PE}$, *then, in addition to P.3.6.1,* $|\tilde{\theta}(t)| \to 0$ *exponentially fast and there exists an optimal value of* γ *for which the rate of parametric convergence is maximum.*

The proof of Property P.3.6.2 can be found in [26, Sect. 7.4].

It is worth noting that in linear state error models the exponentially decaying disturbance ϵ does not violate the stability of the closed-loop system. As it was demonstrated by Example 1.6, this is not the case for nonlinear systems, where special design tools must be employed to remove the negative influence of the exponentially decaying disturbance. One of such design tools is called *nonlinear damping* that is discussed in the next section.

3.3.2 Nonlinear Damping

The adaptation algorithm (3.64) can be considered as a type of *dynamic* nonlinear feedback allowing us to cope with constant uncertainties θ. In this section, we demonstrate that one can guarantee some desirable properties for uncertain systems, namely, boundedness of all closed-loop signals and exponential convergence of the state error to a residual set by means of *static* nonlinear feedback called *nonlinear damping* initially proposed in [4, 6] and further developed in [14, 18].

Following [18, Sect. 2.5.1], we generalize the class of uncertainties considering the vector θ as a vector of unknown uniformly bounded functions $\theta = \theta(e, u, t)$ such that $|\theta(e, u, t)| \leq \bar{\theta}$, where $\bar{\theta}$ is a positive constant. Then the properties of nonlinear damping as a design tool are established by the following proposition [18, Lemma 2.26], [9, Theorem 6.3].

Proposition 3.7 *If Assumption 3.1 holds,* $\omega \in \mathcal{L}$ *and* $\epsilon(t) \equiv 0$, *then the control law* (3.62) *with the static feedback*

$$\hat{\theta} = \mu\omega(e,t)\frac{\partial V_e}{\partial e}(e)\phi(e), \tag{3.75}$$

where $\mu > 0$ is some constant gain, when applied to the plant (3.56) with $\theta = \theta(e, u, t)$, $|\theta(e, u, t)| \leq \bar{\theta}$ renders the closed-loop system ISS with respect to the input $\theta(e, u, t)$ and guarantees the following properties for any initial condition $e(0)$:

P.3.7.1. *e is bounded and exponentially converges to the residual set*

$$D_e = \left\{ e : V_e \leq \frac{c_2}{4\mu c_4}\bar{\theta}^2 \right\}; \tag{3.76}$$

with the rate faster than $\mathrm{e}^{-\frac{c_4}{c_2}t}$;

P.3.7.2. *if $\theta = \theta(t)$ exponentially decays, then e exponentially converges to the origin.*

Proof Substituting (3.62) and (3.75) into (3.56), we obtain the model of the closed-loop system

$$\dot{e} = f_\Sigma(e) + \phi(e)\left(\theta^\top \omega(e, t) - \mu|\omega(e, t)|^2 \frac{\partial V_e(e)}{\partial e}\phi(e)\right). \tag{3.77}$$

Then the time derivative of the Lyapunov function $V_e(e)$ along the solutions of (3.77), in view of (3.61), takes the form

$$\begin{aligned} \dot{V}_e &= \frac{\partial V_e}{\partial e} f_\Sigma + \frac{\partial V_e}{\partial e}\phi\theta^\top\omega - \mu|\omega|^2 \left|\frac{\partial V_e}{\partial e}\phi\right|^2 \\ &\leq -c_4|e|^2 + \left|\frac{\partial V_e}{\partial e}\phi\right| |\omega||\bar{\theta}| - \mu|\omega|^2 \left|\frac{\partial V_e}{\partial e}\phi\right|^2. \end{aligned}$$

Completing the squares, in view of (3.59), we obtain

$$\begin{aligned} \dot{V}_e &\leq -c_4|e|^2 - \left(\sqrt{\mu}|\omega|\left|\frac{\partial V_e}{\partial e}\phi\right| - \frac{1}{2\sqrt{\mu}}|\bar{\theta}|\right)^2 + \frac{1}{4\mu}|\bar{\theta}|^2 \leq -c_4|e|^2 + \frac{1}{4\mu}|\bar{\theta}|^2 \\ &\leq -\frac{c_4}{c_2}V_e + \frac{1}{4\mu}\bar{\theta}^2. \end{aligned}$$

The latter means that e is bounded and exponentially converges to the residual set (3.76). This completes the proof of Property P.3.7.1.

To prove P.3.7.2, let us assume that θ satisfies the equation $\dot{\theta} = G\theta$, where the matrix $G \in \mathbb{R}^{q\times q}$ is Hurwitz. Choose the Lyapunov function

$$V(e, \theta) = V_e(e) + c_\theta \theta^\top P\theta,$$

where $c_\theta = 1 + \frac{1}{4\mu}$, $P = P^\top \succ 0$ is the matrix satisfying the equation $G^\top P + PG = -I$. Then the time derivative of V takes the form

$$\dot{V} \leq -c_4|e|^2 + \left|\frac{\partial V_e}{\partial e}\phi\right| |\omega||\theta| - \mu|\omega|^2 \left|\frac{\partial V_e}{\partial e}\phi\right|^2 - c_\theta|\theta|^2$$
$$= -c_4|e|^2 - \left(\sqrt{\mu}|\omega| \left|\frac{\partial V_e}{\partial e}\phi\right| - \frac{1}{2\sqrt{\mu}}|\theta|\right)^2 - |\theta|^2 \leq -c_4|e|^2 - |\theta|^2 \leq -c_5 V,$$

where $c_5 = \min\left\{\frac{c_4}{c_2}, \frac{1}{c_\theta \lambda_p}\right\}$, λ_p is the maximum eigenvalue of P. Hence, it follows from the latter inequality that $e(t)$ exponentially converges to the origin $e = 0$. This completes the proof of Property P.3.7.2. □

Remark 3.6 Replacing (3.75) in (3.62), we obtain the control law

$$u = u_s(e) - \mu|\omega(e,t)|^2 \frac{\partial V_e(e)}{\partial e}\phi(e), \tag{3.78}$$

where the second term involving $|\omega(e,t)|^2$ represents the nonlinear damping term. In Proposition 3.7, we introduce the expression (3.75) for the methodological purposes in order to demonstrate that the nonlinear damping term is constructed by replacing the "pure" integral feedback (3.64) by the static one (3.75) of the same structure. ■

In adaptive or robust control, nonlinear damping terms can be used for the following purposes.

First, it can be used in robust control to counteract the bounded *functional* or *signal* uncertainties generally defined as $\theta = \theta(e, u, t)$. In this case, the control law is represented by (3.78). As it was demonstrated by Proposition 3.7, this control law guarantees the boundedness of the state error and its exponential convergence to a residual set. Results of Proposition 3.7 can be extended to certain classes of unbounded functional uncertainties as well as to nonlinear parameterizations [9, Sect. 6.1.1].

Second, the nonlinear damping term can be used together with an adaptation algorithm for *modular design* (see Appendix E) to guarantee input-to-state stabilization of the closed-loop system with respect to the vector of parametric errors regarded as the disturbance input. Indeed, consider the adjustable controller involving additional nonlinear damping term

$$u = u_s(e) - \hat{\theta}^\top \omega(e,t) - \mu|\omega(e,t)|^2 \frac{\partial V_e(e)}{\partial e}\phi(e), \tag{3.79}$$

where $\mu > 0$, and $\hat{\theta}$ is the vector of adjustable parameters generated by an adaptation algorithm. Then the closed-loop state error model is given by

$$\dot{e} = f_\Sigma(e) + \phi(e)\left(\tilde{\theta}^\top \omega(e,t) - \mu|\omega(e,t)|^2 \frac{\partial V_e}{\partial e}(e)\phi(e)\right)$$

that is identical to (3.77) with substitution $\tilde{\theta}$ instead of θ. Therefore, if $\tilde{\theta}$ is bounded (that can be guaranteed by an adaptation algorithm), then, in view of Proposition 3.7, the controller (3.79) ensures the boundedness of e and its exponential convergence to a residual set. As a result, this property can be used for independent design of a nonlinear stabilizing controller and an adaptation algorithm. Such a design is called *modular design* and used as a special technique of backstepping [18, Chaps. 5, 6 and 9]. In this book, we combine modular backstepping with nonlinear swapping and high-order tuners (see Appendix E).

Third, a nonlinear damping term can be used in adaptive control to counteract the influence of exponentially decaying terms caused by, for example, auxiliary filters used for plant model and/or exosignal parameterization. As it was demonstrated by Example 1.6, in nonlinear systems, even exponentially vanishing disturbances can cause unbounded growth of the closed-loop signals. To avoid this undesirable effect, the nonlinear damping term must be added in an adaptive controller. For the plant (3.56), the adjustable controller with the damping term is defined as

$$u = u_s - \hat{\theta}^\top \omega(e,t) - \mu \frac{\partial V_e(e)}{\partial e} \phi(e), \tag{3.80}$$

where $\hat{\theta}$ is the vector of adjustable parameters generated by the adaptation algorithm (3.64).

The structure of the controller (3.80) can be explained as follows. Consider a slight modification of the model (3.56) given by

$$\dot{e} = f(e) + \phi(e)\left(u + \theta^\top \omega(e,t)\right),$$

where $\theta = \theta(t) = \epsilon(t)$ is the exponentially decaying term, and $\omega(e,t) \equiv 1$. Then the nonlinear damping term, defined according to Proposition 3.7 as $\mu\omega(e,t)\frac{\partial V_e}{\partial e}(e)\phi(e)$, takes the particular form $\mu\frac{\partial V_e(e)}{\partial e}\phi(e)$. Therefore, the third term in (3.80) provides the nonlinear damping of ϵ.

The properties of the closed-loop system affected by the exponentially decaying disturbance ϵ are formulated as follows.

Proposition 3.8 *If Assumption 3.1 holds and $\omega \in \mathcal{L}$, then the control law (3.80) with the adaptation algorithm (3.64) when applied to the plant (3.56) with the exponentially decaying disturbance ϵ guarantees for any $\gamma > 0$, $\mu > 0$ and for all initial conditions $e(0)$, $\hat{\theta}(0)$ the boundedness of all the closed-loop signals and the convergence $|e(t)| \to 0$ as $t \to \infty$.*

Proof Replacing (3.80) in (3.56), we obtain the following closed-loop error model:

$$\dot{e} = f_\Sigma(e) + \phi(e)\left(\tilde{\theta}^\top \omega(e,t) - \mu \frac{\partial V_e(e)}{\partial e}\phi(e) + \epsilon\right), \tag{3.81}$$

$$\dot{\tilde{\theta}} = -\gamma\omega(e,t)\frac{\partial V_e}{\partial e}(e)\phi(e). \tag{3.82}$$

Then the time derivative of the Lyapunov function

$$V(e, \tilde{\theta}, \epsilon) = V_e(e) + \frac{1}{2\gamma}\tilde{\theta}^\top \tilde{\theta} + \frac{1}{4\mu}\int_t^\infty \epsilon^2(\tau)d\tau$$

along the solutions of (3.81), (3.82) takes the form

$$\begin{aligned}\dot{V} &= \frac{\partial V_e}{\partial e} f_\Sigma - \mu \left|\frac{\partial V_e}{\partial e}\phi\right|^2 + \frac{\partial V_e}{\partial e}\phi\epsilon - \frac{1}{4\mu}\epsilon^2 \\ &\le -c_4|e|^2 - \mu\left|\frac{\partial V_e}{\partial e}\phi\right|^2 + \left|\frac{\partial V_e}{\partial e}\phi\right||\epsilon| - \frac{1}{4\mu}|\epsilon|^2.\end{aligned}$$

Completing the squares, we obtain

$$\dot{V} \le -c_4|e|^2 - \left(\sqrt{\mu}\left|\frac{\partial V_e}{\partial e}\phi\right| - \frac{1}{2\sqrt{\mu}}|\epsilon|\right)^2 \le -c_4|e|^2.$$

The latter means that e and $\hat{\theta}$ are bounded, and $|e(t)| \to 0$ as $t \to \infty$. □

3.4 Algorithms for Dynamic Error Model with Accessible Output

Now, we consider the state error model of the form

$$\dot{e} = f(e) + \phi(e)\left(\theta^\top \omega(e,t) + u\right), \quad \varepsilon = h(e), \tag{3.83}$$

where the state error e is not accessible for measurements, the output $\varepsilon \in \mathbb{R}$ and regressor ω are measured, and $h(e)$ is a smooth function such that $h(0) = 0$. In this case, an adaptation algorithm can involve the measured signals only and takes the form

$$\dot{\hat{\theta}} = \Psi_\varepsilon(\varepsilon, \omega). \tag{3.84}$$

Then the design problem is to define a function Ψ_ε such that all the closed-loop signals are bounded and $\varepsilon(t) \to 0$ as $t \to \infty$. This problem will be called *adaptive output stabilization*.

3.4.1 Adaptive Output Stabilization of Strictly Passive Systems

An adaptation algorithm of the form (3.64) can be directly applied to the model (3.83) if the following assumption holds (see, for example, [9, Sect. 6.2.1]).

Assumption 3.2 The exists a smooth function $V_e(e) \in \mathcal{K}_\infty$ such that

$$\frac{\partial V_e(e)}{\partial e} f(e) = -S(e), \tag{3.85}$$
$$\frac{\partial V_e(e)}{\partial e} \phi(e) = h(e), \tag{3.86}$$

where $S(e) \in \mathcal{K}$.[4]

In view of Theorem B.2, the condition (3.85) means that the origin of the system $\dot{e} = f(e)$ is globally uniformly asymptotically stable, and, as a result, the stabilizing component $u_s(e)$ is not necessary in the control (3.62). The condition (3.86), in turn, implies that in the adaptation law (3.64) the measured signal ε can be used instead of the term $\frac{\partial V_e}{\partial e}(e)\phi(e)$. Therefore, the adaptive controller (3.62), (3.64) can be represented in the physically implementable form

$$u = -\hat{\theta}^\top \omega(e, t), \quad \dot{\hat{\theta}} = \gamma \omega(e, t)\varepsilon. \tag{3.87}$$

The properties of the closed-loop error model are formulated in the following proposition.

Proposition 3.9 *If Assumption 3.2 holds and $\omega \in \mathcal{L}$, then the adaptive controller (3.87) when applied to the plant (3.83) guarantees for any $\gamma > 0$ and for all initial conditions $e(0)$, $\hat{\theta}(0)$ that all the closed-loop signals are bounded and $|e(t)| \to 0$ as $t \to \infty$.*

Proof In this case, the closed-loop error model is given by

$$\dot{e} = f(e) + \phi(e)\tilde{\theta}^\top \omega(e, t), \quad \dot{\tilde{\theta}} = -\gamma \omega(e, t)\varepsilon. \tag{3.88}$$

Then the time derivative of the Lyapunov function (3.65) along the solutions of the closed-loop error model (3.88) takes the form $\dot{V} = -S(e) < 0$, and, therefore, e and $\hat{\theta}$ are bounded. Since $S(e) \in \mathcal{K}$ and $h(e)$ is a smooth function such that $h(0) = 0$, from Theorem B.2 we conclude that $S(e(t))$, $|e(t)|$, $\varepsilon(t) \to 0$ as $t \to \infty$. □

From the practical point of view, it is important to understand how restrictive Assumption 3.2 is. As it was shown in [10, 24], a function $V_e(e)$ satisfying Assumption 3.2 exists if and only if the system

[4] For details about $\mathcal{K}$ and $\mathcal{K}_\infty$ functions, see Appendix B.

$$\dot{e} = f(e) + \phi(e)u, \quad \varepsilon = h(e) \tag{3.89}$$

is *strictly passive*.[5] In particular, it means that the origin of $\dot{e} = f(e)$ is globally uniformly asymptotically stable [18, Lemma D.3] and the system (3.89) has uniform relative degree equal to one. Therefore, Assumption 3.2 is very restrictive and can be satisfied for a narrow class of nonlinear systems.

Now, consider the problem of output adaptive stabilization for the linear state error model

$$\dot{e} = Ae + b\left(\theta^\top \omega + u\right), \quad \varepsilon = c^\top e, \tag{3.90}$$

where instead of the full state error e the output error ε is measured, and the matrix A and the vectors b, c have appropriate dimensions. If we choose $V_e(e) = \frac{1}{2}e^\top Pe$, where $P = P^\top$ is a positive definite matrix, then the Eqs. (3.85) and (3.86) are reduced to the following linear matrix equations:

$$A^\top P + PA = -Q, \quad Pb = c, \tag{3.91}$$

where $Q = Q^\top \succ 0$ is some preselected matrix. As a result, the adaptive controller can be defined as

$$u = -\hat{\theta}^\top \omega, \quad \dot{\hat{\theta}} = \gamma \omega \varepsilon. \tag{3.92}$$

The properties of the system closed by the designed control can be established with the use of the Lyapunov function

$$V(e, \tilde{\theta}) = \frac{1}{2}e^\top Pe + \frac{1}{2\gamma}\tilde{\theta}^\top \tilde{\theta}$$

whose time derivative along the solutions of the closed-loop system takes the form

$$\dot{V} = \frac{1}{2}e^\top (A^\top P + PA)e + \tilde{\theta}^\top \omega e^\top Pb - \tilde{\theta}^\top \omega \varepsilon = -e^\top Qe \leq 0.$$

According to the Yakubovich–Kalman lemma, there exists a positive definite matrix P satisfying simultaneously both Eq. (3.91) iff the transfer function $W(s) = c^\top (sI - A)^{-1} b$ is *strictly positive real* (SPR).[6] In particular, the SPR property implies that $W(s)$ is asymptotically stable, minimum phase and has relative degree one.

Therefore, we conclude that Assumption 3.2 is very restrictive and is not satisfied in the majority of practical applications due to so-called structural obstacle consisting in high relative degree of the error model. To overcome this obstacle and extend the

---

[5] For rigorous definition of strictly passive systems, see, for example, [18, Definition D.2], [9, Definition 2.29].

[6] For rigorous definition of strictly positive real transfer function, see, for example, [26, Sects. 2.5, 2.6], [12, Sect. 3.5.1], [31].

class of uncertain plants, for which implementable output-feedback algorithms of adaptation can be designed, Assumption 3.2 must be removed by means of more sophisticated design tools. In the next section, we will apply one of such design tools employing the *swapping-based* adaptation and the concept of *augmented error* [22, 26].

3.4.2 Swapping-Based Adaptation

Now, we consider the linear state error model (3.90) and make the following assumption.

Assumption 3.3 In the model (3.90), the matrices A, b and c are known and constitute the controllable and observable triplet (A, b, c). The matrix A is Hurwitz.

In contrast to Assumption 3.2, the latter assumption does not require the SPR property of the transfer function $W(s) = c^{\top}(sI - A)^{-1}b$. In particular, the relative degree of $W(s)$ can be greater than one.

According to the certainty equivalence design, we use the controller

$$u = -\hat{\theta}^{\top}\omega, \tag{3.93}$$

where the vector of adjustable parameters $\hat{\theta}$ is generated by an adaptation algorithm of the form

$$\dot{\hat{\theta}} = \Psi_{\varepsilon}(\varepsilon, \omega). \tag{3.94}$$

Then the closed-loop error model is given by

$$\dot{e} = Ae + b\tilde{\theta}^{\top}\omega, \quad \varepsilon = c^{\top}e, \tag{3.95}$$

$$\dot{\tilde{\theta}} = -\gamma\Psi_{\varepsilon}(\varepsilon, \omega). \tag{3.96}$$

The Eq. (3.95) can be rewritten in the input-output form

$$\varepsilon = W(s)\left[\tilde{\theta}^{\top}\omega\right] + \epsilon, \tag{3.97}$$

where $\epsilon = c^{\top}e^{At}e(0)$ exponentially decays.

The concept of the augmented error introduced by Monopoli in [22] consists in a certain transformation of the model (3.97) to a static form like (3.12). This form is suitable for design of stable output-feedback algorithms of adaptation. The tool of this transformation is often called *swapping scheme* (or *linear swapping scheme*).

Lemma 3.1 (Transfer function swapping.) *Along with* (3.97), *we define the* swapping term

$$\zeta = W(s)\left[\hat{\theta}^\top \omega\right] - \hat{\theta}^\top \omega_w, \tag{3.98}$$

where

$$\omega_w = W(s)[\omega] \tag{3.99}$$

is the filtered regressor, and introduce the augmented error

$$\bar{\varepsilon} = \varepsilon + \zeta. \tag{3.100}$$

Then the following equality is valid:

$$\bar{\varepsilon} = \tilde{\theta}^\top \omega_w + \epsilon. \tag{3.101}$$

Proof Taking into account the properties of linear systems and the fact that θ is constant, we get

$$\begin{aligned}\bar{\varepsilon} &= \varepsilon + \zeta = W(s)\left[(\theta - \hat{\theta})^\top \omega\right] + \epsilon + W(s)\left[\hat{\theta}^\top \omega\right] - \hat{\theta}^\top W(s)\,[\omega] \\ &= \theta^\top W(s)[\omega] - W(s)\left[\hat{\theta}^\top \omega\right] + W(s)\left[\hat{\theta}^\top \omega\right] - \hat{\theta}^\top W(s)\,[\omega] + \epsilon \\ &= \tilde{\theta}^\top \omega_w + \epsilon.\end{aligned}$$

□

The value of Lemma 3.1 is as follows. The augmented error is represented by the static model (3.101) (instead of the dynamic model (3.97)) that allows us to use relatively simple basic adaptation algorithms (see Sect. 3.2). However, the augmented error $\bar{\varepsilon}$ is different from ε, and, hence, a more detailed signal analysis is needed to prove the stability of the closed-loop system and the convergence of the actual output error ε to zero.

The next lemma demonstrates a state-space implementation of the swapping scheme and reveals some important properties of the swapping term ζ used later in the proof of the output error convergence [9, Lemma 6.1].

Lemma 3.2 (State-space swapping.) *Along with* (3.97), *we consider the* swapping filters[7]

$$\dot{\Omega}_w = A\Omega_w + b\omega^\top, \quad \omega_w^\top = c^\top \Omega_w, \tag{3.102}$$

$$\dot{\zeta}_w = A\zeta_w - \Omega_w \dot{\hat{\theta}}, \qquad \zeta = c^\top \zeta_w \tag{3.103}$$

[7] In (3.103), $\dot{\hat{\theta}}$ is replaced by the analytical expression $\Psi_\varepsilon(\varepsilon, \omega)$ from the adaptation algorithm (3.94).

with arbitrary initial conditions $\Omega_w(0)$ *and* $\zeta_w(0)$. *Define the* augmented error *as*

$$\bar{\varepsilon} = \varepsilon + \zeta. \tag{3.104}$$

Then the following equality is valid:

$$\bar{\varepsilon} = \tilde{\theta}^\top \omega_w + \bar{\epsilon}, \tag{3.105}$$

where $\bar{\epsilon} = c^\top \epsilon_w$ *exponentially decays, and* ϵ_w *satisfies the equation* $\dot{\epsilon}_w = A\epsilon_w$ *with* $\epsilon_w(0) = e(0) + \zeta_w(0) - \Omega_w(0)\hat{\theta}(0)$.

Proof Calculating the time derivative of $\epsilon_w = e + \zeta_w - \Omega_w\tilde{\theta}$, in view of (3.95), (3.102), (3.103), we have

$$\begin{aligned}\dot{\epsilon}_w &= Ae + b\tilde{\theta}^\top\omega + A\zeta_w - \Omega_w\dot{\hat{\theta}} - A\Omega_w\tilde{\theta} - b\omega^\top\tilde{\theta} + \Omega_w\dot{\hat{\theta}} \\ &= A(e + \zeta_w - \Omega_w\tilde{\theta}) = A\epsilon_w.\end{aligned}$$

Then from $e + \zeta_w = \Omega_w\tilde{\theta} + \epsilon_w$ and $\bar{\varepsilon} = c^\top(e + \zeta_w) = c^\top(\Omega_w\tilde{\theta} + \epsilon_w)$ we directly obtain (3.105). □

Corollary 3.4 *The difference between the output error* ε *and the augmented error* $\bar{\varepsilon}$ *is given by*

$$\zeta = \bar{\varepsilon} - \varepsilon = -W_\zeta(s)\left[W_\Omega(s)\left[\omega^\top\right]\dot{\hat{\theta}}\right]$$

and

$$\tilde{\theta}^\top W(s)[\omega] - W(s)\left[\tilde{\theta}^\top\omega\right] = -W_\zeta(s)\left[W_\Omega(s)\left[\omega^\top\right]\dot{\hat{\theta}}\right] + \bar{\epsilon}_w,$$

where $W_\zeta(s) = c^\top(sI - A)^{-1}$, $W_\Omega(s) = (sI - A)^{-1}b$,

$$\zeta = -W_\zeta(s)\left[W_\Omega(s)\left[\omega^\top\right]\dot{\hat{\theta}}\right]$$

is the swapping term, $\bar{\epsilon}_w = \epsilon - \bar{\epsilon}$.

The value of Lemma 3.2 consists in the following two facts. First, Lemma 3.2 and Corollary 3.4 explicitly demonstrate that the swapping term ζ is proportional to the adaptation rate of the adjustable parameters $\dot{\hat{\theta}}$. If the adaptation rate tends to zero and the regressor ω is bounded, then the swapping term tends to zero and the augmented error $\bar{\varepsilon}$ tends to the real output error ε. Second, the state-space swapping can be extended to certain classes of nonlinear systems [18, Appendix F].

The convergence properties ensured by the gradient algorithm of adaptation with augmented error are defined as follows.

Proposition 3.10 *Let Assumption 3.3 hold. Then the control law (3.93) with the adaptation algorithm*

$$\dot{\hat{\theta}} = \gamma \omega_w \bar{\varepsilon}, \tag{3.106}$$

where $\bar{\varepsilon}$ and ω_w are defined by (3.98) and (3.99), respectively, when applied to the state error model (3.90) guarantees for all $\gamma > 0$ and for any initial conditions $e(0)$, $\hat{\theta}(0)$ the following properties:

P.3.10.1. *if $\omega \in \mathcal{PC}_{[0,\infty)}$, then $\hat{\theta} \in \mathcal{L}_\infty$, $\bar{\varepsilon} \in \mathcal{L}_2$;*

P.3.10.2. *if $\omega, \dot{\omega} \in \mathcal{L}_\infty$, then $\varepsilon, \hat{\theta}, \dot{\hat{\theta}}, \in \mathcal{L}_\infty$, $\dot{\hat{\theta}} \in \mathcal{L}_2$, and $\varepsilon(t)$, $|\dot{\hat{\theta}}(t)| \to 0$ as $t \to \infty$;*

P.3.10.3. *if $\omega \in \mathcal{PE}$, then, in addition to P.3.10.2, $|\tilde{\theta}(t)|$ tends to zero exponentially and there exists an optimal γ for which the rate of convergence is maximum.*

Proof We directly obtain Property P.3.10.1 from Property P.3.1.1 of Proposition 3.1 with substitution $\varepsilon = \bar{\varepsilon}$. If $\omega, \dot{\omega} \in \mathcal{L}_\infty$, by virtue of P.3.1.2, we obtain that $\bar{\varepsilon}(t)$, $|\dot{\hat{\theta}}(t)| \to 0$ as $t \to \infty$. Since $|\dot{\hat{\theta}}(t)| \to 0$ and ω is bounded, then from Corollary 3.4 we obtain that $\bar{\varepsilon}(t) \to \varepsilon(t)$. As a result, $\varepsilon(t) \to 0$ as $t \to \infty$. Other properties directly follow from Proposition 3.1 with substitution $\varepsilon = \bar{\varepsilon}$. □

As seen from the proof of Proposition 3.10, the boundedness of the regressor ω plays key role in order to prove that $|\dot{\hat{\theta}}(t)| \to 0$ and, as a result, $(\bar{\varepsilon}(t) - \varepsilon(t)) \to 0$. However, in many cases, the boundedness of the regressor ω cannot be established a priori. For example, in the problems of output adaptive control of plants with unknown parameters (see Chap. 5), regressor depends on the plant output and is not known to be bounded a priori. Hence, the adaptation algorithm must be modified for rigorous proof of stability of the closed-loop system and asymptotic convergence of the actual output error to zero. One of such modifications includes *normalizing factor* [2, 26]. Algorithm of adaptation with the augmented error and normalization has the form

$$\dot{\hat{\theta}} = \gamma \frac{\omega_w \bar{\varepsilon}}{1 + \omega_w^\top \omega_w}. \tag{3.107}$$

Proposition 3.11 *Instead of (3.106), we consider the algorithm of adaptation with normalization (3.107). Then, additionally to the convergence properties established by Proposition 3.10, the algorithm (3.107) guarantees for $\omega \in \mathcal{PC}_{[0,\infty)}$ that*

$$\dot{\hat{\theta}} \in \mathcal{L}_\infty \quad \text{and} \quad \frac{\bar{\varepsilon}}{\sqrt{1 + \omega_w^\top \omega_w}}, \ \dot{\hat{\theta}} \in \mathcal{L}_2.$$

Proof ([26, Sect. 5.5]) Substituting (3.105) into (3.107) and neglecting the exponentially decaying term $\bar{\epsilon}$, we derive the following parametric error model:

$$\dot{\tilde{\theta}} = -\gamma \frac{\omega_w \omega_w^\top}{1 + \omega_w^\top \omega_w} \tilde{\theta}. \tag{3.108}$$

Therefore, Proposition 3.11 is proved by following the same arguments as in the proof of Proposition 3.2 with the use of $\bar{\varepsilon}$ and ω_w instead of ε and ω, respectively.

□

Since the model of the augmented error (3.105) has the form of the static canonical error model (3.12), one can combine the swapping technique with DRE as well as with MRE to obtain the adaptation algorithms with improved parametric convergence.

Corollary 3.5 *Define $\bar{H}(s) = [1, H_1(s), \ldots, H_p(s)]^\top$, where $p \geq q - 1$, and $H_i(s)$ are distinct PASMP transfer functions. Then, for the dynamically extended augmented error*

$$\varepsilon_H = \bar{H}(s)\left[\varepsilon + W(s)\left[\hat{\theta}^\top \omega\right]\right] - \bar{H}(s)\left[\omega_w^\top\right]\hat{\theta},$$

the following equality is valid:

$$\varepsilon_H = W_w^\top \tilde{\theta} + \bar{H}(s)[\epsilon],$$

where $W_w^\top = \bar{H}(s)\left[\omega_w^\top\right]$.

Corollary 3.6 *Define $L(s)$ in the form of (3.44). Then, for the memory extended augmented error*

$$\varepsilon_L = L(s)\left[\omega_w\left(\varepsilon + W(s)\left[\theta^\top \omega\right]\right)\right] - L(s)\left[\omega_w \omega_w^\top\right]\hat{\theta},$$

the following equality is valid:

$$\varepsilon_L = \Omega_w \tilde{\theta} + L(s)[\omega_w \epsilon],$$

where $\Omega_w = L(s)\left[\omega_w \omega_w^\top\right]$.

By now, the swapping technique is well-developed design tool with many applications to different control problems including MRAC of SISO and MIMO systems, adaptive modular backstepping, adaptive control with input delay, etc. In Appendix D, we present different swapping schemes used in this book.

3.5 Robust Algorithms of Adaptation

The algorithms of adaptation introduced in Sects. 3.2, 3.3 and 3.4 are designed and analyzed under assumption that the disturbance ϵ is either absent or exponentially decays to zero. However, as it is demonstrated by Example 1.3, even asymptotically (but not exponentially) vanishing unmodelled disturbances can cause unbounded parametric drift. Such undesirable effect can be demonstrated even for much more simpler case of a static error model with single adjustable parameter.

Example 3.2 ([12, Sect. 8.3.1.]) Consider the error model $\varepsilon = \omega\tilde{\theta} + \delta$ with bounded irregular disturbance δ and the gradient algorithm of adaptation

$$\dot{\hat{\theta}} = \gamma\omega\varepsilon,$$

where $\tilde{\theta} = \theta - \hat{\theta}$, $\omega(t) = (1+t)^{-\frac{1}{2}}$, $\theta = 2$, $\gamma = 1$, $\hat{\theta}(0) = 1$, and

$$\delta(t) = (1+t)^{-\frac{1}{4}}\left(\frac{5}{4} - 2(1+t)^{-\frac{1}{4}}\right) \to 0 \text{ as } t \to \infty.$$

In this case, in spite of the fact that the disturbance δ asymptotically vanishes, we obtain unbounded parametric drift and $\hat{\theta}(t) = (1+t)^{\frac{1}{4}} \to \infty$ as $t \to \infty$. ∎

In order to avoid this unacceptable phenomenon, several robust modifications of the basic algorithms of adaptation were proposed in literature [12, Chap. 8], [21], [26, Chap. 8], [30, Chap. 5]. In this book, we use the *continuous switching σ-modification* [11], [12, Sect. 8.4.1]. Applied to the static error model, such σ-modifications of the basic adaptation algorithms are defined by the following two propositions.

Proposition 3.12 *Consider the output error model*

$$\varepsilon = \tilde{\theta}^{\top}\omega + \delta, \tag{3.109}$$

where the output error $\varepsilon \in \mathbb{R}$ and the regressor $\omega : \mathbb{R}_+ \to \mathbb{R}^q$ are measured, $\tilde{\theta} = \theta - \hat{\theta}$ is the parametric error, $\hat{\theta}$ is the vector of adjustable parameters, θ is a vector of unknown parameters, and δ is a bounded disturbance such that $|\delta(t)| \le \bar{\delta}$ for all $t \ge 0$, and $\bar{\delta} > 0$ is a constant. Then the robust gradient algorithm of adaptation

$$\dot{\hat{\theta}} = \gamma\omega\varepsilon - \gamma\sigma(\hat{\theta})\hat{\theta}, \tag{3.110}$$

where

$$\sigma(\hat{\theta}) = \begin{cases} 0, & |\hat{\theta}| < \theta^*, \\ \frac{|\hat{\theta}|}{\theta^*} - 1, & \theta^* \le |\hat{\theta}| \le 2\theta^*, \\ 1, & |\hat{\theta}| > 2\theta^*, \end{cases} \tag{3.111}$$

and θ^* is a positive constant, guarantees that:

P.3.12.1. *if $\omega \in \mathcal{PC}_{[0,\infty)}$, then $\hat{\theta}$ is bounded and $\tilde{\theta}$ exponentially converges to the residual set*

$$D_{\tilde{\theta}} = \left\{\tilde{\theta} : |\tilde{\theta}|^2 \le \frac{1}{c_6}\left(\frac{1}{c_6}|\theta|^2 + \bar{\delta}^2 + 2c_7\right)\right\}, \tag{3.112}$$

where

$$c_7 = c_6 \left(|\theta| + (1 + c_6)\theta^*\right)^2, \tag{3.113}$$

and $0 < c_6 < 1$;

P.3.12.2. *if* $\theta^* > |\theta|$ *and* $\omega \in \mathcal{L}_\infty$, *then* $\varepsilon, \hat{\theta}, \dot{\hat{\theta}} \in \mathcal{L}_\infty$, *and both* ε *and* $|\dot{\hat{\theta}}|$ *are of the order of* $\bar{\delta}$ *in the mean square sense, i.e.,* ε, $|\dot{\hat{\theta}}| \in \mathcal{MS}(\bar{\delta})$;

P.3.12.3. *if* $\theta^* > |\theta|$ *and* δ *exponentially decays, then the algorithm* (3.110), (3.111) *has the same properties as the unmodified algorithm* (3.15) *established by Proposition* 3.1.

Proof From (3.110), in view of the equality $\dot{\hat{\theta}} = -\dot{\tilde{\theta}}$, we obtain the following model of the parametric errors:

$$\dot{\tilde{\theta}} = -\gamma\omega\varepsilon + \gamma\sigma(\hat{\theta})\hat{\theta}. \tag{3.114}$$

Then, calculating the time derivative of the Lyapunov function $V(\tilde{\theta}) = \frac{1}{2\gamma}\tilde{\theta}^\top\tilde{\theta}$, we have

$$\begin{aligned} \dot{V} &= -\tilde{\theta}^\top\omega\varepsilon + \sigma(\hat{\theta})\tilde{\theta}^\top\hat{\theta} = -\varepsilon^2 + \varepsilon\delta + \sigma(\hat{\theta})\tilde{\theta}^\top\hat{\theta} \\ &\le -\varepsilon^2 + \bar{\delta}|\varepsilon| + \sigma\tilde{\theta}^\top\hat{\theta} \qquad (3.115) \\ &\le -\frac{1}{2}\varepsilon^2 - \frac{1}{2}\left(\varepsilon - \bar{\delta}\right)^2 + \frac{1}{2}\bar{\delta}^2 + \sigma\tilde{\theta}^\top\hat{\theta} \\ &\le -\frac{1}{2}\varepsilon^2 + \frac{1}{2}\bar{\delta}^2 + \sigma\tilde{\theta}^\top\hat{\theta}. \qquad (3.116) \end{aligned}$$

Because

$$\sigma\tilde{\theta}^\top\hat{\theta} = -\sigma|\tilde{\theta}|^2 + \sigma\tilde{\theta}^\top\theta, \tag{3.117}$$

we have

$$\dot{V} \le -\frac{1}{2}\varepsilon^2 - \sigma|\tilde{\theta}|^2 + \frac{1}{2}\bar{\delta}^2 + \sigma\tilde{\theta}^\top\theta.$$

It can be verified that

$$-\sigma(\hat{\theta})|\tilde{\theta}|^2 \le -c_6|\tilde{\theta}|^2 + c_7. \tag{3.118}$$

Then

$$\begin{aligned} \dot{V} &\le -\frac{1}{2}\varepsilon^2 - c_6|\tilde{\theta}|^2 + |\tilde{\theta}||\theta| + \frac{1}{2}\bar{\delta}^2 + c_7 \\ &\le -\frac{1}{2}\varepsilon^2 - \frac{c_6}{2}|\tilde{\theta}|^2 - \left(\sqrt{\frac{c_6}{2}}|\tilde{\theta}| - \frac{1}{\sqrt{2c_6}}|\theta|\right)^2 + \frac{1}{2c_6}|\theta|^2 + \frac{1}{2}\bar{\delta}^2 + c_7 \end{aligned}$$

$$\leq -\frac{1}{2}\varepsilon^2 - \frac{c_6}{2}|\tilde{\theta}|^2 + \frac{1}{2c_6}|\theta|^2 + \frac{1}{2}\bar{\delta}^2 + c_7$$
$$\leq -c_6\gamma V + \frac{1}{2c_6}|\theta|^2 + \frac{1}{2}\bar{\delta}^2 + c_7.$$

Therefore, $\hat{\theta}$ is bounded and exponentially converges to the residual set $D_{\tilde{\theta}}$ (3.112). Property P.3.7.1 is proved.

If $\omega \in \mathcal{L}_\infty$, then from the boundedness of $\hat{\theta}$ and ϵ we directly obtain the boundedness of ε and $|\dot{\hat{\theta}}|$. Now, integrating (3.116), we have [12, Sect. 8.4.1]

$$-2\int_t^{t+T} \sigma\tilde{\theta}^\top\hat{\theta}d\tau + \int_t^{t+T} \varepsilon^2 d\tau \leq \bar{\delta}^2 T + c_8, \tag{3.119}$$

where $c_8 = 2\sup_{t\geq 0}(V(t) - V(t+T))$, $T \geq 0$. Since $\theta^* \geq |\theta|$, we have $\sigma\tilde{\theta}^\top\hat{\theta} \leq 0$. Therefore, from (3.119) we obtain that $\varepsilon \in \mathcal{MS}(\bar{\delta})$. Since ω is bounded, from (3.110) we can conclude that $|\dot{\hat{\theta}}| \in \mathcal{MS}(\bar{\delta})$. This completes the proof of Property P.3.12.2.

To prove Property P.3.12.3, consider the Lyapunov function

$$V(\tilde{\theta}) = \frac{1}{2}\tilde{\theta}^\top\tilde{\theta} + \int_t^\infty \delta^2(\tau)d\tau$$

whose time derivative along the solutions of (3.114) takes the form

$$\dot{V} = -\varepsilon^2 + \varepsilon\delta - \delta^2 + \sigma\tilde{\theta}^\top\hat{\theta}.$$

Because $\theta^* \geq |\theta|$, we have $\sigma\tilde{\theta}^\top\hat{\theta} \leq 0$ for any $\hat{\theta}$. Therefore,

$$\dot{V} \leq -\varepsilon^2 + \varepsilon\delta - \delta^2 = -\frac{1}{2}\varepsilon^2 - \frac{1}{2}(\varepsilon - \delta)^2 - \frac{1}{2}\delta^2 \leq -\frac{1}{2}\varepsilon^2.$$

The latter inequality is in the form of (3.20) and enables us to prove Property P.3.12.3. □

Proposition 3.13 *Consider the output error model* (3.109) *with* $|\delta(t)| < \bar{\delta}$ *for all* $t \geq 0$, *where* $\bar{\delta} > 0$ *is a constant. Then the robust algorithm of adaptation with MRE*

$$\dot{\hat{\theta}} = \gamma\left(y_L - \Omega\hat{\theta}\right) - \sigma(\hat{\theta})\hat{\theta}, \tag{3.120}$$

where y_L, Ω *and* $\sigma(\hat{\theta})$ *are defined by* (3.37), (3.39) *and* (3.111), *ensures that:*

P.3.13.1. *if* $\omega \in \mathcal{PC}_{[0,\infty)}$, *then* $\hat{\theta}$ *is bounded and* $\tilde{\theta}$ *exponentially converges to the residual set* (3.112);

P.3.13.2. *if* $\theta^* > |\theta|$ *and* $\omega \in \mathcal{L}_\infty$, *then* $\varepsilon, \hat{\theta}, \dot{\hat{\theta}}, \in \mathcal{L}_\infty$, *and* ε, $|\dot{\hat{\theta}}| \in \mathcal{MS}(\bar{\delta})$;

P.3.13.3. *if* $\theta^* > |\theta|$ *and* δ *exponentially decays, then the adaptation algorithm* (3.120) *has the same properties as the algorithm* (3.45) *established by Proposition 3.4.*

Proof In this case, the parametric error model takes the form

$$\dot{\tilde{\theta}} = -\gamma\Omega\tilde{\theta} - \gamma L(s)[\omega\delta] + \sigma(\hat{\theta})\hat{\theta},$$

and the time derivative of the Lyapunov function $V = \frac{1}{2\gamma}\tilde{\theta}^\top\tilde{\theta}$ takes the form

$$\dot{V} = -\tilde{\theta}^\top\Omega\tilde{\theta} - \tilde{\theta}^\top L(s)[\omega\delta] + \sigma\tilde{\theta}^\top\hat{\theta}.$$

Then, using the Schwartz inequality as in the proof of Corollary 3.3, we can demonstrate the validity of the following inequality (for $b_0 = \prod_{i=1}^{p} a_i$):

$$\dot{V} \le -\tilde{\theta}^\top\Omega\tilde{\theta} + \left(\tilde{\theta}^\top\Omega\tilde{\theta}\right)^{\frac{1}{2}}\bar{\delta} + \sigma\tilde{\theta}^\top\hat{\theta}.$$

Denoting $\varepsilon_\Omega = \left(\tilde{\theta}^\top\Omega\tilde{\theta}\right)^{\frac{1}{2}}$, we can write

$$\dot{V} \le -\varepsilon_\Omega^2 + \bar{\delta}|\varepsilon_\Omega| + \sigma\tilde{\theta}^\top\hat{\theta}.$$

The latter expression is in the form of (3.115) that allows us to extend the results of Proposition 3.12 to the algorithm of adaptation with DRE (3.120). □

Now, consider particular case of the nonlinear plant (3.56), where the external disturbance $\delta \in \mathbb{R}^n$ enters the plant linearly

$$\dot{e} = f(e) + \phi(e)\left(u + \theta^\top\omega(e,t)\right) + \delta. \tag{3.121}$$

Proposition 3.14 *Let Assumption 3.1 be valid, $\omega \in \mathcal{L}$ and $|\delta(t)| \le \bar{\delta}$, where $\bar{\delta} > 0$ is a constant. Then the control law (3.62) with the adaptation algorithm*

$$\dot{\hat{\theta}} = \gamma\omega(e,t)\frac{\partial V_e(e)}{\partial e}\phi(e) - \gamma\sigma(\hat{\theta})\hat{\theta}, \tag{3.122}$$

where $\sigma(\hat{\theta})$ is defined by (3.111), when applied to the plant (3.121) guarantees for any $\gamma > 0$ and for all initial conditions $e(0)$, $\hat{\theta}(0)$ that:

P.3.14.1. *e, $\hat{\theta} \in \mathcal{L}_\infty$, and e, $\tilde{\theta}$ exponentially converge to the residual set*

$$D_{e,\tilde{\theta}} = \left\{e, \tilde{\theta} : V(e,\tilde{\theta}) \le \frac{1}{c_8}\left(\frac{c_3^2}{2c_4}\bar{\delta}^2 + \frac{1}{2c_6}|\theta|^2 + c_7\right)\right\}, \tag{3.123}$$

where $0 < c_6 < 1$, c_7 is defined by (3.113), and $c_8 = \min\{\frac{c_4}{2c_2}, \gamma c_6\}$;

P.3.14.2. *if $\theta^* > |\theta|$, then e is of the order of $\bar{\delta}$ in the mean square sense, i.e., $|e| \in \mathcal{MS}(\bar{\delta})$;*

P.3.14.3. *if $\theta^* > |\theta|$ and δ exponentially decays, then $|e(t)| \to 0$ as $t \to \infty$.*

Proof Substituting (3.62) into (3.121) and differentiating $\tilde{\theta}$, in view of (3.122), we can derive the following closed-loop error model:

$$\dot{e} = f_\Sigma(e) + \phi(e)\tilde{\theta}^\top \omega(e,t) + \delta, \tag{3.124}$$

$$\dot{\tilde{\theta}} = -\gamma\omega(e,t)\frac{\partial V_e(e)}{\partial e}\phi(e) + \gamma\sigma(\hat{\theta})\hat{\theta}. \tag{3.125}$$

Calculating the time derivative of the Lyapunov function (3.65) along the solutions of (3.124), (3.125), we obtain

$$\dot{V} \leq -c_4|e|^2 + \sigma\tilde{\theta}^\top\hat{\theta} + \frac{\partial V_e}{\partial e}\delta \leq -c_4|e|^2 + \sigma\tilde{\theta}^\top\hat{\theta} + c_3|e|\bar{\delta}.$$

Completing the squares, we obtain

$$\dot{V} \leq -\frac{c_4}{2}|e|^2 - \left(\sqrt{\frac{c_4}{2}}|e| - \frac{c_3}{\sqrt{2c_4}}\bar{\delta}\right)^2 + \frac{c_3^2}{2c_4}\bar{\delta}^2 + \sigma\tilde{\theta}^\top\hat{\theta}$$

$$\leq -\frac{c_4}{2}|e|^2 + \frac{c_3^2}{2c_4}\bar{\delta}^2 + \sigma\tilde{\theta}^\top\hat{\theta}. \tag{3.126}$$

Then, in view of (3.117) and (3.118), we have

$$\dot{V} \leq -\frac{c_4}{2}|e|^2 - c_6|\tilde{\theta}|^2 + \frac{c_3^2}{2c_4}\bar{\delta}^2 + |\theta||\tilde{\theta}| + c_7,$$

and, therefore,

$$\dot{V} \leq -\frac{c_4}{2}|e|^2 - \frac{c_6}{2}|\tilde{\theta}|^2 - \left(\sqrt{\frac{c_6}{2}}|\tilde{\theta}| - \frac{1}{\sqrt{2c_6}}|\theta|\right)^2 + \frac{1}{2c_6}|\theta|^2 + \frac{c_3^2}{2c_4}\bar{\delta}^2 + c_7$$

$$\leq -c_8 V + \frac{c_3^2}{2c_4}\bar{\delta}^2 + \frac{1}{2c_6}|\theta|^2 + c_7. \tag{3.127}$$

The latter means that e and $\hat{\theta}$ are bounded and exponentially converge to the residual set (3.123). Further analysis demonstrates that all the closed-loop signals are bounded. Part P.3.14.1 is proved.

To prove Property P.3.14.2, we integrate the inequality (3.126) and use the arguments similar to those from the proof of Property P.3.12.2.

To prove part P.3.14.3, we use the Lyapunov function

$$V_\epsilon = V_e(e) + \frac{1}{2\gamma}\tilde{\theta}^\top\tilde{\theta} + \frac{c_3^2}{c_4}\int_t^\infty |\epsilon(\tau)|^2 d\tau,$$

and note that for $\theta^* > |\theta|$ we have $\sigma(\hat{\theta})\tilde{\theta}^\top\hat{\theta} \leq 0$ for all $\hat{\theta}$. Therefore,

$$
\begin{aligned}
\dot{V}_{\epsilon} &\leq -c_4|e|^2 - \frac{c_3^2}{c_4}|\delta|^2 + c_3|e||\delta| \\
&\leq -\frac{c_4}{2}|e|^2 - \frac{c_3^2}{2c_4}|\delta|^2 - \left(\sqrt{\frac{c_4}{2}}|e| - \frac{c_3}{\sqrt{2c_4}}|\delta|^2\right)^2 \\
&\leq -\frac{c_4}{2}|e|^2 - \frac{c_3^2}{2c_4}|\delta|^2,
\end{aligned}
$$

that enables us to prove Property P.3.14.3. □

Now, consider more general form of nonlinear plant (3.56), where the disturbance δ enters the plant nonlinearly, i.e., is multiplied by the nonlinear term $\phi(e)$.

Proposition 3.15 *Let Assumption 3.1 be valid, $\omega \in \mathcal{L}$ and $|\delta(t)| \leq \bar{\delta}$, where $\bar{\delta} > 0$ is a constant. Then the control law*

$$
u = u_s - \hat{\theta}^{\top}\omega(e,t) - \mu\frac{\partial V_e}{\partial e}\phi(e) \tag{3.128}
$$

with the adaptation algorithm (3.122), (3.111) when applied to the plant (3.56) guarantees for any $\gamma > 0$ and for all initial conditions $e(0)$, $\hat{\theta}(0)$ that:

P.3.15.1. *e, $\hat{\theta} \in \mathcal{L}_{\infty}$, and e, $\tilde{\theta}$ exponentially converge to the residual set*

$$
D_{e,\tilde{\theta}} = \left\{ e, \tilde{\theta} : V(e,\tilde{\theta}) \leq \frac{1}{c_9}\left(\frac{1}{4}\bar{\delta}^2 + \frac{1}{2c_6}|\theta|^2 + c_7\right)\right\}, \tag{3.129}
$$

where $0 < c_6 < 1$, c_7 is defined by (3.113), and $c_9 = \min\{\frac{c_4}{c_2}, \gamma c_6\}$;

P.3.15.2. *if $\theta^* > |\theta|$, then e is of the order of $\bar{\delta}$ in the mean square sense, i.e., $|e| \in \mathcal{MS}(\bar{\delta})$;*

P.3.15.3. *if $\theta^* > |\theta|$ and δ exponentially decays, then $|e(t)| \to 0$ as $t \to \infty$.*

Proof Substituting (3.128) into (3.56) and differentiating $\tilde{\theta}$ in view of (3.122), we can derive the following closed-loop error model:

$$
\dot{e} = f_{\Sigma}(e) + \phi(e)\tilde{\theta}^{\top}\omega(e,t) - \phi(e)\frac{\partial V_e}{\partial e}\phi(e) + \phi(e)\delta, \tag{3.130}
$$

$$
\dot{\tilde{\theta}} = -\gamma\omega(e,t)\frac{\partial V_e}{\partial e}(e)\phi(e) + \gamma\sigma(\hat{\theta})\hat{\theta}. \tag{3.131}
$$

Calculating the time derivative of the Lyapunov function (3.65) along the solutions of (3.130), (3.131), we obtain

$$\begin{aligned}\dot{V} &\leq -c_4|e|^2 + \left|\frac{\partial V}{\partial e}\phi\right|\bar{\delta} - \left|\frac{\partial V}{\partial e}\phi\right|^2 + \sigma\tilde{\theta}^\top\hat{\theta} \\ &= -c_4|e|^2 - \left(\left|\frac{\partial V}{\partial e}\phi\right| - \frac{1}{2}\bar{\delta}\right)^2 + \frac{1}{4}\bar{\delta}^2 + \sigma\tilde{\theta}^\top\hat{\theta} \qquad (3.132) \\ &\leq -c_4|e|^2 + \frac{1}{4}\bar{\delta}^2 + \sigma\tilde{\theta}^\top\hat{\theta}. \qquad (3.133)\end{aligned}$$

Then, in view of (3.117) and (3.118), we have

$$\dot{V} \leq -c_4|e|^2 + \frac{1}{4}\bar{\delta}^2 - c_6|\tilde{\theta}|^2 + |\theta||\tilde{\theta}| + c_7,$$

and, therefore,

$$\begin{aligned}\dot{V} &\leq -c_4|e|^2 + \frac{1}{4}\bar{\delta}^2 - \frac{c_6}{2}|\tilde{\theta}|^2 - \left(\sqrt{\frac{c_6}{2}}|\tilde{\theta}| - \frac{1}{\sqrt{2c_6}}|\theta|\right)^2 + \frac{1}{2c_6}|\theta|^2 + c_7 \\ &\leq -c_9 V + \frac{1}{4}\bar{\delta}^2 + \frac{1}{2c_6}|\theta|^2 + c_7. \qquad (3.134)\end{aligned}$$

The latter means that e and $\hat{\theta}$ are bounded and exponentially converge to the residual set (3.129). Further analysis demonstrates that all the closed-loop signals are bounded. Part P.3.15.1 is proved.

Properties P.3.15.2 and P.3.15.3 are proved by following the same arguments as in the proof of Proposition 3.14. □

References

1. Aranovskiy, S., Bobtsov, A., Ortega, R., Pyrkin, A.: Performance enhancement of parameter estimator via dynamic regressor extension and mixing. IEEE Trans. Autom. Control **62**(7), 3546–3550 (2017)
2. Barabanov, N.E., Ortega, R.: Is normalization necessary for stable model reference adaptive control? IEEE Trans. Autom. Control **50**(9), 1384–1390 (2005)
3. Barabanov, N.E., Ortega, R.: On global asymptotic stability of $\dot{x} = \phi(t)\phi^\top(t)x$ with $\phi(t)$ bounded and not persistently exciting. Syst. Control Lett. **109**, 24–29 (2017)
4. Barmish, B.R., Corless, M., Leitmann, G.: A new class of stabilizing controllers for uncertain dynamical systems. SIAM J. Control Optim. **21**, 246–255 (1983)
5. Chong, E.K.P., Zak, S.H.: An Introduction to Optimization, 3rd edn. Wiley, New York (2008)
6. Corless, M., Leitmann, G.: Continuous state feedback guaranteeing uniform ultimate boundedness for uncertain dynamical systems. IEEE Trans. Autom. Control **26**, 1139–1144 (1981)
7. Gerasimov, D.N., Nikiforov, V.O.: On key properties of the Lion's and Kreisselmeier's adaptation algorithms. Int. J. Adapt. Control Signal Proc. (2021). https://doi.org/10.1002/acs.3311
8. Fletcher, R.: Practical Methods of Optimization, 2nd edn. Wiley, New York (2000)
9. Fradkov, A.L., Miroshnik, I.V., Nikiforov, V.O.: Nonlinear and Adaptive Control of Complex Systems. Kluwer Academic Publishers, Dordrecht (1999)

10. Hill, D.J., Moylan, P.J.: The stability of nonlinear dissipative systems. IEEE Trans. Autom. Control **21**(5), 708–711 (1976)
11. Ioannou, P., Tskalis, K.: A robust direct adaptive controller. IEEE Trans. Autom. Control **31**(11), 1033–1043 (1986)
12. Ioannou, P., Sun, J.: Robust Adaptive Control. Prentice-Hall, Englewood Cliffs (1996)
13. Kaczorek, T., Sajewski, L.: The Realization Problem for Positive and Fractional Systems. Springer International Publishing, Switzerland (2014)
14. Kanellakopoulos, I., Kokotović, P., Morse, A.S.: A toolkit for nonlinear feedback design. Syst. Control Lett. **18**(2), 83–92 (1992)
15. Khalil, H.K.: Nonlinear Control. Prentice-Hall, Englewood Cliffs (2015)
16. Kreisselmeier, G.: Adaptive observers with exponential rate of convergence. IEEE Trans. Autom. Control **22**(1), 2–8 (1977)
17. Kreisselmeier, G., Joos, D.: Rate of convergence in model reference adaptive control. IEEE Trans. Autom. Control **27**(3), 710–713 (1982)
18. Krstić, M., Kanellakopoulos, I., Kokotović, P.: Nonlinear and Adaptive Control Design. Wiley, New York (1995)
19. Lion, P.M.: Rapid identification of linear and nonlinear systems. AIAA J. **5**(10), 1835–1842 (1967)
20. Ljung, L.: System Identification. Theory for the User, 2nd edn. Prentice Hall, Upper Saddle River (1999)
21. de Mathelin, M., Lozano, R.: Robust adaptive identification of slowly time-varying parameters with bounded disturbances. Automatica **35**, 1291–1305 (1999)
22. Monopoli, R.V.: Model reference adaptive control with an augmented signal. IEEE Trans. Autom. Control **19**(5), 474–484 (1974)
23. Morse, A.S.: High-order parameter tuners for adaptive control of nonlinear systems. In: Isidori, A., Tarn, T.J. (eds.) Systems, Models and Feedback: Theory and Applications, pp. 339–364. Birkhauser, Basel (1992)
24. Moylan, P.J.: Implication of passivity in a class of nonlinear systems. IEEE Trans. Autom. Control **19**(4), 373–381 (1974)
25. Narendra, K.S., Tripathi, S.S., Luders, G., Kudva, P.: Adaptive control using Lyapunov's direct method, Technical Report No. CT-43, Becton Center, Yale University, New Haven, CT (1971)
26. Narendra, K.S., Annaswamy, A.M.: Stable Adaptive Systems. Prentice Hall, Englewood Cliffs (1989)
27. Nikiforov, V.O.: Robust high-order tuner of simplified structure. Automatica **35**(8), 1409–1415 (1999)
28. Ortega, R.: On Morse's new adaptive controller: parameter convergence and transient performance. IEEE Trans. Autom. Control **38**(8), 1191–1202 (1993)
29. Ortega, R., Nikiforov, V., Gerasimov, D.: On modified parameter estimators for identification and adaptive control: a unified framework and some new schemes. Annu. Rev. Control **50**, 278–293 (2020)
30. Sastry, S., Bodson, M.: Adaptive Control: Stability, Convergence and Robustness. Prentice-Hall, Englewood Cliffs (1989)
31. Yakubovich, V.A.: The solution of certain matrix inequalities in automatic control theory. Soviet Math. AMS **3**, 620–623 (1962)

Chapter 4
Adaptive Regulation in Systems with Known Parameters

Abstract The chapter presents controllers providing adaptive reference tracking and adaptive disturbance rejection for the plants with known parameters. In other words, the chapter presents design methods applied to the plants with known parameters in order to solve control Problems 1.2, 1.3 and 1.4 formulated in Sect. 1.3.3. The employed design methods are based on the combined use of the exosignal observers introduced in Sects. 2.3, 2.4 and the basic adaptation algorithms presented in Chap. 3. The designed adaptive controllers provide complete rejection of the external disturbance and/or asymptotic reference tracking when the exosignals are modelled by linear exosystems of known order but with unknown parameters. In this case, the adaptation counteracts the exosignal uncertainty. Section 4.1 presents adaptive controllers for LTI SISO plants, while Sects. 4.2 and 4.3 extend the obtained solutions to certain classes of LTI MIMO plants as well as to nonlinear plants.

4.1 LTI SISO Plants

In this section, the class of LTI SISO plants with known parameters is considered. For methodological purposes, we start with separate design of reference tracking controllers and disturbance rejection ones. The simplest case of the state-feedback control under disturbance matching condition is considered first. Then the approach will be extended to the cases of unmatched disturbance, output-feedback adaptive control, and simultaneous disturbance rejection and reference tracking.

4.1.1 Reference Tracking

Consider the LTI plant

$$\dot{x} = Ax + bu, \quad y = c^{\top}x, \tag{4.1}$$

V. Nikiforov and D. Gerasimov, *Adaptive Regulation*, Lecture Notes in Control and Information Sciences 491, https://doi.org/10.1007/978-3-030-96091-9_4

where $x \in \mathbb{R}^n$ is the state, $y \in \mathbb{R}$ is the output (regulated) variable, $u \in \mathbb{R}$ is the control, while A, b and c are known and have appropriate dimensions.

The considered control problem is to design a state-feedback (or output-feedback) controller providing the boundedness of all the closed-loop signals and driving the tracking error to zero according to the equality

$$\lim_{t\to\infty} (y(t) - g(t)) = 0, \tag{4.2}$$

where the reference g satisfies Assumption 2.1 (with substitution $\upsilon = g$) and is accessible for measurements. In other words, we consider Problem 1.2 of adaptive reference tracking formulated in Sect. 1.3.3.

We accept the following assumption regarding the plant (4.1).

Assumption 4.1 The plant (4.1) is such that:

A.4.1.1. the triplet (A, b, c) is controllable and observable;

A.4.1.2. all the zeros of the transfer function $W_u(s) = c^\top(sI - A)^{-1}b$ have negative real parts.

Remark 4.1 In view of Lemma 1.1, it is enough to require that none of the zeros of $W_u(s)$ coincides with an eigenvalue of Γ. At the same time, the eigenvalues of Γ are unknown. Therefore, we strengthen the requirement for $W_u(s)$ and assume that all its zeros have negative real parts. The latter guarantees that they do not coincide with the eigenvalues of Γ which have zero real parts in accordance with Assumption 2.1. ■

First, we stabilize the plant with the use of the controller

$$u = -k^\top x + u_t, \tag{4.3}$$

where the vector of feedback gains k is selected so that the matrix of the closed-loop system $A_\Sigma = A - bk^\top$ is Hurwitz [2, Sect. 8.2], while the *tracking component* u_t will be defined later. Substituting (4.3) into (4.1), we obtain

$$\dot{x} = A_\Sigma x + bu_t, \qquad y = c^\top x. \tag{4.4}$$

Remark 4.2 If the state x is inaccessible for measurements, its estimate $\hat{x}$ can be generated by a state observer and used in (4.3) instead of x [2, Sect. 8.5]. ■

Second, we design the adaptive tracking component u_t. Below, we present two different approaches to this design problem. One of them utilizes the concept of augmented error, while the second one uses estimates of the HOTD of the reference signal.

4.1.1.1 Adaptive Controller Using Augmented Error

In view of Lemma 2.1, the exosystem generating the reference g can be represented in the canonical form

$$\dot{\bar{\xi}} = (G + l\theta^{\top})\bar{\xi}, \tag{4.5}$$

$$g = \theta^{\top}\bar{\xi}, \tag{4.6}$$

where $\bar{\xi} \in \mathbb{R}^q$ is the state of the canonical form with a certain initial value $\bar{\xi}(0)$, $\theta \in \mathbb{R}^q$ is an unknown constant vector, $G \in \mathbb{R}^{q\times q}$ is an arbitrary designer-chosen Hurwitz matrix, and the vector l is such that the pair (G, l) is controllable.

It is easy to see that the exosystem (4.5), (4.6) is given by the form (1.64) with $\xi_g = \bar{\xi}$, $\Gamma_g = G + l\theta^{\top}$, $Q_g = \theta^{\top}$. Therefore, in accordance with Lemmas 1.1, 1.2, there exists a solution pair $M_g \in \mathbb{R}^{n\times q}$ and $\theta_g = \bar{\theta}_g \in \mathbb{R}^q$ of the matrix equation (1.66) such that in the new coordinates $e = x - M_g\bar{\xi}$, $\varepsilon = y - g$ the plant model (4.4) takes the form

$$\dot{e} = A_{\Sigma}e + b(\theta_g^{\top}\bar{\xi} + u_t), \qquad \varepsilon = c^{\top}e \tag{4.7}$$

with a new vector on unknown parameters $\theta_g \in \mathbb{R}^q$, the unmeasured state e (since the matrix M_g is unknown) and the measured output $\varepsilon = y - g$.

Now, using statement L.2.3.2 of Lemma 2.3, we can write

$$\bar{\xi} = \hat{\xi} + \epsilon_{\xi}, \tag{4.8}$$

where ϵ_{ξ} exponentially decays, and the regressor estimate $\hat{\xi}$ is generated by the reference observer

$$\dot{\hat{\xi}} = G\hat{\xi} + lg \tag{4.9}$$

with the pair (G, l) from the canonical form of the exosystem (4.5). It is worth stressing that the filter (4.9) is implementable since its input g is measured.

The model (4.7) and the equality (4.8) motivate the following choice of the tracking component

$$u_t = -\hat{\theta}_g^{\top}\hat{\xi}, \tag{4.10}$$

where $\hat{\theta}_g$ is the vector of adjustable parameters. Substituting (4.10) and (4.8) into (4.7) and rewriting the plant in the input–output form, we obtain the following closed-loop error model:

$$\varepsilon = W_{\Sigma}(s)\left[\tilde{\theta}_g^{\top}\hat{\xi}\right] + \epsilon, \tag{4.11}$$

where the transfer function $W_\Sigma(s) = c^\top(sI - A_\Sigma)^{-1}b$ is asymptotically stable, $\tilde{\theta}_g = \theta_g - \hat{\theta}_g$ is the vector of parametric errors, and $\epsilon = W_\Sigma(s)\left[\theta_g^\top \epsilon_\xi\right] + c^\top e^{A_\Sigma t} e(0)$ exponentially decays.

The model (4.11) is in the form of the canonical error model (3.97) that belongs to the class of linear dynamic error models with measured output. Therefore, to complete design of the adaptive controller, we can use one of the basic algorithms of adaptation employing the augmented error (see Sect. 3.4.2).

The gradient algorithm of adaptation. By applying the swapping scheme from Lemma 3.1, we construct the algorithm of adaptation

$$\dot{\hat{\theta}}_g = \gamma \xi_w \bar{\varepsilon}, \tag{4.12}$$

where $\gamma > 0$ is the adaptation gain,

$$\xi_w = W_\Sigma(s)\left[\hat{\xi}\right] \tag{4.13}$$

is the filtered regressor, and $\bar{\varepsilon}$ is the augmented error defined as

$$\bar{\varepsilon} = \varepsilon + W_\Sigma(s)\left[\hat{\xi}^\top \hat{\theta}_g\right] - \xi_w^\top \hat{\theta}_g. \tag{4.14}$$

In view of Lemma 3.1, for the augmented error $\bar{\varepsilon}$, we have

$$\bar{\varepsilon} = \xi_w^\top \tilde{\theta}_g + \epsilon,$$

and, therefore, the algorithm of adaptation (4.12)–(4.14) yields the closed-loop parametric error model of the form

$$\dot{\tilde{\theta}}_g = -\gamma \xi_w \xi_w^\top \tilde{\theta}_g - \gamma \xi_w \epsilon.$$

The properties of the closed-loop system are established as follows.

Proposition 4.1 *If Assumption 2.1 (with $\upsilon = g$) and Assumption 4.1 hold, then the adaptive controller (4.3), (4.10) with the reference observer (4.9) and the adaptation algorithm (4.12)–(4.14) when applied to the plant (4.1) provides:*

P.4.1.1. *the boundedness of all the closed-loop signals and asymptotic tracking (4.2);*

P.4.1.2. *if $\hat{\xi} \in \mathcal{PE}$, then, in addition to P.4.1.1, $|\tilde{\theta}_g(t)| \to 0$ exponentially fast and there exists an optimal value of γ for which the rate of parametric convergence is maximum.*

Proof In accordance with Lemma 2.3, $\hat{\xi}$ is bounded. Therefore, the proposition directly follows from Properties P.3.10.2 and P.3.10.3 of Proposition 3.10. □

Below, we combine the swapping technique with DRE and MRE in order to design adaptation algorithms with improved parametric convergence (see Corollaries 3.5, 3.6).

The algorithm with DRE. Define the transfer function vector $\bar{H}(s) = [1,\ H_1(s), \ldots, H_p(s)]^\top$, where $p \geq (q-1)$, $H_i(s)$ are distinct designer-chosen PASMP transfer function ($i = 1, 2, \ldots, p$) and introduce the dynamically extended augmented error

$$\bar{\varepsilon}_g = \bar{H}(s)[\varepsilon_w] - \Xi^\top \hat{\theta}_g \in \mathbb{R}^{p+1} \tag{4.15}$$

and the dynamically extended regressor

$$\Xi^\top = \bar{H}(s)\left[\xi_w^\top\right] = \begin{bmatrix} \xi_w^\top \\ H_1(s)\left[\xi_w^\top\right] \\ \vdots \\ H_p(s)\left[\xi_w^\top\right] \end{bmatrix} \in \mathbb{R}^{p+1\times q}, \tag{4.16}$$

where

$$\varepsilon_w = \varepsilon + W_\Sigma(s)\left[\hat{\xi}^\top \hat{\theta}_g\right], \quad \xi_w = W_\Sigma(s)\left[\hat{\xi}\right]. \tag{4.17}$$

In view of Corollary 3.5, we obtain that

$$\bar{\varepsilon}_g = \Xi^\top \tilde{\theta}_g + \epsilon_w, \tag{4.18}$$

where $\epsilon_w = \bar{H}(s)[\epsilon]$ exponentially decays.

Then the adaptation algorithm with DRE is defined as

$$\dot{\hat{\theta}}_g = \gamma \Xi \bar{\varepsilon}_g, \quad \gamma > 0. \tag{4.19}$$

In this case, the parametric error model takes the form

$$\dot{\tilde{\theta}}_g = -\gamma \Xi \Xi^\top \tilde{\theta}_g - \gamma \Xi \epsilon_w.$$

Proposition 4.2 *If Assumption 2.1 (with $\upsilon = g$) and Assumption 4.1 hold, then the adaptive controller (4.3), (4.10) with the reference observer (4.9) and the adaptation algorithm (4.15)–(4.19) when applied to the plant (4.1) provides:*

P.4.2.1. *the boundedness of all the closed-loop signals and asymptotic tracking (4.2);*

P.4.2.2. *if $\lambda_\Xi(t) \notin \mathcal{L}_1$, where $\lambda_\Xi(t)$ is the minimum eigenvalue of $\Xi(t)\Xi^\top(t)$, then, in addition to P.4.2.1, $|\tilde{\theta}_g(t)| \to 0$ as $t \to \infty$;*

P.4.2.3. *if $\hat{\xi} \in \mathcal{PE}$, then, in addition to P.4.2.1, $|\tilde{\theta}_g(t)| \to 0$ exponentially fast and the rate of parametric convergence can be increased by increasing γ.*

Proof Since $\hat{\xi}$ and Ξ are bounded, then, in view of Corollary 3.5, the proposition directly follows from Properties P.3.4.2, P.3.4.3, and P.3.4.4 of Proposition 3.3 and Corollary 3.2. □

The algorithm with MRE is defined as

$$\dot{\hat{\theta}}_g = \gamma \left(L(s) \left[\xi_w \left(\varepsilon + W_\Sigma(s) \left[\hat{\xi}^\top \hat{\theta}_g \right] \right) \right] - \Omega \hat{\theta}_g \right), \quad \gamma > 0, \tag{4.20}$$

where $L(s)$ is a designer-chosen transfer function of the form (3.44), and $\Omega = L(s)\left[\xi_w \xi_w^\top\right]$. Then, in view of Corollary 3.6, we obtain

$$\dot{\tilde{\theta}}_g = -\gamma \Omega \tilde{\theta}_g - \gamma L(s)[\xi_w \epsilon], \tag{4.21}$$

where ϵ exponentially decays.

Proposition 4.3 *If Assumption 2.1 (with $\upsilon = g$) and Assumption 4.1 hold, then the adaptive controller (4.3), (4.10) with the reference observer (4.9) and the adaptation algorithm (4.20) when applied to the plant (4.1) provides:*

P.4.3.1. *the boundedness of all the closed-loop signals and asymptotic tracking (4.2);*

P.4.3.2. *if $\lambda_\Omega(t) \notin \mathcal{L}_1$, where $\lambda_\Omega(t)$ is the minimum eigenvalue of $\Omega(t)$, then, in addition to P.4.3.1, $|\tilde{\theta}_g(t)| \to 0$ as $t \to \infty$;*

P.4.3.3. *if $\hat{\xi} \in \mathcal{PE}$, then, in addition to P.4.3.1, $|\tilde{\theta}_g(t)| \to 0$ exponentially fast and the rate of parametric convergence can be increased by increasing γ.*

Proof Since $\hat{\xi}$ is bounded, in view of Corollary 3.6, the proposition directly follows from Properties P.3.4.2, P.3.4.3, and P.3.4.4 of Proposition 3.4 and Corollary 3.3. □

Example 4.1 Consider the plant (4.1) with

$$A = \begin{bmatrix} 0 & 1 \\ 0 & 0 \end{bmatrix}, \quad b = \begin{bmatrix} 0 \\ 1 \end{bmatrix}, \quad c = \begin{bmatrix} 30 \\ 0 \end{bmatrix}. \tag{4.22}$$

Let it be known a priori that the reference g can be modelled by the second-order exosystem. In view of Assumption 2.1 (see also Example 2.6), it means that the reference g can be either a sinusoidal signal with unknown amplitude, frequency and phase or an unknown constant. In order to emphasise the adaptive properties of the designed controller, we consider a priori unknown reference g with variable waveform (see Fig. 4.1)

$$g(t) = \begin{cases} 1 & \text{if } \ 0 \le t \le 9 \text{ sec}, \\ \sin\left(\dfrac{\pi t}{2}\right) & \text{if } \ t > 9 \text{ sec}. \end{cases} \tag{4.23}$$

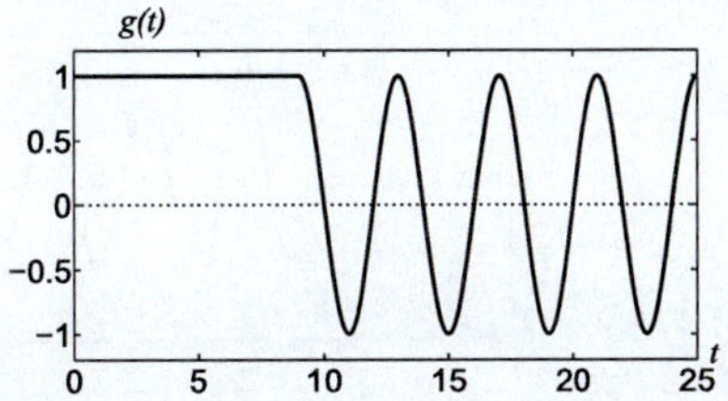

Fig. 4.1 Evolution of the reference signal (4.23)

We apply the control (4.3) with $k = [30,\ 11]^\top$. The tracking component u_t is given by (4.10), where the estimate $\hat{\xi}$ is generated by the second-order filter (4.9) with

$$G = \begin{bmatrix} 0 & 1 \\ -20 & -9 \end{bmatrix}, \quad l = \begin{bmatrix} 0 \\ 20 \end{bmatrix}. \tag{4.24}$$

The vector of adjustable parameters $\hat{\theta}_g$ is calculated using one of the following three adaptation algorithms:

- the gradient algorithm (4.12);
- the algorithm (4.19) with DRE and $p = 1$, $H_1(s) = \frac{1}{0.25s+1}$;
- the algorithm (4.20) with MRE and $L(s) = \frac{1}{0.2s+1}$.

All simulations are carried out with zero initial conditions $x(0)$, $\hat{\xi}(0)$, $\hat{\theta}_g(0)$, and for different adaptation gains $\gamma = 0.1,\ 1,\ 100$.

Simulation results of the closed-loop system with the gradient algorithm of adaptation (see Proposition 4.1) are presented in Fig. 4.2 and demonstrate asymptotic tracking of a priori unknown reference with variable waveform. However, as it is seen from the plots, increase of the adaptation gain γ accelerates the transients up to a certain rate only. Further increase of γ degrades the transient performance.

Simulation results of the closed-loop system with algorithm of adaptation involving DRE (see Proposition 4.2) are presented in Fig. 4.3, while Fig. 4.4 demonstrates the transients in the closed-loop system with the adaptation algorithm involving MRE (see Proposition 4.3). As shown by Figs. 4.3 and 4.4, the plant output asymptotically tracks a priori unknown reference (4.23), and, in contrast to the results presented in Fig. 4.2, the transients can be systematically accelerated by increasing γ. ■

4.1.1.2 Adaptive Controller Using the Estimates of the Reference HOTD

Now, we consider an alternative approach for design of the control tracking component u_t avoiding employment of the augmented error.

First, we construct an adaptive observer for the HOTD of the reference g. Substituting (4.8) into (4.6), we obtain

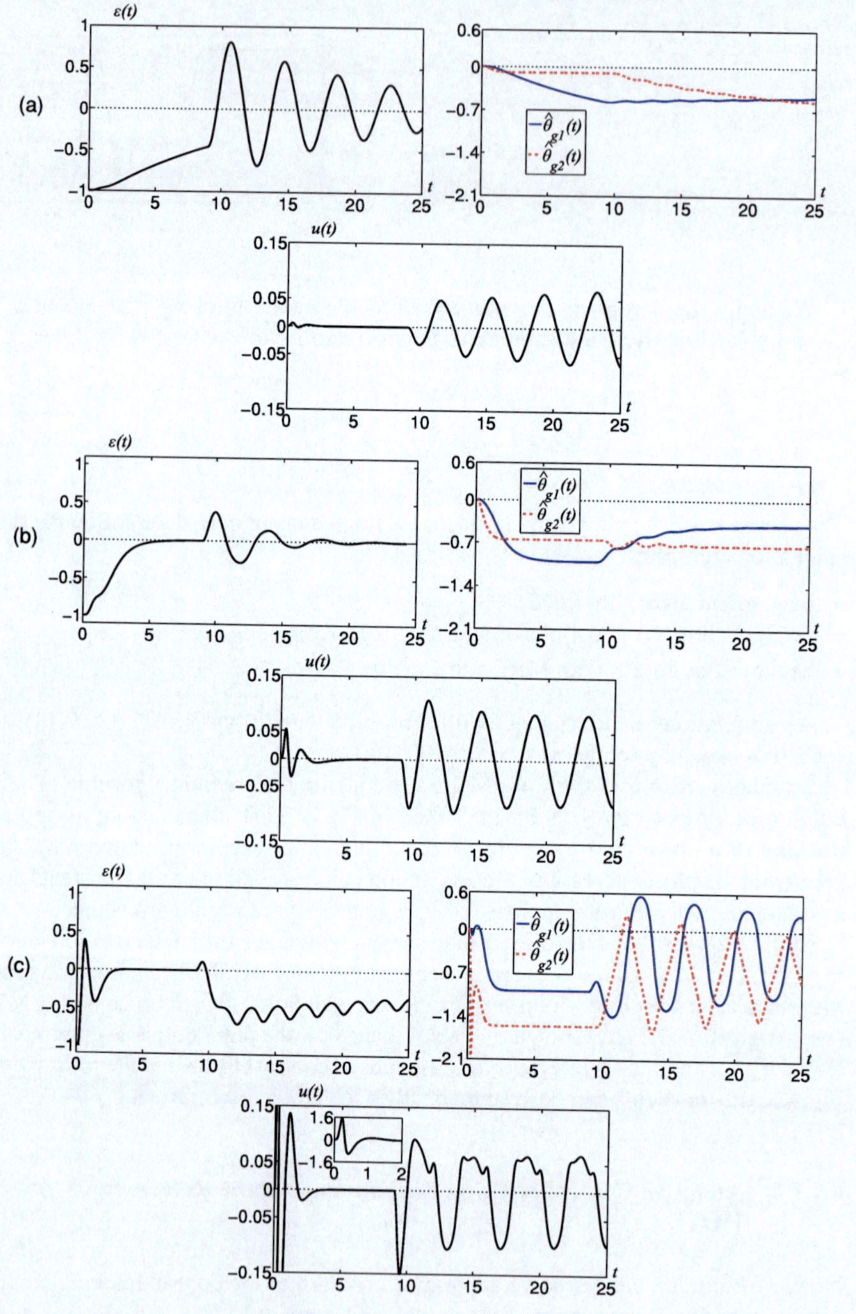

Fig. 4.2 Example 4.1: adaptive tracking of the reference (4.23) in the system with the gradient adaptation algorithm and: **a** $\gamma = 0.1$; **b** $\gamma = 1$; **c** $\gamma = 100$

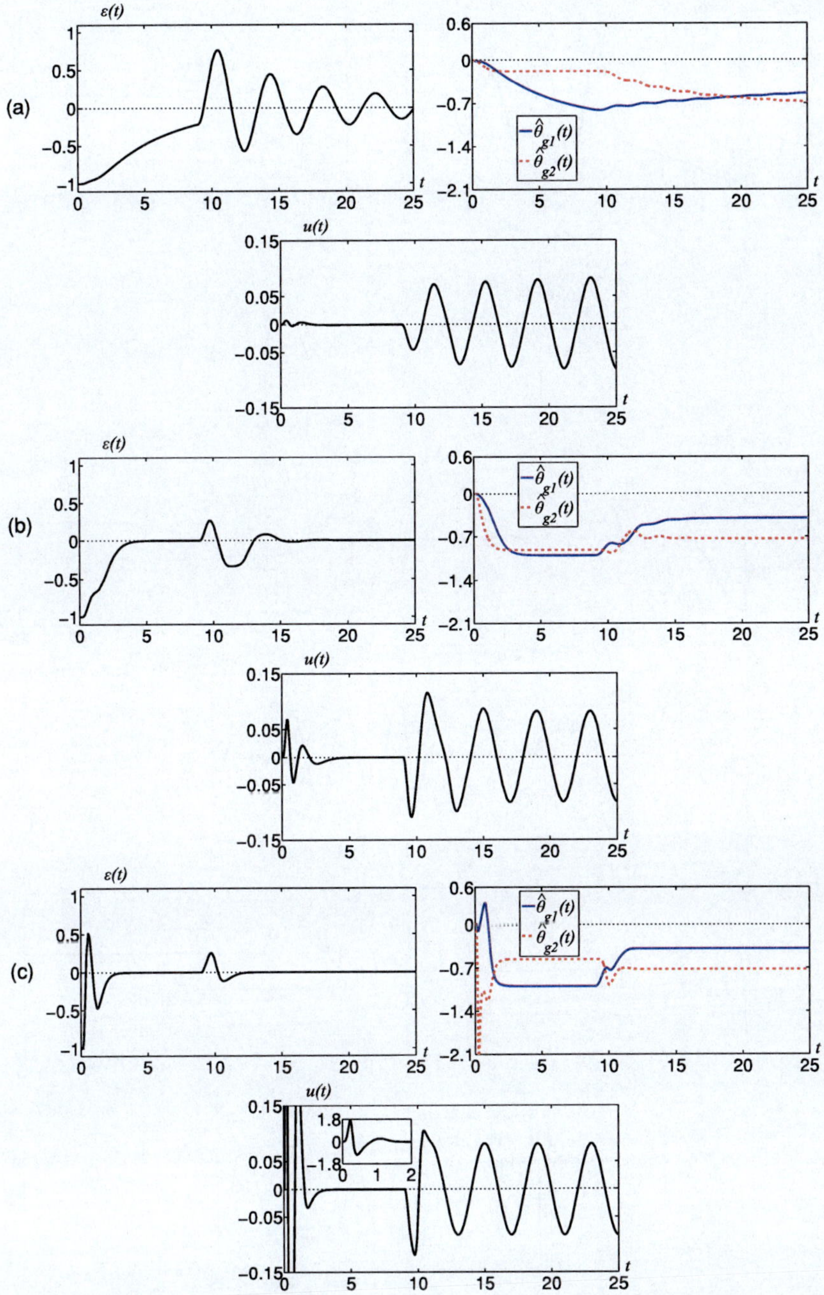

Fig. 4.3 Example 4.1: adaptive tracking of the reference (4.23) in the system with the algorithm of adaptation with DRE and: **a** $\gamma = 0.1$; **b** $\gamma = 1$; **c** $\gamma = 100$

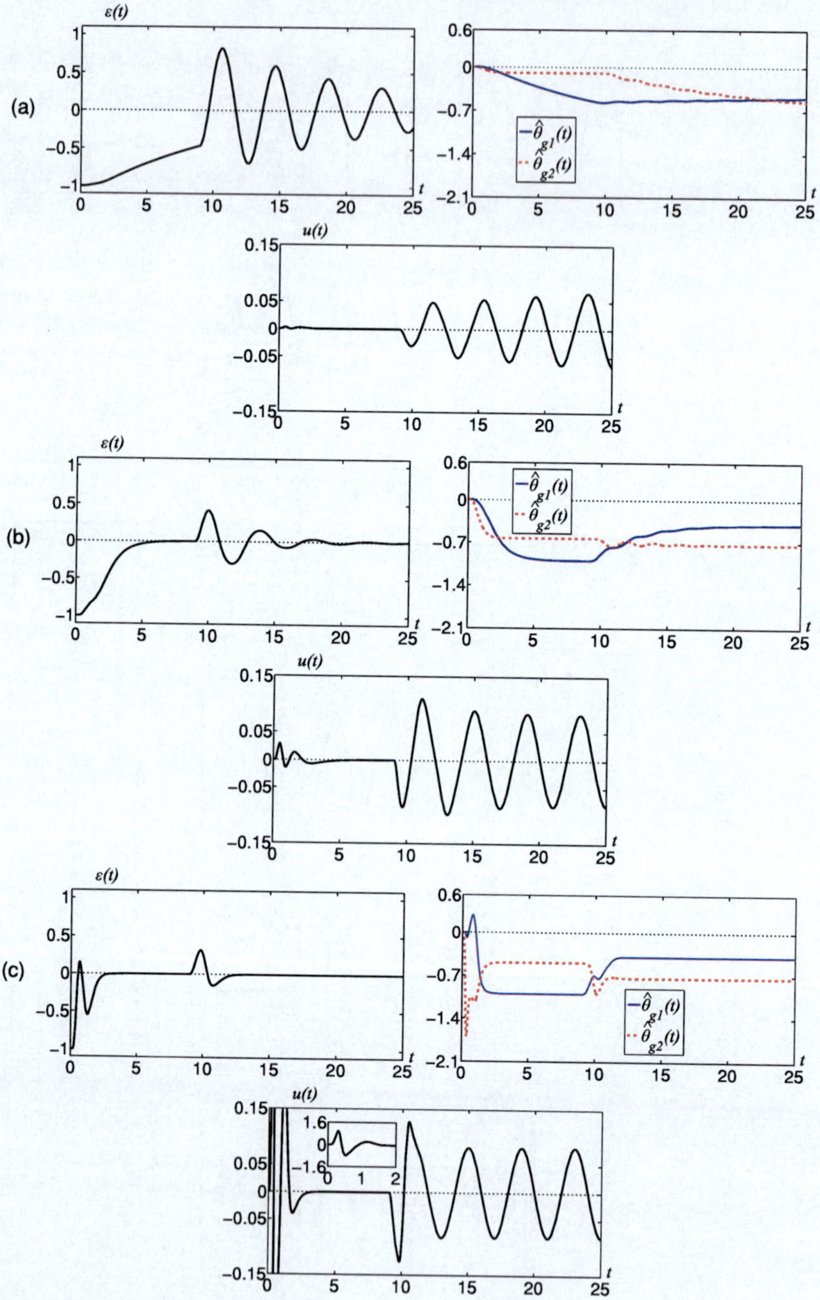

Fig. 4.4 Example 4.1: adaptive tracking of the reference (4.23) in the system with the algorithm of adaptation with MRE with: **a** $\gamma = 0.1$; **b** $\gamma = 1$; **c** $\gamma = 100$

$$g = \theta^\top \hat{\xi} + \epsilon, \tag{4.25}$$

where $\epsilon = \theta^\top \epsilon_\xi$ exponentially decays. Then, in view of Corollary 2.2, together with the filter (4.9), we consider the estimates of the reference and its HOTD defined by the following equations:

$$\hat{g}_i = \hat{\theta}^\top \left(G + l\hat{\theta}^\top\right)^i \hat{\xi}, \quad i \in \mathbb{N}_0, \tag{4.26}$$

where $\hat{g}_i$ is the estimate of the ith time derivative of the reference $g^{(i)}$, $\hat{\theta} \in \mathbb{R}^q$ is the vector of adjustable parameters generated by the gradient algorithm of adaptation

$$\dot{\hat{\theta}} = \gamma \hat{\xi}(g - \hat{\theta}^\top \hat{\xi}), \tag{4.27}$$

and $\gamma > 0$ is the adaptation gain.

Lemma 4.1 ([11]) *Under Assumption 2.1 (with $\upsilon = g$), the adaptive observer (4.9), (4.26), (4.27) provides the boundedness of $\hat{\theta}$ and asymptotic convergence*

$$\lim_{t\to\infty} \left(g^{(i)}(t) - \hat{g}_i(t)\right) = 0, \quad i \in \mathbb{N}_0. \tag{4.28}$$

Proof Define the estimation errors $\tilde{g}_i = g^{(i)} - \hat{g}_i$. In view of (4.25), we can see that the algorithm of adaptation (4.27) yields the following model of parametric errors:

$$\dot{\tilde{\theta}} = -\gamma \hat{\xi}\hat{\xi}^\top \tilde{\theta} - \gamma \hat{\xi} \epsilon,$$

where $\tilde{\theta} = \theta - \hat{\theta}$. Note that $\hat{\xi}$ is bounded. Then from Proposition 3.1, we immediately conclude that $\hat{\theta}$ is bounded and $|\dot{\hat{\theta}}(t)| \to 0$, $\tilde{g}_0(t) \to 0$ as $t \to \infty$. As a result, the limit of the integral

$$\lim_{t\to\infty} \int_0^t \dot{\tilde{g}}_0(\tau) d\tau = \lim_{t\to\infty} \tilde{g}_0(t) - \tilde{g}_0(0) = \tilde{g}_0(0)$$

exists and is finite, and, in view of the uniform continuity of $\ddot{\tilde{g}}_0(t) = \ddot{g}(t) - \hat{g}_2(t)$, from the Barbalat's lemma (see Lemma C.1), we obtain that $\dot{\tilde{g}}_0(t) \to 0$ as $t \to \infty$. Then, calculating the time derivative of $\tilde{g}_0 = g - \hat{g}_0 = g - \hat{\theta}^\top \hat{\xi} - \epsilon$, in view of (4.9) and (4.26), we obtain

$$\begin{aligned}
\dot{\tilde{g}}_0 &= \dot{g} - \dot{\hat{\theta}}^\top \hat{\xi} - \hat{\theta}^\top \left(G\hat{\xi} + lg\right) - \dot{\epsilon} \\
&= \dot{g} - \dot{\hat{\theta}}^\top \hat{\xi} - \hat{\theta}^\top \left(G + l\hat{\theta}^\top\right) \hat{\xi} - \hat{\theta}^\top l(\tilde{g}_0 - \epsilon) - \dot{\epsilon} \\
&= \dot{g} - \hat{g}_1 - \dot{\hat{\theta}}^\top \hat{\xi} - \hat{\theta}^\top l(\tilde{g}_0 + \epsilon) - \dot{\epsilon}.
\end{aligned}$$

Since $|\dot{\tilde{\theta}}(t)|$, $\tilde{g}_0(t)$, $\dot{\tilde{g}}_0(t)$, $\epsilon(t)$, $\dot{\epsilon}(t) \to 0$ as $t \to \infty$, from the latter equality, we see that

$$\lim_{t\to\infty} \tilde{g}_1(t) = \lim_{t\to\infty} \left(\dot{g}(t) - \hat{g}_1(t)\right) = 0.$$

Then, proceeding the proof in the recursive manner with the use of the same arguments as above, we can prove (4.28) for any $i > 1$. □

The algorithm (4.27) is the standard **gradient algorithm of adaptation**. To improve the parametric convergence, we apply the algorithms with DRE or MRE represented in Sects. 3.2.2 and 3.2.3, respectively.

To derive **the algorithm with DRE**, we introduce $p \geq (q-1)$ distinct PASMP transfer functions $H_i(s)$, compose the transfer function vector $\bar{H}(s) = \left[1, H_1(s), \ldots, H_p(s)\right]^\top$ and form the dynamically extended reference

$$\bar{g} = \bar{H}(s)[g] = \left[g, H_1(s)[g], \ldots, H_p(s)[g]\right]^\top, \quad \bar{g} \in \mathbb{R}^{p+1} \tag{4.29}$$

together with the dynamically extended regressor

$$\Xi^\top = \bar{H}(s)\left[\hat{\xi}^\top\right] = \begin{bmatrix} \hat{\xi}^\top \\ H_1(s)\left[\hat{\xi}^\top\right] \\ \vdots \\ H_p(s)\left[\hat{\xi}^\top\right] \end{bmatrix}. \tag{4.30}$$

Then the adaptation algorithm with DRE is given by the expression

$$\dot{\hat{\theta}} = \gamma \Xi \left(\bar{g} - \Xi^\top \hat{\theta}\right), \quad \gamma > 0 \tag{4.31}$$

that yields the following model of the parametric errors:

$$\dot{\tilde{\theta}} = -\gamma \Xi \Xi^\top \tilde{\theta} - \gamma \Xi \epsilon_w,$$

where ϵ_w exponentially decays.

The algorithm with MRE is given by

$$\dot{\hat{\theta}} = \gamma \left(L(s)\left[\hat{\xi} g\right] - \Omega \hat{\theta}\right), \quad \gamma > 0, \tag{4.32}$$

where $L(s)$ is a designer-chosen transfer function of the form (3.44), $\Omega = L(s)\left[\hat{\xi}\hat{\xi}^\top\right]$. In this case, the model of parametric errors takes the form

$$\dot{\tilde{\theta}} = -\gamma \Omega \tilde{\theta} - \gamma L(s)\left[\hat{\xi}\epsilon\right],$$

where ϵ exponentially decays.

Taking Propositions 3.2–3.4 into account and using the same approach as in the proof of Lemma 4.1, we can show that the algorithms (4.31) and (4.32) provide asymptotic convergence (4.28). Under some additional conditions being similar to the second and third conditions of Propositions 4.2 and 4.3, the adaptation algorithms with extended regressors (4.31) and (4.32) provide acceleration of the parametric convergence by increasing γ.

Example 4.2 Let the reference g, whose HOTD are to be estimated, be the single harmonic $g(t) = \sin(3t)$ with unknown frequency, amplitude and phase. We apply the second-order filter (4.9) with G and l defined by (4.24). In this case, the estimates (4.26) for $i = 0, 1, 2$ take the form

$$\hat{g}_0 = \hat{\theta}_1\hat{\xi}_1 + \hat{\theta}_2\hat{\xi}_2, \quad \hat{g}_1 = \hat{\theta}_1\hat{\xi}_2 + \hat{\theta}_2 r_1, \quad \hat{g}_2 = \hat{\theta}_1 r_1 + \hat{\theta}_2 r_2, \tag{4.33}$$

where

$$r_1 = 20\hat{g}_0 - 20\hat{\xi}_1 - 9\hat{\xi}_2, \quad r_2 = 20\hat{g}_1 - 20\hat{\xi}_2 - 9r_1.$$

We choose $p = 1$ and $H_1(s) = \frac{1}{0.25s+1}$ for the adaptation algorithm with DRE (4.31) and $L(s) = \frac{1}{0.2s+1}$ for the adaptation algorithm with MRE (4.32). Simulation results are presented in Fig. 4.5 and demonstrate the potentials for transient acceleration of the algorithms with extended regressors achieved by increasing the adaptation gain γ (see Fig. 4.5b, c). At the same time, in the system with the gradient algorithm increasing of the adaptation gain γ from 10 to 100 degrades the transient performance (see Fig. 4.5a). ■

Now, we design the tracking component u_t. To this end, substituting (4.3) into (4.1), we obtain the following input-output model of the tracking error:

$$\varepsilon = y - g = W_\Sigma(s)[u_t] - g + \epsilon, \tag{4.34}$$

where $\epsilon = c^\top e^{A_\Sigma t}x(0)$ exponentially decays. If the transfer function $W_\Sigma(s)$ is given by

$$W_\Sigma(s) = \frac{\beta(s)}{\alpha(s)} = \frac{b_m s^m + b_{m-1}s^{m-1} + \cdots + b_0}{s^n + a_{n-1}s^{n-1} + \cdots + a_0},$$

then $W_\Sigma^{-1}(s)$ can be represented in the form

$$W_\Sigma^{-1}(s) = \frac{1}{\beta(s)}\sum_{i=0}^{n} a_i s^i$$

with $a_n = 1$.

Therefore, the tracking error model (4.34) can be rewritten as

$$\varepsilon = y - g = W_\Sigma(s)\left[u_t - \frac{1}{\beta(s)}\sum_{i=0}^{n} a_i g^{(i)}\right] + \epsilon. \tag{4.35}$$

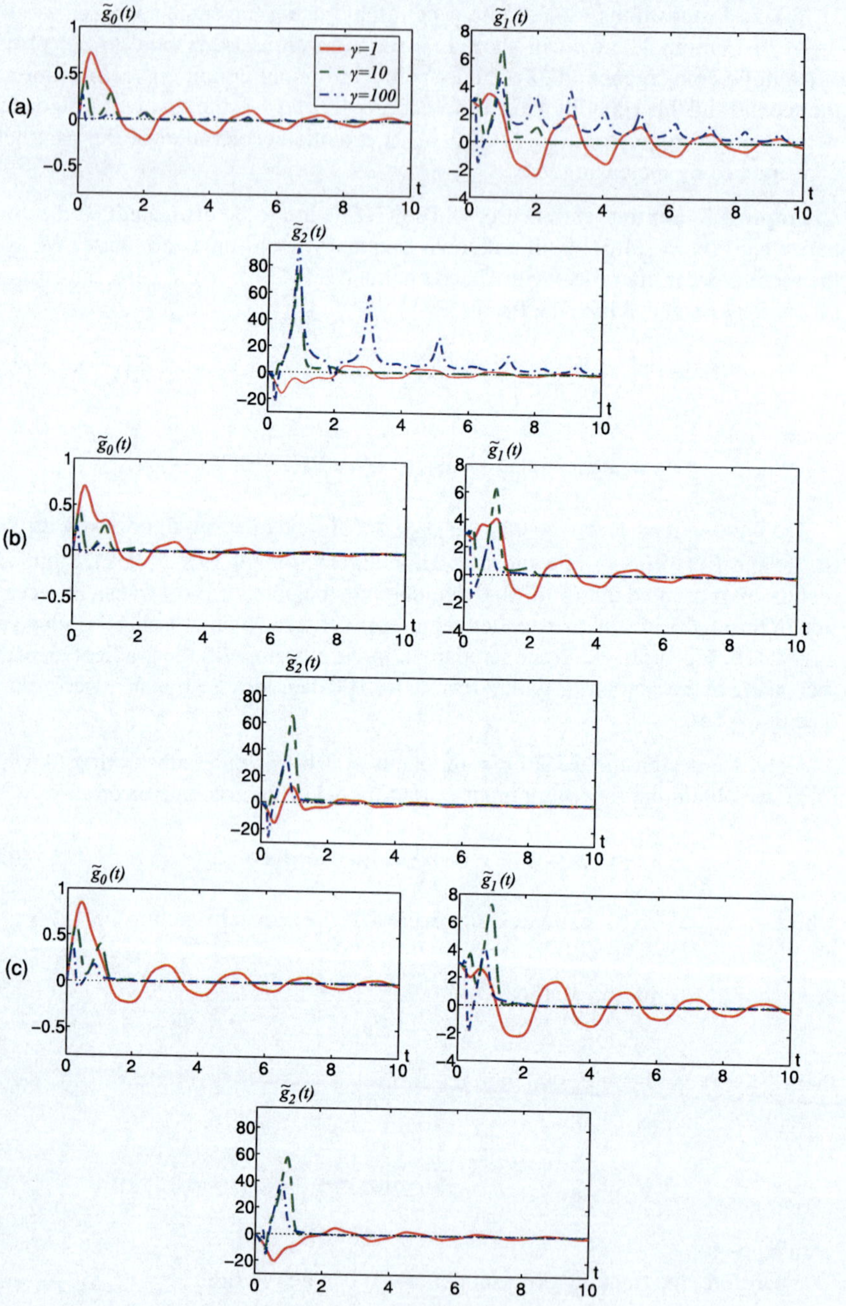

Fig. 4.5 Example 4.2: adaptive estimation of HOTD of the reference $g(t) = \sin(3t)$ by applying: **a** the gradient-based algorithm; **b** the algorithm with DRE; **c** the algorithm with MRE

Since by Assumption 4.1.2 $\beta(s)$ is Hurwitz, then the tracking component

$$u_t = \frac{1}{\beta(s)}\left[a_0 g + \sum_{i=1}^{n} a_i \hat{g}_i\right] \tag{4.36}$$

is applicable and implementable, where $\hat{g}_i$ is the estimate of the HOTD $g^{(i)}$ generated by the HOTD estimator (4.26) with the adaptation algorithm (4.27) (or (4.31), or (4.32)).

Remark 4.3 The expression (4.36) requires the HOTD $\hat{g}_i$ of up to the nth order. With a special factorization of $W_\Sigma^{-1}(s)$, the required highest order of the estimate $\hat{g}_i$ can be decreased down to $n - m$. However, we use (4.36) due to its analytical simplicity. ■

Substituting (4.36) into (4.35), we obtain the closed-loop error model

$$\varepsilon = W_\Sigma(s)\left[\sum_{i=1}^{n} a_i \tilde{g}_i\right] + \epsilon, \tag{4.37}$$

where $\tilde{g}_i = g^{(i)} - \hat{g}_i$ are the estimation errors. By virtue of Lemma 4.1, $\tilde{g}_i(t) \to 0$ as $t \to \infty$. Therefore, we have proved the following statement.

Proposition 4.4 *If Assumption 2.1 (with $\upsilon = g$) and Assumption 4.1 hold, then the adaptive controller (4.3), (4.36) with the reference observer (4.9), the HOTD estimator (4.26), and the adaptation algorithm (4.27) (or (4.31), or (4.32)) when applied to the plant (4.1) provides the boundedness of all the closed-loop signals and asymptotic tracking (4.2).*

Under some additional conditions being similar to the second and third conditions of Propositions 4.2 and 4.3, the adaptation algorithms with extended regressors (4.31) and (4.32) provide acceleration of the parametric convergence by increasing γ.

Example 4.3 Consider the plant (4.1) with the matrices (4.22) and the reference $g(t) = \sin(3t)$. We use the second-order reference observer (4.9) with G and l defined by (4.24). Selecting $k = [-30, -11]^\top$, we design the control law

$$u = -30x_1 - 11x_2 + \frac{1}{30}\left(30g + 11\hat{g}_1 + \hat{g}_2\right),$$

where the estimates $\hat{g}_1$ and $\hat{g}_2$ are given by (4.33).

Simulation results of the closed-loop systems with the algorithms of adaptation (4.27), (4.31), and (4.32) are presented in Fig. 4.6 and demonstrate asymptotic tracking of a priori unknown reference. At the same time, it is shown that the transients can be accelerated by increasing the adaptation gain γ in the systems with the adaptation algorithms employing the extended regressor. ■

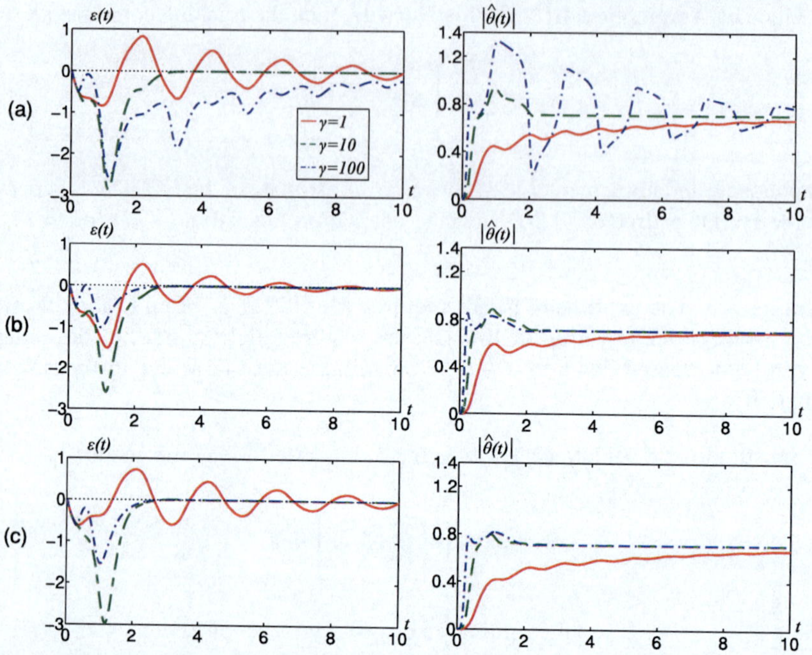

Fig. 4.6 Example 4.3: adaptive tracking of the reference $g(t) = \sin(3t)$ in the systems using estimates of the HOTD of the reference signal and: **a** the gradient algorithm; **b** the algorithm with DRE; **c** the algorithm with MRE

4.1.2 Compensation of Matched Disturbance

Let us consider the LTI plant

$$\dot{x} = Ax + b(u + \upsilon), \tag{4.38}$$

where $x \in \mathbb{R}^n$ is the measured state, $u \in \mathbb{R}$ is the control, $\upsilon \in \mathbb{R}$ is the unmeasured disturbance satisfying Assumption 2.1, while A and b are known and have appropriate dimensions. The pair (A, b) is assumed to be controllable.

The considered problem is to design a state-feedback controller ensuring the boundedness of all the closed-loop signals and asymptotic convergence of the state to zero according to the equality

$$\lim_{t\to\infty} |x(t)| = 0. \tag{4.39}$$

In other words, we consider Problem 1.3 formulated in Sect. 1.3.3 and resolve it for the particular case of matched disturbance.

First of all, we stabilize the plant with the use of the controller

$$u = -k^\top x - u_\upsilon, \tag{4.40}$$

where the vector of feedback gains k is selected so that the matrix of the closed-loop system $A_\Sigma = A - bk^\top$ is Hurwitz [2, Sect. 8.2], while the *compensating component* u_υ will be defined later. Substituting (4.40) into (4.38), we obtain

$$\dot{x} = A_\Sigma x + b(u_\upsilon + \upsilon). \tag{4.41}$$

Then, applying Lemma 2.4 to the model (4.41), we represent the disturbance in the form of linear regression

$$\upsilon = \theta^\top \hat{\xi} + \epsilon, \tag{4.42}$$

where $\theta \in \mathbb{R}^q$ is a vector of unknown parameters, ϵ is the exponentially decaying term and $\hat{\xi} \in \mathbb{R}^q$ is the vector generated by the disturbance observer

$$\hat{\xi} = \eta + Nx, \tag{4.43}$$
$$\dot{\eta} = G\eta + (GN - NA_\Sigma)x - Nbu_\upsilon \tag{4.44}$$

with the matrix $N \in \mathbb{R}^{q \times n}$ satisfying the equation

$$Nb = l \tag{4.45}$$

and the pair (G, l) from the canonical form of the exosystem (2.15).

Substituting (4.42) into (4.41), we obtain the plant model of the form

$$\dot{x} = A_\Sigma x + b\left(u + \theta^\top \hat{\xi} + \epsilon\right). \tag{4.46}$$

The latter equation motivates the following choice of the compensating component:

$$u_\upsilon = -\hat{\theta}^\top \hat{\xi}, \tag{4.47}$$

where the vector of adjustable parameters $\hat{\theta}$ is generated by an adaptation algorithm to be defined later. Note that under Assumption 2.1 the regressor estimate $\hat{\xi}$ is bounded.

Substitution of (4.47) into (4.46) gives the following closed-loop error model:

$$\dot{x} = A_\Sigma x + b\left(\tilde{\theta}^\top \hat{\xi} + \epsilon\right), \tag{4.48}$$

where $\tilde{\theta} = \theta - \hat{\theta}$ is the vector of parametric errors. The model (4.48) is in the form of the canonical error model (3.72) that belongs to the class of linear dynamic error models with measured state. Therefore, to complete the design of the adaptive controller, we can use the following basic algorithm of adaptation defined by Proposition 3.6:

$$\dot{\hat{\theta}} = \gamma \hat{\xi} b^\top P x, \tag{4.49}$$

where $\gamma > 0$ is the adaptation gain, $P \in \mathbb{R}^{n \times n}$ is the solution of the Lyapunov matrix equation $A_\Sigma^\top P + P A_\Sigma = -Q$, $Q = Q^\top \succ 0$.

Then the properties of the closed-loop system are established as follows.

Proposition 4.5 *If Assumption 2.1 holds and the pair* (A, b) *is controllable, then the adaptive controller* (4.40), (4.47) *with the disturbance observer* (4.43)–(4.45) *and the adaptation algorithm* (4.49) *when applied to the plant* (4.38) *provides:*

P.4.5.1. *the boundedness of all the closed-loop signals and asymptotic convergence* (4.39)*;*

P.4.5.2. *if* $\hat{\xi} \in \mathcal{PE}$*, then, in addition to P.4.5.1,* $|\tilde{\theta}(t)| \to 0$ *exponentially fast and there exists an optimal value of* γ *for which the rate of parametric convergence is maximum.*

The proposition directly follows from Proposition 3.6.

Example 4.4 Let us consider the second-order plant (4.38) with

$$A = \begin{bmatrix} 0 & 1 \\ 0 & 0 \end{bmatrix}, \quad b = \begin{bmatrix} 0 \\ 1 \end{bmatrix}.$$

Let the disturbance υ be modelled by the exosystem (2.1), (2.2) of the second order. In this case, we choose the second-order disturbance observer (4.43)–(4.45) and use G and l defined by (4.24). We also select

$$N = \begin{bmatrix} 0 & 0 \\ 0 & 20 \end{bmatrix}, \quad k = \begin{bmatrix} 1.5 \\ 2 \end{bmatrix}, \quad Q = \begin{bmatrix} 3 & 0 \\ 0 & 2 \end{bmatrix}.$$

Solving the Lyapunov equation $A_\Sigma^\top P + P A_\Sigma = -Q$, we obtain

$$P = \begin{bmatrix} 3.5 & 1 \\ 1 & 1 \end{bmatrix}.$$

The vector of adjustable parameters $\hat{\theta}$ is generated by the Lyapunov-based algorithm (4.49).

We consider a priori unknown disturbance υ with variable waveform

$$\upsilon(t) = \begin{cases} 1 & \text{if} \quad 0 \le t \le 17 \text{ sec}, \\ \sin\left(\dfrac{\pi t}{2}\right) & \text{if} \quad t > 17 \text{ sec}. \end{cases} \tag{4.50}$$

All simulations are carried out with zero initial conditions $x(0)$, $\hat{\xi}(0)$, $\hat{\theta}(0)$, and with different adaptation gains $\gamma = 1,\ 2,\ 100$.

Simulation results of the closed-loop systems with the Lyapunov-based algorithm of adaptation (see Proposition 4.5) are presented in Fig. 4.7. It is seen from the presented results that, despite the influence of a priori unknown disturbance and even the step change of its waveform, the proposed adaptive controller provides complete compensation of the disturbance, and the state x converges to zero asymptotically.

■

4.1.3 *Compensation of Unmatched Disturbance*

In this section, we present approaches to the state-feedback and output-feedback adaptive compensation of unmatched disturbances.

4.1.3.1 Plants with Measured State

Consider the LTI plant

$$\dot{x} = Ax + bu + d\upsilon, \quad y = c^\top x + d_n\upsilon, \tag{4.51}$$

where $x \in \mathbb{R}^n$ is the measured state, $y \in \mathbb{R}$ is the output (regulated) variable, $u \in \mathbb{R}$ is the control, $\upsilon \in \mathbb{R}$ is the unmeasured disturbance satisfying Assumption 2.1, A, b, d, c are known and have appropriate dimensions, while a constant d_n is possibly unknown.

Additionally, we accept the following assumption regarding the plant (4.51).

Assumption 4.2[1] The plant (4.1) is such that:

A.4.2.1. the triplet (A, b, c) is controllable and observable;

A.4.2.2. all the zeros of the transfer function $W_u(s) = c^\top(sI - A)^{-1}b$ have negative real parts.

The considered control problem is to design a state-feedback controller that will provide the boundedness of all the closed-loop signals and asymptotic convergence

$$\lim_{t\to\infty} y(t) = 0. \tag{4.52}$$

In other words, we consider Problem 1.3 formulated in Sect. 1.3.3 and solve it for the case of unmatched disturbance and for plants with measured state.

Again, as the first step, we stabilize the plant (4.51) with the use of the controller

$$u = -k^\top x + u_\upsilon, \tag{4.53}$$

[1] This is Assumption 2.4 repeated here for the sake of convenience. See also Remark 2.8.

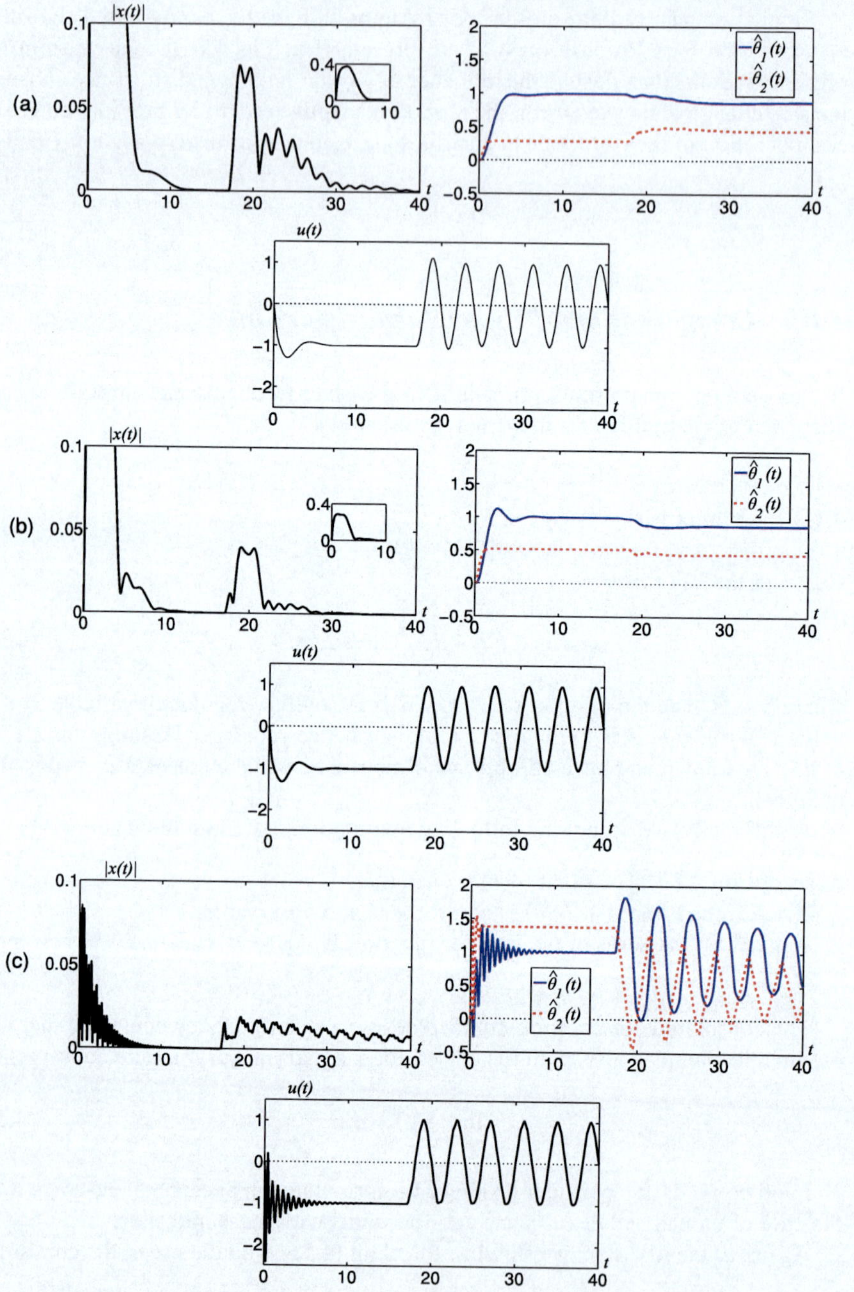

Fig. 4.7 Example 4.4: adaptive state-feedback compensation of the matched disturbance (4.50) in the system closed by the Lyapunov-based algorithm of adaptation with: **a** $\gamma = 1$; **b** $\gamma = 2$; **c** $\gamma = 100$

where the vector of feedback gains k is selected so that the matrix of the closed-loop system $A_\Sigma = A - bk^\top$ is Hurwitz, while the compensating component u_υ will be defined later. Substituting (4.53) into (4.51), we obtain

$$\dot{x} = A_\Sigma x + bu_\upsilon + d\upsilon, \quad y = c^\top x + d_n\upsilon. \tag{4.54}$$

Then, applying Proposition 2.2 to the model (4.54), we employ the observer (4.43), (4.44) with the matrix N such that

$$Nd = l \tag{4.55}$$

and represent the plant subjected to the external disturbance in the form

$$\dot{e} = A_\Sigma e + b\left(u_\upsilon + \theta_\upsilon^\top\hat{\xi} + \epsilon\right), \quad y = c^\top e, \tag{4.56}$$

where $e \in \mathbb{R}^n$ is inaccessible state error, $\theta_\upsilon \in \mathbb{R}^q$ is a vector of unknown parameters, $\hat{\xi}$ is generated by (4.43)–(4.55) and ϵ exponentially decays. In the model (4.56), the disturbance $\bar{\upsilon} = \theta_\upsilon^\top\hat{\xi}$ is matched with the control u.

In view of (4.56), we choose the following compensating component:

$$u_\upsilon = -\hat{\theta}_\upsilon^\top\hat{\xi}, \tag{4.57}$$

where $\hat{\theta}_\upsilon \in \mathbb{R}^q$ is the vector of adjustable parameters generated by an adaptation algorithm defined later. Substituting (4.57) into (4.56), we obtain the closed-loop error model in the state-space form

$$\dot{e} = A_\Sigma e + b\left(\tilde{\theta}_\upsilon^\top\hat{\xi} + \epsilon\right), \quad y = c^\top e \tag{4.58}$$

or in the input–output form

$$y = W_\Sigma(s)\left[\tilde{\theta}_\upsilon^\top\hat{\xi}\right] + \epsilon_y, \tag{4.59}$$

where the transfer function $W_\Sigma(s) = c^\top(sI - A_\Sigma)^{-1}b$ is asymptotically stable, $\tilde{\theta}_\upsilon = \theta_\upsilon - \hat{\theta}_\upsilon$, and $\epsilon_y = W_\Sigma(s)[\epsilon] + c^\top e^{A_\Sigma t}e(0)$ exponentially decays.

The model (4.59) is in the form of the canonical error model (3.97) that belongs to the class of linear dynamic error models with measured output. Therefore, we can use Proposition 3.10 and Corollaries 3.5, 3.6 to derive a gradient swapping-based algorithm of adaptation as well as algorithms with DRE and MRE.

The gradient algorithm of adaptation is given by

$$\dot{\hat{\theta}}_\upsilon = \gamma\xi_w\bar{y}, \quad \gamma > 0, \tag{4.60}$$

where $\bar{y}$ is the augmented output defined as

$$\bar{y} = y + W_\Sigma(s)\left[\hat{\theta}_\upsilon^\top \hat{\xi}\right] - \xi_w^\top \hat{\theta}_\upsilon, \tag{4.61}$$

and

$$\xi_w = W_\Sigma(s)\left[\hat{\xi}\right] \tag{4.62}$$

is the filtered regressor.

In view of Lemma 3.1, we can show that, for the augmented output, we have

$$\bar{y} = \tilde{\theta}_\upsilon^\top \xi_w + \epsilon_y, \tag{4.63}$$

and the algorithm (4.60) yields the following parametric error model:

$$\dot{\hat{\theta}}_\upsilon = -\gamma \xi_w \xi_w^\top \tilde{\theta}_\upsilon - \gamma \xi_w \epsilon_y.$$

The algorithm with DRE is defined as

$$\dot{\hat{\theta}}_\upsilon = \gamma \Xi_y y_H, \quad \gamma > 0, \tag{4.64}$$

where

$$y_H = \bar{H}(s)\left[\bar{y}\right] - \Xi_y^\top \hat{\theta}_\upsilon, \quad \Xi_y^\top = \bar{H}(s)\left[\xi_w^\top\right], \tag{4.65}$$

$$\bar{y} = y + W_\Sigma(s)\left[\hat{\theta}_\upsilon^\top \hat{\xi}\right], \quad \xi_w = W_\Sigma(s)\left[\hat{\xi}\right], \tag{4.66}$$

$\bar{H}(s) = \left[1,\ H_1(s), \ldots, H_p(s)\right]^\top$, $H_i(s)$ are designer-chosen distinct PASMP transfer functions, and $p \geq (q-1)$. Then, in view of Corollary 3.5, we can show that

$$y_H = \Xi_y^\top \tilde{\theta}_\upsilon + \bar{\epsilon}_y,$$

where $\bar{\epsilon}_y = H(s)[\epsilon_y]$, and the algorithm (4.64)–(4.66) yields the parametric error model

$$\dot{\hat{\theta}}_\upsilon = -\gamma \Xi_y \Xi_y^\top \tilde{\theta}_\upsilon - \gamma \Xi_y \bar{\epsilon}_y.$$

The algorithm with MRE is defined as

$$\dot{\hat{\theta}}_\upsilon = \gamma\left(L(s)\left[\xi_w\left(y + W_\Sigma(s)\left[\hat{\xi}^\top \hat{\theta}_\upsilon\right]\right)\right] - \Omega \hat{\theta}_\upsilon\right), \quad \gamma > 0, \tag{4.67}$$

where $L(s)$ is a designer-chosen transfer function of the form (3.44), and $\Omega = L(s)\left[\xi_w \xi_w^\top\right]$. Then, in view of Corollary 3.5, we can shown that

$$\dot{\hat{\theta}}_\upsilon = -\gamma \Omega \tilde{\theta}_\upsilon - \gamma L(s)\left[\xi_w \epsilon_y\right].$$

The main properties of the closed-loop systems with the adaptive controller employing different algorithms of adaptation are established by the following statement.[2]

Proposition 4.6 *If Assumptions 2.1 and 4.2 hold, then the adaptive controller* (4.53), (4.57) *with the disturbance observer* (4.43)–(4.55) *and the adaptation algorithm* (4.60)–(4.62) *(or* (4.64)–(4.66), *or* (4.67)*) when applied to the plant* (4.51) *provides the boundedness of all the closed-loop signals and asymptotic convergence of y to zero in accordance with* (4.52).

Proposition 4.6 follows from Proposition 3.10 and Corollaries 3.5, 3.6. Under some additional conditions being similar to the second and the third conditions of Propositions 4.2 and 4.3, the adaptation algorithms with the extended regressors (4.64) and (4.67) provide acceleration of parametric convergence by increasing γ.

Example 4.5 Consider the second-order plant (4.51) with

$$A = \begin{bmatrix} 0 & 1 \\ 0 & 0 \end{bmatrix}, \; b = \begin{bmatrix} 0 \\ 1 \end{bmatrix}, \; c = \begin{bmatrix} 30 \\ 0 \end{bmatrix}, \; d = \begin{bmatrix} 1 \\ 2 \end{bmatrix},$$

and unknown $d_n = 3$. Let the disturbance υ be modelled by the exosystem (2.1), (2.2) of the second order. We use the second-order disturbance observer (4.43)–(4.55) with G and l defined by (4.24). We also select

$$N = \begin{bmatrix} 0 & 0 \\ 0 & 10 \end{bmatrix}, \; k = \begin{bmatrix} 30 \\ 11 \end{bmatrix}.$$

The vector of adjustable parameters $\hat{\theta}_\upsilon$ is calculated using one of the following three adaptation algorithms:

- the gradient-based algorithm (4.60);
- the algorithm (4.64) with DRE and $p = 1$, $H_1(s) = \frac{1}{0.25s+1}$;
- the algorithm (4.67) with MRE and $L(s) = \frac{1}{0.2s+1}$.

All simulations are carried out with zero initial conditions $x(0)$, $\hat{\xi}(0)$, $\hat{\theta}_\upsilon(0)$, and with different gains $\gamma = 0.1, \; 1, \; 100$.

Simulations of the closed-loop systems employing the gradient algorithm of adaptation, the adaptation algorithm with DRE and the algorithm of adaptation with MRE are presented in Figs. 4.8, 4.9 and 4.10, respectively. To emphasize the adaptive properties of the controller designed, we use the discontinuously changing disturbance (4.50).

Simulation results demonstrate complete rejection of a priori unknown disturbance with variable waveform and asymptotic convergence of the output y to zero. Moreover, the adaptation algorithms with DRE and MRE allow one to accelerate parametric convergence by increasing the adaptation gain γ. ■

[2] An alternative approach to the state-feedback compensation of unmatched disturbances in LTI SISO plants is presented in [10].

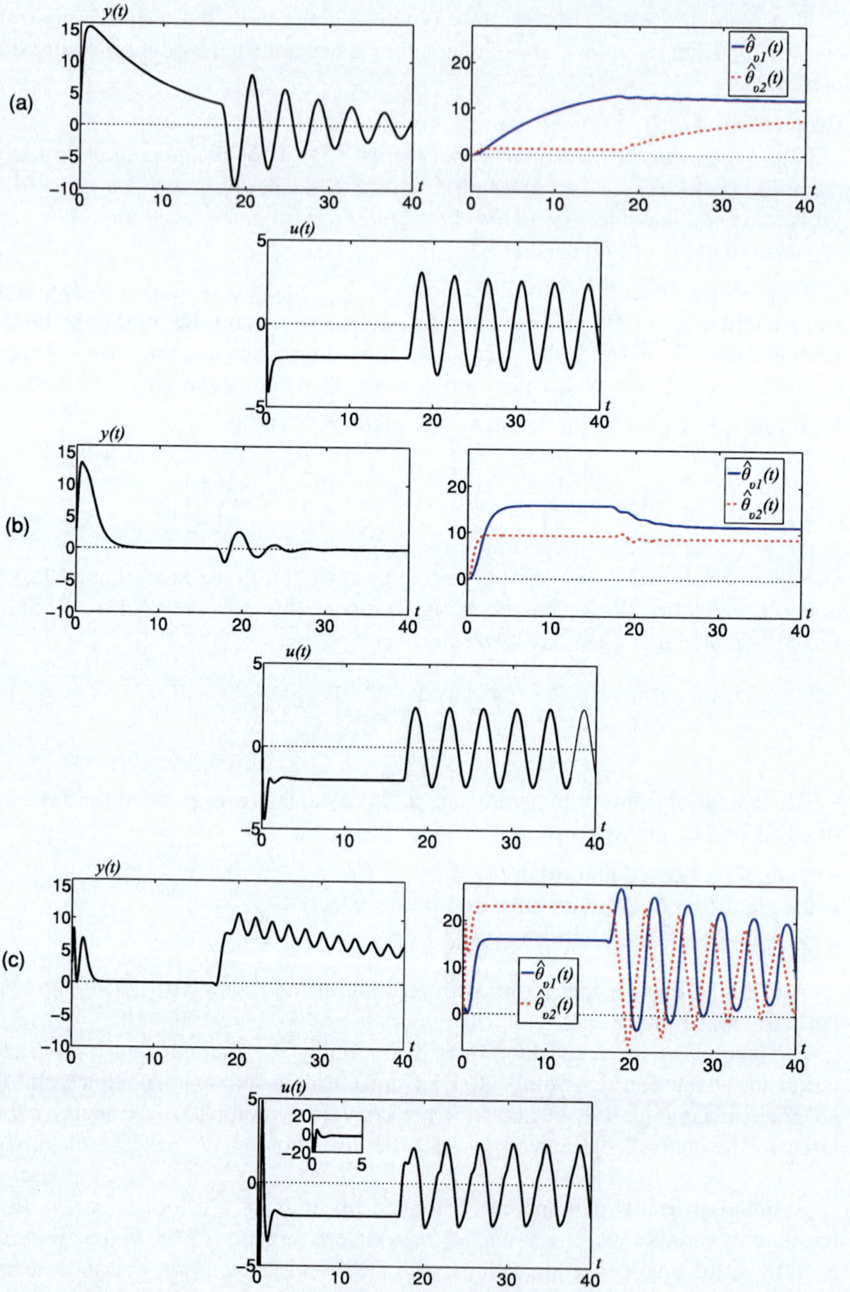

Fig. 4.8 Example 4.5: adaptive state-feedback compensation of the unmatched disturbance (4.50) in the system closed by the gradient-based algorithm of adaptation with: **a** $\gamma = 0.1$; **b** $\gamma = 1$; **c** $\gamma = 100$

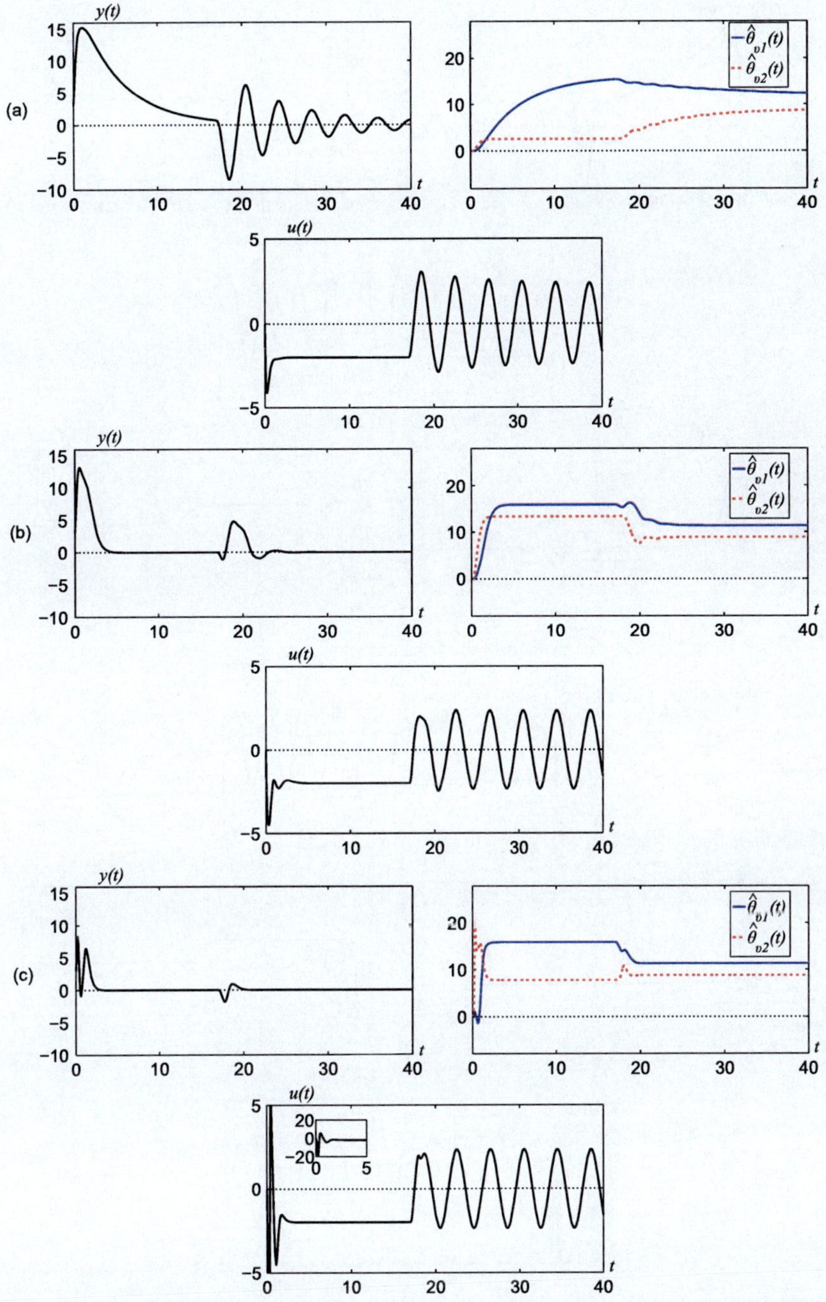

Fig. 4.9 Example 4.5: adaptive state-feedback compensation of the unmatched disturbance (4.50) in the system closed by the algorithm of adaptation with DRE and with: **a** $\gamma = 0.1$; **b** $\gamma = 1$; **c** $\gamma = 100$

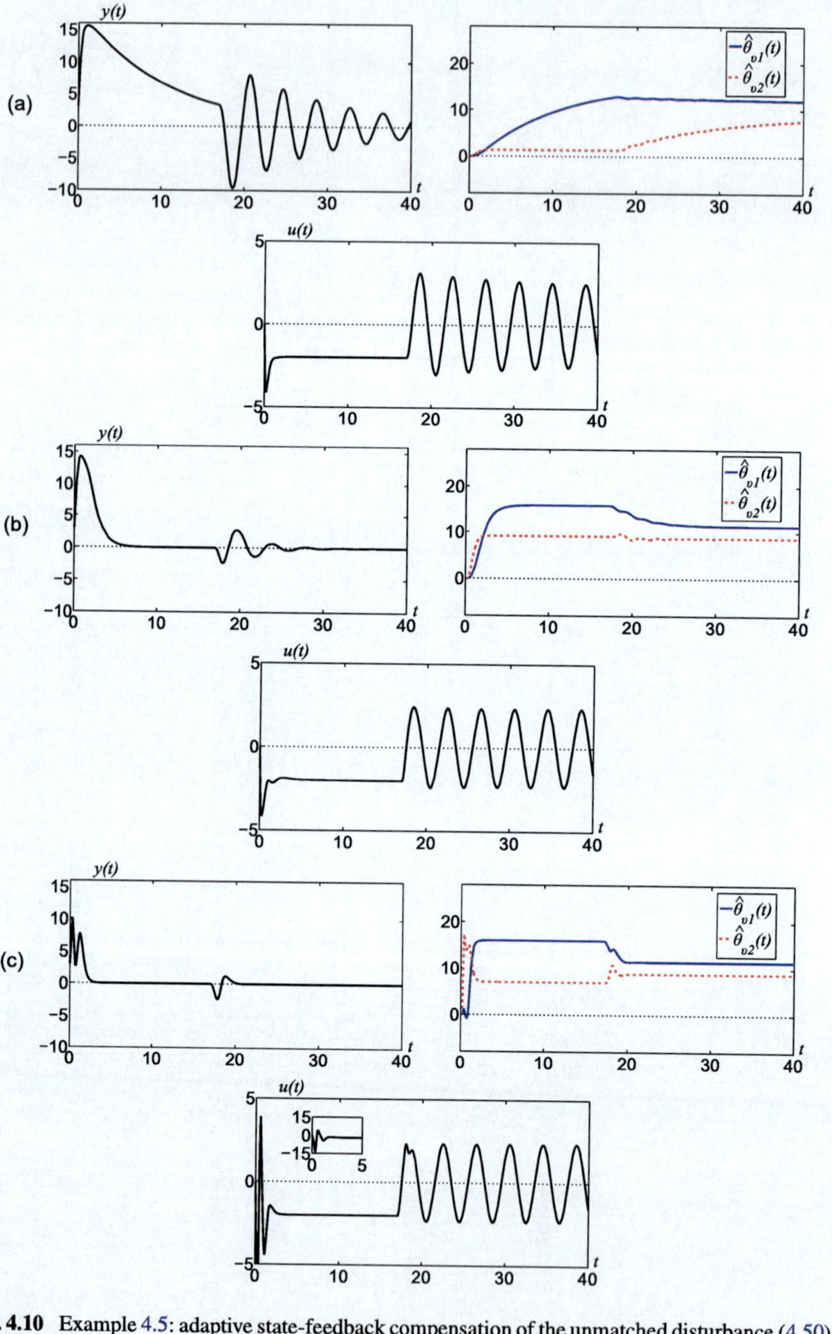

Fig. 4.10 Example 4.5: adaptive state-feedback compensation of the unmatched disturbance (4.50) in the system closed by the algorithm of adaptation with MRE with: **a** $\gamma = 0.1$; **b** $\gamma = 1$; **c** $\gamma = 100$

4.1.3.2 Plants with Measured Output

Consider the plant (4.51), where the state x is inaccessible for measurements. The control objective is to design an output-feedback adaptive controller providing the boundedness of all the closed-loop signals and driving the plant output to zero according to (4.52). Additionally, we assume that Assumption 4.2 holds.[3]

Following the main stream of the book, we consider Problem 1.3 formulated in Sect. 1.3.3 and solve it for the case of unmatched disturbance and for the plants with measured output.

Below, we present two different approaches to the considered problem, the first of which is based on direct application of the certainty equivalence principle, while the second one uses adaptive backstepping design procedure.

Adaptive disturbance compensation based on the certainty equivalence principle.[4] Applying the result of Proposition 2.3, we employ the following disturbance observer:

$$\dot{\hat{e}} = A\hat{e} + bu + l_e \left(y - c^\top \hat{e}\right), \tag{4.68}$$

$$\dot{\hat{\xi}}_f = G\hat{\xi}_f + l \left(y - c^\top \hat{e}\right), \tag{4.69}$$

where the vector of feedback gains l_e is such that the matrix $A_e = A - l_e c^\top$ is Hurwitz [2, Sect. 8.4], while the pair (G, l) is from the exosystem canonical form (2.15). Then the plant model can be represented as

$$\dot{e} = Ae + b\left(u + \bar{\theta}^\top \hat{\xi}_f + \epsilon\right), \quad y = c^\top e, \tag{4.70}$$

where $e \in \mathbb{R}^n$ is inaccessible state error, $\bar{\theta} \in \mathbb{R}^q$ is a vector of unknown parameters and ϵ exponentially decays. This model determines the following structure of adjustable controller:

$$u = u_s - \hat{\theta}^\top \hat{\xi}_f, \tag{4.71}$$

where u_s is the stabilizing component to be defined later, $\hat{\theta}$ is the vector of adjustable parameters. Substituting (4.71) into (4.70), we obtain

$$\dot{e} = Ae + b\left(u_s + \tilde{\theta}^\top \hat{\xi}_f + \epsilon\right), \quad y = c^\top e, \tag{4.72}$$

where $\tilde{\theta} = \bar{\theta} - \hat{\theta}$.

[3] A particular case of a stable LTI SISO plant is considered in [1]. An alternative approach based on Youla–Kucera parameterization is proposed in [17]. The case when all exosystem oscillatory modes are assumed to be excited by the initial condition is considered in [8, 9]. A particular case of a plant with output disturbance only is considered in [16].

[4] An alternative approach to the output-feedback compensation of unmatched disturbances in LTI SISO plants is presented in [10].

At first glance, the model (4.72) belongs to the class of canonical dynamic error models with measured output (see Sect. 3.4). However, the model (4.72) cannot be directly used for design of an adaptation algorithm since the matrix A is possibly unstable. Therefore, we use a special swapping scheme that is applicable to unstable plants (see Lemma D.3). In this case, the swapping scheme is defined by the following expressions:

$$\dot{\hat{e}}_\xi = A\hat{e}_\xi - l_e\left(y + c^\top\hat{e}_\xi\right) + b\left(\hat{\theta}^\top\hat{\xi}_f - u_s\right), \tag{4.73}$$

$$\bar{y} = y + c^\top\hat{e}_\xi - \hat{\theta}^\top\hat{\xi}_e, \tag{4.74}$$

where

$$\hat{\xi}_e = W_e(s)\left[\hat{\xi}_f\right] \tag{4.75}$$

is the new filtered regressor, the vector of feedback gains l_e is such that the matrix $A_e = A - l_e c^\top$ is Hurwitz, and $W_e(s) = c^\top(sI - A_e)^{-1}b$ is the asymptotically stable transfer function.

In view of Lemma D.3, we immediately obtain that

$$\bar{y} = \tilde{\theta}^\top\hat{\xi}_e + \epsilon_y, \tag{4.76}$$

where ϵ_y exponentially decays. The model (4.76) is in the form of the canonical error model (3.12) that belongs to the class of static error models. Therefore, we can use different types of basic adaptation algorithms introduced in Sect. 3.2. In particular, we can use the gradient algorithm of adaptation

$$\dot{\hat{\theta}} = \gamma\hat{\xi}_e\bar{y}, \qquad \gamma > 0. \tag{4.77}$$

Finally, we define the stabilizing component as

$$u_s = -k^\top\bar{e}, \tag{4.78}$$

where $\bar{e}$ is the state of the filter

$$\dot{\bar{e}} = A\bar{e} + bu_s + l_e\left(y - c^\top\bar{e}\right), \tag{4.79}$$

and k is the vector of feedback gains selected so that the matrix $A_\Sigma = A - bk^\top$ is Hurwitz.

Proposition 4.7 *If Assumptions 2.1 and 4.2 hold, then the adjustable controller (4.71) with the disturbance observer (4.68), (4.69), the swapping scheme (4.73), (4.74), the adaptation algorithm (4.77), and the stabilizing component (4.78), (4.79) when applied to the plant (4.51) provides the boundedness of all the closed-loop signals and drives the plant output to zero in accordance with (4.52).*

Proof By Proposition 2.3, it follows that the regressor $\hat{\xi}_f$ and, therefore, $\hat{\xi}_e$ are bounded. Then from Property P.3.1.2 of Proposition 3.1, we conclude that the algorithm of adaptation (4.77) ensures that $\bar{y},\ \hat{\theta} \in \mathcal{L}_\infty$, while $|\dot{\hat{\theta}}(t)| \to 0$ and $\bar{y}(t) \to 0$ as $t \to \infty$. Since $\tilde{\theta}^\top(t)\hat{\xi}_e(t) \to 0$ (see (4.76)), $|\dot{\hat{\theta}}(t)| \to 0$, and $\hat{\xi}_f,\ \hat{\xi}_e \in \mathcal{L}_\infty$, it follows from Corollary 3.4 that

$$\tilde{\theta}^\top(t)\hat{\xi}_e(t) - W_e(s)\left[\tilde{\theta}^\top\hat{\xi}_f(t)\right] \to 0 \quad \text{as} \quad t \to \infty,$$

and, therefore, $\tilde{\theta}^\top(t)\hat{\xi}_f(t) \to 0$ as $t \to \infty$. Introducing the state estimation error $\tilde{e} = e - \bar{e}$ and calculating its time derivative, in view of (4.72) and (4.79), we obtain

$$\dot{\tilde{e}} = A_e\tilde{e} + b\left(\tilde{\theta}^\top\hat{\xi}_f + \epsilon\right). \tag{4.80}$$

Replacing $\bar{e} = e - \tilde{e}$ in (4.78) and substituting the obtained expression for u_s into (4.72), we get the following model of the closed-loop system:

$$\dot{e} = A_\Sigma e + b\left(k^\top\tilde{e} + \tilde{\theta}^\top\hat{\xi}_f + \epsilon\right). \tag{4.81}$$

Since $\tilde{\theta}^\top(t)\hat{\xi}_f(t) \to 0$, it follows from (4.80) that $|\tilde{e}(t)| \to 0$, and from (4.81) that $|e(t)| \to 0$ and $y(t) \to 0$ as $t \to \infty$. Further analysis of the corresponding equations demonstrates the boundedness of all the closed-loop signals. □

Again, using the results of Sects. 3.2.2 and 3.2.3, we can construct adaptation algorithms with improved parametric convergence. In particular, we can define the extended output $y_H \in \mathbb{R}^{p+1}$ and the extended regressor $\Xi_e \in \mathbb{R}^{q\times p+1}$ as

$$y_H = \bar{H}(s)\left[y_w\right] - \Xi_e^\top\hat{\theta}, \quad \Xi_e^\top = \bar{H}(s)\left[\xi_e^\top\right], \tag{4.82}$$

$$y_w = y + c^\top\hat{e}_\xi, \tag{4.83}$$

where $\bar{H}(s) = \left[1,\ H_1(s), \dots, H_p(s)\right]^\top$, $p \geq (q-1)$, $H_i(s)$ are distinct designer-chosen PASMP transfer functions, $i = 1, 2, \dots, p$, $\hat{e}_\xi$ is the state of the filter (4.73). Then the adaptation algorithm with DRE takes the form

$$\dot{\hat{\theta}} = \gamma\Xi_e y_H, \quad \gamma > 0. \tag{4.84}$$

The adaptation algorithm with MRE is

$$\dot{\hat{\theta}} = \gamma\left(L(s)\left[\hat{\xi}_e\bar{y}\right] - \Omega\hat{\theta}\right), \quad \gamma > 0, \tag{4.85}$$

where $L(s)$ is a designer-chosen transfer function of the form (3.44), and $\Omega = L(s)\left[\hat{\xi}_e\hat{\xi}_e^\top\right]$.

Example 4.6 Let us consider the plant and the disturbance used in Example 4.5 with exception that the plant state is not accessible for measurement. The vector of feedback gains k in (4.78) is selected as $k = [30,\ 11]^\top$, the matrix G and the vector l of the filter (4.69) are defined by (4.24). For the filters (4.68), (4.73), (4.79), we choose the vector $l_e = \left[\frac{13}{30}, \frac{42}{30}\right]^\top$ and get the transfer function

$$W_e(s) = \frac{30}{s^2 + 13s + 42}.$$

The vector of adjustable parameters $\hat{\theta}$ is generated by one of the following three adaptation algorithms:

- the gradient algorithm (4.77);
- the algorithm (4.84) with DRE and $p = 1$, $H_1(s) = \frac{1}{0.25s+1}$;
- the algorithm (4.85) with MRE and $L(s) = \frac{1}{0.2s+1}$.

All simulations are carried out with the variable disturbance (4.50), for zero initial conditions $x(0)$, $\hat{\xi}_f(0)$, $\hat{e}(0)$, $\hat{e}_\xi(0)$, $\bar{e}(0)$, $\hat{\theta}(0)$ and with different gains $\gamma = 0.1,\ 1,\ 100$.

Simulation results of the closed-loop systems employing different algorithms of adaptation introduced above are presented in Figs. 4.8, 4.9 and 4.10.

The results demonstrate complete rejection of unknown disturbance generated by a second-order uncertain exosystem and asymptotic convergence of the output y to zero (Figs. 4.11, 4.12, and 4.13). Again, the adaptation algorithms with DRE and MRE allow one to accelerate the parametric convergence by increasing the adaptation gain γ. ■

Adaptive disturbance compensation based on backstepping. To compensate the unmatched disturbance, we can use an alternative approach based on the modular adaptive backstepping with high-order tuners (see Appendix E).

First, for the design purposes, we write out the corresponding transfer functions of the plant (4.51)

$$W_u(s) = c^\top(sI - A)^{-1}b = \frac{\beta(s)}{\alpha(s)}, \quad W_v(s) = d_n + c^\top(sI - A)^{-1}d = \frac{\delta(s)}{\alpha(s)}$$

with the polynomials

$$\alpha(s) = s^n + a_{n-1}s^{n-1} + \cdots + a_0, \quad \beta(s) = b_m s^m + b_{m-1}s^{m-1} + \cdots + b_0$$

and define their known parameters $a_{n-1}, \ldots, a_0$ and $b_m, \ldots, b_0$.

Further, using Proposition 2.4, we obtain the following parameterized model:

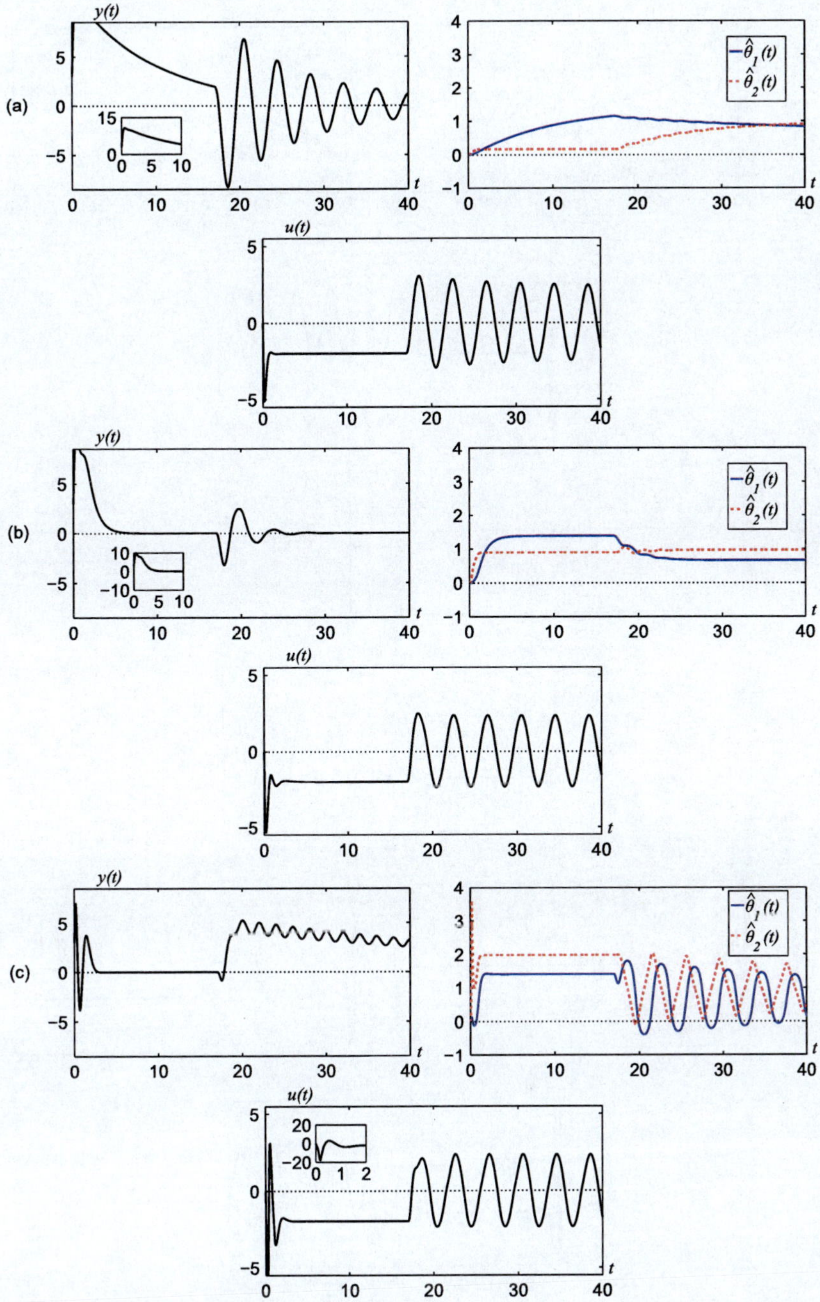

Fig. 4.11 Example 4.6: adaptive output-feedback compensation of the unmatched disturbance (4.50) in the system closed by the gradient adaptation algorithm with: **a** $\gamma = 0.1$; **b** $\gamma = 1$; **c** $\gamma = 100$

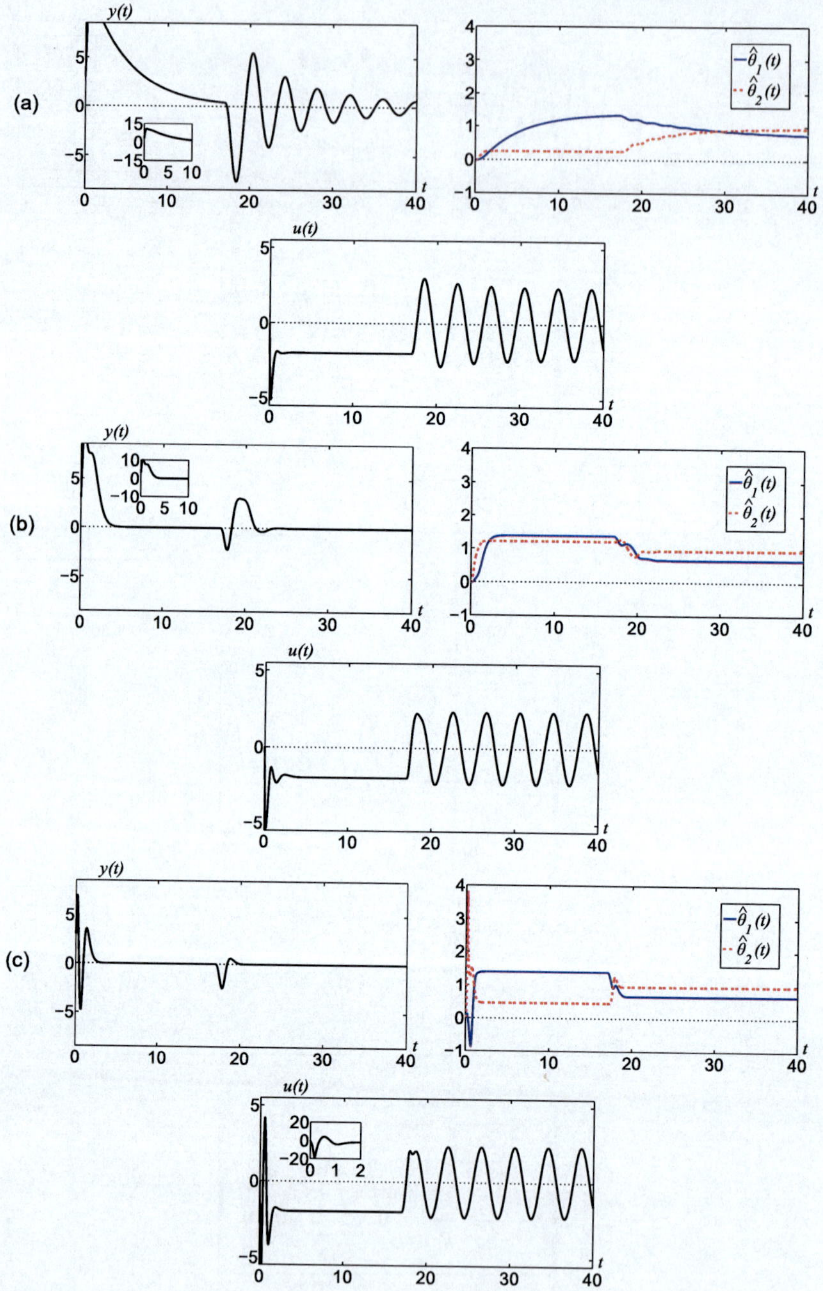

Fig. 4.12 Example 4.6: adaptive output-feedback compensation of the unmatched disturbance (4.50) in the system closed by the augmented error adaptation algorithm with DRE and with: **a** $\gamma = 0.1$; **b** $\gamma = 1$; **c** $\gamma = 100$

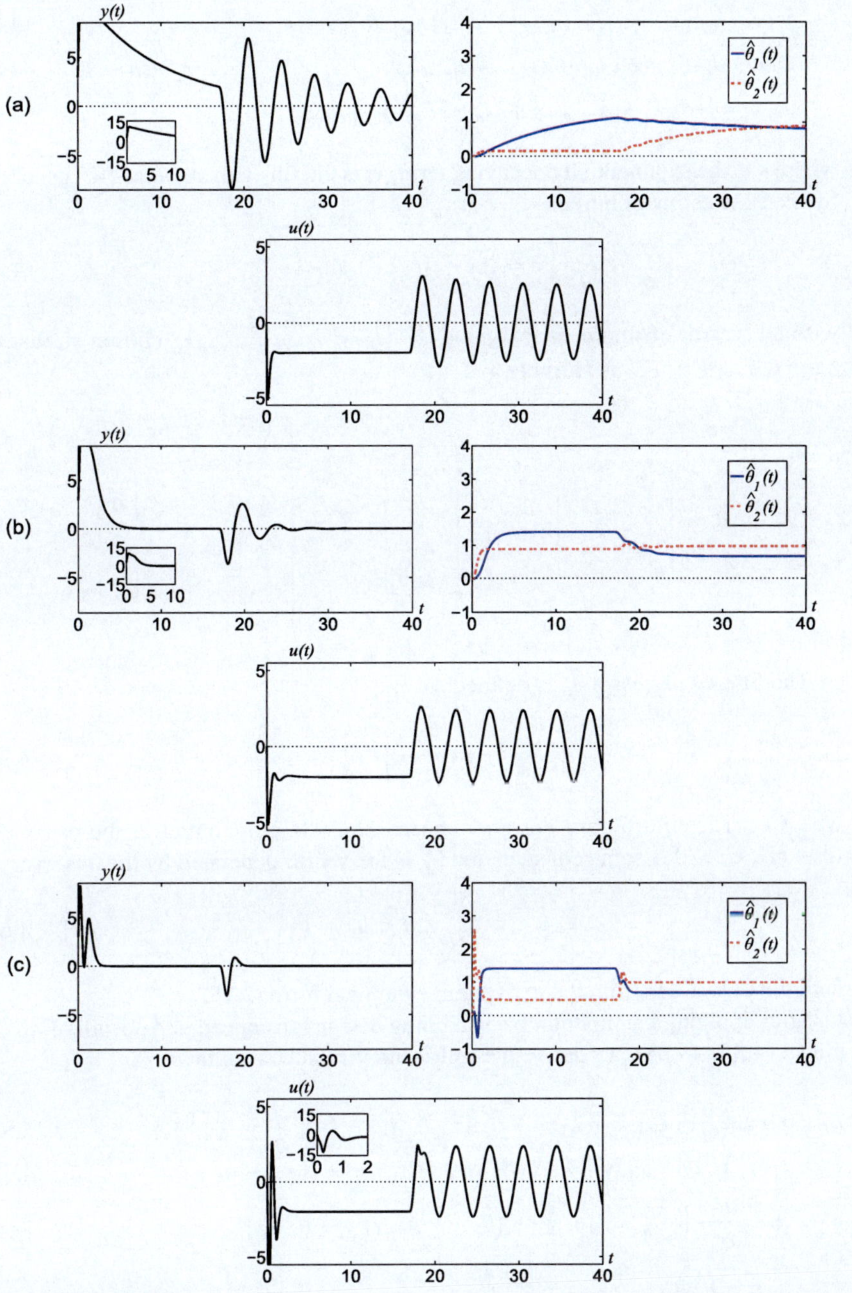

Fig. 4.13 Example 4.6: adaptive output-feedback compensation of the unmatched disturbance (4.50) in the system closed by the augmented error adaptation algorithm with MRE with: **a** $\gamma = 0.1$; **b** $\gamma = 1$; **c** $\gamma = 100$

$$\dot{y} = -\bar{l}_{n-1}\hat{x}_1 + \hat{x}_2 + \left(\bar{l}_{n-1} - a_{n-1}\right) y + \theta^\top \dot{\hat{\xi}}_\kappa + \epsilon, \tag{4.86}$$

$$\dot{\hat{x}}_i = -\bar{l}_{n-i}\hat{x}_1 + \hat{x}_{i+1} + \left(\bar{l}_{n-i} - a_{n-i}\right) y, \qquad i = 2, 3, \ldots, n-m-1, \tag{4.87}$$

$$\dot{\hat{x}}_{n-m} = -\bar{l}_m\hat{x}_1 + \hat{x}_{n-m+1} + \left(\bar{l}_m - a_m\right) y + b_m u, \tag{4.88}$$

where ϵ is the exponentially decaying term, $\hat{x}_i$ is the ith element of $\hat{x} \in \mathbb{R}^n$ generated by the state estimation filter

$$\dot{\hat{x}} = \bar{A}\hat{x} + \left(\bar{l} - \bar{a}\right) y + \bar{b}u \tag{4.89}$$

with the vector of the feedback gains $\bar{l} = \left[\bar{l}_{n-1}, \bar{l}_{n-2}, \ldots, \bar{l}_0\right]^\top$ chosen so that the matrix $\bar{A} = E - \bar{l}\bar{c}^\top$ is Hurwitz and

$$E = \begin{bmatrix} 0 & \\ \vdots & I_{n-1} \\ 0 & \cdots 0 \end{bmatrix}, \quad \bar{a} = \begin{bmatrix} a_{n-1} \\ \vdots \\ a_1 \\ a_0 \end{bmatrix}, \quad \bar{b} = \begin{bmatrix} 0 \\ \vdots \\ 0 \\ b_m \\ \vdots \\ b_0 \end{bmatrix}, \quad \bar{c} = \begin{bmatrix} 1 \\ 0 \\ \vdots \\ 0 \end{bmatrix}.$$

The filtered regressor $\hat{\xi}_\kappa$ is defined as

$$\hat{\xi}_\kappa = \frac{\kappa_0}{\kappa(s)}\left[\hat{\xi}_f\right], \tag{4.90}$$

where $\kappa(s)$ is an arbitrary designer-chosen Hurwitz polynomial of the order $r \geq (n-m)$, $\kappa_0 > 0$ is some constant and $\hat{\xi}_f$ is the vector generated by the observer

$$\dot{\hat{\xi}}_f = G\hat{\xi}_f + l\left(y - \bar{c}^\top\hat{x}\right) \tag{4.91}$$

with the pair (G, l) from the exosystem canonical form (2.15).

Now, applying the modular backstepping design introduced in Appendix E to the model (4.86)–(4.88), we derive the following virtual control laws:

$$\alpha_1(y, \hat{x}_1, \hat{\theta}, \hat{\xi}_\kappa) = -c_1 z_1 + \bar{l}_{n-1}\hat{x}_1 - \left(\bar{l}_{n-1} - a_{n-1}\right) y - \hat{\theta}^\top \dot{\hat{\xi}}_\kappa, \tag{4.92}$$

$$\begin{aligned} \alpha_i(y, \bar{x}_i, \bar{\theta}_{i-1}, \bar{\xi}_i) &= -z_{i-1} - c_i z_i + \bar{l}_{n-i}\hat{x}_1 - \left(\bar{l}_{n-i} - a_{n-i}\right) y \\ &+ \frac{\partial \alpha_{i-1}}{\partial y}\left(-\bar{l}_{n-1}\hat{x}_1 + \hat{x}_2 + \left(\bar{l}_{n-1} - a_{n-1}\right) y + \hat{\theta}^\top \dot{\hat{\xi}}_\kappa\right) \\ &+ \sum_{j=1}^{i-1}\left(\frac{\partial \alpha_{i-1}}{\partial \hat{\theta}^{(j-1)}}\hat{\theta}^{(j)} + \frac{\partial \alpha_{i-1}}{\partial \hat{\xi}_\kappa^{(j)}}\hat{\xi}_\kappa^{(j+1)}\right. \end{aligned} \tag{4.93}$$

$$+\frac{\partial\alpha_{i-1}}{\partial\hat{x}_j}\left(-\bar{l}_{n-j}\hat{x}_1+\hat{x}_{j+1}+\left(\bar{l}_{n-j}-a_{n-j}\right)y\right)\bigg),\quad i=2,3,\ldots,n-m,$$

where $z_1 = y$, $z_i = \hat{x}_i - \alpha_{i-1}$ are the state errors, $c_i > 0$ are the design parameters, $\hat{\theta} \in \mathbb{R}^q$ is the vector of adjustable parameters and

$$\hat{\xi}_\kappa^{(j+1)} = \frac{\kappa_0 s^{j+1}}{\kappa(s)}\left[\hat{\xi}_f\right],\quad j = 0, 1, 2, \ldots, n-m-1 \tag{4.94}$$

are measured, $\bar{x}_i = (\hat{x}_1, \hat{x}_2, \ldots, \hat{x}_i)$, $\bar{\xi}_i = (\dot{\hat{\xi}}_\kappa, \ddot{\hat{\xi}}_\kappa, \ldots, \hat{\xi}_\kappa^{(i)})$, and $\bar{\theta}^{(i)} = (\hat{\theta}, \dot{\hat{\theta}}, \ldots, \hat{\theta}^{(i)})$ are the sets of corresponding variables. The procedure is finished by design of the actual control law

$$u = \frac{1}{b_m}\left(\alpha_{n-m}(y, \bar{x}_{n-m}, \bar{\theta}_{n-m-1}, \bar{\xi}_{n-m}) - \hat{x}_{n-m+1}\right). \tag{4.95}$$

The control law (4.92)–(4.95) when applied to (4.86)–(4.88) results in the following closed-loop state error model:

$$\dot{z} = A_z z + b_z \dot{\hat{\xi}}_\kappa^\top \tilde{\theta} + b_z \epsilon, \tag{4.96}$$

$$y = z_1, \tag{4.97}$$

where $z = [z_1,\ z_2, \ldots,\ z_{n-m}]^\top$ is the vector of the state errors,

$$A_z = \begin{bmatrix} -c_1 & 1 & 0 & \cdots & 0 \\ -1 & -c_2 & 1 & \ddots & \vdots \\ 0 & -1 & \ddots & \ddots & 0 \\ \vdots & \ddots & \ddots & \ddots & 1 \\ 0 & \cdots & 0 & -1 & -c_{n-m} \end{bmatrix},\quad b_z = \begin{bmatrix} 1 \\ -\frac{\partial\alpha_1}{\partial y} \\ -\frac{\partial\alpha_2}{\partial y} \\ \vdots \\ -\frac{\partial\alpha_{n-m-1}}{\partial y} \end{bmatrix}.$$

Remark 4.4 Since in the model (4.96) the b_z is constant (all the partial derivatives $\frac{\partial\alpha_{i-1}}{\partial y}$ are constants), while $\dot{\hat{\xi}}_\kappa$ is known to be bounded a priori, the ISS property of the closed-loop system is guaranteed by linear feedback terms $-c_i z_i$, and no additional nonlinear damping terms are needed. ■

Remark 4.5 In a certain sense, the state error model (4.96), (4.97) is not complete because it does not contain additional m equations describing zero dynamics. However, by virtue of Assumption 4.2.2, if we stabilize the model (4.96), (4.97), we will provide the boundedness of all the closed-loop signals [6, Sect. 2.2.3]. ■

To derive an adaptation algorithm, we apply the nonlinear swapping scheme proposed in [6, Chap. 9] (see Lemma D.9). To this end, we introduce the swapping filters

$$\dot{\zeta}_0 = A_z\zeta_0 + b_z\dot{\hat{\xi}}_\kappa^\top\hat{\theta}, \tag{4.98}$$

$$\dot{\Omega}_\theta^\top = A_z\Omega_\theta^\top + b_z\dot{\hat{\xi}}_\kappa^\top \tag{4.99}$$

with the states $\zeta_0 \in \mathbb{R}^{n-m}$, $\Omega_\theta \in \mathbb{R}^{q\times n-m}$ and define the *augmented state error*

$$\bar{z} = z + \zeta_0. \tag{4.100}$$

Then, in view of Lemma D.9, we have

$$\bar{z} = \Omega_\theta^\top\theta + \epsilon_z, \tag{4.101}$$

where Ω_θ is bounded (since b_z is constant and $\dot{\hat{\xi}}_\kappa$ is bounded), and ϵ_z is the exponentially decaying term satisfying the equation $\dot{\epsilon}_z = A_z\epsilon_z + b_z\epsilon$.

Remark 4.6 To reduce the effect caused by the initial condition $\epsilon_z(0)$ (not completely due to the influence of ϵ), we select the initial conditions of the filters (4.98), (4.99) as follows: $\zeta_0(0) = -z(0)$, $\Omega_\theta(0) = O$. Hence, $\epsilon_z(0) = z(0) + \zeta_0(0) - \Omega_\theta^\top(0)\theta = 0$. ■

It is necessary to note that the control laws (4.93), (4.95) involve the HOTD of adjustable parameters $\hat{\theta}^{(i)}$ up to the order $n-m-1$. At the same time, the static linear regression (4.101) allows one to employ the adaptation algorithm with MRE, special realization of which can generate the HOTD of the adjustable parameters (see Sect. 3.2.3). According to the approach, such an algorithm of adaptation is given by

$$\dot{\hat{\theta}} = \gamma\left(\bar{z}_L - \bar{\Omega}_\theta\hat{\theta}\right), \quad \gamma > 0, \tag{4.102}$$

where $\bar{z}_L = L(s)\,[\Omega_\theta\bar{z}]$, $\bar{\Omega}_\theta = L(s)\left[\Omega_\theta\Omega_\theta^\top\right]$, $L(s)$ is a designer-chosen transfer function of the form (3.44) with a relative degree greater or equal to $n-m-2$. Then, in view of (4.101), the adaptation algorithm (4.102) yields the following closed-loop parametric error model:

$$\dot{\tilde{\theta}} = -\gamma\bar{\Omega}_\theta\tilde{\theta} - \gamma L(s)\left[\Omega_\theta\epsilon_z\right]. \tag{4.103}$$

Since the relative degree of the transfer function $L(s)$ is greater or equal to $n-m-2$, the HOTD of $\bar{z}_L$ and $\bar{\Omega}_\theta$ are physically implementable. As a result, the adaptation algorithm (4.102) allows us to calculate the HOTD of adjustable parameters $\hat{\theta}$ up to the $(n-m-1)$st order via the following expressions [4]:

$$\hat{\theta}^{(i)} = \gamma\left(\bar{z}_H^{(i-1)} - \sum_{j=0}^{i-1} C_j^{i-1}\bar{\Omega}_\theta^{(i-j-1)}\hat{\theta}^{(j)}\right), \tag{4.104}$$

where $i = 2, 3, \ldots, n-m-1$, and C_j^i are the binomial coefficients.

The main properties of the closed-loop adaptive system are established as follows.

Proposition 4.8 *If Assumptions 2.1 and 4.2 hold, then the adjustable controller (4.92)–(4.95) with the state estimation filter (4.89), disturbance observer (4.91), the swapping scheme (4.98)–(4.100), and the adaptation algorithm (4.102), (4.104) when applied to the plant (4.51) provides the boundedness of all the closed-loop signals and drives the plant output to zero according to (4.52).*

Proof From Proposition E.1, we directly obtain that $z, \hat{\theta}, \dot{\hat{\theta}}, \ldots, \hat{\theta}^{(n-m-1)}$ are bounded and $|z(t)| \to 0$ as $t \to \infty$. Now, taking Assumption 4.2 into account and using the same arguments as in [6, Sect. 10.2.2], we can prove that all the closed-loop signals are bounded. □

Example 4.7 Let us consider the third-order plant (4.51) with

$$A = \begin{bmatrix} 0\,1\,0 \\ 0\,0\,1 \\ 0\,0\,0 \end{bmatrix}, \quad b = \begin{bmatrix} 0 \\ 0 \\ 1 \end{bmatrix}, \quad c = \begin{bmatrix} 1 \\ 0 \\ 0 \end{bmatrix}, \quad d = \begin{bmatrix} 1 \\ 0 \\ 0 \end{bmatrix}, \quad d_3 = 1,$$

and the disturbance υ given by the function (4.50). Selecting the state estimation filter (4.89) with $\bar{l}_0 = 24,\ \bar{l}_1 = 26,\ \bar{l}_2 = 9$ and applying the backstepping procedure, we design the virtual control laws $\alpha_1,\ \alpha_2, \alpha_3$ and the actual control $u = \alpha_3$ with $c_1 = c_2 = c_3 = 1$. The regressor $\dot{\hat{\xi}}_\kappa$ and its time derivatives are calculated via the equalities

$$\hat{\xi}_\kappa^{(i)} = \frac{8s^i}{s^3 + 6s^2 + 12s + 8}\left[\hat{\xi}_f\right], \quad i = 1, 2, 3,$$

where $\hat{\xi}_f$ is generated by the observer (4.91) with G and l defined by (4.24).

The vector of adjustable parameters $\hat{\theta}$ and its HOTD up to the second order are calculated via the algorithm with MRE with $L(s) = \frac{1}{s+1}$:

$$\dot{\hat{\theta}} = \gamma\left(\frac{1}{s+1}[\Omega_\theta(z + \zeta_0)] - \frac{1}{s+1}\left[\Omega_\theta\Omega_\theta^\top\right]\hat{\theta}\right) \tag{4.105}$$

and the expression

$$\ddot{\hat{\theta}} = -\left(I_2 + \gamma\frac{1}{s+1}\left[\Omega_\theta\Omega_\theta^\top\right]\right)\dot{\hat{\theta}} + \gamma\Omega_\theta\left(z + \zeta_0 - \Omega_\theta^\top\hat{\theta}\right), \tag{4.106}$$

where Ω_θ and ζ_0 are generated by the swapping filters (4.98), (4.99), $z = [z_1, z_2, z_3]^\top = \left[y, \hat{x}_2 - \alpha_1, \hat{x}_3 - \alpha_2\right]^\top, \hat{x}_2, \hat{x}_3$ are calculated by means of the disturbance observer, and $\gamma = 1$.

All simulations are carried out with zero initial conditions for x, $\hat{x}$, $\hat{\xi}_f$, ζ_0, Ω_θ, and $\hat{\theta}$.

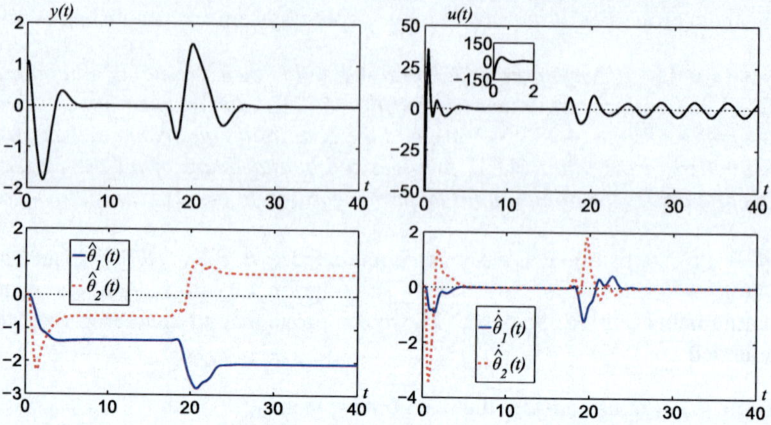

Fig. 4.14 Example 4.7: adaptive output-feedback compensation of the unmatched disturbance (4.50) in the system closed by the backstepping controller with the adaptation algorithm with MRE (4.105), (4.106)

Simulations results of the closed-loop systems employing the algorithm of adaptation with MRE are presented in Fig. 4.14.

As seen form the presented plots, a priori unknown disturbance generated by a second-order uncertain exosystem is completely compensated, while the output y is driven to zero. ■

4.1.4 Regulation in LTI SISO Plants

Now, we demonstrate how the problems of adaptive reference tracking and adaptive disturbance rejection can be resolved simultaneously by applying the above techniques.

Consider the plant (4.51), where the state x is inaccessible for measurements. The control problem consists in design of an output-feedback control ensuring the boundedness of all the closed-loop signals and achievement of the control objective

$$\lim_{t\to\infty} (g(t) - y(t)) = 0, \tag{4.107}$$

where g is the reference signal. In addition to Assumption 4.2, we make the following one.

Assumption 4.3 The disturbance υ and the reference g are such that:

A.4.3.1. they can be modelled as the outputs of the exosystems

$$\dot{\xi}_g = \Gamma_g \xi_g, \quad g = h_g^\top \xi_g, \tag{4.108}$$

$$\dot{\xi}_\upsilon = \Gamma_\upsilon \xi_\upsilon, \quad \upsilon = h_\upsilon^\top \xi_\upsilon, \tag{4.109}$$

where $\xi_g \in \mathbb{R}^{q_g}$, $\xi_\upsilon \in \mathbb{R}^{q_\upsilon}$, and the constant matrices Γ_g, Γ_υ with the vectors h_g, h_υ have appropriate dimensions.

A.4.3.2 both the exosystems (4.108) and (4.109) satisfy Assumption 2.1.

In other words, we consider Problem 1.4 of adaptive output regulation formulated in Sect. 1.3.3 and resolve it for the case of unmatched disturbances affecting LTI SISO plants with measured outputs.

The design procedure contains the following steps. First, as a technical intermediate result, we represent the exosystem (4.108) in the autonomous canonical form (see Lemma 2.1)

$$\dot{\bar{\xi}}_g = \left(G_g + l_g \bar{h}_g^\top\right) \bar{\xi}_g, \quad g = \bar{h}_g^\top \bar{\xi}_g, \tag{4.110}$$

where $\bar{\xi}_g \in \mathbb{R}^{q_g}$ is the state, $\bar{h}_g \in \mathbb{R}^{q_g}$ is a vector of unknown parameters, G_g is an arbitrary designer-chosen Hurwitz matrix, and (G_g, l_g) is the controllable pair.

Then we represent the plant model in the terms of corresponding state and output errors with matched disturbance. To this end, we rewrite the plant model in the form

$$\dot{x} = Ax + bu + dh_\upsilon^\top \xi_\upsilon, \quad y = c^\top x + d_n h_\upsilon^\top \xi_\upsilon. \tag{4.111}$$

It is easy to see that the model (4.111) together with the exosystems (4.109), (4.110) and the tracking error

$$\varepsilon = y - g = c^\top x + d_n h_\upsilon^\top \xi_\upsilon - \bar{h}_g^\top \bar{\xi}_g$$

exactly duplicates the form of (1.48), (1.49) and (1.64), (1.65) (with appropriate change of notations). Therefore, according to Lemmas 1.1 and 1.2:

(1) there exists solution pairs (M_g, θ_g) and $(M_\upsilon, \theta_\upsilon)$ of the matrix equations (1.66), (1.67) (with $\Theta_g = \theta_g^\top$ and $\Theta_\upsilon = \theta_\upsilon^\top$) such that for the state error $e_g = x - M\xi$, we have

$$\dot{e}_g = Ae_g + b\left(u + \theta_g^\top \bar{\xi}_g + \theta_\upsilon^\top \xi_\upsilon\right), \quad \varepsilon = c^\top e_g, \tag{4.112}$$

where $M = \left[M_g,\ M_\upsilon\right]$, e_g is inaccessible for measurements, $\xi = \left[\bar{\xi}_g^\top,\ \xi_\upsilon^\top\right]^\top$, $\theta_\upsilon \in \mathbb{R}^{q_\upsilon}$ and $\theta_g \in \mathbb{R}^{q_g}$ are vectors of unknown parameters, and $\varepsilon = y - g$ is the measured output error;

(2) there exists a solution pair M_υ and $\Theta_\upsilon = \theta_\upsilon^\top$ of the matrix equation (1.67) such that for the state error $e_\upsilon = x - M_\upsilon \xi_\upsilon$ we have

$$\dot{e}_\upsilon = Ae_\upsilon + b\left(u + \theta_\upsilon^\top \xi_\upsilon\right), \quad y = c^\top e_\upsilon, \tag{4.113}$$

where e_υ is inaccessible for measurements.

In the expressions (4.112) and (4.113), the signal $\bar{\upsilon} = \theta_\upsilon^\top \xi_\upsilon$ can be considered as the matched disturbance. Below, the model (4.113) will be used for design of a disturbance observer, while the model (4.112) will be used for design of the overall adaptive controller and algorithms of adaptation.

Further, we construct observers for the inaccessible regressor $\bar{\xi}_g$ and matched disturbance $\bar{\upsilon}$. In view of (4.110), we can use statement L.2.3.2. of Lemma 2.3 and represent $\bar{\xi}_g$ as

$$\bar{\xi}_g = \hat{\xi}_g + \epsilon_g, \tag{4.114}$$

where ϵ_g exponentially decays, and $\hat{\xi}_g$ is the state of the implementable filter

$$\dot{\hat{\xi}}_g = G_g \hat{\xi}_g + l_g g \tag{4.115}$$

with an arbitrary initial condition $\hat{\xi}_g(0)$.

In view of (4.113), we use the results of Proposition 2.3 and employ the disturbance observer

$$\dot{\hat{e}} = A\hat{e} + bu + l_e \left(y - c^\top \hat{e}\right), \tag{4.116}$$

$$\dot{\hat{\xi}}_f = G_\upsilon \hat{\xi}_f + l_\upsilon \left(y - c^\top \hat{e}\right), \tag{4.117}$$

where the vector of feedback gains l_e is such that the matrix $A_e = A - l_e c^\top$ is Hurwitz, G_υ is an arbitrary designer-chosen Hurwitz matrix, and the pair (G_υ, l_υ) is controllable. Therefore (see Proposition 2.3), we can show that the matched disturbance can be given by

$$\bar{\upsilon} = \bar{\theta}^\top \hat{\xi}_f + \epsilon_\upsilon, \tag{4.118}$$

where $\bar{\theta}$ is a new vector of unknown parameters and ϵ_υ exponentially decays.

Substituting (4.114) and (4.118) into (4.112), we obtain the following error model:

$$\dot{e} = Ae + b\left(u + \theta^\top \hat{\xi} + \epsilon\right), \quad \varepsilon = c^\top e, \tag{4.119}$$

where $\theta = [\theta_g^\top, \ \bar{\theta}^\top]^\top$ is the aggregated vector of unknown parameters, $\hat{\xi} = [\hat{\xi}_g^\top, \ \hat{\xi}_f^\top]^\top$ is the aggregated regressor and $\epsilon = \theta_g^\top \epsilon_g + \epsilon_\upsilon$ exponentially decays.

We see that the model (4.119) exactly repeats the form of (4.70) (with the use of corresponding notations). Therefore, to complete the design procedure we can use the results of Sect. 4.1.3.2 and Proposition 4.7. In this way, the adaptive controller simultaneously resolving the reference tracking and disturbance rejection problems consists of:

- the reference observer (4.115);
- the disturbance observer (4.116), (4.117);
- the adjustable controller with the state estimation filter

$$u = -k^{\top}\bar{e} - \hat{\theta}^{\top}\hat{\xi}, \tag{4.120}$$

$$\dot{\bar{e}} = A\bar{e} - bk^{\top}\bar{e} + l_e\left(\epsilon - c^{\top}\bar{e}\right), \tag{4.121}$$

where the vectors of feedback gains k and l_e are such that the matrices $A_{\Sigma} = A - bk^{\top}$ and $A_e = A - l_e c^{\top}$ are Hurwitz, $\hat{\theta} \in \mathbb{R}^{q_g+q_v}$ is the vector of adjustable parameters;
- the swapping scheme

$$\dot{\hat{e}}_{\xi} = A\hat{e}_{\xi} - l_e\left(\varepsilon + c^{\top}\hat{e}_{\xi}\right) + b\left(\hat{\theta}^{\top}\hat{\xi} - u_s\right), \tag{4.122}$$

$$\bar{\varepsilon} = \varepsilon + c^{\top}\hat{e}_{\xi} - \hat{\theta}^{\top}\hat{\xi}_e, \tag{4.123}$$

where $\hat{\xi}_e = W_e(s)[\hat{\xi}]$ and $W_e(s) = c^{\top}(sI - A_e)^{-1}b$;
- one of the following adaptation algorithms:
the gradient adaptation algorithm

$$\dot{\hat{\theta}} = \gamma\hat{\xi}_e\bar{\varepsilon}, \qquad \gamma > 0; \tag{4.124}$$

the adaptation algorithm with DRE

$$\dot{\hat{\theta}} = \gamma\Xi\bar{\varepsilon}_H, \quad \gamma > 0, \tag{4.125}$$

where

$$\bar{\varepsilon}_H = \bar{H}(s)\left[\hat{\epsilon}\right] - \Xi^{\top}\hat{\theta}, \quad \Xi^{\top} = \bar{H}(s)\left[\hat{\xi}_e^{\top}\right], \quad \hat{\varepsilon} = \varepsilon + c^{\top}\hat{e}_{\xi}, \tag{4.126}$$

$\bar{H}(s) = \left[1, H_1(s), \ldots, H_p(s)\right]^{\top}$, $H_i(s)$ are distinct designer-chosen PASMP transfer functions, $p \geq (q_g + q_v - 1)$;
the adaptation algorithm with MRE

$$\dot{\hat{\theta}} = \gamma\left(L(s)\left[\hat{\xi}_e\hat{\varepsilon}\right] - \Omega\hat{\theta}\right), \quad \gamma > 0, \tag{4.127}$$

where $L(s)$ is a designer-chosen transfer function of the form (3.44), $\hat{\varepsilon}$ is defined in (4.126) and $\Omega = L(s)\left[\hat{\xi}_e\hat{\xi}_e^{\top}\right]$.

Finally, using the same arguments as in Sects. 4.1.1.1 and 4.1.3.2, we can prove that under Assumptions 4.2 and 4.3 the adaptive controller (4.120) with the reference observer (4.115), the disturbance observer (4.116), (4.117), the state estimation filter (4.121), the swapping scheme (4.122), (4.123), and the gradient adaptation algorithm (4.124) (or algorithm with DRE (4.125), or algorithm with MRE (4.127)) when applied to the plant (4.51) provides the boundedness of all the closed-loop signals and achievement of the control objective (4.107).

Example 4.8 Let us consider the plant defined in Example 4.6 assuming that the state x is not accessible for measurements. Let the reference g and the disturbance υ be given by the functions $g(t) = 2 + \sin(3t + 4)$ and

$$\upsilon(t) = \begin{cases} 1 & \text{if } \ 0 \le t \le 33 \text{ sec}, \\ \sin\left(\dfrac{\pi t}{2}\right) & \text{if } \ t > 33 \text{ sec}. \end{cases} \tag{4.128}$$

The parameters of these functions, i.e., the amplitudes, phases, frequencies and biases, are unknown.

To design the reference and disturbance observers, we use the third and the second-order exosystems, respectively, and select the matrices

$$G_g = \begin{bmatrix} 0 & 1 & 0 \\ 0 & 0 & 1 \\ -24 & -26 & -9 \end{bmatrix}, \quad l_g = \begin{bmatrix} 0 \\ 0 \\ 24 \end{bmatrix}, \quad G_\upsilon = \begin{bmatrix} 0 & 1 \\ -20 & -9 \end{bmatrix}, \quad l_\upsilon = \begin{bmatrix} 0 \\ 20 \end{bmatrix},$$

$$l_e = \frac{1}{30}\begin{bmatrix} 13 \\ 42 \end{bmatrix}, \quad A_e = A - l_e c^\top = \begin{bmatrix} -13 & 1 \\ -42 & 0 \end{bmatrix}.$$

For the stabilizing component, we select $k = [-30, \ -11]^\top$.

The vector of adjustable parameters $\hat{\theta}$ is generated by one of the following three adaptation algorithms with $\gamma = 1$:

- the gradient algorithm (4.124);
- the algorithm (4.125) with DRE and $p = 4$, $H_1(s) = \frac{3}{0.25s+1}$, $H_2(s) = \frac{9}{0.5s+1}$, $H_3(s) = \frac{18}{0.75s+1}$, $H_4(s) = \frac{36}{s+1}$;
- the algorithm (4.127) with MRE and $L(s) = \frac{1}{2s+1}$.

All simulations are carried out with zero initial conditions $x(0)$, $\hat{\xi}_g(0)$, $\hat{\xi}_\upsilon(0)$, $\hat{e}(0)$, $\hat{\xi}_f(0)$, $\bar{e}(0)$, $\hat{e}_\xi(0)$, and $\hat{\theta}(0)$.

Simulations of the closed-loop systems employing the gradient algorithm of adaptation (4.124), the adaptation algorithm with DRE (4.125) and the algorithm of adaptation with MRE (4.127) are presented in Figs. 4.15, 4.16, and 4.17, respectively, and demonstrate asymptotic tracking of a priori uncertain reference g with complete compensation of the uncertain external disturbance with variable waveform. ■

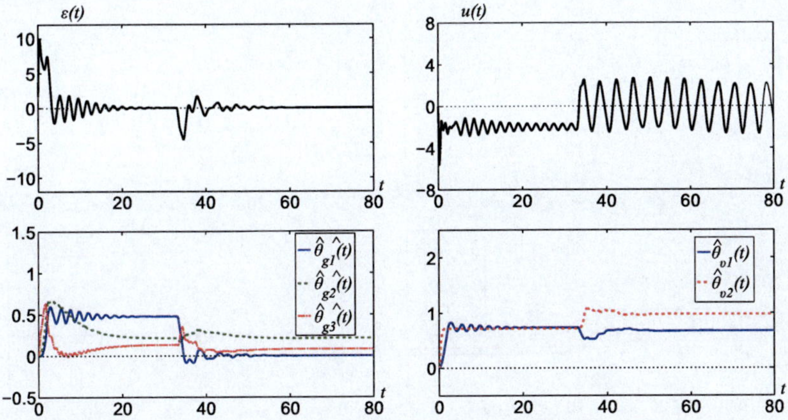

Fig. 4.15 Example 4.8: adaptive output-feedback reference tracking with compensation of the unmatched disturbance in the system closed by the gradient adaptation algorithm

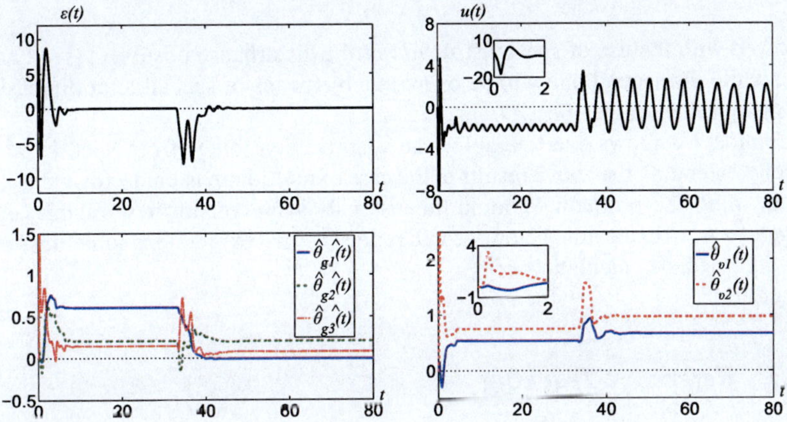

Fig. 4.16 Example 4.8: adaptive output-feedback reference tracking with compensation of the unmatched disturbance in the system closed by the adaptation algorithm with DRE

4.2 LTI MIMO Plants

In this section, we demonstrate how the results obtained for scalar exosignals can be extended to the case of multidimensional ones. We will consider two different classes of the multidimensional exosignals generated by the single exosystem (2.21) and by the set of individual exosystems (2.32).

To cope with multidimensional exosignals, one needs to solve the following two technical problems: construction of multidimensional exosignal observers, and extension of the augmented error technique to the MIMO systems. The first problem

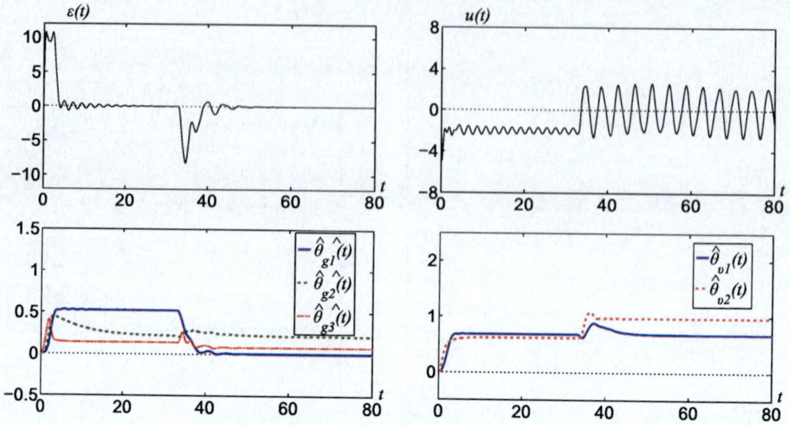

Fig. 4.17 Example 4.8: adaptive output-feedback reference tracking with compensation of the unmatched disturbance in the system closed by the adaptation algorithm with MRE

is solved with the use of special structure of the disturbance observers (see Lemma 2.2), while the second one will be overcome by means of special multidimensional swapping schemes (see Sect. D.2).

Lemma 2.2 allows one to employ the adaptive controllers from Sect. 4.1 almost directly extending the above results to the case of multidimensional exosignals keeping in mind the problem of multidimensional swapping. In view of this almost straightforward extension, below we will represent final expressions sometimes omitting intermediate calculations.

4.2.1 Reference Tracking

Consider the LTI MIMO plant

$$\dot{x} = Ax + Bu, \quad y = Cx, \tag{4.129}$$

where $x \in \mathbb{R}^n$ is the state, $y \in \mathbb{R}^\rho$ is the set of output variables ($\rho \le n$), $u \in \mathbb{R}^\rho$ is the vector of control signals, A, B, and C are the known matrices of appropriate dimensions. We make the following assumption regarding the plant (4.129).

Assumption 4.4 The plant (4.129) is such that:

A.4.4.1. the triplet (A, B, C) is controllable and observable;

A.4.4.2. the matrix C has full row rank, while the matrix B has full column rank;

A.4.4.3. the transfer function matrix $W_u(s) = C(sI - A)^{-1}B$ is invertable, and its transmission zeros have negative real parts.

We consider Problem 1.2 of adaptive multidimensional reference tracking formulated in Sect. 1.3.3. The problem consists in design of a state-feedback (or output-feedback) controller ensuring the boundedness of all the closed-loop signals and achievement of the control objective

$$\lim_{t\to\infty} |y(t) - g(t)| = 0, \tag{4.130}$$

where $g \in \mathbb{R}^\rho$ is the measured reference.

Additionally, we assume that the reference g satisfies Assumption 2.3 (with substitution $\upsilon = g$), i.e., its each component g_i is modelled by the individual exosystem (2.32). Then, in view of Corollary 2.1 and Lemma 2.3, we form the following model of the reference:

$$g = \Theta_d \hat{\xi} + \epsilon_g, \tag{4.131}$$

where ϵ_g is the exponentially decaying term, $\Theta_d \in \mathbb{R}^{\rho\times q}$ is an unknown constant matrix, and $\hat{\xi} \in \mathbb{R}^q$ is the regressor generated by the reference observer

$$\dot{\hat{\xi}} = G_d \hat{\xi} + L_d g. \tag{4.132}$$

The matrices G_d, Θ_d and L_d are block-diagonal of the form (2.36). It is obvious that the observer (4.132) can be splitted into ρ independent ones:

$$\dot{\hat{\xi}}_i = G_i \hat{\xi}_i + l_i g_i, \quad i = 1, 2, \ldots, \rho, \tag{4.133}$$

where the pairs (G_i, l_i) are defined in Corollary 2.1, $\hat{\xi} = [\hat{\xi}_1^\top, \hat{\xi}_2^\top, \ldots, \hat{\xi}_\rho^\top]^\top$, $\hat{\xi}_i \in \mathbb{R}^{q_i}$, and $\sum_{i=1}^{\rho} q_i = q$.

As in Sect. 4.1, we first stabilize the plant by applying the controller

$$u = -Kx + u_t, \tag{4.134}$$

where the matrix of feedback gains K is such that $A_\Sigma = A - BK$ is Hurwitz [2, Sect. 8.6], and u_t is the *tracking component* to be defined later. If the state x is not accessible for measurements, then in (4.134) instead of x it is used the estimate $\hat{x}$ generated by a state observer [2, Sect. 8.8].

Further, we design the adaptive tracking component u_t. As for SISO plant, below we present two different approaches to this design problem. One of them utilizes the concept of the augmented error, while the second one uses estimates of the HOTD of the reference signal.

4.2.1.1 Adaptive Controller Using Augmented Error

By the same arguments as in Sect. 4.1, we can demonstrate that there exists a solution pair (M_g, Θ_g) of the matrix equation (1.66) such that for the plant (4.129), we can write the following error model:

$$\dot{e} = A_\Sigma e + B\left(u_t + \Theta_g \hat{\xi} + \epsilon_g\right), \quad \varepsilon = Ce, \tag{4.135}$$

where the state error $e = x - M_g\hat{\xi}$ is inaccessible for measurement, while the output error $\varepsilon = y - g$ is measured. Note that in general case the matrix Θ_g does not have block-diagonal structure in spite of the fact that the matrix Θ_d does.

Using the tracking component

$$u_t = -\hat{\Theta}_g \hat{\xi}, \tag{4.136}$$

where $\hat{\Theta}_g \in \mathbb{R}^{\rho\times q}$ is the matrix of adjustable parameters, we obtain the following closed-loop error model:

$$\varepsilon = W_\Sigma(s)\left[\tilde{\Theta}_g \hat{\xi}\right] + \epsilon, \tag{4.137}$$

where the transfer function matrix $W_\Sigma(s) = C(sI - A_\Sigma)^{-1}B$ is asymptotically stable and minimum phase,[5] $\tilde{\Theta}_g = \Theta_g - \hat{\Theta}_g$ is the matrix of parametric errors, and $\epsilon = W_\Sigma[\epsilon_g] + Ce^{A_\Sigma t}e(0)$ exponentially decays.

By Θ_{gi} and $\hat{\Theta}_{gi}$, we denote the ith raw of Θ_g and $\hat{\Theta}_g$, respectively, and compose the vector of unknown parameters

$$\vartheta_g = \left[\Theta_{g1},\ \Theta_{g2}, \ldots,\ \Theta_{g\rho}\right]^\top$$

with the vector of adjustable parameters

$$\hat{\vartheta}_g = \left[\hat{\Theta}_{g1},\ \hat{\Theta}_{g2}, \ldots,\ \hat{\Theta}_{g\rho}\right]^\top.$$

Then, using the MIMO swapping scheme (see Sect. D.2.1), we construct the following three algorithms of adaptation.

The gradient algorithm of adaptation is defined as

$$\dot{\hat{\vartheta}}_g = \gamma \Xi_w \bar{\varepsilon}, \quad \gamma > 0, \tag{4.138}$$

where

[5] The state feedback in (4.134) has no effect on the transmission zeros of the closed-loop system [2, Sect. 8.6.4]. Therefore, the transfer function matrix $W_\Sigma(s)$ is minimum phase under Assumption A.4.4.3.

$$\Xi_w^\top = \begin{bmatrix} W_{\Sigma 1,1}(s)\left[\hat{\xi}^\top\right] \dots W_{\Sigma 1,\rho}(s)\left[\hat{\xi}^\top\right] \\ \vdots \quad \ddots \quad \vdots \\ W_{\Sigma \rho,1}(s)\left[\hat{\xi}^\top\right] \dots W_{\Sigma \rho,\rho}(s)\left[\hat{\xi}^\top\right] \end{bmatrix} \tag{4.139}$$

is the $\rho \times q\rho$ matrix regressor, $W_{\Sigma i,j}(s)$ are the entries of the transfer function matrix $W_\Sigma(s)$, and

$$\bar{\varepsilon} = \varepsilon + W_\Sigma(s)\left[\hat{\Theta}_g \hat{\xi}\right] - \Xi_w^\top \hat{\vartheta}. \tag{4.140}$$

In this case, for the output augmented error, we have

$$\bar{\varepsilon} = \Xi_w^\top \tilde{\vartheta} + \epsilon,$$

and the closed-loop model of the parametric errors takes the form

$$\dot{\tilde{\vartheta}}_g = -\gamma \Xi_w \Xi_w^\top \tilde{\vartheta}_g - \gamma \Xi_w \epsilon,$$

where $\tilde{\vartheta}_g = \vartheta_g - \hat{\vartheta}_g$.

Then, with the use of standard arguments, it can be proved that the adaptive controller (4.134), (4.136) with the reference observer (4.132) and the adaptation algorithm (4.138)–(4.140) when applied to the plant (4.129) provides the boundedness of all the closed-loop signals and the asymptotic reference tracking according to (4.130).

Now, we demonstrate that the results of Sects. 3.2.2 and 3.2.3 can be directly extended to the multidimensional case to derive adaptation algorithms with improved parametric convergence.

The algorithm with DRE. Define the dynamically extended augmented error $\bar{\varepsilon}_H \in \mathbb{R}^{\rho(p+1)}$ and the dynamically extended regressor $\bar{\Xi}_w \in \mathbb{R}^{q\rho \times \rho(p+1)}$:

$$\bar{\varepsilon}_H = \begin{bmatrix} \hat{\varepsilon} - \Xi_w^\top \hat{\vartheta}_g \\ H_1(s)\left[\hat{\varepsilon}\right] - \bar{\Xi}_{w1}^\top \hat{\vartheta}_g \\ \vdots \\ H_p(s)\left[\hat{\varepsilon}\right] - \bar{\Xi}_{wp}^\top \hat{\vartheta}_g \end{bmatrix}, \quad \bar{\Xi}_w^\top = \begin{bmatrix} \Xi_w^\top \\ \bar{\Xi}_{w1}^\top \\ \vdots \\ \bar{\Xi}_{wp}^\top \end{bmatrix},$$

where $p \geq (q-1)$, $H_i(s)$ are distinct designer-chosen PASMP transfer functions, $\hat{\varepsilon} = [\hat{\varepsilon}_1, \hat{\varepsilon}_2, \dots, \hat{\varepsilon}_\rho]^\top$ is given by

$$\hat{\varepsilon} = \varepsilon + W_\Sigma(s)\left[\hat{\Theta}_g \hat{\xi}\right], \tag{4.141}$$

$$\bar{\Xi}_{wi}^\top = H_i(s)\left[\Xi_w^\top\right], \qquad i = 1, 2, \dots, p, \tag{4.142}$$

and $\Xi_w^\top$ is defined by (4.139).

Then the adaptation algorithm with DRE takes the form

$$\dot{\hat{\vartheta}}_g = \gamma \bar{\Xi}_w \bar{\varepsilon}_H, \quad \gamma > 0 \tag{4.143}$$

and yields the following closed-loop parametric error model:

$$\dot{\tilde{\vartheta}}_g = -\gamma \bar{\Xi}_w \bar{\Xi}_w^\top \tilde{\vartheta}_g - \gamma \bar{\Xi}_w \epsilon_H,$$

where ϵ_H exponentially decays.

The algorithm with MRE is defined as

$$\dot{\hat{\vartheta}}_g = \gamma \left(L(s) \left[\Xi_w \hat{\varepsilon} \right] - \Omega_\Xi \hat{\vartheta}_g \right), \quad \gamma > 0, \tag{4.144}$$

where $L(s)$ is a designer-chosen transfer function of the form (3.44), $\hat{\varepsilon}$ is defined by (4.141), and $\Omega_\Xi = L(s) \left[\Xi_w \Xi_w^\top \right]$. The algorithm yields the parametric error model

$$\dot{\tilde{\vartheta}}_g = -\gamma \Omega_\Xi \tilde{\vartheta}_g - \gamma L(s) \left[\Xi_w \epsilon_\Omega \right],$$

where ϵ_Ω exponentially decays.

Example 4.9 Consider the two-input and two-output unstable plant (4.129) of the third order with

$$A = \begin{bmatrix} 0 & 1 & 0 \\ 0.1 & -0.9 & 0 \\ 0 & 0 & 0 \end{bmatrix}, \quad B = \begin{bmatrix} 0 & 0 \\ 1 & 0 \\ 0 & 1 \end{bmatrix}, \quad C = \begin{bmatrix} 1 & 0 & 0 \\ 1 & 0 & 1 \end{bmatrix},$$

and the transfer function matrix

$$W_u(s) = \begin{bmatrix} \frac{1}{s^2+0.9s-0.1} & 0 \\ \frac{1}{s^2+0.9s-0.1} & \frac{1}{s} \end{bmatrix}. \tag{4.145}$$

Let the reference be given by

$$g(t) = \begin{bmatrix} g_1(t) \\ g_2(t) \end{bmatrix} = \begin{bmatrix} \beta_2 \sin(wt + \varphi) \\ \beta_1 \end{bmatrix},$$

where $\beta_1 = 2$, $\beta_2 = 3$, $w = 4$, and $\varphi = 5$ are unknown. Therefore, we use two observers (4.133). The first one is of the second order, and the second one is of the first order. For the observers, we select the following parameters:

$$G_1 = \begin{bmatrix} 0 & 1 \\ -20 & -9 \end{bmatrix}, \quad l_1 = \begin{bmatrix} 0 \\ 20 \end{bmatrix}, \quad G_2 = -5, \quad l_2 = 5. \tag{4.146}$$

To design the stabilizing component, we calculate the matrix

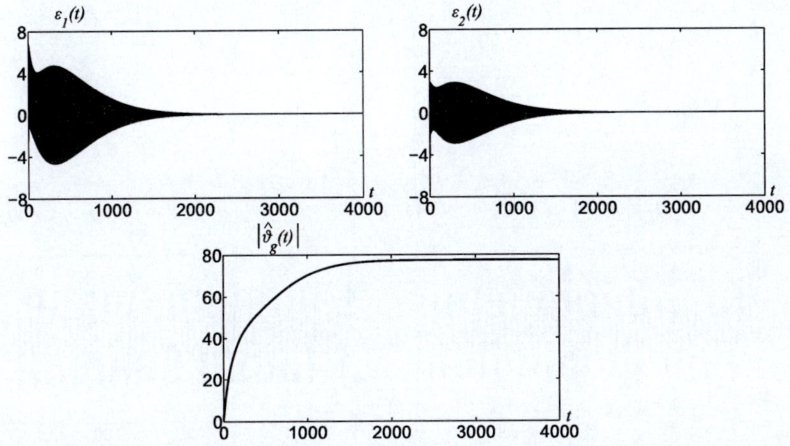

Fig. 4.18 Example 4.9: adaptive reference tracking in the MIMO system closed by the gradient-based adaptation algorithm

$$K = \begin{bmatrix} 5.685 & 8.246 & 20.732 \\ -1.197 & -1.049 & -0.146 \end{bmatrix} \tag{4.147}$$

so that the matrix $A_\Sigma = A - BK$ has the eigenvalues $\lambda\{A_\Sigma\} = \{-2, -3, -4\}$. Hence, the transfer function matrix $W_\Sigma(s)$ is given by

$$W_\Sigma(s) = \frac{1}{s^3 + 9s^2 + 26s + 24} \begin{bmatrix} s - 0.146 & -20.732 \\ 2.049s + 1.051 & s^2 + 9.146s - 15.146 \end{bmatrix}. \tag{4.148}$$

It is worth noting that the both transfer function matrices (4.145) and (4.148) have no transmission zeros.

The vector of adjustable parameters $\hat{\vartheta}_g$ is generated by one of the following three algorithms of adaptation:

- the gradient algorithm (4.138) with $\gamma = 4$;
- the algorithm (4.143) with DRE and $p = 2$, $H_1(s) = \frac{2}{s+1}$, $H_2(s) = \frac{2}{0.2s+1}$, $\gamma = 200$;
- the algorithm (4.144) with MRE and $L(s) = \frac{1}{s+1}$, $\gamma = 200$.

The adaptation gain of the gradient algorithm is selected smaller than the gains of the algorithms with DRE and MRE since for $\gamma = 200$ the gradient algorithm provides inappropriately slow convergence.

All simulations are carried out with zero initial conditions $x(0)$, $\hat{\xi}(0)$, and $\hat{\vartheta}_g(0)$.

Simulation results for the closed-loop systems employing the gradient algorithm of adaptation (4.138), the adaptation algorithm with DRE (4.143) and the algorithm of adaptation with MRE (4.144) are presented in Figs. 4.18, 4.19, and 4.20, respectively. Here we draw the readers' attention to different time scales used to demonstrate the convergence of the gradient scheme.

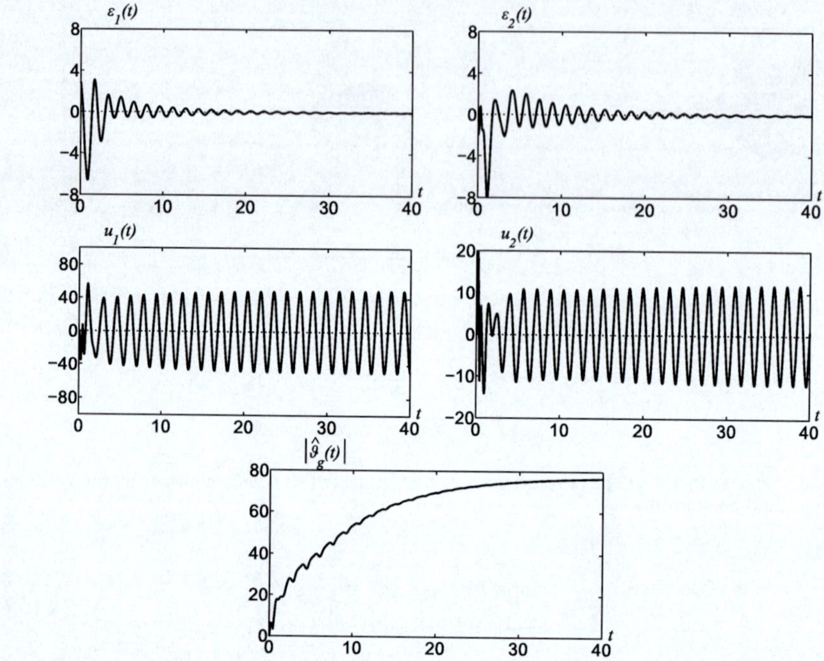

Fig. 4.19 Example 4.9: adaptive reference tracking in the MIMO system closed by the adaptation algorithm with DRE

The results demonstrate asymptotic tracking of a priori uncertain multidimensional reference g. In contrast to the gradient adaptation algorithm, the algorithms with DRE and MRE demonstrate dramatically faster convergence. ■

4.2.1.2 Adaptive Controller Using Estimates of the Reference HOTD

Since the matrix Θ_d in (4.131) has block-diagonal structure (see Corollary 2.1), we can use the approach from Sect. 4.1.1.2 applying it to each element of g.

Together with the observers (4.133) generating the estimates $\hat{\xi}_i \in \mathbb{R}^{q_i}$, we define the estimates of the reference and its HOTD as

$$\hat{g}_{j,i} = \hat{\theta}_i^\top \left(G_i + l_i \hat{\theta}_i^\top\right)^j \hat{\xi}_i, \quad i = 1, 2, \ldots, \rho, \quad j \in \mathbb{N}_0, \qquad (4.149)$$

where $\hat{g}_{j,i}$ is the estimate of the jth-order time derivative of the ith element of g (i.e., the estimate of $g_i^{(j)}$), $\hat{\theta}_i \in \mathbb{R}^{q_i}$ is the ith vector of adjustable parameters generated by one of the three algorithms of adaptation presented below.

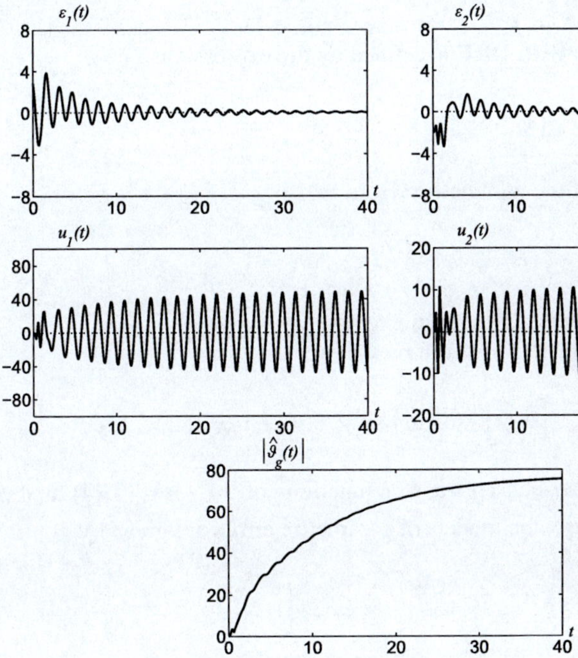

Fig. 4.20 Example 4.9: adaptive reference tracking in the MIMO system closed by the adaptation algorithm with MRE

The gradient algorithm of adaptation is given by

$$\dot{\hat{\theta}}_i = \gamma_i \hat{\xi}_i \left(g_i - \hat{\theta}_i^\top \hat{\xi}_i \right), \qquad \gamma_i > 0, \qquad i = 1, 2, \ldots, \rho. \tag{4.150}$$

To derive the algorithm with DRE, we introduce the transfer function vector $\bar{H}_i(s) = \left[1,\ H_{i,1}(s), \ldots, H_{i,p_i}(s)\right]^\top$ with distinct PASMP transfer functions $H_{i,j}(s)$ and form the dynamically extended reference elements

$$\bar{g}_i = \bar{H}_i(s)[g_i] = \left[g_i,\ H_{i,1}(s)[g_i], \ldots, H_{i,p_i}[g_i]\right]^\top, \quad \bar{g}_i \in \mathbb{R}^{p_i+1}$$

together with the dynamically extended matrix regressors

$$\varXi_i^\top = \bar{H}_i(s)\left[\xi_i^\top\right] = \begin{bmatrix} \hat{\xi}_i^\top \\ H_{i,1}(s)\left[\hat{\xi}_i^\top\right] \\ \vdots \\ H_{i,p_i}(s)\left[\hat{\xi}_i^\top\right] \end{bmatrix}, \tag{4.151}$$

where $i = 1, 2, \ldots, \rho$, $j = 1, 2, \ldots, p_i$, $p_i \geq (q_i - 1)$.

Then **the algorithm with DRE** is defined by the expressions

$$\dot{\hat{\theta}}_i = \gamma_i \Xi_i \left(\bar{g}_i - \Xi_i^\top \hat{\theta}_i\right), \quad \gamma_i > 0, \quad i = 1, 2, \ldots \rho \tag{4.152}$$

that yield the models of the parametric errors

$$\dot{\tilde{\theta}}_i = -\gamma_i \Xi_i \Xi_i^\top \tilde{\theta}_i - \gamma_i \Xi_i \epsilon_{wi}, \quad i = 1, 2, \ldots \rho,$$

where ϵ_{wi} are the exponentially decaying terms.

The algorithm with MRE is given by the expressions

$$\dot{\hat{\theta}}_i = \gamma_i \left(L_i(s)\left[\hat{\xi}_i g_i\right] - \Omega_i \hat{\theta}_i\right), \quad \gamma_i > 0, \quad i = 1, 2, \ldots, \rho, \tag{4.153}$$

where $L_i(s)$ are designer-chosen transfer functions of the form (3.44), and $\Omega_i = H_i(s)\left[\hat{\xi}_i \hat{\xi}_i^\top\right]$. In this case, the models of parametric errors are given by

$$\dot{\tilde{\theta}}_i = -\gamma_i \Omega_i \tilde{\theta}_i - \gamma_i L_i(s)\left[\hat{\xi}_i \epsilon_i\right],$$

where ϵ_i exponentially decay.

Taking Corollaries 3.1–3.3 into account and using the same approach as in the proof of Lemma 4.1, we can show that the adaptive observer (4.133), (4.149) with the adaptation algorithms (4.150) (or (4.151), or (4.152)) provides the boundedness of $\hat{\theta}_i$ and the convergence

$$\lim_{t \to \infty} \left(g_i^{(j)}(t) - \hat{g}_{j,i}(t)\right) = 0 \quad \text{for} \quad i = 1, 2, \ldots, \rho, \quad j \in \mathbb{N}_0.$$

Under some additional conditions being similar to the second and third conditions of Propositions 4.2 and 4.3, the adaptation algorithms with extended regressors (4.152) and (4.153) provide acceleration of the parametric convergence by increasing γ.

Further, since, by Assumption A.4.4.3, $W_u(s)$ is invertable and minimum phase, while the state feedback in (4.134) has no effect on the transmission zeros [2, Sect. 8.6.4], then $W_\Sigma(s)$ is also invertable and minimum phase. Hence, we can define

$$W_\Sigma^{-1}(s) = \frac{1}{\det\{W_\Sigma(s)\}}\operatorname{adj}\{W_\Sigma(s)\} = \frac{1}{\beta(s)}\begin{bmatrix} \lambda_{1,1}(s) & \ldots & \lambda_{1,\rho}(s) \\ \vdots & \ddots & \vdots \\ \lambda_{\rho,1}(s) & \ldots & \lambda_{\rho,\rho}(s) \end{bmatrix},$$

where the Hurwitz polynomial $\beta(s)$ contains the transmission zeros of $W_\Sigma(s)$, $\lambda_{k,i}(s) = \sum_{j=0}^{n-1} \lambda_{k,i_j} s^j$ are the polynomials with known constant coefficients, $k = 1, 2, \ldots, \rho$, $i = 1, 2, \ldots, \rho$.

Then, following the approach introduced in Sect. 4.1.1.2, we define the tracking component u_t of the controller (4.134) as

$$u_t = \frac{1}{\beta(s)} \begin{bmatrix} \sum_{i=1}^{\rho} \left(\lambda_{1,i_0} g_i + \sum_{j=1}^{n-1} \lambda_{1,i_j} \hat{g}_{j,i} \right) \\ \sum_{i=1}^{\rho} \left(\lambda_{2,i_0} g_i + \sum_{j=1}^{n-1} \lambda_{2,i_j} \hat{g}_{j,i} \right) \\ \vdots \\ \sum_{i=1}^{\rho} \left(\lambda_{\rho,i_0} g_i + \sum_{j=1}^{n-1} \lambda_{\rho,i_j} \hat{g}_{j,i} \right) \end{bmatrix}, \tag{4.154}$$

where the estimates $\hat{g}_{j,i}$ are generated by the adaptive observer (4.133), (4.149) with the adaptation algorithm (4.150) (or (4.152), or (4.153)).

It can be proved that the adjustable controller (4.134), (4.154) with the adaptive observer (4.133), (4.149) and the adaptation algorithm (4.150) (or (4.152), or (4.153)) when applied to the plant (4.129) provides the boundedness of all the closed-loop signals and the asymptotic reference tracking (4.130).

Example 4.10 Let us consider the plant and the reference defined in Example 4.9. For the matrix of feedback gains K (4.147), the transfer function matrix $W_\Sigma(s)$ is given by (4.148). Hence,

$$W_\Sigma^{-1}(s) = \begin{bmatrix} s^2 + 9.146\,s - 15.146 & 20.731 \\ 2.049s - 1.051 & s - 0.146 \end{bmatrix},$$

where $\beta(s) = 1$. As a result, the tracking component takes the form

$$u_t = \begin{bmatrix} u_{t1} \\ u_{t2} \end{bmatrix} = \begin{bmatrix} \hat{g}_{2,1} + 9.146\hat{g}_{1,1} - 15.146 g_1 + 20.731 g_2 \\ -2.049\hat{g}_{1,1} - 1.051 g_1 + \hat{g}_{1,2} - 0.146 g_2 \end{bmatrix},$$

where $\hat{g}_{1,1}$ and $\hat{g}_{2,1}$ are the estimates of the first- and second-order time derivatives of the first reference element g_1, while $\hat{g}_{1,2}$ is the estimate of the first-order time derivative of the second reference element g_2. It is worth noting that since the second component g_2 is known a priori to be a constant, its time derivative is identically zero. Therefore, we set $\hat{g}_{1,2}$ to zero.

We use the observer (4.133) with G_1, G_2, l_1, l_2 given by (4.146) and calculate the estimates (4.149) as follows[6]:

$$\hat{g}_i = \hat{\theta}_i^\top \hat{\xi}_i, \quad \hat{g}_{1,i} = \hat{\theta}_i^\top \left(G_i \hat{\xi}_i + l_i \hat{g}_i \right), \quad \hat{g}_{2,i} = \hat{\theta}_i^\top \left(G_i^2 \hat{\xi}_i + G_i l_i \hat{g}_i + l_i \hat{g}_{1,i} \right), \quad i = 1, 2.$$

[6] For the convenience of notation, we introduce $\hat{g}_i = \hat{\theta}_i^\top \hat{\xi}_i$.

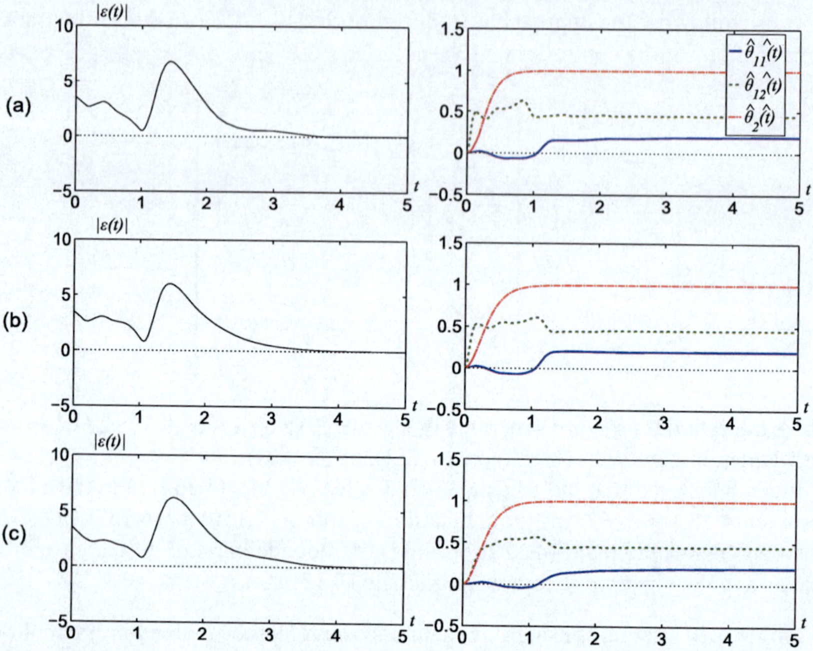

Fig. 4.21 Example 4.10: adaptive tracking of the reference $g(t) = [3\sin(4t+5),\ 2]^\top$ in the MIMO system using estimates of the HOTD of the reference signal and: **a** the gradient-based algorithm; **b** the algorithm with DRE; **c** the algorithm with MRE

The adjustable parameters $\hat{\theta}_1 \in \mathbb{R}^2$ and $\hat{\theta}_2 \in \mathbb{R}$ are generated by one of the following three algorithms of adaptation with $\gamma_1 = \gamma_2 = 1$:

- the gradient algorithm (4.150);
- the algorithm (4.152) with DRE and $p_1 = 1$, $H_{1,1}(s) = \frac{1}{0.2s+1}$, $p_2 = 0$;
- the algorithm (4.153) with MRE and $L_1(s) = \frac{1}{0.2s+1}$, $L_2(s) = 1$.

Simulation results obtained for zero initial conditions $x(0)$, $\hat{\xi}(0)$, $\hat{\theta}_{1,1}(0)$, $\hat{\theta}_{1,2}(0)$, and $\hat{\theta}_2(0)$ are presented in Fig. 4.21 and demonstrate asymptotic tracking of a priori unknown multidimensional reference signal. ■

4.2.2 *Compensation of Unmatched Disturbance*

In this section, we consider the problem of compensation of multidimensional disturbances affecting LTI MIMO plants with measured state. Omitting the case of matched disturbance considered for SISO systems, we present more general solutions for unmatched disturbance.

Consider the LTI MIMO plant

$$\dot{x} = Ax + Bu + D\upsilon, \quad y = Cx + R\upsilon, \tag{4.155}$$

where $x \in \mathbb{R}^n$ is the state accessible for measurements, $y \in \mathbb{R}^m$ is the vector of output variables ($m \leq n$), $u \in \mathbb{R}^m$ is the vector of control signals, $\upsilon \in \mathbb{R}^\rho$ is the disturbance satisfying Assumption 2.2, A, B, D, and C are the known matrices of appropriate dimensions, and $R \in \mathbb{R}^{m\times\rho}$ is unknown.

The considered control problem is to design a state-feedback controller providing the boundedness of all the closed-loop signals and ensuring the control objective

$$\lim_{t\to\infty} |y(t)| = 0. \tag{4.156}$$

We also accept the following assumptions.

Assumption 4.5 The plant (4.155) is such that:
A.4.5.1. the triplet (A, B, C) is observable and controllable;
A.4.5.2. the matrix C has full row rank, while the matrix B has full column rank;
A.4.5.3. all the transmission zeros of the transfer function matrix $W_u(s) = C(sI - A)^{-1}B$ have negative real parts.

We use the controller

$$u = -Kx + u_\upsilon, \tag{4.157}$$

where the matrix of feedback gains K is such that the matrix $A_\Sigma = A - BK$ is Hurwitz [2, Sect. 8.6], while the compensating component u_υ will be defined later.

Further, taking Lemma 2.2 into account and extending the results of Sect. 4.1.3.1 to the case of MIMO plants, we form the following regressor estimate:

$$\hat{\xi} = \eta + Nx, \tag{4.158}$$

where $\eta \in \mathbb{R}^q$ is the state of the filter

$$\dot{\eta} = G\eta + (GN - NA_\Sigma)x - NBu_\upsilon \tag{4.159}$$

with an arbitrary initial condition $\eta(0)$, the matrix $N \in \mathbb{R}^{q\times n}$ satisfying the condition

$$ND = L, \tag{4.160}$$

and the matrices G and L given by (2.25), (2.26).

Remark 4.7 In the particular case, when each element υ_i of the multidimensional disturbance υ is modelled by an individual exosystem (2.32), the matrices G and L can be represented in the block-diagonal form (2.36), and the equations (4.158)–(4.160) can be splitted into ρ independent ones:

$$\hat{\xi}_i = \eta_i + N_i x, \quad \dot{\eta}_i = G_i \eta_i + (G_i N_i - N_i A_\Sigma)x - N_i B u_\upsilon,$$

where the matrices $N_i \in \mathbb{R}^{q_i \times n}$ are such that

$$N_i D = L_{0i},$$

and the matrix

$$L_{0i} = \left[0_{q_i}, \ldots, 0_{q_i}, l_i, 0_{q_i}, \ldots, 0_{q_i}\right], \quad i = 1, 2, \ldots, \rho$$

contains the vector l_i as the ith column, while 0_{q_i} is the q_ith-dimensional zero vector. ■

Using, in view of Lemma 2.2, the same approach as in the proof of Lemma 2.4, we can demonstrate that

$$\upsilon = \Theta \hat{\xi} + \epsilon,$$

where $\Theta \in \mathbb{R}^{\rho \times q}$ is a matrix of unknown parameters, and ϵ exponentially decays.

Therefore, we can directly extend Proposition 2.2 to the case of multidimensional exosignal and represent the plant (4.155) in the form

$$\dot{e} = A_\Sigma e + B\left(u_\upsilon + \Theta_\upsilon \hat{\xi} + \epsilon_\upsilon\right), \quad y = Ce, \tag{4.161}$$

where $\Theta_\upsilon \in \mathbb{R}^{m \times q}$ is a new matrix of unknown parameters, and ϵ_υ is the exponentially decaying term. Choosing the compensating component as

$$u_\upsilon = -\hat{\Theta}_\upsilon \hat{\xi}, \tag{4.162}$$

where $\hat{\Theta}_\upsilon \in \mathbb{R}^{m \times q}$ is the matrix of adjustable parameters, we obtain the following closed-loop error model:

$$y = W_\Sigma(s)\left[\tilde{\Theta}_\upsilon \hat{\xi}\right] + \epsilon_y, \tag{4.163}$$

where $\tilde{\Theta}_\upsilon = \Theta_\upsilon - \hat{\Theta}_\upsilon$ is the matrix of parametric errors, the transfer function matrix $W_\Sigma(s) = C(sI - A_\Sigma)^{-1}B$ is asymptotically stable and minimum phase, and ϵ_y exponentially decays.

The model (4.163) is identical to (4.137) (with substitution of y instead of ε and $\tilde{\Theta}_\upsilon$ instead of $\tilde{\Theta}_g$). Therefore, we can directly employ adaptation algorithms like (4.138), (4.143), or (4.144) with the use of appropriate notations.

Finally, using the same arguments as in Sect. 4.1.3.2, we can prove that under Assumptions 2.2 and 4.5 the adaptive controller (4.157) and (4.162) with the disturbance observer (4.158)–(4.160) and the adaptation algorithm of the form (4.138) (or (4.143), or (4.144)) when applied to the plant (4.155) provides the boundedness of all the closed-loop signals and fulfilment of the control objective (4.156).

4.2.3 Regulation in LTI MIMO Plants

Now, we demonstrate how in the case of multidimensional exosignals the problems of adaptive reference tracking and adaptive disturbance rejection can be solved simultaneously by applying the above techniques [14].[7] In other words, we will solve Problem 1.4 of the adaptive output regulation from Sect. 1.3.3 formulated for known LTI MIMO plants and multidimensional exosignals.

Consider the plant (4.155) with measured state x. The control problem consists in design of a state-feedback controller providing the boundedness of all the closed-loop signals and achievement of the control objective

$$\lim_{t\to\infty} |y(t) - g(t)| = 0, \tag{4.164}$$

where $g \in \mathbb{R}^m$ is the multidimensional reference signal.

Additionally to Assumption 4.5, we make the following one.

Assumption 4.6 The exosignals υ and g are such that:

A.4.6.1. the disturbance $\upsilon \in \mathbb{R}^\rho$ and the reference $g \in \mathbb{R}^m$ can be modelled as the outputs of the exosystems

$$\dot{\xi}_g = \Gamma_g \xi_g, \quad g = Q_g \xi_g, \tag{4.165}$$

$$\dot{\xi}_\upsilon = \Gamma_\upsilon \xi_\upsilon, \quad \upsilon = H_\upsilon \xi_\upsilon, \tag{4.166}$$

where $\xi_g \in \mathbb{R}^{q_g}$, $\xi_\upsilon \in \mathbb{R}^{q_\upsilon}$, and constant matrices Γ_g, Γ_υ, Q_g, H_υ have appropriate dimensions;

A.4.6.2. both of the exosystems (4.165) and (4.166) satisfy Assumption 2.2.

As it was demonstrated in Sects. 4.2.1 and 4.2.2, Lemma 2.2 allows one to extend straightforwardly the techniques derived for SISO plants with scalar exosignals to the MIMO plants with multidimensional exosignals. Therefore, following Sect. 4.1.4, we will present below only the key design steps omitting the most of intermediate derivations.

First, we stabilize the plant with the use of the controller

$$u = -Kx + u_c, \tag{4.167}$$

where the matrix of feedback gains $K \in \mathbb{R}^{m\times n}$ is such that $A_\Sigma = A - BK$ is Hurwitz, and the compensating component u_c will be defined later.

In view of Lemma 2.2, we represent the exosystems (4.165) and (4.166) in the autonomous canonical forms

$$\dot{\bar{\xi}}_g = (G_g + L_g \bar{Q}_g)\bar{\xi}_g, \quad g = \bar{Q}_g \bar{\xi}_g, \tag{4.168}$$

$$\dot{\bar{\xi}}_\upsilon = (G_\upsilon + L_\upsilon \bar{H}_\upsilon)\bar{\xi}_\upsilon, \quad \upsilon = \bar{H}_\upsilon \bar{\xi}_\upsilon, \tag{4.169}$$

[7] An alternative approach applicable to right-invertible MIMO LTI plants is considered in [15].

where $\bar{\xi}_g$ and $\bar{\xi}_\upsilon$ are the states of the canonical forms, $\bar{Q}_g$ and $\bar{H}_\upsilon$ are unknown matrices and the pairs (G_g, L_g) and (G_υ, L_υ) are defined by (2.25), (2.26) (with appropriate change of notations).

Then, by Lemmas 1.1 and 1.2, we conclude that there exist solution pairs (M_g, Θ_g) and $(M_\upsilon, \Theta_\upsilon)$ of the matrix equations (1.66), (1.67) such that, for the state error $e = x - M\xi$, we obtain the following model:

$$\dot{e} = A_\Sigma e + B\left(u_c + \Theta_g\bar{\xi}_g + \Theta_\upsilon\bar{\xi}_\upsilon\right), \quad \varepsilon = Ce, \tag{4.170}$$

where e is inaccessible for measurements, $M = [M_g,\ M_\upsilon]$, $\xi = \left[\bar{\xi}_g^\top,\ \bar{\xi}_\upsilon^\top\right]^\top$, the output error $\varepsilon = y - g$ is measured, $\Theta_g \in \mathbb{R}^{m\times q_g}$, $\Theta_\upsilon \in \mathbb{R}^{m\times q_\upsilon}$ are matrices with unknown parameters.

To estimate inaccessible states $\bar{\xi}_g$ and $\bar{\xi}_\upsilon$, we rewrite the exosystems (4.165) and (4.166) in the canonical form with input

$$\dot{\bar{\xi}}_g = G_g\bar{\xi}_g + L_g g, \quad g = \bar{Q}_g\bar{\xi}_g,$$
$$\dot{\bar{\xi}}_\upsilon = G_\upsilon\bar{\xi}_\upsilon + L_\upsilon\upsilon, \quad \upsilon = \bar{H}_\upsilon\bar{\xi}_\upsilon.$$

The latter allows us to directly extend the results of Lemmas 2.3 and 2.5 to the case of multidimensional exosignals and to use the reference observer

$$\dot{\hat{\xi}}_g = G_g\hat{\xi}_g + L_g g \tag{4.171}$$

and the disturbance observer

$$\hat{\xi}_\upsilon = \eta + Nx, \quad \dot{\eta} = G_\upsilon\eta + (G_\upsilon N - NA_\Sigma)x - NBu_c \tag{4.172}$$

with arbitrary initial conditions $\hat{\xi}_g(0)$ and $\eta(0)$. The matrix $N \in \mathbb{R}^{q_\upsilon\times n}$ is selected so that

$$ND = L_\upsilon. \tag{4.173}$$

Then the error model takes the form

$$\dot{e} = A_\Sigma e + B\left(u_c + \Theta\hat{\xi} + \epsilon\right), \quad \varepsilon = Ce, \tag{4.174}$$

where $\hat{\xi} = [\hat{\xi}_g^\top,\ \hat{\xi}_\upsilon^\top]^\top$ is the aggregated regressor, $\Theta = \left[\Theta_g, \Theta_\upsilon\right]$ is the aggregated matrix of unknown parameters, and ϵ is the exponentially decaying term.

We choose the compensating component in the form

$$u_c = -\hat{\Theta}\hat{\xi}, \tag{4.175}$$

where $\hat{\Theta} \in \mathbb{R}^{m\times(q_g+q_v)}$ is the matrix of adjustable parameters.Then, substituting (4.175) into (4.174), we show that the closed-loop plant model in the coordinates e and ε takes the form

$$\dot{e} = A_\Sigma e + B\left(\tilde{\Theta}\hat{\xi} + \epsilon\right), \quad \varepsilon = Ce,$$

where $\tilde{\Theta} = \Theta - \hat{\Theta}$ is the matrix of parametric errors. To design adaptation algorithms used for calculation of $\hat{\Theta}$, we involve the MIMO swapping scheme (see Lemma D.4 from Appendix D). We define the augmented output error

$$\bar{\varepsilon} = \varepsilon + W_\Sigma(s)\left[\hat{\Theta}\hat{\xi}\right] - \Xi^\top\hat{\vartheta}, \tag{4.176}$$

where

$$\Xi^\top = \begin{bmatrix} W_{\Sigma 1,1}(s)\left[\hat{\xi}^\top\right] & \dots & W_{\Sigma 1,m}(s)\left[\hat{\xi}^\top\right] \\ \vdots & \ddots & \vdots \\ W_{\Sigma m,1}(s)\left[\hat{\xi}^\top\right] & \dots & W_{\Sigma m,m}(s)\left[\hat{\xi}^\top\right] \end{bmatrix}$$

is the $m \times (q_g + q_v)m$ matrix regressor, $W_\Sigma(s) = C(sI - A_\Sigma)^{-1}B$ is the asymptotically stable transfer function matrix, $\hat{\vartheta} =\leq [\hat{\Theta}_1, \hat{\Theta}_2, \dots, \hat{\Theta}_m]^\top$ is the $(q_g + q_v)m$-dimensional vector of adjustable parameters corresponding to the vector of unknown parameters $\vartheta = \left[\Theta_1, \Theta_2, \dots, \Theta_\rho\right]^\top$, $\hat{\Theta}_j$, Θ_j are the jth rows of the matrices $\hat{\Theta}$ and Θ, respectively.

Applying the results of Lemma D.4, we get the static augmented error model

$$\bar{\varepsilon} = \Xi^\top\tilde{\vartheta} + \epsilon_\varepsilon, \tag{4.177}$$

where ϵ_ε is the exponentially decaying term. The model is used for design of the following adaptation algorithms:

The gradient adaptation algorithm

$$\dot{\hat{\vartheta}} = \gamma\Xi\bar{\varepsilon}, \quad \gamma > 0; \tag{4.178}$$

The adaptation algorithm with DRE

$$\dot{\hat{\vartheta}} = \gamma\bar{\Xi}\varepsilon_H, \quad \gamma > 0, \tag{4.179}$$

where

$$\varepsilon_H = \begin{bmatrix} \hat{\varepsilon} - \Xi^\top \hat{\vartheta} \\ H_1(s)\,[\hat{\varepsilon}] - \bar{\Xi}_1^\top \hat{\vartheta} \\ \vdots \\ H_p(s)\,[\hat{\varepsilon}] - \bar{\Xi}_p^\top \hat{\vartheta} \end{bmatrix}, \quad \bar{\Xi}^\top = \begin{bmatrix} \Xi^\top \\ \bar{\Xi}_1^\top \\ \vdots \\ \bar{\Xi}_p^\top \end{bmatrix},$$

$$\hat{\varepsilon} = \varepsilon + W_\Sigma(s)\left[\hat{\Theta}\hat{\xi}\right], \qquad \bar{\Xi}_j = H_j(s)\,[\Xi], \; j = 1, 2, \ldots, p, \tag{4.180}$$

$\bar{H}(s) = \left[1, H_1(s), \ldots, H_p(s)\right]^\top$, $H_i(s)$ are distinct designer-chosen PASMP transfer functions, $p \geq (q_g + q_\upsilon - 1)$;

The adaptation algorithm with MRE

$$\dot{\hat{\vartheta}} = \gamma\left(L(s)\left[\Xi\hat{\varepsilon}\right] - \Omega\hat{\vartheta}\right), \quad \gamma > 0, \tag{4.181}$$

where $L(s)$ is a designer-chosen transfer function of the form (3.44), $\hat{\varepsilon}$ is defined by (4.180), and $\Omega = L(s)\left[\Xi\Xi^\top\right]$.

Now, using the same arguments as in Sect. 4.2.2, we can prove that under Assumptions 4.5 and 4.6 the adaptive controller (4.167) with the compensating component (4.175), the reference observer (4.171), the disturbance observer (4.172), (4.173), the swapping scheme (4.176), the gradient adaptation algorithm (4.178) (or algorithm with DRE (4.179), or algorithm with MRE (4.181)) when applied to the plant (4.155) provides the boundedness of all the closed-loop signals and fulfilment of the control objective (4.164).

Example 4.11 Consider the two-input and two-output unstable plant (4.155) with the matrices

$$A = \begin{bmatrix} 0 & 1 & 0 \\ 0.1 & -0.9 & 0 \\ 0 & 0 & 1 \end{bmatrix}, \quad B = \begin{bmatrix} 0 & 0 \\ 1 & 0 \\ 0 & 1 \end{bmatrix}, \quad C = \begin{bmatrix} 1 & 0 & 0 \\ 1 & 0 & 1 \end{bmatrix}, \quad D = \begin{bmatrix} 1 & 1 \\ 0 & 1 \\ 1 & 0 \end{bmatrix}, \quad R = \begin{bmatrix} 1 & 2 \\ 2 & 3 \end{bmatrix}.$$

To stabilize the plant, we calculate the matrix

$$K = \begin{bmatrix} 8.629 & 9.298 & 28.099 \\ -1.218 & -1.057 & -0.198 \end{bmatrix}$$

so that the matrix $A_\Sigma = A - BK$ has the eigenvalues $\lambda\{A_\Sigma\} = \{-2, -3, -4\}$, and, hence, the transfer function matrix $W_\Sigma(s)$ is given by

$$W_\Sigma(s) = \frac{1}{s^3 + 9s^2 + 26s + 24}\begin{bmatrix} s - 1.198 & -28.099 \\ 2.057s + 0.02 & s^2 + 10.198s - 19.57 \end{bmatrix}.$$

Let it be known a priori that the reference $g \in \mathbb{R}^2$ and the disturbance $\upsilon \in \mathbb{R}^2$ are modelled by the third-order exosystems (4.165) and (4.166), respectively. Hence, we select

$$G_g = G_\upsilon = \begin{bmatrix} 0 & 1 & 0 \\ 0 & 0 & 1 \\ -6 & -11 & -6 \end{bmatrix}, \quad L_g = L_\upsilon = \begin{bmatrix} 0 & 0 \\ 0 & 0 \\ 6 & 6 \end{bmatrix}.$$

The vector of adjustable parameters $\hat{\vartheta}$ is generated by one of the following three algorithms of adaptation:

- the gradient algorithm (4.178) with $\gamma = 5$;
- the algorithm (4.179) with DRE and $p = 5$, $H_1(s) = \frac{1}{4s+1}$, $H_2(s) = \frac{1}{2s+1}$, $H_3(s) = \frac{1}{s+1}$, $H_4(s) = \frac{2}{s+2}$, $H_5(s) = \frac{3}{s+3}$, $\gamma = 500$;
- the algorithm (4.181) with MRE with $L(s) = \frac{1}{2s+1}$, $\gamma = 500$.

It worth noting that the adaptation gain of the gradient-based algorithm is selected smaller than the gains of the algorithms with DRE and MRE since for $\gamma = 500$ the gradient-based algorithm provides inappropriately slow convergence.

All simulations are carried out with zero initial conditions for x, $\hat{\xi}_g$, $\hat{\xi}_f$, and $\hat{\vartheta}$. In simulation, the reference g and the disturbance υ are given by the functions

$$g(t) = \begin{bmatrix} g_1(t) \\ g_2(t) \end{bmatrix} = \begin{bmatrix} \beta_{g1,1} + \beta_{g1,2} \sin(w_g t + \varphi_g) \\ \beta_{g2,1} \end{bmatrix},$$

$$\upsilon(t) = \begin{bmatrix} \upsilon_1(t) \\ \upsilon_2(t) \end{bmatrix} = \begin{bmatrix} \beta_{\upsilon 1,1} \\ \beta_{\upsilon 2,1} + \beta_{\upsilon 2,2} \sin(w_\upsilon t + \varphi_\upsilon) \end{bmatrix},$$

respectively. The parameters of these functions $\beta_{g1,1} = 6, \beta_{g1,2} = 5, w_g = 1, \varphi_g = 4$, $\beta_{g2,1} = 3$, $\beta_{\upsilon 1,1} = 4$, $\beta_{\upsilon 2,1} = -4$, $\beta_{\upsilon 2,2} = 5$, $w_\upsilon = 2$, and $\varphi_\upsilon = 1$ are unknown.

Simulation results for the closed-loop systems employing the gradient algorithm of adaptation (4.178), the adaptation algorithm with DRE (4.179), and the algorithm of adaptation with MRE (4.181) are presented in Figs. 4.22, 4.23, and 4.24, respectively. We draw the readers' attention to different time scales in the plots of $\hat{\vartheta}(t)$ used to demonstrate the convergence of the gradient-based scheme.

The results demonstrate asymptotic tracking of a priori uncertain multidimensional reference g despite the influence of uncertain multidimensional disturbance. In contrast to the gradient adaptation algorithm, the algorithms with DRE and MRE demonstrate dramatically faster parametric convergence. ■

4.3 Disturbance Compensation in Nonlinear Plants

In this section, we demonstrate how the above control techniques of adaptive regulation can be extended to certain classes of nonlinear plants. We start again with consideration of a general form of nonlinear plants and then apply these techniques to the globally defined canonical normal form, strict-feedback canonical form and output-feedback canonical form.

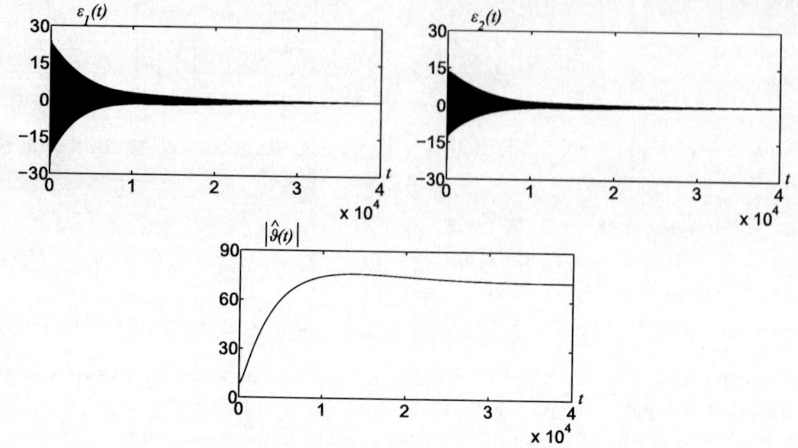

Fig. 4.22 Example 4.11: adaptive output regulation in the LTI MIMO system closed by the gradient adaptation algorithm

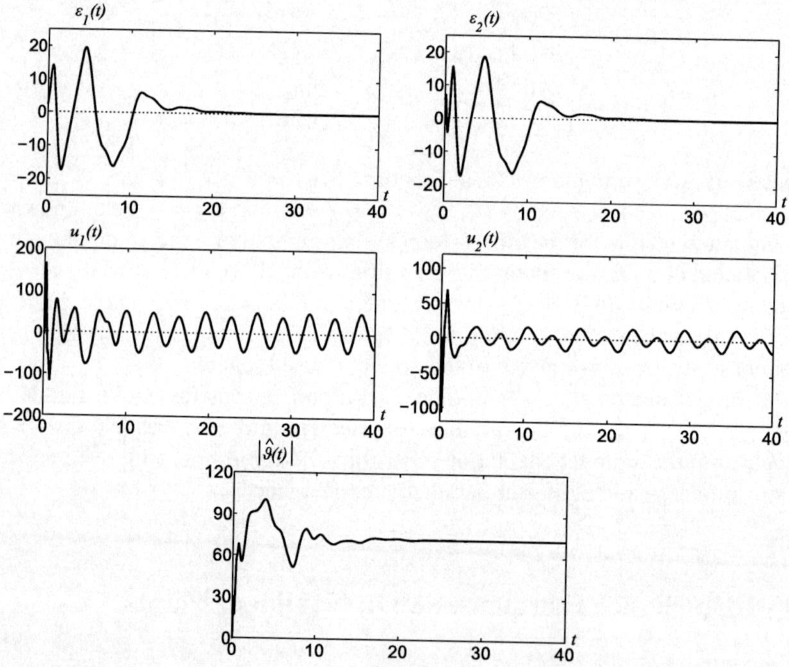

Fig. 4.23 Example 4.11: adaptive output regulation in the LTI MIMO system closed by the adaptation algorithm with DRE

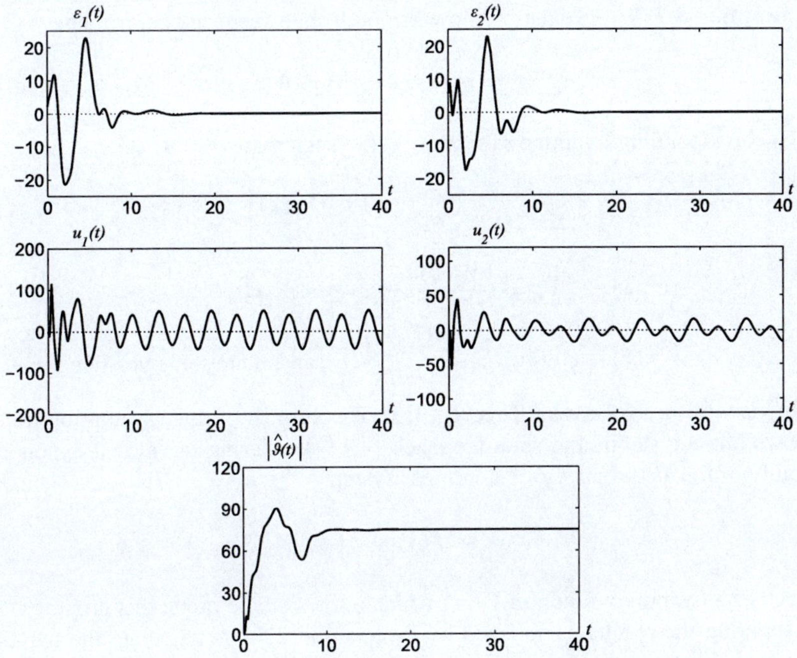

Fig. 4.24 Example 4.11: adaptive output regulation in the LTI MIMO system closed by the adaptation algorithm with MRE

4.3.1 Plant in the General Form

Consider the nonlinear plant

$$\dot{x} = f(x) + \phi(x)(u + \upsilon), \tag{4.182}$$

where $x \in \mathbb{R}^n$ is the measured state with the initial condition $x(0)$, $f(x)$ and $\phi(x)$ are the smooth vector functions, υ is the external disturbance satisfying Assumption 2.1. We also assume that $f(x)$ and $\phi(x)$ are such that for any $x(0)$ and admissible u and υ there exists a unique solution $x(t, x(0))$ defined for all $t \geq 0$.

The control problem is to design a state-feedback controller providing the boundedness of all the closed-loop signals and achievement of the control objective

$$\lim_{t\to\infty} |x(t)| = 0. \tag{4.183}$$

In order to focus the main attention on external disturbance rejection, we make the following assumption.

Assumption 4.7 There exists a known smooth state-feedback control law

$$u = u_s(x), \quad u_s(0) = 0 \tag{4.184}$$

and a corresponding Lyapunov function $V_x(x)$ such that

$$c_1|x|^2 \le V_x(x) \le c_2|x|^2, \quad \left|\frac{\partial V_x(x)}{\partial x}\right| \le c_3|x|,$$
$$\dot{V}_x = \frac{\partial V_x(x)}{\partial x} f_\Sigma(x) \le -c_4|x|^2,$$

where $f_\Sigma(x) = f(x) + \phi(x)u_s(x)$, and c_1, c_2, c_3, and c_4 are some positive constants.

In view of the Krasovskii Theorem (see Theorem B.3), this assumption implies that we know a stabilizing state-feedback (4.184) ensuring the global exponential stability of the disturbance free nonlinear system

$$\dot{x} = f(x) + \phi(x)u$$

as well as a Lyapunov function $V_x(x)$ which allows one to prove this property.

Applying the results established by Proposition 2.5, we calculate the regressor estimate via the equality

$$\hat{\xi} = \eta + \pi(x), \tag{4.185}$$

where the auxiliary vector $\eta \in \mathbb{R}^q$ is generated by the nonlinear filter

$$\dot{\eta} = G\eta + G\pi(x) - \frac{\partial \pi(x)}{\partial x}\left(f(x) + \phi(x)u\right), \tag{4.186}$$

the vector function $\pi(x) \in \mathbb{R}^q$ satisfies the PDE

$$\frac{\partial \pi(x)}{\partial x}\phi(x) = l, \tag{4.187}$$

while the pair (G, l) is from the exosystem canonical form (2.15). Then the plant subjected to the external disturbance can be represented in the following form (see (2.89) with $d(x) = \phi(x)$):

$$\dot{x} = f(x) + \phi(x)\left(u + \Theta^\top \hat{\xi} + \epsilon\right), \tag{4.188}$$

where $\Theta \in \mathbb{R}^q$ is a vector of unknown parameters, $\epsilon = \Theta^\top \epsilon_\xi$ is the exponentially decaying term, while ϵ_x satisfies the equation $\dot{\epsilon}_\xi = G\epsilon_\xi$.

It is worth noting that in the case of nonlinear systems we cannot neglect the decaying term ϵ (see Example 1.6). Therefore, a control law must involve a special term called *nonlinear damping term* that will counteract harmful influence of the additive disturbance $\phi(x)\epsilon$.

We consider the *Lyapunov-based* adjustable controller (see Sect. 3.3.2)

$$u = u_s(x) - \hat{\theta}^\top \hat{\xi} - \mu \frac{\partial V_x(x)}{\partial x} \phi(x), \tag{4.189}$$

where $\mu > 0$ is a constant gain, $\hat{\theta}$ is the vector of the adjustable parameters generated by the algorithm of adaptation

$$\dot{\hat{\theta}} = \gamma \hat{\xi} \frac{\partial V_x(x)}{\partial x} \phi(x) \tag{4.190}$$

with the adaptation gain $\gamma > 0$. The controller (4.189) is called Lyapunov-based because all its components (i.e., the stabilizing component $u_s(x)$, the damping term $\frac{\partial V_x(x)}{\partial x}\phi(x)$ and the adaptation algorithm (4.190)) are designed with the use of appropriate Lyapunov function [6, Sect. 1.4.4].

After replacement of (4.189) into (4.188), the closed-loop error model takes the form

$$\dot{x} = f_\Sigma(x) + \phi(x)\left(\tilde{\theta}^\top \hat{\xi} + \theta^\top \epsilon_\xi - \mu \frac{\partial V_x(x)}{\partial x} \phi(x)\right), \tag{4.191}$$

$$\dot{\tilde{\theta}} = -\gamma \hat{\xi} \frac{\partial V_x(x)}{\partial x} \phi(x), \tag{4.192}$$

$$\dot{\epsilon}_\xi = G \epsilon_\xi, \tag{4.193}$$

where $\tilde{\theta} = \theta - \hat{\theta}$ is the vector of parametric errors.

To prove the stability properties of the closed-loop error model (4.191)–(4.193), we use the Lyapunov function

$$V(x, \tilde{\theta}, \epsilon_\xi) = V_x(x) + \frac{1}{2\gamma} \tilde{\theta}^\top \tilde{\theta} + \frac{1 + |\theta|^2}{4\mu} \epsilon_\xi^\top P \epsilon_\xi$$

with the positive definite matrix P satisfying the matrix equation $G^\top P + PG = -I$. Then, calculating the time derivative of $V(x, \tilde{\theta}, \epsilon_\xi)$ along the solutions of (4.191)–(4.193), in view of Assumption 4.7, we have

$$\begin{aligned}
\dot{V}(x, \tilde{\theta}, \epsilon_\xi) &\le -c_4|x|^2 + \frac{\partial V_x(x)}{\partial x} \phi(x) \tilde{\theta}^\top \hat{\xi} + \frac{\partial V_x(x)}{\partial x} \phi(x) \theta^\top \epsilon_\xi - \mu \left| \frac{\partial V_x(x)}{\partial x} \phi(x) \right|^2 \\
&- \tilde{\theta}^\top \hat{\xi} \frac{\partial V_x(x)}{\partial x} \phi(x) - \frac{1 + |\theta|^2}{4\mu} |\epsilon_\xi|^2 \le -c_4|x|^2 \\
&+ \left| \frac{\partial V_x(x)}{\partial x} \phi(x) \right| |\theta| |\epsilon_\xi| - \mu \left| \frac{\partial V_x(x)}{\partial x} \phi(x) \right|^2 - \frac{1 + |\theta|^2}{4\mu} |\epsilon_\xi|^2.
\end{aligned}$$

Completing the squares, we obtain

$$\dot{V}(x,\tilde{\theta},\epsilon_\xi) \le -c_4|x|^2 - \frac{1}{4\mu}|\epsilon_\xi|^2 - \left(\sqrt{\mu}\left|\frac{\partial V_x(x)}{\partial x}\phi(x)\right| - \frac{|\theta|}{2\sqrt{\mu}}|\epsilon_\xi|\right)^2$$
$$\le -c_4|x|^2 - \frac{1}{4\mu}|\epsilon_\xi|^2.$$

The latter inequality means that the origin of the closed-loop error model (4.191)–(4.193) is globally uniformly stable, and $|x(t)| \to 0$ as $t \to 0$ (see Theorems B.1, B.4). Then, with the use of standard arguments, the boundedness of all the closed-loop signals can be shown. Therefore, we have proved the following statement.

Proposition 4.9 *If Assumptions 2.1 and 4.7 hold, then the adaptive controller (4.189) with the disturbance observer (4.185)–(4.187) and the adaptation algorithm (4.190) when applied to the plant (4.182) provides the boundedness of all the closed-loop signals and achievement of the control objective (4.183).*

4.3.2 Plant in the Normal Canonical Form

Now, we assume that Assumption 2.5 holds, and the nonlinear plant (4.182) is strongly minimum phase and possesses the globally defined normal form

$$\dot{z} = f_0(z,\chi), \tag{4.194}$$
$$\dot{\chi} = E\chi + \varsigma_\rho\left(\phi_\chi(z,\chi) + \phi_u(z,\chi)(u+\upsilon)\right), \quad y = \chi_1, \tag{4.195}$$

where $z \in \mathbb{R}^{n-\rho}$ and $\chi \in \mathbb{R}^\rho$ are defined by the nonlinear change of coordinates

$$\begin{bmatrix} z \\ \chi \end{bmatrix} = \Pi(x), \tag{4.196}$$

$\rho < n$ is the uniform relative degree, $f_0(z,\chi)$, $\phi_\chi(z,\chi)$, $\phi_u(z,\chi)$ are the smooth functions, and[8]

$$E = \begin{bmatrix} 0 & \\ \vdots & I_{\rho-1} \\ 0 \cdots 0 & \end{bmatrix}, \quad \varsigma_\rho = \begin{bmatrix} 0 \\ \vdots \\ 0 \\ 1 \end{bmatrix}.$$

The control problem is to design a state-feedback controller that guarantees the boundedness of the closed-loop signals and drives the plant output to zero, i.e., $\lim_{t\to\infty} y(t) = 0$.

Using the results established by Proposition 2.6, we introduce the regressor estimate

[8] For more details, see Sect. 2.4.2.2.

$$\hat{\xi} = \eta + \pi(z, \chi), \tag{4.197}$$

where the vector $\eta \in \mathbb{R}^q$ is generated by the filter

$$\begin{aligned}\dot{\eta} = G\eta + G\pi(z, \chi) - \frac{\partial \pi(z, \chi)}{\partial \chi}\left(E\chi + \varsigma_\rho \phi_\chi(z, \chi)\right) \\ - lu - \frac{\partial \pi(z, \chi)}{\partial z} f_0(z, \chi),\end{aligned} \tag{4.198}$$

the vector function $\pi(z, \chi)$ satisfies the PDE

$$\frac{\partial \pi(z, \chi)}{\partial \chi} \varsigma_\rho \phi_u(z, \chi) = l, \tag{4.199}$$

and the pair (G, l) is from the exosystem canonical form (2.15). Then the equation (4.195) takes the form

$$\dot{\chi} = E\chi + \varsigma_\rho \left(\phi_\chi(z, \chi) + \phi_u(z, \chi)\left(u + \theta^\top \hat{\xi} + \epsilon\right)\right), \tag{4.200}$$

where $\theta \in \mathbb{R}^q$ is a vector of unknown parameters, $\epsilon = \theta^\top \epsilon_\xi$ exponentially decays, while ϵ_ξ obeys the equation $\dot{\epsilon}_\xi = G\epsilon_\xi$.

Analysis of (4.200) leads to the following structure of adjustable controller employing additional nonlinear damping term that counteracts the influence of the exponentially decaying estimation error ϵ (see Proposition 3.8):

$$u = -\frac{1}{\phi_u(z, \chi)}\left(k^\top \chi + \phi_\chi(z, \chi)\right) - \hat{\theta}^\top \hat{\xi} - \mu \chi^\top P \varsigma_\rho \phi_u(z, \chi), \tag{4.201}$$

where $\hat{\theta}$ is the vector of adjustable parameters, k is a vector of feedback gains such that the matrix $A_\Sigma = E - \varsigma_\rho k^\top$ is Hurwitz, the positive definite matrix P is the solution of the matrix equation $A_\Sigma^\top P + PA_\Sigma = -I$, and $\mu > 0$.

To adjust the vector $\hat{\theta}$, we use the standard algorithm of adaptation (see Propositions 3.5 and 3.8)

$$\dot{\hat{\theta}} = \gamma \phi_u(z, \chi) \hat{\xi} \varsigma_\rho^\top P \chi, \quad \gamma > 0. \tag{4.202}$$

Substituting (4.201) into (4.200), we derive the following closed-loop error model:

$$\dot{z} = f_0(z, \chi), \tag{4.203}$$

$$\dot{\chi} = A_\Sigma \chi + \varsigma_\rho \phi_u(z, \chi)\left(\tilde{\theta}^\top \hat{\xi} + \theta^\top \epsilon_\xi - \mu \chi^\top P \varsigma_\rho \phi_u(z, \chi)\right), \tag{4.204}$$

$$\dot{\tilde{\theta}} = -\gamma \phi_u(z, \chi) \hat{\xi} \varsigma_\rho^\top P \chi, \tag{4.205}$$

$$\dot{\epsilon}_\xi = G\epsilon_\xi, \tag{4.206}$$

where $\tilde{\theta} = \theta - \hat{\theta}$ is the vector of parametric errors.

To prove stability properties of (4.203)–(4.206), we use the Lyapunov function

$$V(\chi, \tilde{\theta}, \epsilon_\xi) = \chi^\top P \chi + \frac{1}{2\gamma}\tilde{\theta}^\top \tilde{\theta} + \frac{1 + |\theta|^2}{4}\epsilon_\xi^\top P_\epsilon \epsilon_\xi,$$

where the positive definite matrix P_ϵ is the solution of the matrix equation $G^\top P_\epsilon + P_\epsilon G = -I$. Then, with the use of the same approach as in Sect. 4.3.1, we can show that the time derivative of the Lyapunov function $V(\chi, \tilde{\theta}, \epsilon_\xi)$ along the solutions of the error model (4.203)–(4.206) satisfies the inequality

$$\dot{V}(\chi, \tilde{\theta}, \epsilon_\xi) \le -|\chi|^2 - \frac{1}{4\mu}|\epsilon_\xi|^2.$$

The latter means that the equilibrium $(\chi, \tilde{\theta}, \epsilon_\xi) = (0, 0, 0)$ is globally uniformly stable and $|\chi(t)| \to 0$ as $t \to 0$ (see Theorems B.1, B.4). Since, by Assumption 2.5, the subsystem (4.203) is ISS, then $|z(t)| \to 0$, and, as a result, $|x(t)| \to 0$ as $t \to \infty$. With the use of additional arguments, we can prove the boundedness of all the closed-loop signals. Therefore, we have proved the following properties of the closed-loop adaptive system [12].

Proposition 4.10 *Let Assumptions 2.1 and 2.5 hold, and a nonlinear change of coordinates (4.196) and a function $\pi(z, \chi)$ satisfying the equation (4.199) exist. Then the adaptive controller (4.201) with the disturbance observer (4.197)–(4.199) and the adaptation algorithm (4.202) when applied to the plant (4.182) provides the boundedness of all the closed-loop signals and drives the plant output to zero, i.e.,* $\lim_{t\to\infty} y(t) = 0$.

Example 4.12 Consider the plant [13, Example 4.3]

$$\dot{x}_1 = x_1^2 + x_2, \tag{4.207}$$

$$\dot{x}_2 = \sin(x_2) + \left(2 + x_1^2 + \cos(x_3)\right)(u + \upsilon), \tag{4.208}$$

$$\dot{x}_3 = -x_3^3 + (x_1^2 + x_2)x_3, \tag{4.209}$$

$$y = x_1 \tag{4.210}$$

subjected to unmeasured disturbance $\upsilon(t) = \beta \sin(wt + \phi)$ with a priori unknown parameters $\beta = 2$, $w = 1$, and $\phi = 3$.

It can be shown that this plant satisfies Assumption 2.5 and in the new coordinates

$$z = x_3, \quad \chi_1 = x_1, \quad \chi_2 = x_1^2 + x_2$$

it takes the form (4.194), (4.195) with

$$f_0(z, \chi) = -z^3 + z\chi_2, \ \phi_\chi(z, \chi) = 2\chi_1\chi_2 + \sin(\chi_2 - \chi_1^2), \ \phi_u(z, \chi) = 2 + \chi_1^2 + \cos(z).$$

Since we know a priori that the disturbance υ is a sinusoid, we construct the second-order disturbance observer and choose

$$G = \begin{bmatrix} 0 & 1 \\ -12 & -7 \end{bmatrix}, \quad l = \begin{bmatrix} 0 \\ 12 \end{bmatrix}.$$

For the selected l, the PDE (4.199) takes the form

$$(2 + \chi_1^2 + \cos(z)) \frac{\partial \pi(z, \chi)}{\partial \chi_2} = 12,$$

and the function $\pi(z, \chi)$ can be chosen as

$$\pi(z, \chi) = \begin{bmatrix} \pi_1(z, \chi) \\ \pi_2(z, \chi) \end{bmatrix} = \begin{bmatrix} 0 \\ \frac{12\chi_2}{2+\chi_1^2+\cos(z)} \end{bmatrix}.$$

In this case, the nonlinear observer (4.197), (4.198) is given by

$$\begin{aligned}
&\hat{\xi}_1 = \eta_1, \quad \hat{\xi}_2 = \eta_2 + \pi_2, \\
&\dot{\eta}_1 = \eta_2 + \pi_2, \\
&\dot{\eta}_2 = -12\eta_1 - 7\eta_2 - 7\pi_2 + \frac{12}{2 + \chi_1^2 + \cos(z)} \left(\frac{2\chi_1\chi_2^2 - \chi_2 f_0 \sin(z)}{2 + \chi_1^2 + \cos(z)} - \phi_\chi \right) - 12u.
\end{aligned}$$

Taking $k = [2, 3]^\top$, we calculate the matrices A_Σ and P as

$$A_\Sigma = \begin{bmatrix} 0 & 1 \\ -2 & -3 \end{bmatrix}, \quad P = \begin{bmatrix} 1.25 & 0.25 \\ 0.25 & 0.25 \end{bmatrix}.$$

Then the adaptive controller takes the form

$$\begin{aligned}
u &= -\frac{1}{2 + \chi_1^2 + \cos(z)} \left(2\chi_1 + 3\chi_2 + 2\chi_1\chi_2 + \sin(\chi_2 - \chi_1^2)\right) \qquad (4.211)\\
&\quad - \hat{\theta}_1\hat{\xi}_1 - \hat{\theta}_2\hat{\xi}_2 - 0.25\mu \left(2 + \chi_1^2 + \cos(z)\right) (\chi_1 + \chi_2), \\
\dot{\hat{\theta}}_1 &= 0.25\gamma \left(2 + \chi_1^2 + \cos(z)\right) (\chi_1 + \chi_2) \hat{\xi}_1, \\
\dot{\hat{\theta}}_2 &= 0.25\gamma \left(2 + \chi_1^2 + \cos(z)\right) (\chi_1 + \chi_2) \hat{\xi}_2,
\end{aligned}$$

where $\hat{\xi}_1$, $\hat{\xi}_2$ are the elements of the vector $\hat{\xi}$.

Simulation results demonstrating adaptive compensation of a priori unknown disturbance υ are presented in Fig. 4.25. In simulation, we set the initial conditions for x, η, $\hat{\theta}$ to zero and select $\gamma = 1$, $\mu = 1$. ■

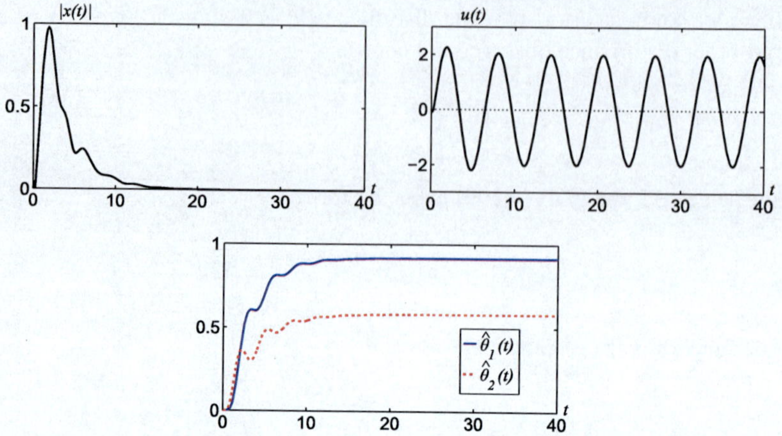

Fig. 4.25 Example 4.12: adaptive state-feedback compensation of the matched disturbance in the nonlinear system transformable into the normal canonical form

4.3.3 *Plant with Unmatched Disturbance and Measured State*

Now, we consider the problem of adaptive control for the class of nonlinear plants representable in (or transformable into) the *strict-feedback canonical form* with external disturbance

$$\dot{x}_i = x_{i+1} + \phi_i(x_1, x_2, \ldots, x_i) + \bar{d}_{n-i}^\top \upsilon, \quad i = 1, 2, \ldots, n-1, \tag{4.212}$$

$$\dot{x}_n = \phi_n(x) + \phi_u(x)u + \bar{d}_0^\top \upsilon, \tag{4.213}$$

$$y = x_1, \tag{4.214}$$

where $\phi_i(x_1, x_2, \ldots, x_i)$ and $\phi_u(x)$ are the known smooth scalar functions, $\bar{d}_j \in \mathbb{R}^\rho$ ($j = 0, 1, \ldots, n-1$) are the known vectors. It is assumed that the function $\phi_u(z, \chi)$ is bounded away from zero by some constant ϕ_0, i.e., $\phi_u(z, \chi) \geq \phi_0 > 0$. It is also assumed that the multidimensional disturbance $\upsilon \in \mathbb{R}^\rho$ satisfies Assumption 2.2.

The control problem is to design such a state-feedback controller that ensures the boundedness of all the closed-loop signals and the convergence $y(t) \to 0$ as $t \to 0$.

In contrast to the general form (4.182) and to the normal form (4.194), (4.195), in (4.212), (4.213) the disturbance is not matched with the control. Hence, it is suggested to use a backstepping control procedure for complete disturbance compensation. Here we apply the *modular backstepping with high-order tuners* described in Appendix E.

Together with the representation (4.212)–(4.214), we also use the matrix form

$$\dot{x} = Ex + \phi(x) + \varsigma_n \phi_u(x)u + D\upsilon, \tag{4.215}$$

$$y = \bar{c}^\top x, \tag{4.216}$$

where

$$E = \begin{bmatrix} 0 & \\ \vdots & I_{n-1} \\ 0 & \cdots 0 \end{bmatrix}, \ \phi(x) = \begin{bmatrix} \phi_1(x_1) \\ \phi_2(x_1, x_2) \\ \vdots \\ \phi_n(x_1, x_2, \ldots, x_n) \end{bmatrix}, \ \varsigma_n = \begin{bmatrix} 0 \\ 0 \\ \vdots \\ 1 \end{bmatrix},$$

$$D = \begin{bmatrix} \bar{d}_{n-1}^\top \\ \vdots \\ \bar{d}_1^\top \\ \bar{d}_0^\top \end{bmatrix}, \ \bar{c} = \begin{bmatrix} 1 \\ 0 \\ \vdots \\ 0 \end{bmatrix}.$$

Based on the results of Proposition 2.7, we represent the plant in the parameterized form

$$\dot{x}_i = x_{i+1} + \phi_i(x_1, x_2, \ldots, x_i) + \theta^\top \bar{D}_{n-i}\hat{\xi}_\kappa + \bar{d}_{n-i}^\top \epsilon_\kappa, \tag{4.217}$$

$$\dot{x}_n = \phi_n(x) + \phi_u(x)u + \theta^\top \bar{D}_0\hat{\xi}_\kappa + \bar{d}_0^\top \epsilon_\kappa, \tag{4.218}$$

$$y = x_1, \tag{4.219}$$

where $i = 1, 2, \ldots, n-1$, $\theta \in \mathbb{R}^{q\rho}$ is a vector of unknown parameters,

$$\bar{D}_i = \begin{bmatrix} d_{i,1}I_q \\ d_{i,2}I_q \\ \vdots \\ d_{i,\rho}I_q \end{bmatrix}$$

is the $q\rho \times q$ matrix containing the elements of the vectors $\bar{d}_i$ denoted by $d_{i,j}$ ($i = 0, 1, \ldots, n-1, j = 1, 2, \ldots, \rho$), ϵ_κ is the exponentially decaying term. The filtered regressor $\hat{\xi}_\kappa$ is defined as

$$\hat{\xi}_\kappa = \frac{\kappa_0}{\kappa(s)}\left[\hat{\xi}\right], \tag{4.220}$$

where $\kappa(s)$ is a designer-chosen Hurwitz polynomial of degree no less than $n-1$, $\kappa_0 > 0$ is a constant, and the regressor estimate $\hat{\xi}$ is generated by the disturbance observer

$$\hat{\xi} = \eta + Nx, \quad \dot{\eta} = G\eta + GNx - N\left(Ex + \phi(x)\right) - N\varsigma_n\phi_u(x)u \tag{4.221}$$

with a matrix $N \in \mathbb{R}^{q\times n}$ satisfying the equality

$$ND = L \tag{4.222}$$

and the pair (G, L) defined by (2.25), (2.26).

Here we recall that the filtering (4.220) is required for simple implementation of the HOTD $\hat{\xi}_\kappa^{(i)}$ $(i = 1, 2, \ldots, n-1)$ appearing in the steps of the backstepping procedure.

Now, applying the modular backstepping introduced in Appendix E to the model (4.217), (4.218), we derive the virtual control laws

$$\alpha_1(x_1, \hat{\theta}, \hat{\xi}_\kappa) = -(c_1 + s_1) z_1 - \phi_1 - \hat{\xi}_\kappa^\top \bar{D}_{n-1}^\top \hat{\theta}, \tag{4.223}$$

$$\begin{aligned}\alpha_i(\bar{x}_i, \bar{\theta}_{i-1}, \bar{\xi}_{i-1}) = &-z_{i-1} - (c_i + s_i) z_i - \phi_i \\ &- \hat{\xi}_\kappa^\top \left(\bar{D}_{n-i}^\top - \sum_{j=1}^{i-1} \bar{D}_{n-j}^\top \frac{\partial \alpha_{i-1}}{\partial x_j} \right) \hat{\theta} \\ &+ \sum_{j=1}^{i-1} \left(\frac{\partial \alpha_{i-1}}{\partial \hat{\theta}^{(j-1)}} \hat{\theta}^{(j)} + \frac{\partial \alpha_{i-1}}{\partial \hat{\xi}_\kappa^{(j-1)}} \hat{\xi}_\kappa^{(j)} + \frac{\partial \alpha_{i-1}}{\partial x_j} (x_{j+1} + \phi_j) \right),\end{aligned} \tag{4.224}$$

where $i = 2, 3, \ldots, n$, $z_1 = y$, $z_i = x_i - \alpha_{i-1}$ are the state errors, $c_i > 0$ are the design parameters, $\hat{\theta} \in \mathbb{R}^q$ is the vector of adjustable parameters,

$$s_1 = \mu_{1,1} \left| \bar{D}_{n-1} \hat{\xi}_\kappa \right|^2 + \mu_{1,2} \left| \bar{d}_{n-1} \right|^2, \tag{4.225}$$

$$\begin{aligned} s_i = \mu_{i,1} &\left| \left(\bar{D}_{n-i} - \sum_{j=1}^{i-1} \bar{D}_{n-j} \frac{\partial \alpha_{i-1}}{\partial x_j} \right) \hat{\xi}_\kappa \right|^2 \\ &+ \mu_{i,2} \left| \bar{d}_{n-i} - \sum_{j=1}^{i-1} \bar{d}_{n-j} \frac{\partial \alpha_{i-1}}{\partial x_j} \right|^2 \end{aligned} \tag{4.226}$$

are the damping terms with constant parameters $\mu_{i,1}$, $\mu_{i,2} > 0$,

$$\hat{\xi}_\kappa^{(j)} = \frac{\kappa_0 s^j}{\kappa(s)} \left[\hat{\xi} \right] \tag{4.227}$$

are the measured regressor HOTD, $\bar{x}_i = (x_1, x_2, \ldots, x_i)$, $\bar{\xi}_i = (\hat{\xi}_\kappa, \dot{\hat{\xi}}_\kappa, \ldots, \hat{\xi}_\kappa^{(i)})$, and $\bar{\theta}^{(i)} = (\hat{\theta}, \dot{\hat{\theta}}, \ldots, \hat{\theta}^{(i)})$ are the sets of corresponding variables. Finally, we design the actual control law

$$u = \frac{1}{\phi_u(x)} \alpha_n(x, \bar{\theta}_{n-1}, \bar{\xi}_{n-1}). \tag{4.228}$$

Remark 4.8 Since the vector $\hat{\xi}_\kappa$ is bounded a priori, while $\bar{d}_i$ and $\bar{D}_i$ consist of the same elements, the coefficients $\mu_{j,1}$ $(j = 0, 1, \ldots, n)$ in the damping terms (4.225),

(4.226) can be zeroed. In this case, the first damping term s_1 will be a constant and can be removed by adding to the constant c_1. However, we use the most general structure of $s_1,\ s_i$ to use the general proof of the closed-loop stability presented in Appendix E. ■

The control law (4.223)–(4.228) when applied to the plant (4.212)–(4.214) results in the following closed-loop error model:

$$\dot{z} = A_z(z, x, \hat{\theta}, \hat{\xi}_\kappa)z + W_\theta^\top(z, x, \hat{\theta}, \hat{\xi}_\kappa)\tilde{\theta} + W_\epsilon^\top(z, x, \hat{\theta}, \hat{\xi}_\kappa)\epsilon_\kappa, \tag{4.229}$$

$$y = z_1, \tag{4.230}$$

where $z = [z_1,\ z_2, \dots,\ z_n]^\top$ is the vector of the state errors,

$$A_z = \begin{bmatrix} -c_1 - s_1 & 1 & 0 & \cdots & 0 \\ -1 & -c_2 - s_2 & 1 & \ddots & \vdots \\ 0 & -1 & \ddots & \ddots & 0 \\ \vdots & \ddots & \ddots & \ddots & 1 \\ 0 & \cdots & 0 & -1 & -c_n - s_n \end{bmatrix},$$

$$W_\theta^\top = \begin{bmatrix} \hat{\xi}_\kappa^\top \bar{D}_{n-1}^\top \\ \hat{\xi}_\kappa^\top \left(\bar{D}_{n-2}^\top - \bar{D}_{n-1}^\top \frac{\partial \alpha_1}{\partial x_1}\right) \\ \hat{\xi}_\kappa^\top \left(\bar{D}_{n-3}^\top - \bar{D}_{n-1}^\top \frac{\partial \alpha_2}{\partial x_1} - \bar{D}_{n-2}^\top \frac{\partial \alpha_2}{\partial x_2}\right) \\ \vdots \\ \hat{\xi}_\kappa^\top \left(\bar{D}_0^\top - \sum_{j=1}^{n-1} \bar{D}_{n-j}^\top \frac{\partial \alpha_{n-1}}{\partial x_j}\right) \end{bmatrix},$$

$$W_\epsilon^\top = \begin{bmatrix} \bar{d}_{n-1}^\top \\ \bar{d}_{n-2}^\top - \bar{d}_{n-1}^\top \frac{\partial \alpha_1}{\partial x_1} \\ \bar{d}_{n-3}^\top - \bar{d}_{n-1}^\top \frac{\partial \alpha_2}{\partial x_1} - \bar{d}_{n-2}^\top \frac{\partial \alpha_2}{\partial x_2} \\ \vdots \\ \bar{d}_0^\top - \sum_{j=1}^{n-1} \bar{d}_{n-j}^\top \frac{\partial \alpha_{n-1}}{\partial x_j} \end{bmatrix}.$$

Now, to derive an adaptation algorithm, we apply the nonlinear swapping scheme [6, Chap. 9] (see Lemma D.9)

$$\dot{\zeta}_0 = A_z\zeta_0 + W_\theta^\top \hat{\theta}, \tag{4.231}$$

$$\dot{\Omega}_\theta^\top = A_z\Omega_\theta^\top + W_\theta^\top \tag{4.232}$$

with the states $\zeta_0 \in \mathbb{R}^n$ and $\Omega_\theta \in \mathbb{R}^{n\times q}$ and define the *augmented state error* as

$$\bar{z} = z + \zeta_0. \tag{4.233}$$

In view of Lemma D.9, we have

$$\bar{z} = \Omega_\theta^\top \theta + \epsilon_z, \tag{4.234}$$

where ϵ_z is the exponentially decaying term satisfying the equation $\dot{\epsilon}_z = A_z\epsilon_z + W_\epsilon^\top \epsilon_\kappa$. Note that Ω_θ is bounded due to the structure of the matrix A_z with the damping terms s_i, and its boundedness is irrespective of the boundedness of the matrix W_θ.

Remark 4.9 To reduce the effect caused by the initial condition $\epsilon_z(0)$ (not completely due to the influence of ϵ_κ), we select the initial conditions of the filters (4.231), (4.232) as follows: $\zeta_0(0) = -z(0)$, $\Omega_\theta(0) = 0$. Hence, $\epsilon_z(0) = z(0) + \zeta_0(0) - \Omega_\theta^\top(0)\theta = 0$. ■

It is worth noting that the control laws (4.224), (4.228) involve the HOTD of adjustable parameters $\hat{\theta}^{(i)}$ up to the order $n-1$. At the same time, the static linear regression (4.234) allows one to employ the following special realization of the adaptation algorithm with MRE to calculate $\hat{\theta}^{(i)}$ (see Sect. 3.2.3):

$$\dot{\hat{\theta}} = \gamma\left(\bar{z}_\Omega - \bar{\Omega}_\theta^\top \hat{\theta}\right), \tag{4.235}$$

where $\bar{z}_\Omega = L(s)\left[\Omega_\theta \bar{z}\right]$, $\bar{\Omega}_\theta = L(s)\left[\Omega_\theta \Omega_\theta^\top\right]$, $L(s)$ is a designer-chosen transfer function of the form (3.44) with a relative degree greater or equal to $n-2$, and $\gamma > 0$ is the adaptation gain. Then, in view of (4.234), the adaptation algorithm (4.235) yields the closed-loop parametric error model

$$\dot{\tilde{\theta}} = -\gamma\bar{\Omega}_\theta\tilde{\theta} - \gamma L(s)\left[\Omega_\theta \epsilon_z\right]. \tag{4.236}$$

Since the relative degree of the transfer function $L(s)$ is greater to $n-1$, the HOTD of $\bar{z}_\Omega$ and $\bar{\Omega}_\theta$ are physically implementable. As a result, the adaptation algorithm (4.235) allows us to calculate the HOTD of adjustable parameters $\hat{\theta}$ up to the $(n-1)$st order via the following expressions [4]:

$$\hat{\theta}^{(i)} = \gamma\left(\bar{z}_\Omega^{(i-1)} - \sum_{j=0}^{i-1} C_j^{i-1}\bar{\Omega}_\theta^{(i-j-1)}\hat{\theta}^{(j)}\right), \tag{4.237}$$

where $i = 2, 3, \ldots, n-1$ and C_j^i are the binomial coefficients.

The main properties of the closed-loop system are established by the following statement that directly follows from Proposition E.1.

Proposition 4.11 *If Assumption 2.1 holds, then the adjustable controller (4.223)–(4.228) with the disturbance observer (4.220)–(4.222), the swapping scheme (4.231)–(4.233), and the adaptation algorithm (4.235), (4.237) when applied to the plant (4.212)–(4.214) provides the boundedness of all the closed-loop signals and drives the plant output to zero, i.e.,* $\lim_{t\to\infty} y(t) = 0$.

Example 4.13 Consider the plant

$$\dot{x}_1 = x_2 + x_1^2 + \upsilon_1, \tag{4.238}$$
$$\dot{x}_2 = x_3 + \upsilon_2, \tag{4.239}$$
$$\dot{x}_3 = (2 + \cos(x_2))\, u + 2\upsilon_1 + 3\upsilon_2, \tag{4.240}$$
$$y = x_1 \tag{4.241}$$

affected by the unmeasured disturbance $\upsilon(t) = [\upsilon_1(t),\ \upsilon_2(t)]^\top = [\beta_1 \sin(wt + \phi), \beta_2]^\top$ with unknown parameters $\beta_1 = 2$, $\beta_2 = 1$, $w = 1$, and $\phi = 3$. The structure of the plant corresponds to the form (4.215), (4.216), where $n = 3$, $x = [x_1,\ x_2,\ x_3]^\top$, $\phi = [\phi_1,\ \phi_2,\ \phi_3]^\top$, $\phi_1(y) = x_1^2$, $\phi_2(x_1, x_2) = \phi_3(x) = 0$, $\phi_u(x) = 2 + \cos(x_2)$,

$$\bar{D} = \begin{bmatrix} \bar{d}_2^\top \\ \bar{d}_1^\top \\ \bar{d}_0^\top \end{bmatrix} = \begin{bmatrix} 1 & 0 \\ 0 & 1 \\ 2 & 3 \end{bmatrix}.$$

Due to prior knowledge about the number and waveforms of modes in the disturbance υ, we construct the third-order disturbance observer (4.220)–(4.222) with

$$G = \begin{bmatrix} 0 & 1 & 0 \\ 0 & 0 & 1 \\ -6 & -11 & -6 \end{bmatrix}, \quad L = \begin{bmatrix} 0 & 0 \\ 0 & 0 \\ 12 & 12 \end{bmatrix}, \quad N = \begin{bmatrix} 0 & 0 & 0 \\ 0 & 0 & 0 \\ 12 & 12 & 0 \end{bmatrix},$$

and

$$\hat{\xi}_\kappa = \frac{8}{s^2 + 6s + 8}\left[\hat{\xi}\right].$$

Applying the backstepping procedure, we design the virtual control laws α_1, α_2, α_3, and the actual control

$$u = \frac{1}{2 + \cos(x_2)}\alpha_2$$

with $c_1 = c_2 = c_3 = 1$, $\mu_{1,2} = \mu_{2,2} = \mu_{3,2} = 0.001$. The parameters $\mu_{1,1}$, $\mu_{2,1}$, $\mu_{3,1}$, in view of Remark 4.8, are set to zero.

The vector of the adjustable parameters $\hat{\theta}$ and its HOTD up to the second order are calculated via the algorithm of adaptation with MRE with $L(s) = \frac{1}{s+1}$

$$\dot{\hat{\theta}} = \gamma \left(\frac{1}{s+1}\left[\Omega_\theta(z + \zeta_0)\right] - \frac{1}{s+1}\left[\Omega_\theta \Omega_\theta^\top\right] \hat{\theta} \right) \tag{4.242}$$

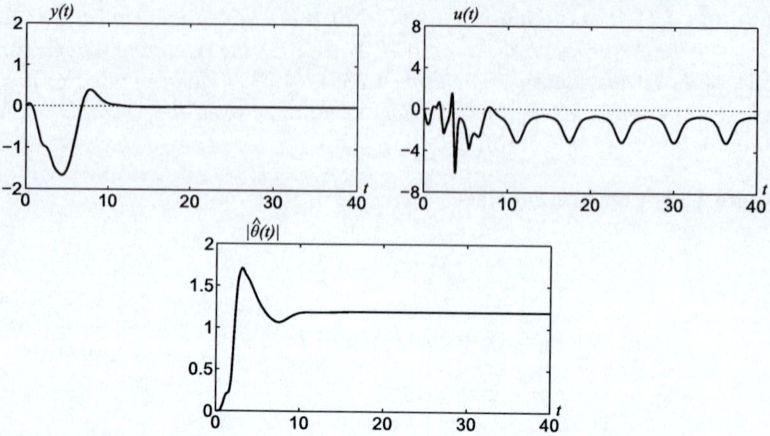

Fig. 4.26 Example 4.13: adaptive backstepping state-feedback compensation of unmatched disturbance in the nonlinear system

and the expression

$$\ddot{\hat{\theta}} = -\left(I_6 + \gamma \frac{1}{s+1}\left[\Omega_\theta \Omega_\theta^\top\right]\right)\dot{\hat{\theta}} + \gamma \Omega_\theta (z + \zeta_0 - \Omega_\theta^\top \hat{\theta}), \qquad (4.243)$$

where Ω_θ and ζ_0 are generated by the swapping filters (4.231), (4.232), $z = [z_1, z_2, z_3]^\top = [y, x_2 - \alpha_1, x_3 - \alpha_2]^\top$, and $\gamma = 1$.

All simulations are carried out with zero initial conditions $x(0), \eta(0), \zeta_0(0), \Omega_\theta(0)$, and $\hat{\theta}(0)$

Simulation results are presented in Fig. 4.26 and demonstrate complete compensation of the influence of the multidimensional unmatched disturbance on the plant output which is driven to zero. ■

4.3.4 *Plant with Unmatched Disturbance and Measured Output*

In this section, we consider the problem of adaptive control for nonlinear plants with the state inaccessible for measurement. Let the plant be representable in (or transformable into) the *output-feedback* canonical form with external disturbance

$$\dot{x} = Ex + \phi(y) + \bar{b}\phi_u(y)u + \bar{d}\upsilon, \qquad (4.244)$$

$$y = \bar{c}^\top x + d_n \upsilon, \qquad (4.245)$$

where the state $x \in \mathbb{R}^n$ is inaccessible for measurements,

$$E = \begin{bmatrix} 0 & \\ \vdots & I_{n-1} \\ 0 \cdots 0 & \end{bmatrix}, \quad \phi = \begin{bmatrix} \phi_1 \\ \phi_2 \\ \vdots \\ \phi_n \end{bmatrix}, \quad \bar{b} = \begin{bmatrix} 0 \\ \vdots \\ 0 \\ b_m \\ \vdots \\ b_1 \\ b_0 \end{bmatrix}, \quad \bar{d} = \begin{bmatrix} d_{n-1} \\ \vdots \\ d_1 \\ d_0 \end{bmatrix}, \quad \bar{c} = \begin{bmatrix} 1 \\ 0 \\ \vdots \\ 0 \end{bmatrix},$$

$\phi_i(y)$ and $\phi_u(y)$ are the known smooth scalar functions, b_j are the known constant parameters, while d_{i-1} are possibly unknown constant parameters ($i = 1, 2, \ldots, n$, $j = 0, 1, \ldots, m$). It is assumed that the function $\phi_u(y)$ is bounded away from zero by some constant ϕ_0, i.e., $\phi_u(y) \geq \phi_0 > 0$.

The class of the considered plants is restricted by Assumption 2.6.

The control problem is to design a state-feedback controller that guarantees the boundedness of all the closed-loop signals and provides the convergence $y(t) \to 0$ as $t \to 0$.[9]

Applying Proposition 2.8, we use the following parameterized model:

$$\dot{y} = -\bar{l}_{n-1}\hat{x}_1 + \hat{x}_2 + \bar{l}_{n-1}y + \phi_1(y) + \theta^\top \dot{\hat{\xi}}_\kappa + \bar{\epsilon}_\kappa, \tag{4.246}$$

$$\dot{\hat{x}}_i = -\bar{l}_{n-i}\hat{x}_1 + \hat{x}_{j+1} + \bar{l}_{n-i}y + \phi_i(y), \qquad i = 2, 3, \ldots, n-m-1, \tag{4.247}$$

$$\dot{\hat{x}}_{n-m} = -\bar{l}_m\hat{x}_1 + \hat{x}_{n-m+1} + \bar{l}_m y + \phi_m(y) + b_m\phi_u(y)u, \tag{4.248}$$

where $\bar{\epsilon}_\kappa$ is the exponentially decaying term, $\hat{\xi}_\kappa$ is the filtered regressor defined as

$$\hat{\xi}_\kappa = \frac{\kappa_0}{\kappa(s)}\left[\hat{\xi}_f\right], \tag{4.249}$$

$\kappa(s)$ is a designer-chosen Hurwitz polynomial of a degree no less than $n - m$ and $\kappa_0 > 0$ is a constant. The regressor estimate $\hat{\xi}_f$ is generated by the following disturbance observer:

$$\dot{\hat{\xi}}_f = G\hat{\xi}_f + l(y - \bar{c}^\top \hat{x}), \tag{4.250}$$

where the pair (G, l) is from the exosystem canonical form (2.15), while the state estimate $\hat{x}$ is generated by the state observer

$$\dot{\hat{x}} = \bar{A}\hat{x} + \bar{l}y + \phi(y) + \bar{b}\phi_u(y)u \tag{4.251}$$

---

[9] A similar approach based on backstepping design is proposed in [3]. Extension to the class of nonminimum phase nonlinear plants is presented in [7]. An approach based on adaptive disturbance observer is presented in [5].

with the vector of the feedback gains $\bar{l} = [\bar{l}_{n-1},\ \bar{l}_{n-2}, \ldots,\ \bar{l}_0]^\top$ selected so that the matrix $\bar{A} = E - \bar{l}\bar{c}^\top$ is Hurwitz. In (4.246)–(4.248), $\hat{x}_j$ $(j = 1, 2, \ldots, n)$ are the elements of the vector $\hat{x}$.

Since the uncertainties are not matched with the control in the model (4.246)–(4.248), we apply the modular backstepping with the high-order tuners described in Appendix E in order to design an appropriate adaptive controller.[10]

The filtering (4.249) is required for implementation of the HOTD $\hat{\xi}_\kappa^{(i)}$ appearing in the steps of the backstepping procedure ($i = 1, 2, \ldots, n - m$).

Now, applying to the model (4.246)–(4.248) the modular backstepping introduced in Appendix E, we derive the virtual control laws

$$\alpha_1(y, \hat{x}_1, \hat{\theta}, \dot{\hat{\xi}}_\kappa) = -(c_1 + s_1)z_1 + \bar{l}_{n-1}\left(\hat{x}_1 - y\right) - \phi_1(y) - \hat{\theta}^\top \dot{\hat{\xi}}_\kappa, \tag{4.252}$$

$$\begin{aligned}\alpha_i(y, \bar{x}_i, \bar{\theta}_{i-1}, \bar{\xi}_i) &= -z_{i-1} - (c_i + s_i)z_i + \bar{l}_{n-i}\left(\hat{x}_1 - y\right) - \phi_i(y) \\ &+ \frac{\partial \alpha_{i-1}}{\partial y}\left(-\bar{l}_{n-1}\hat{x}_1 + \hat{x}_2 + \bar{l}_{n-1}y + \phi_1(y) + \hat{\theta}^\top \dot{\hat{\xi}}_\kappa\right) \\ &+ \sum_{j=1}^{i-1}\left(\frac{\partial \alpha_{i-1}}{\partial \hat{\theta}^{(j-1)}}\hat{\theta}^{(j)} + \frac{\partial \alpha_{i-1}}{\partial \hat{\xi}_\kappa^{(j)}}\hat{\xi}_\kappa^{(j+1)}\right. \\ &+ \left.\frac{\partial \alpha_{i-1}}{\partial \hat{x}_j}\left(-\bar{l}_{n-j}\hat{x}_1 + \hat{x}_{j+1} + \bar{l}_{n-j}y + \phi_j(y)\right)\right), \quad i = 2, 3, \ldots, n - m,\end{aligned} \tag{4.253}$$

where $z_1 = y$, $z_i = \hat{x}_i - \alpha_{i-1}$ are the state errors, $c_i > 0$ are the design parameters, $\hat{\theta} \in \mathbb{R}^q$ is the vector of adjustable parameters,

$$s_1 = \mu_{1,1}\left|\dot{\hat{\xi}}_\kappa\right|^2 + \mu_{1,2}, \quad s_i = \mu_{i,1}\left|\frac{\partial \alpha_{i-1}}{\partial y}\dot{\hat{\xi}}_\kappa\right|^2 + \mu_{i,2}\left|\frac{\partial \alpha_{i-1}}{\partial y}\right|^2 \tag{4.254}$$

are the damping terms with constant parameters $\mu_{i,1}, \mu_{i,2} > 0$, $\bar{x}_i = (\hat{x}_1, \hat{x}_2, \ldots, \hat{x}_i)$, $\bar{\xi}_i = (\dot{\hat{\xi}}_\kappa, \ddot{\hat{\xi}}_\kappa, \ldots, \hat{\xi}_\kappa^{(i)})$, and $\bar{\theta}^{(i)} = (\hat{\theta}, \dot{\hat{\theta}}, \ldots, \hat{\theta}^{(i)})$ are the sets of corresponding variables.

Remark 4.10 Since the vector $\dot{\hat{\xi}}_\kappa$ is known to be bounded a priori, the coefficients $\mu_{j,1}$ $(j = 0, 1, \ldots, n)$ in the damping terms (4.254) can be zeroed. In this case, the first damping term s_1 will be constant and can be removed by adding to the constant c_1. However, we use the most general structure of s_1, s_i to use the proof of the closed-loop stability presented in Appendix E. ■

The procedure is finished by design of the actual control law

$$u = \frac{1}{b_m\phi_u(y)}\left(\alpha_{n-m}\left(y, \bar{x}_{n-m}, \bar{\theta}_{n-m-1}, \bar{\xi}_{n-m}\right) - \hat{x}_{n-m+1}\right) - \hat{x}_{n-m+1}\right). \tag{4.255}$$

[10] An alternative approach employing adaptive backstepping with tuning functions is presented in [13, Sect. 4.2.2].

The control law (4.252)–(4.255) when applied to (4.246)–(4.248) results in the following closed-loop state error model:

$$\dot{z} = A_z(z, y, \hat{\theta}, \dot{\hat{\xi}}_\kappa)z + b_z(z, y, \hat{\theta}, \dot{\hat{\xi}}_\kappa)\dot{\hat{\xi}}_\kappa^\top \tilde{\theta} + b_z(z, y, \hat{\theta}, \dot{\hat{\xi}}_\kappa)\bar{\epsilon}_\kappa, \tag{4.256}$$

$$y = \bar{c}^\top z, \tag{4.257}$$

where $z = \left[z_1, z_2, \ldots, z_{n-m}\right]^\top$ is the vector of the state errors,

$$A_z = \begin{bmatrix} -c_1 - s_1 & 1 & 0 & \cdots & 0 \\ -1 & -c_2 - s_2 & 1 & \ddots & \vdots \\ 0 & -1 & \ddots & \ddots & 0 \\ \vdots & \ddots & \ddots & \ddots & 1 \\ 0 & \cdots & 0 & -1 & -c_{n-m} - s_{n-m} \end{bmatrix}, \quad b_z = \begin{bmatrix} 1 \\ -\frac{\partial \alpha_1}{\partial y} \\ -\frac{\partial \alpha_2}{\partial y} \\ \vdots \\ -\frac{\partial \alpha_{n-m-1}}{\partial y} \end{bmatrix}.$$

Remark 4.11 The state error model (4.256), (4.257) is not complete because it does not contain additional m equations describing zero dynamics. However, by virtue of Assumption 2.6, if we stabilize the model (4.256), (4.257), we provide the boundedness of all the closed-loop signals [6, Sect. 2.2.3]. ■

To obtain a linear regression and derive an adaptation algorithm for parameters tuning, we introduce the swapping filters [6, Chap. 9] (see Lemma D.9)

$$\dot{\zeta}_0 = A_z\zeta_0 + b_z\dot{\hat{\xi}}_\kappa^\top \hat{\theta}, \tag{4.258}$$

$$\dot{\Omega}_\theta^\top = A_z\Omega_\theta^\top + b_z\dot{\hat{\xi}}_\kappa^\top, \tag{4.259}$$

where $\zeta_0 \in \mathbb{R}^{n-m}$, $\Omega_\theta \in \mathbb{R}^{n-m\times q}$ are the filters states, and define the *augmented output error* as

$$\bar{z} = z + \zeta_0. \tag{4.260}$$

Then, in view of Lemma D.9, we have

$$\bar{z} = \Omega_\theta^\top \theta + \epsilon_z, \tag{4.261}$$

where Ω_θ is bounded independently of the boundedness of $b_z\dot{\hat{\xi}}_\kappa^\top$ due to the structure of the matrix A_z with the damping terms s_i, while ϵ_z is the exponentially decaying term satisfying the equation $\dot{\epsilon}_z = A_z\epsilon_z + w_\epsilon\bar{\epsilon}_\kappa$.

Remark 4.12 To reduce the effect caused by the initial condition $\epsilon_z(0)$ (not completely due to the influence of $\bar{\epsilon}_\kappa$), we select the initial conditions of the filters (4.258), (4.259) as follows: $\zeta_0(0) = -z(0)$, $\Omega_\theta(0) = 0$. Hence, $\epsilon_z(0) = z(0) + \zeta_0(0) - \Omega_\theta^\top(0)\theta = 0$. ■

It is worth noting that the control laws (4.253), (4.255) involve the HOTD of the adjustable parameters $\hat{\theta}^{(i)}$ up to the order $n - m - 1$. At the same time, the static linear regression (4.261) allows one to employ the following special realization of the adaptation algorithm with MRE to calculate $\hat{\theta}^{(i)}$ (see Sect. 3.2.3):

$$\dot{\hat{\theta}} = \gamma \left(\bar{z}_\Omega - \bar{\Omega}_\theta \hat{\theta} \right), \tag{4.262}$$

where $\bar{z}_\Omega = L(s)\left[\Omega_\theta \bar{z}\right]$, $\bar{\Omega}_\theta = L(s)\left[\Omega_\theta \Omega_\theta^\top\right]$, $L(s)$ is a designer-chosen transfer function of the form (3.44) with a relative degree greater or equal to $n - m - 2$, and $\gamma > 0$ is the adaptation gain. Then, in view of (4.234), the adaptation algorithm (4.235) yields the closed-loop parametric error model

$$\dot{\tilde{\theta}} = -\gamma \bar{\Omega}_\theta \tilde{\theta} - \gamma L(s) \left[\Omega_\theta^\top \epsilon_z\right]. \tag{4.263}$$

Since the relative degree of the transfer function $L(s)$ is greater or equal to $n - m - 1$, the HOTD of $\bar{z}_\Omega$ and $\bar{\Omega}_\theta$ are physically implementable. As a result, the adaptation algorithm (4.262) allows us to calculate the HOTD of adjustable parameters $\hat{\theta}$ up to the $(n - m - 1)$st order via the following expressions [4]:

$$\hat{\theta}^{(i)} = \gamma \left(\bar{z}_\Omega^{(i-1)} - \sum_{j=0}^{i-1} C_j^{i-1} \bar{\Omega}_\theta^{(i-j-1)} \hat{\theta}^{(j)} \right), \tag{4.264}$$

where $i = 2, 3, \ldots, n - m - 1$ and C_j^i are the binomial coefficients.

The main properties of the closed-loop system are established by the following statement.

Proposition 4.12 *If Assumptions 2.1 and 2.6 hold, then the adjustable controller (4.252)–(4.255) with the disturbance observer (4.249)–(4.251), the swapping scheme (4.258), (4.260), and the adaptation algorithm (4.262), (4.264) when applied to the plant (4.244), (4.245) provides the boundedness of all the closed-loop signals and drives the plant output to zero, i.e.,* $\lim_{t\to\infty} y(t) = 0$.

Proof From Proposition E.1, we directly obtain that $z, \hat{\theta}, \dot{\hat{\theta}}, \ldots, \hat{\theta}^{(n-m-1)}$ are bounded and $|z(t)| \to 0$ as $t \to \infty$. Now, taking Assumption 2.6 into account and using the same arguments as in [6, Sect. 8.1.3], we can prove that all the closed-loop signals are bounded.

□

Example 4.14 Consider the plant

$$\dot{x}_1 = x_2 + y\sin(y) + 3\upsilon, \tag{4.265}$$

$$\dot{x}_2 = x_3, \tag{4.266}$$

$$\dot{x}_3 = x_4 + (1 + y^2)u, \tag{4.267}$$

$$\dot{x}_4 = 2(1 + y^2)u + \upsilon, \tag{4.268}$$

$$y = x_1 + 2\upsilon \tag{4.269}$$

with unmeasured state x affected by the unmeasured disturbance $\upsilon(t) = \beta\sin(wt + \phi)$ with unknown parameters $\beta = 2$, $w = 1$, and $\phi = 3$.

The structure of the plant corresponds to the form (4.244), (4.245), where $n = 4$, $m = 1$, $x = [x_1,\ x_2,\ x_3,\ x_4]^\top$, $\bar{b} = [0,\ 0,\ 2,\ 1]^\top$, $\bar{d} = [3,\ 0,\ 0,\ 1]^\top$, $d_4 = 2$, $\phi = [\phi_1,\ \phi_2,\ \phi_3,\ \phi_4]^\top$, $\phi_1(y) = y\sin(y)$, $\phi_2(y) = \phi_3(y) = \phi_4(y) = 0$, $\phi_u(y) = 1 + y^2$.

Since the polynomial $b_1 s + b_0 = 2s + 1$ is Hurwitz, the plant (4.265)–(4.269) satisfies Assumption 2.6.

Due to prior knowledges about the disturbance υ, we construct the second-order disturbance observer (4.249)–(4.251) with

$$G = \begin{bmatrix} 0 & 1 \\ -12 & -7 \end{bmatrix},\quad l = \begin{bmatrix} 0 \\ 12 \end{bmatrix},\quad \bar{l} = \begin{bmatrix} 10 \\ 35 \\ 50 \\ 24 \end{bmatrix},$$

and

$$\hat{\xi}_\kappa = \frac{6}{s^3 + 6s + 11s^2 + 6}\left[\hat{\xi}_f\right].$$

Applying the backstepping procedure consisting of $n - m = 3$ steps, we design the virtual control laws α_1, α_2, α_3, and the actual control

$$u = \frac{1}{1 + y^2}\left(\alpha_3 - \hat{x}_4\right)$$

with $c_1 = c_2 = c_3 = 1$, $\mu_{1,2} = \mu_{2,2} = \mu_{3,2} = 0.001$. The parameters $\mu_{1,1}$, $\mu_{2,1}$, $\mu_{3,1}$, in view of Remark 4.10, are set to zero.

The adjustable parameters $\hat{\theta}$ and their HOTD up to the second order are calculated using the algorithm with MRE with $L(s) = \frac{1}{s+1}$

$$\dot{\hat{\theta}} = \gamma\left(\frac{1}{s+1}\left[\Omega_\theta(z + \zeta_0)\right] - \frac{1}{s+1}\left[\Omega_\theta\Omega_\theta^\top\right]\hat{\theta}\right) \tag{4.270}$$

and the expression

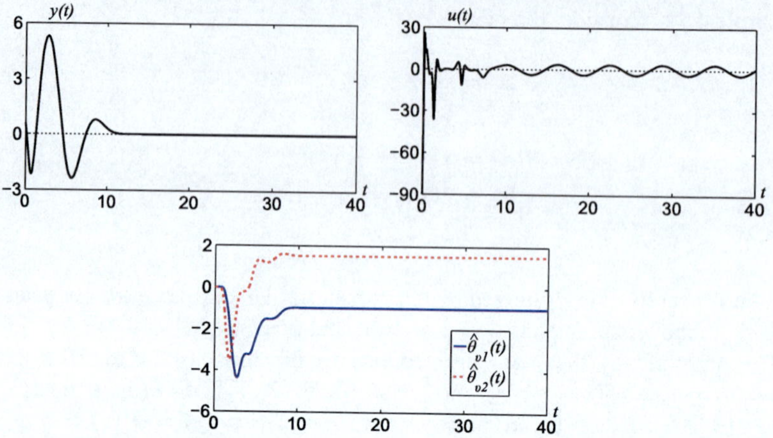

Fig. 4.27 Example 4.14: adaptive backstepping output-feedback compensation of unmatched disturbance in the nonlinear system

$$\ddot{\hat{\theta}} = -\left(I_2 + \gamma \frac{1}{s+1}\left[\Omega_\theta \Omega_\theta^\top\right]\right)\dot{\hat{\theta}} + \gamma \Omega_\theta \left(z + \zeta_0 - \Omega_\theta^\top \hat{\theta}\right), \qquad (4.271)$$

where Ω_θ and ζ_0 are generated by the swapping filters (4.258), (4.259), $z = [z_1, z_2, z_3]^\top = \left[y, \hat{x}_2 - \alpha_1, \hat{x}_3 - \alpha_2\right]^\top$, and $\gamma = 1$. The variables $\hat{x}_2$ and $\hat{x}_3$ are generated by the disturbance observer.

All simulations are carried out with zero initial conditions $x(0)$, $\hat{x}(0)$, $\hat{\xi}_f(0)$, $\zeta_0(0)$, $\Omega_\theta(0)$, and $\hat{\theta}(0)$.

Simulation results are presented in Fig. 4.27 and demonstrate complete compensation of the unmatched disturbance in the ouput of the nonlinear plant (4.265)–(4.269). ■

References

1. Bobtsov, A.A., Pyrkin, A.A.: Compensation of unknown sinusoidal disturbances in linear plants of arbitrary relative degree. Autom. Remote. Control. **70**, 449–456 (2009)
2. Chen, C.-T.: Linear Systems. Theory and Design, 3rd edn. Oxford University Press, NY, Oxford (1999)
3. Ding, Z.: Global stabilization and disturbance suppression of a class of nonlinear systems with uncertain internal model. Automatica **39**, 471–479 (2003)
4. Gerasimov, D.N., Nikiforov, V.O.: On key properties of the Lion's and Kreisselmeier's adaptation algorithms. Int. J. Adapt Control Signal Process (2021). https://doi.org/10.1002/acs.3311
5. Han, J., Zhang, H., Wang, Y., Liu, X.: Anti-disturbance control for nonlinear system via adaptive disturbance observer. Int. J. Robust Nonlinear Control **27**, 2121–2144 (2017)

6. Krstić, M., Kanellakopoulos, I., Kokotović, P.: Nonlinear and Adaptive Control Design. Wiley, NY (1995)
7. Marino, R., Santosuosso, G.L.: Global compensation of unknown sinusoidal disturbances for a class of nonlinear nonminimum phase systems. IEEE Trans. Autom. Control **50**(11), 1816–1822 (2005)
8. Marino, R., Santosuosso, G.L.: Regulation of linear systems with unknown exosystems of uncertain order. IEEE Trans. Autom. Control **52**(2), 352–359 (2007)
9. Marino, R., Tomei, P.: Output regulation for linear systems via adaptive internal model. IEEE Trans. Autom. Control **48**, 2199–2202 (2003)
10. Nikiforov, V.O.: Adaptive servocompensation of input disturbances. In: Proceedings of 13th World Congress of IFAC, San Francisco, USA, vol. 29, no. 1, pp. 5114–5119 (1996)
11. Nikiforov, V.O.: Adaptive servomechanism controller with an implicit reference model. Int. J. Control **68**(2), 277–286 (1997)
12. Nikiforov, V.O.: Adaptive nonlinear servocompensation of unknown external disturbances. In: 14th World Congress of IFAC, vol. I, Beijing, P.R. China, pp. 283–289 (1999)
13. Nikiforov, V.O.: Adaptive and Robust Control with Compensation of External Disturbance. Nauka, St.-Petersburg (2003). (in Russian)
14. Nikiforov, V.O., Paramonov, A.V., Gerasimov, D.N.: Adaptive control algorithms in MIMO systems with control delay. Autom. Remote. Control. **81**, 1107–1124 (2020)
15. Pyrkin, A., Isidori, A.: Adaptive output regulation of right-invertible MIMO LTI systems, with application to vessel motion control. Eur. J. Control **46**, 63–79 (2019)
16. Serrani, A.: Rejection of harmonic disturbances at the controller input via hybrid adaptive external model. Automatica **42**, 1977–1985 (2006)
17. Wang, J., Aranovskiy, S.V., Bobtsov, A.A., Pyrkin, A.A.: Compensating for a multisinusoidal disturbance based on Youla-Kucera parametrization. Autom. Remote. Control. **78**(9), 1559–1571 (2017)

Chapter 5
Adaptive Regulation in Systems with Unknown Parameters

Abstract The chapter presents controllers providing adaptive reference tracking and adaptive disturbance rejection for plants with unknown parameters. In other words, it extends the results of Chap. 4 to different classes of uncertain plants. In this case, the adaptation counteracts the influence of both the exosignal as well as the plant uncertainties. The introduced design methods are based on the combined use of the exosignal observers designed in Sect. 2.5 and the basic adaptation algorithms presented in Chap. 3. Section 5.1 presents adaptive controllers for LTI SISO plants, while Sect. 5.2 presents solutions for certain classes of nonlinear plants.

5.1 LTI SISO Plants

As in Chap. 4, we start with the class of linear plants. However, in contrast to Sect. 4.1, the linear plants considered here have unknown parameters. For methodological purposes, we start again with separate design of reference tracking controllers and disturbance rejection ones. We consider first the simplest case of state-feedback control of the plants with matched disturbance and then extend the approach for the cases of unmatched disturbance, output-feedback adaptive control and simultaneous disturbance rejection with reference tracking.

5.1.1 Reference Tracking

Let us consider the LTI SISO plant

$$y = b_m \frac{\beta(s)}{\alpha(s)}[u], \tag{5.1}$$

where $y \in \mathbb{R}$ is the regulated variable, $u \in \mathbb{R}$ is the control, $\alpha(s)$ and $\beta(s)$ are monic polynomials with unknown constant parameters. Regarding the plant (5.1) we make the following standard assumption [7, 8, 15].

V. Nikiforov and D. Gerasimov, *Adaptive Regulation*, Lecture Notes in Control and Information Sciences 491, https://doi.org/10.1007/978-3-030-96091-9_5

Assumption 5.1 *The plant (5.1) is such that:*

A.5.1.1. *the degrees n and m of the polynomials $\alpha(s)$ and $\beta(s)$ are known, the relative degree $n - m \geq 1$;*

A.5.1.2. *the high-frequency gain b_m is unknown, but the sign of b_m is known. Without loss of generality, we assume that $b_m > 0$;*

A.5.1.3. *the polynomial $\beta(s)$ is Hurwitz.*

The control problem is to design an output-feedback controller providing the boundedness of all the closed-loop signals and asymptotic tracking of the plant output according to the equality

$$\lim_{t\to\infty} (y(t) - g(t)) = 0, \tag{5.2}$$

where the reference g is measured and satisfies Assumption 2.1 (with substitution $\upsilon = g$).

First, using Lemma 2.3, we represent g in the form of linear regression

$$g = \theta^\top \hat{\xi} + \epsilon_g, \tag{5.3}$$

where $\theta \in \mathbb{R}^q$ is a vector of unknown parameters, ϵ_g is the exponentially decaying term, $\hat{\xi} \in \mathbb{R}^q$ is the vector generated by the observer

$$\dot{\hat{\xi}} = G\hat{\xi} + lg \tag{5.4}$$

with the pair (G, l) from the exosystem canonical form (2.15) and arbitrary initial condition $\hat{\xi}(0)$.

Second, using the results of Corollary A.5 (with substitutions $\upsilon = g$ in (A.1), $\bar{\xi} = \hat{\xi}$ in (A.2) and $\bar{\xi}_w = \hat{\xi}_w$ in (A.8)), we obtain (neglecting exponentially decaying term)

$$g = \bar{\theta}^\top \hat{\xi}_w = \bar{\theta}^\top \frac{\bar{\alpha}_0}{\bar{\alpha}(s)} \left[\hat{\xi}\right] = \frac{\bar{\alpha}_0}{\bar{\alpha}(s)} \left[\bar{\theta}^\top \hat{\xi}\right], \tag{5.5}$$

where $\bar{\theta} \in \mathbb{R}^q$ is a new vector of unknown parameters, $\hat{\xi}_w = \frac{\bar{\alpha}_0}{\bar{\alpha}(s)}[\hat{\xi}]$ is the filtered regressor, $\bar{\alpha}(s)$ is an arbitrary designer-chosen Hurwitz polynomial of the order $n - m$, $\bar{\alpha}_0 > 0$ is some constant. Finally, we can write

$$g = \frac{\bar{\alpha}_0}{\bar{\alpha}(s)} [r(t)], \tag{5.6}$$

where $r = \bar{\theta}^\top \hat{\xi}$.

Third, considering $\frac{\bar{\alpha}_0}{\bar{\alpha}(s)}$ as a virtual reference model and $r = \bar{\theta}^\top \hat{\xi}$ as its virtual reference input, we can reduce the formulated problem to the standard problem of *model reference adaptive control* (MRAC) [16, 20]. In order to get a solution to the MRAC problem, one can apply well-known approaches based on augmented error

[13, 15], high-order tuners [14, 19, 22] or adaptive backstepping [8, Sect. 10]. Here we use the augmented error approach and apply the following technical lemma that establishes a suitable parameterization of the tracking error $\varepsilon = y - g$ for the case of the known reference input r [13, 15].

Lemma 5.1 *Together with the plant (5.1) and the reference model (5.6) with the known reference input r, consider the auxiliary filters*

$$\dot{\bar{\chi}}_1 = \bar{\Lambda}\bar{\chi}_1 + \varsigma_{n-1}u, \tag{5.7}$$

$$\dot{\bar{\chi}}_2 = \bar{\Lambda}\bar{\chi}_2 + \varsigma_{n-1}y, \tag{5.8}$$

where $\bar{\chi}_1 \in \mathbb{R}^{n-1}$ and $\bar{\chi}_2 \in \mathbb{R}^{n-1}$ are the states of the filters with arbitrary initial conditions and a Hurwitz designer-chosen matrix

$$\bar{\Lambda} = \begin{bmatrix} 0 & 1 & 0 & \cdots & 0 \\ 0 & 0 & 1 & \ddots & 0 \\ \vdots & \vdots & \ddots & \ddots & \vdots \\ 0 & 0 & 0 & \cdots & 1 \\ -\lambda_0 & -\lambda_1 & -\lambda_2 & \cdots & -\lambda_{n-2} \end{bmatrix}.$$

Then there exists a constant vector $\psi \in \mathbb{R}^{2n}$ such that

$$\varepsilon = b_m \frac{\bar{\alpha}_0}{\bar{\alpha}(s)} \left[\psi^\top \omega_\psi + u\right] + \epsilon, \tag{5.9}$$

where ϵ is the exponentially decaying term, $\omega_\psi \in \mathbb{R}^{2n}$ is the regressor defined as

$$\omega_\psi = \left[\bar{\chi}_1^\top, \bar{\chi}_2^\top, y, r\right]^\top. \tag{5.10}$$

However, in the case considered, the reference input $r = \bar{\theta}^\top \hat{\xi}$ is not measured, because the vector $\bar{\theta}$ is unknown. As a result, we use the following corollary of the above lemma.

Corollary 5.1 *For the tracking error $\varepsilon = y - g$, where g satisfies Assumption 2.1 (with substitution $\upsilon = g$), the following model is valid:*

$$\varepsilon = b_m \frac{\bar{\alpha}_0}{\bar{\alpha}(s)} \left[\vartheta^\top \omega + u\right] + \epsilon, \tag{5.11}$$

where $\vartheta \in \mathbb{R}^{2n+q-1}$ is a vector of unknown parameters, ϵ is the exponentially decaying term, $\omega \in \mathbb{R}^{2n+q-1}$ is the regressor given by

$$\omega = \left[\bar{\chi}_1^\top, \bar{\chi}_2^\top, y, \hat{\xi}^\top\right]^\top. \tag{5.12}$$

The corollary is proved by substitution of $r = \bar{\theta}^\top \hat{\xi}$ into (5.10) and aggregation of the corresponding vectors.

For the control law selected as

$$u = -\hat{\vartheta}^\top \omega, \tag{5.13}$$

where $\hat{\vartheta} \in \mathbb{R}^{2n+q-1}$ is the vector of adjustable parameters, the closed-loop error model takes the form

$$\varepsilon = b_m \frac{\bar{\alpha}_0}{\bar{\alpha}(s)} \left[\tilde{\vartheta}^\top \omega\right] + \epsilon, \tag{5.14}$$

in which $\tilde{\vartheta} = \vartheta - \hat{\vartheta}$ is the vector of parametric errors.

Extending the results from Sect. 3.4.2 to the error model with unknown high-frequency gain (see [15, Sect. 5.4.6], [6, Lemma 6.4]), we define the *augmented error*

$$\bar{\varepsilon} = \varepsilon + \hat{b}_m \zeta, \tag{5.15}$$

$$\zeta = \frac{\bar{\alpha}_0}{\bar{\alpha}(s)} \left[\hat{\vartheta}^\top \omega\right] - \hat{\vartheta}^\top \bar{\omega}, \tag{5.16}$$

where $\bar{\omega} = \frac{\bar{\alpha}_0}{\bar{\alpha}(s)}[\omega]$ is the filtered regressor, $\hat{b}_m$ is the additional adjustable parameter. In this case, the augmented error model takes the form

$$\bar{\varepsilon} = b_m \tilde{\vartheta}^\top \bar{\omega} - \tilde{b}_m \zeta + \epsilon, \tag{5.17}$$

where $\tilde{b}_m = b_m - \hat{b}_m$. For the tuning of the controller parameters, we apply the normalized adaptation laws [15, Sect. 5.4.6]

$$\dot{\hat{\vartheta}} = \gamma_1 \frac{\bar{\omega}\bar{\varepsilon}}{1 + \bar{\omega}^\top \bar{\omega}}, \quad \dot{\hat{b}}_m = -\gamma_2 \frac{\zeta \bar{\varepsilon}}{1 + \bar{\omega}^\top \bar{\omega}}, \tag{5.18}$$

where $\gamma_1 > 0$ and $\gamma_2 > 0$ are the adaptation gains. The normalization is used to ensure the boundedness of the closed-loop signals in the system since the vector ω depends on the output y, the boundedness of which is not established a priori. It is known (see Sect. 3.4.2, [15, Sect. 5.5.2], [7, Sect. 6.5.3]) that the adaptation laws (5.18) when applied to the error model (5.17) provide the boundedness of $\hat{\vartheta}$, $\hat{b}_m$, $\bar{\varepsilon}$, ε and drive ε to zero. Therefore, we can prove the following proposition.

Proposition 5.1 *If Assumptions 2.1 (with $\upsilon = g$) and 5.1 hold, then the adaptive controller (5.13) with the reference observer (5.4), the auxiliary filters (5.7), (5.8), the swapping scheme (5.15), (5.16), and the adaptation algorithms (5.18) when applied to the plant (5.1) provides the boundedness of all the closed-loop signals and the asymptotic reference tracking (5.2).*

Example 5.1 Consider the unstable relative-degree-two plant

$$y = \frac{b_0}{s^2 + a_1 s + a_0}[u], \tag{5.19}$$

where $b_0 = 3, a_1 = 0.9$, and $a_0 = -0.1$ are unknown. Let a priori unknown reference g be modelled by a linear exosystem of the second order, i.e., it can be represented by a sinusoidal signal with unknown amplitude, frequency, and phase or by an unknown constant. Therefore, for design of the reference observer, we select

$$G = \begin{bmatrix} 0 & 1 \\ -20 & -9 \end{bmatrix}, \quad l = \begin{bmatrix} 0 \\ 20 \end{bmatrix}.$$

To obtain the filtered regressor $\bar{\omega}$, we design the parameterization filters (5.7), (5.8) with $\bar{\chi}_1, \ \bar{\chi}_2 \in \mathbb{R}^1$, $\bar{\Lambda} = -5$, construct the regressor $\omega = [\bar{\chi}_1, \bar{\chi}_2, \hat{\xi}^\top, y]^\top \in \mathbb{R}^5$ and apply the transfer function $\frac{\bar{\alpha}_0}{\bar{\alpha}(s)}$ with the polynomial $\bar{\alpha}(s) = s^2 + 5s + 6$ (since $n - m = 2$) and $\bar{\alpha}_0 = 6$. In simulations, we stress that the proposed adaptive controller is able to track any reference generated by uncertain exosystem of the second order. To show this property, we simulate the reference with variable form

$$g(t) = \begin{cases} 1 & \text{if} \quad 0 \le t \le 45 \text{ s}, \\ \sin\left(\dfrac{\pi t}{2}\right) & \text{if} \quad t > 45 \text{ s}. \end{cases} \tag{5.20}$$

The initial conditions of $\hat{\vartheta}, \ \hat{b}_0, \ \bar{\chi}_1, \ \bar{\chi}_2$ are set to zero. Simulation results are presented in Fig. 5.1 and demonstrate asymptotic tracking of both types of a priori unknown reference signals generated by the uncertain exosystem of the second order. ■

5.1.2 *Compensation of Matched Disturbance*

Let us consider the LTI SISO plant

$$\dot{x} = Ax + b_M(\psi_u u + \upsilon), \tag{5.21}$$

where $x \in \mathbb{R}^n$ is the measured state, $u \in \mathbb{R}$ is the control, $\upsilon \in \mathbb{R}$ is the inaccessible for measurement disturbance satisfying Assumption 2.1, b_M is the known vector, and ψ_u is an unknown parameter. We make also the following assumptions regarding the plant (5.21).

Assumption 5.2 *The plant is such that:*

5.2.1. *the matrix A can be represented in the form*

$$A = A_M + b_M \psi^\top, \tag{5.22}$$

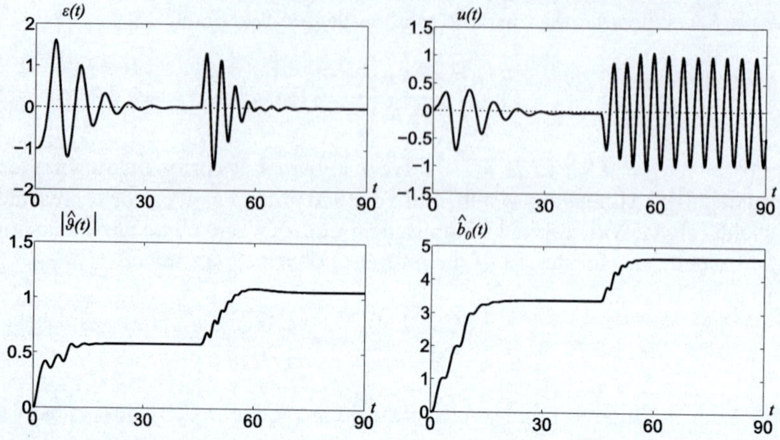

Fig. 5.1 Example 5.1: adaptive tracking of the output of the plant (5.19) to the reference (5.20) in the system closed by the algorithms of adaptation (5.18) with $\gamma_1 = 0.5$, $\gamma_2 = 10$

where $\psi \in \mathbb{R}^n$ *is an unknown vector,* $A_M \in \mathbb{R}^{n\times n}$ *is the known matrix;*

5.2.2. *the pair* (A_M, b_M) *is controllable;*

5.2.3. *the sign of* ψ_u *is known. Without loss of generality, we assume that* $\psi_u > 0$.

The considered control problem is to design a state-feedback controller ensuring the boundedness of all the closed-loop signals and asymptotic convergence of the state to zero:

$$\lim_{t\to\infty} |x(t)| = 0. \tag{5.23}$$

Based on the results established by Lemma 2.7, we use the filters

$$\dot{\eta}_0 = G\eta_0 + (GN - NA_M)x, \tag{5.24}$$

$$\dot{\eta}_i = G\eta_i - lx_i, \qquad i = 1, 2, \dots, n, \tag{5.25}$$

$$\dot{\eta}_u = G\eta_u - lu, \tag{5.26}$$

where $\eta_0, \eta_i, \eta_u \in \mathbb{R}^q$ are the states of the filters with arbitrary initial conditions, $N \in \mathbb{R}^{q\times n}$ is a matrix satisfying the equality

$$Nb_M = l, \tag{5.27}$$

and (G, l) is the pair from the exosystem canonical form (2.15) [21].

Then, as it is shown by Proposition 2.9, the parameterized plant model takes the form

$$\dot{x} = A_M x + b_M \left(\vartheta^\top \omega + \psi_u u + \epsilon\right)$$

or

$$\dot{x} = A_M x + \psi_u b_M \left(\bar{\vartheta}^\top \omega + u + \bar{\epsilon}\right), \tag{5.28}$$

where

$$\bar{\vartheta} = \frac{\vartheta}{\psi_u} = \frac{1}{\psi_u}\left[\psi^\top, \theta^\top, \psi_1\theta^\top, \psi_2\theta^\top, \ldots, \psi_n\theta^\top, \psi_u\theta^\top\right]^\top$$

is the aggregated vector of unknown parameters of dimension $n + (n+2)q$, ω is the regressor given by

$$\omega = \left[x^\top, (\eta_0 + Nx)^\top, \eta_1^\top, \eta_2^\top, \ldots, \eta_n^\top, \eta_u^\top\right]^\top, \tag{5.29}$$

η_0, η_i, η_u are generated by the filters (5.24)–(5.26), $\epsilon = \theta^\top \epsilon_\xi$ and $\bar{\epsilon} = \frac{\epsilon}{\psi_u}$ are the exponentially decaying terms, ϵ_ξ is the vector satisfying the equation $\dot{\epsilon}_\xi = G\epsilon_\xi$.

The model (5.28) motivates the following structure of the adjustable controller:

$$u = -k^\top x - \hat{\vartheta}^\top \omega, \tag{5.30}$$

where the vector of feedback gains k is chosen so that the matrix $A_\Sigma = A_M - b_M k^\top$ is Hurwitz, $\hat{\vartheta}$ is the vector of adjustable parameters. Substituting (5.30) into (5.28), we obtain

$$\dot{x} = A_\Sigma x + \psi_u b_M \left(\tilde{\vartheta}^\top \omega + \bar{\epsilon}\right), \tag{5.31}$$

where $\tilde{\vartheta} = \bar{\vartheta} - \hat{\vartheta}$. Then, for the tuning of parameters $\hat{\vartheta}$, we can use the Lyapunov-based algorithm of adaptation (see Remark 3.5)

$$\dot{\hat{\vartheta}} = \gamma \omega b_M^\top P x, \tag{5.32}$$

where $\gamma > 0$ is the adaptation gain, $P = P^\top \succ 0$ is the solution of the Lyapunov matrix equation $A_\Sigma^\top P + P A_\Sigma = -Q$ with arbitrary matrix $Q = Q^\top \succ 0$. Then, with the use of the Lyapunov function

$$V(x, \tilde{\vartheta}, \epsilon_\xi) = x^\top P x + \frac{\psi_u}{\gamma}\tilde{\vartheta}^\top \tilde{\vartheta} + \frac{\lambda_Q}{2\psi_u^2 \left|b_M^\top P\right|^2 |\theta|^2}\epsilon_\xi^\top P_\xi \epsilon_\xi,$$

where the matrix $P_\xi = P_\xi^\top \succ 0$ is the solution of the Lyapunov matrix equation $G^\top P_\xi + P_\xi G = -I$, λ_Q is the minimum eigenvalue of Q, we can prove the following proposition [17, 20].

Proposition 5.2 *If Assumptions 2.1 and 5.2 hold, then the adaptive controller (5.30) with the filters (5.24)–(5.26), and the adaptation algorithms (5.32) when applied to the plant (5.21) provides the boundedness of all the closed-loop signals and achievement of the control objective (5.23).*

Example 5.2 Consider the third-order unstable plant

$$\dot{x}_1 = -2x_1 - x_2 - \psi_3 x_3 - \psi_u u + \upsilon, \tag{5.33}$$

$$\dot{x}_2 = 4x_1 + 2x_2, \tag{5.34}$$

$$\dot{x}_3 = (3 + \psi_3)x_3 + \psi_u u + \upsilon. \tag{5.35}$$

In this case, $A = A_M + b_M \psi^\top$, $b = \psi_u b_M$, where

$$A_M = \begin{bmatrix} -2 & -1 & 0 \\ 4 & 2 & 0 \\ 0 & 0 & 3 \end{bmatrix}, \quad b_M = \begin{bmatrix} -1 \\ 0 \\ 1 \end{bmatrix}, \quad \psi = \begin{bmatrix} 0 \\ 0 \\ \psi_3 \end{bmatrix}.$$

The parameters $\psi_3 = 2$, $\psi_u = 3$ are unknown.

Let a priori unknown disturbance υ be modelled by a linear exosystem of the second order, i.e., it can be represented by a sinusoidal function with unknown amplitude, frequency and phase or by an unknown constant. For the filters (5.24)–(5.26), we choose

$$G = \begin{bmatrix} 0 & 1 \\ -12 & -7 \end{bmatrix}, \quad l = \begin{bmatrix} 0 \\ 12 \end{bmatrix}, \quad N = \begin{bmatrix} 0 & 0 & 0 \\ 0 & 0 & 12 \end{bmatrix}.$$

To obtain the control with the adaptation algorithms, we select

$$k = [20, 14, 35]^\top, \quad Q = -I_3$$

and get

$$A_\Sigma = A_M - b_M k^\top = \begin{bmatrix} 18 & 13 & 35 \\ 4 & 2 & 0 \\ 20 & -14 & -32 \end{bmatrix}, \quad P = \begin{bmatrix} 6.95 & 6.10 & 7.50 \\ 6.10 & 5.82 & 6.53 \\ 7.50 & 6.53 & 8.22 \end{bmatrix},$$

where the matrix A_Σ has eigenvalues $\lambda\{A_\Sigma\} = \{-2, -4, -6\}$.

To emphasize the fact that the proposed adaptive controller is able to reject any disturbance generated by uncertain exosystem of the second order, we again use the disturbance with variable waveform

$$\upsilon(t) = \begin{cases} 1 & \text{if} \quad 0 \leq t \leq 18 \text{ sec}, \\ \sin\left(\dfrac{\pi t}{2}\right) & \text{if} \quad t > 18 \text{ sec}. \end{cases} \tag{5.36}$$

In simulation, we use zero initial conditions $\hat{\vartheta}(0)$, $\eta_0(0)$, $\eta_1(0)$, $\eta_2(0)$, $\eta_3(0)$, and $\eta_u(0)$. The adaptation gain is given by $\gamma = 100$.

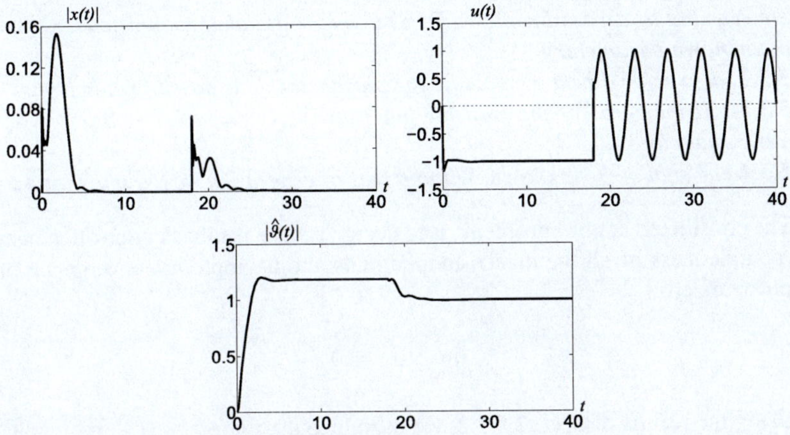

Fig. 5.2 Example 5.2: adaptive state-feedback compensation of the matched disturbance (5.36) affecting the linear plant with unknown parameters

Simulation results are presented in Fig. 5.2 and demonstrate complete compensation of a priori unknown disturbance affecting the linear plant with unknown parameters. ■

5.1.3 *Compensation of Unmatched Disturbance*

In this section, we present approaches to the state-feedback and output-feedback adaptive compensation of unmatched disturbances.

5.1.3.1 Plants with Measured State

Now, let us consider the SISO LTI plant

$$\dot{x} = Ax + bu + d\upsilon, \quad y = c^\top x + d_n \upsilon, \tag{5.37}$$

where $x \in \mathbb{R}^n$ is the measured state, $u \in \mathbb{R}$, $y \in \mathbb{R}$ are the input and the output, respectively, $\upsilon \in \mathbb{R}$ is the inaccessible for measurements disturbance satisfying Assumption 2.1. We make the following assumption regarding the plant.

Assumption 5.3 *The plant* (5.37) *is such that:*

5.3.1. *the matrices A and b can be represented in the forms*

$$A = A_M + b_M \psi^\top, \quad b = \psi_u b_M, \tag{5.38}$$

where A_M and b_M are known, $\psi \in \mathbb{R}^n$ is a vector of unknown parameters, and ψ_u is an unknown parameter;

5.3.2. *the vector d is known, while the parameter d_n is possibly unknown;*

5.3.3. *all the zeros of the transfer function $W_M(s) = c^\top(sI - A_M)^{-1}b_M$ have negative real parts;*

5.3.4. *the sign of ψ_u is known. Without loss of generality, we assume that $\psi_u > 0$.*

The considered control problem[1] is to design a state-feedback controller ensuring the boundedness of all the closed-loop signals and asymptotic convergence of the output y to zero:

$$\lim_{t\to\infty} y(t) = 0. \tag{5.39}$$

Using the results of Sect. 2.5.1.2, we introduce the filters

$$\dot{\eta}_0 = G\eta_0 + (GN - NA_\Sigma)x, \tag{5.40}$$

$$\dot{\eta}_i = G\eta_i - Nb_M x_i, \qquad i = 1, 2, \ldots, n, \tag{5.41}$$

$$\dot{\eta}_u = G\eta_u - Nb_M u, \tag{5.42}$$

where $\eta_0, \eta_i, \eta_u \in \mathbb{R}^q$ are the states of the filters with arbitrary initial conditions, the pair (G, l) is from the exosystem canonical form (2.15), $N \in \mathbb{R}^{q\times n}$ is a matrix that satisfies the equality [21]

$$Nd = l. \tag{5.43}$$

Then, according to Proposition 2.10, the parameterized plant model can be represented in the form

$$\dot{e} = A_\Sigma e + b_M(\vartheta^\top\omega + \psi_u u + \epsilon), \quad y = c^\top e$$

or

$$\dot{e} = A_\Sigma e + \psi_u b_M(\bar{\vartheta}^\top\omega + u + \bar{\epsilon}), \quad y = c^\top e, \tag{5.44}$$

where $e \in \mathbb{R}^n$ is the error state (not accessible for measurement), $\bar{\vartheta} \in \mathbb{R}^{n+(n+2)q}$ is the vector of unknown parameters given by

$$\bar{\vartheta} = \frac{\vartheta}{\psi_u} = \frac{1}{\psi_u}\left[\bar{\psi}^\top, \theta^\top, \bar{\psi}_1\theta^\top, \bar{\psi}_2\theta^\top, \ldots, \bar{\psi}_n\theta^\top, \psi_u\theta^\top\right]^\top, \tag{5.45}$$

$\bar{\psi} = \psi + k$, $k \in \mathbb{R}^n$ is the vector chosen so that the matrix $A_\Sigma = A_M - b_M k^\top$ is Hurwitz, ω is the extended regressor defined as

[1] A particular case of the state derivative feedback is considered in [1, 2].

$$\omega = \left[x^\top, (\eta_0 + Nx)^\top, \eta_1^\top, \eta_2^\top, \ldots, \eta_n^\top, \eta_u^\top\right]^\top, \tag{5.46}$$

η_0, η_i, η_u are generated by the filters (5.40)–(5.43), and ϵ, $\bar{\epsilon}$ are the exponentially decaying terms.

Choosing the control law as

$$u = -\hat{\vartheta}^\top \omega, \tag{5.47}$$

where $\hat{\vartheta}$ is the vector of adjustable parameters, we obtain the closed-loop error model

$$\dot{e} = A_\Sigma x + \psi_u b_M \left[\tilde{\vartheta}^\top \omega + \bar{\epsilon}\right], \quad y = c^\top e$$

and (since the state e is not measured) its input–output form

$$y = \psi_u W_\Sigma(s) \left[\tilde{\vartheta}^\top \omega\right] + \epsilon_y \tag{5.48}$$

with the vector of parametric errors $\tilde{\vartheta} = \vartheta - \hat{\vartheta}$, the known asymptotically stable transfer function $W_\Sigma(s) = c^\top (sI - A_\Sigma)^{-1} b_M$, and some exponentially decaying term ϵ_y .

To adjust $\hat{\vartheta}$, one can use, for example, adaptation algorithms with augmented error applied to the error model with an unknown high-frequency gain (for more details, see Sect. 5.1.1, the error model (5.14)). In the considered case, the algorithms are given by the following equations:

$$\dot{\hat{\vartheta}} = \gamma_1 \frac{\bar{\omega} \bar{y}}{1 + \bar{\omega}^\top \bar{\omega}}, \quad \dot{\hat{\psi}}_u = -\gamma_2 \frac{\zeta \bar{y}}{1 + \bar{\omega}^\top \bar{\omega}}, \tag{5.49}$$

where $\gamma_1, \gamma_2 > 0$ are the adaptation gains, $\bar{\omega} = W_\Sigma(s)[\omega]$ is the filtered regressor, $\hat{\psi}_u$ is the additional adjustable parameter, $\bar{y}$ is the augmented output given by

$$\bar{y} = y + \hat{\psi}_u \zeta, \tag{5.50}$$

ζ is the auxiliary signal (the swapping term) defined as

$$\zeta = W_\Sigma(s) \left[\hat{\vartheta}^\top \omega\right] - \hat{\vartheta}^\top \bar{\omega}. \tag{5.51}$$

Substitution of (5.51) and (5.48) into (5.50) gives the static error model

$$\bar{y} = \psi_u \tilde{\vartheta}^\top \bar{\omega} - \tilde{\psi}_u \zeta + \epsilon_y, \tag{5.52}$$

where $\tilde{\psi}_u = \psi_u - \hat{\psi}_u$ is the additional parametric error. This model determines the structure of the adaptation algorithms (5.49). Here we use the normalized adaptation algorithms since the regressor ω depends on y and u and is not known to be bounded a priori.

The model (5.52) with the adaptation algorithms (5.49) is standard for the problems of MRAC [15, Sect. 5.4.6]. Therefore, according to the well-known results (see Proposition 3.10 and [15, Sect. 5.5.2]), we directly obtain the following proposition.

Proposition 5.3 *If Assumptions 2.1 and 5.3 hold, then the adaptive controller (5.47) with the filters (5.40)–(5.43) and the adaptation algorithms (5.49)–(5.51) when applied to the plant (5.37) provides the boundedness of all the closed-loop signals and achievement of the control objective (5.39).*

Example 5.3 Consider the plant

$$\dot{x}_1 = x_2, \tag{5.53}$$

$$\dot{x}_2 = x_3 + 2\upsilon, \tag{5.54}$$

$$\dot{x}_3 = \psi_1 x_1 + \psi_2 x_2 + \psi_3 x_3 + \psi_u u + \upsilon, \tag{5.55}$$

$$y = x_1 + x_2 + 2x_3 + \upsilon. \tag{5.56}$$

It is easy to see that this plant belongs to the class of plants (5.37) with $A = A_M + b_M \psi^\top$, $b = \psi_u b_M$,

$$A_M = \begin{bmatrix} 0\,1\,0 \\ 0\,0\,1 \\ 0\,0\,0 \end{bmatrix}, \quad b_M = \begin{bmatrix} 0 \\ 0 \\ 1 \end{bmatrix}, \quad \psi = \begin{bmatrix} 0 \\ 0 \\ \psi_3 \end{bmatrix},$$

$$c = \begin{bmatrix} 1 \\ 1 \\ 2 \end{bmatrix}, \quad d = \begin{bmatrix} 0 \\ 2 \\ 1 \end{bmatrix}, \quad d_3 = 1,$$

where $\psi = [\psi_1, \psi_2, \psi_3]^\top = [1, 2, 3]^\top$ is the unknown vector, $\psi_u = 1$ is the unknown parameter.

Let, as in Example 5.2, a priori unknown disturbance υ be modelled by a linear exosystem of the second order. Choosing G, l, and N as

$$G = \begin{bmatrix} 0 & 1 \\ -12 & -7 \end{bmatrix}, \quad l = \begin{bmatrix} 0 \\ 12 \end{bmatrix}, \quad N = \begin{bmatrix} 0\,0\,0 \\ 0\,0\,12 \end{bmatrix},$$

we obtain the filters (5.40)–(5.42) with the regressor (5.46). For the selected $k = [20, 14, 35]^\top$, we have

$$W_\Sigma(s) = \frac{2s^2 + s + 1}{s^3 + 12s^2 + 44s + 48}.$$

In simulation, we set the initial conditions of $\hat{\theta}$, $\hat{\psi}_u$, η_0, η_3, and η_u to zero and select the adaptation gains as $\gamma_1 = 15$, $\gamma_2 = 10$. As in the previous examples, the plant is affected by the disturbance with variable waveform

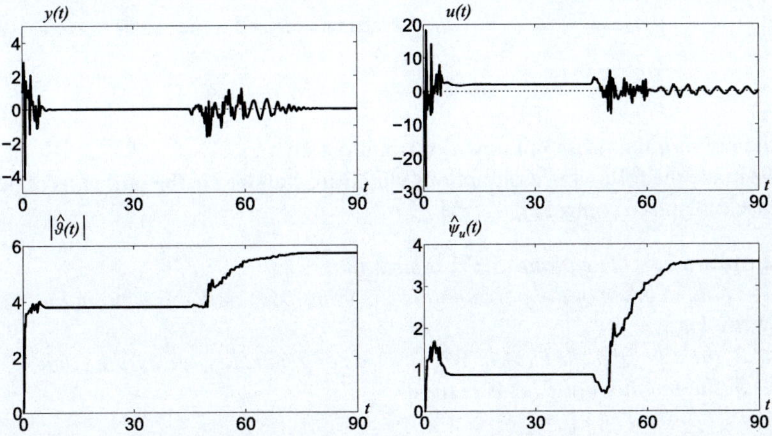

Fig. 5.3 Example 5.3: adaptive state-feedback compensation of the unmatched disturbance (5.57) affecting the linear plant with unknown parameters

$$\upsilon(t) = \begin{cases} 1 & \text{if } \quad 0 \leq t \leq 45 \text{ s}, \\ \sin\left(\dfrac{\pi t}{2}\right) & \text{if } \quad t > 45 \text{ s}. \end{cases} \tag{5.57}$$

Simulation results obtained for closed-loop adaptive system with the adaptation algorithms (5.49) are presented in Fig. 5.3 and demonstrate the ability of the adaptive controller to completely compensate a priori unknown disturbance that belongs to a certain class of time functions. ■

5.1.3.2 Plants with Measured Output

Let us consider the plant (5.37), where A, b, c, d, and d_n are unknown, the triple (A, b, c) is observable and controllable, the state x is not accessible for measurements. The control objective is to design an output-feedback adaptive controller providing the boundedness of all the closed-loop signals and driving the plant output to zero according to the control objective (5.39) [18, 26].[2]

We introduce the transfer functions of the plant (5.37)

$$W_u(s) = c^\top (sI - A)^{-1} b = b_m \frac{\beta(s)}{\alpha(s)}, \quad W_\upsilon(s) = d_n + c^\top (sI - A)^{-1} d = \frac{\delta(s)}{\alpha(s)},$$

where

[2] A particular case of a single sinusoidal matched disturbance and identification-based controller is considered in [3].

$$\alpha(s) = s^n + a_{n-1}s^{n-1} + \cdots + a_0, \quad b_m\bar{\beta}(s) = b_m s^m + b_{m-1}s^{m-1} + \cdots + b_0,$$

$$\delta(s) = d_n s^n + d_{n-1}s^{n-1} + \cdots + d_0$$

are the polynomials with unknown constant parameters.

We make the following assumptions which are standard in the problems of output-feedback adaptive control [7, 8, 15].

Assumption 5.4 *The plant (5.37) is such that:*

5.4.1. *the degrees of the polynomials $\alpha(s)$ and $\bar{\beta}(s)$ denoted by n and m, respectively, are known;*

5.4.2. *the sign of b_m is known. Without loss of generality, we assume that $b_m > 0$;*

5.4.3. *the polynomial $\bar{\beta}(s)$ is Hurwitz.*

Now, applying the results of Sect. 2.5.1.3, we use the filters

$$\dot{\zeta}_n = \bar{A}\zeta_n + \bar{l}y, \tag{5.58}$$

$$\dot{\zeta}_i = \bar{A}\zeta_i + \varsigma_{n-i}y, \qquad i = 0, 1, \ldots, n-1, \tag{5.59}$$

$$\dot{\nu}_i = \bar{A}\nu_i + \varsigma_{n-i}u, \qquad i = 0, 1, \ldots, m, \tag{5.60}$$

$$\dot{\eta}_0 = G\eta_0 + Gly - l\beta_1, \tag{5.61}$$

$$\dot{\eta}_i = G\eta_i - l\bar{\varphi}_i, \qquad i = 1, 2, \ldots, n+m+1, \tag{5.62}$$

where the vector of feedback gains $\bar{l} = \left[\bar{l}_{n-1}, \bar{l}_{n-2}, \ldots, \bar{l}_0\right]^{\top}$ is chosen so that the matrix $\bar{A} = E - \bar{l}\bar{c}^{\top}$ is Hurwitz,

$$E = \begin{bmatrix} 0 & \\ \vdots & I_{n-1} \\ 0 & \cdots\ 0 \end{bmatrix},$$

the pair (G, l) is from the exosystem canonical form (2.15), the function β_1 is given by

$$\beta_1 = -\bar{l}_{n-1}\zeta_{n,1} + \zeta_{n,2} + \bar{l}_{n-1}y, \tag{5.63}$$

the functions $\bar{\varphi}_i$ are the elements of the regressor

$$\begin{aligned} \bar{\varphi} = \big[&-\bar{l}_{n-1}\nu_{m,1} + \nu_{m,2}, -\bar{l}_{n-1}\nu_{m-1,1} + \nu_{m-1,2}, \ldots, \\ &-\bar{l}_{n-1}\nu_{0,1} + \nu_{0,2}, -\bar{l}_{n-1}\zeta_{n-1,1} + \zeta_{n-1,2} + y, \\ &-\bar{l}_{n-1}\zeta_{n-2,1} + \zeta_{n-2,2}, \ldots, -\bar{l}_{n-1}\zeta_{0,1} + \zeta_{0,2}\big]^{\top}. \end{aligned} \tag{5.64}$$

Then, according to the results from Sect. 2.5.1.3, the plant (5.37) affected by the external disturbance admits the following parameterization:

$$\dot{y} = \beta_1 + \vartheta^\top \bar{\omega} + b_m(-\bar{l}_{n-1}v_{m,1} + v_{m,2}) + \epsilon, \tag{5.65}$$

$$\dot{v}_{m,i} = -\bar{l}_{n-i}v_{m,1} + v_{m,i+1}, \qquad i = 2, 3, \ldots, n-m-1, \tag{5.66}$$

$$\dot{v}_{m,n-m} = -\bar{l}_{n-m}v_{m,1} + v_{m,n-m+1} + u, \tag{5.67}$$

where $\vartheta \in \mathbb{R}^{(n+m+1)(q+1)+q}$ is a vector of unknown parameters related to the plant unknown parameters vector $\psi \in \mathbb{R}^{n+m+1}$ and the vector of exosystem unknown parameters $\theta \in \mathbb{R}^q$ via the expressions

$$\vartheta = \left[\psi^\top, \theta^\top, \psi_1\theta^\top, \ldots, \psi_{n+m+1}\theta^\top\right]^\top,$$
$$\psi = \left[b_m, b_{m-1}, \ldots, b_0, -a_{n-1}, -a_{n-2}, \ldots, -a_0\right]^\top,$$

$\bar{\omega} = \left[\bar{\varphi}_0^\top, (\eta_0 + ly)^\top, \eta_1^\top, \ldots, \eta_{n+m+1}^\top\right]^\top$, $\bar{\varphi}_0$ is obtained from $\bar{\varphi}$ (5.64) by zeroing of the first element, i.e.,

$$\bar{\varphi}_0 = \left[0, -\bar{l}_{n-1}v_{m-1,1} + v_{m-1,2}, \ldots, -\bar{l}_{n-1}v_{0,1} + v_{0,2},\right. \tag{5.68}$$
$$\left. -\bar{l}_{n-1}\zeta_{n-1,1} + \zeta_{n-1,2} + y, -\bar{l}_{n-1}\zeta_{n-2,1} + \zeta_{n-2,2}, \ldots, -\bar{l}_{n-1}\zeta_{0,1} + \zeta_{0,2}\right]^\top,$$

β_1 is defined by (5.63), ϵ is the exponentially decaying term.

Equation (5.65) can be represented in the following more compact form used in further calculations:

$$\dot{y} = \beta_1 + \vartheta^\top \omega + \epsilon, \tag{5.69}$$

where $\omega = \left[\bar{\varphi}^\top, (\eta_0 + ly)^\top, \eta_1^\top, \ldots, \eta_{n+m+1}^\top\right]^\top$ is the extended $((n+m+1)(q+1)+q)$-dimensional regressor.

The model (5.65)–(5.67) is used in the backstepping design of virtual control laws (stabilizing functions α_i) and virtual adaptation laws (tuning functions τ_i) as follows (see Sect. E.2):

$$\alpha_1(y, \hat{\vartheta}, \hat{\kappa}_m, \varpi) = -(c_1 + s_1)z_1 + \hat{\kappa}_m\bar{\alpha}_1 + \bar{l}_{n-1}v_{m,1}, \tag{5.70}$$

$$\bar{\alpha}_1(y, \hat{\vartheta}, \varpi) = -\beta_1 - \bar{\omega}^\top\hat{\vartheta}, \tag{5.71}$$

$$\alpha_2(y, \hat{\vartheta}, \hat{\kappa}_m, \varpi) = -(c_2 + s_2)z_2 + \left(\frac{\partial \alpha_1}{\partial y}\omega^\top - \varsigma_1^\top z_1\right)\hat{\vartheta} + \frac{\partial \alpha_1}{\partial \hat{\vartheta}}\tau_2 + \beta_2, \tag{5.72}$$

$$\alpha_i(y, \hat{\vartheta}, \hat{\kappa}_m, \varpi) = -z_{i-1} - (c_i + s_i)z_i + \frac{\partial \alpha_{i-1}}{\partial y}\omega^\top\hat{\vartheta} + \frac{\partial \alpha_{i-1}}{\partial \hat{\vartheta}}\tau_i$$
$$+ \beta_i - \gamma_1 \frac{\partial \alpha_{i-1}}{\partial y}\left(\sum_{j=2}^{i-1} \frac{\partial \alpha_{j-1}}{\partial \hat{\vartheta}} z_j\right)\omega, \quad i = 3, 4, \ldots, n-m, \tag{5.73}$$

where $z_1 = v_{m,1}$, $z_i = v_{m,i} - \alpha_{i-1}$ are the state errors,

$$s_1 = \mu_1, \quad s_i = \mu_i \left(\frac{\partial \alpha_{i-1}}{\partial y} \right)^2, \quad i = 1, 2, \ldots, n - m \tag{5.74}$$

are the damping terms with constant gains $\mu_i > 0$, $c_i > 0$, and $\gamma_1 > 0$ are the design parameters,

$$\tau_1 = \gamma_1 \bar{\omega} z_1, \tag{5.75}$$

$$\tau_2 = \tau_1 - \gamma_1 \left(\frac{\partial \alpha_1}{\partial y} \omega - \varsigma_1 z_1 \right) z_2, \tag{5.76}$$

$$\tau_i = \tau_{i-1} - \gamma_1 \frac{\partial \alpha_{i-1}}{\partial y} \omega z_i, \quad i = 3, 4, \ldots, n - m \tag{5.77}$$

are the tuning functions, $\hat{\vartheta} \in \mathbb{R}^{(n+m+1)(q+1)+q}$ is the vector of adjustable parameters, $\hat{\kappa}_m$ is the additional adjustable parameter corresponding to the unknown constant $\kappa_m = \frac{1}{b_m}$, $\varpi = (\zeta_0, \zeta_2, \ldots, \zeta_n, \nu_0, \nu_1, \ldots, \nu_m, \eta_0, \eta_1, \ldots, \eta_{n+m+1})$ is the set of known variables necessary for control laws design,

$$\begin{aligned}
\beta_2 = {} & \bar{l}_{n-2} \nu_{m,1} + \frac{\partial \alpha_1}{\partial y} \beta_1 + \frac{\partial \alpha_1}{\partial \hat{\kappa}_m} \dot{\hat{\kappa}}_m \\
& + \frac{\partial \alpha_1}{\partial \zeta_n} (\bar{A} \zeta_n + \bar{l} y) + \sum_{r=0}^{n-1} \left(\frac{\partial \alpha_1}{\partial \zeta_r} (\bar{A} \zeta_r + \varsigma_{n-r} y) \right) \\
& + \sum_{r=0}^{m-1} \sum_{j=1}^{2} \left(\frac{\partial \alpha_1}{\partial \nu_{r,j}} (-\bar{l}_{n-j} \nu_{r,1} + \nu_{r,j+1}) \right) + \frac{\partial \alpha_1}{\partial \nu_{m,1}} (-\bar{l}_{n-1} \nu_{m,1} + \nu_{m,2}) \\
& + \frac{\partial \alpha_1}{\partial \eta_0} (G \eta_0 + G l y - l \beta_1) + \sum_{r=1}^{n+m+1} \left(\frac{\partial \alpha_1}{\partial \eta_r} (G \eta_r - l \bar{\varphi}_r) \right),
\end{aligned} \tag{5.78}$$

$$\begin{aligned}
\beta_i = {} & \bar{l}_{n-i} \nu_{m,1} + \frac{\partial \alpha_{i-1}}{\partial y} \beta_1 + \frac{\partial \alpha_{i-1}}{\partial \hat{\kappa}_m} \dot{\hat{\kappa}}_m \\
& + \frac{\partial \alpha_{i-1}}{\partial \zeta_n} (\bar{A} \zeta_n + \bar{l} y) + \sum_{r=0}^{n-1} \left(\frac{\partial \alpha_{i-1}}{\partial \zeta_r} (\bar{A} \zeta_r + \varsigma_{n-r} y) \right) \\
& + \sum_{r=0}^{m-1} \sum_{j=1}^{i} \left(\frac{\partial \alpha_{i-1}}{\partial \nu_{r,j}} (-\bar{l}_{n-j} \nu_{r,1} + \nu_{r,j+1}) \right) + \sum_{j=1}^{i-1} \left(\frac{\partial \alpha_{i-1}}{\partial \nu_{m,j}} (-\bar{l}_{n-j} \nu_{m,1} + \nu_{m,j+1}) \right) \\
& + \frac{\partial \alpha_{i-1}}{\partial \eta_0} (G \eta_0 + G l y - l \beta_1) + \sum_{r=1}^{n+m+1} \left(\frac{\partial \alpha_{i-1}}{\partial \eta_r} (G \eta_r - l \bar{\varphi}_r) \right)
\end{aligned} \tag{5.79}$$

are the known functions.

The actual control law and the actual adaptation algorithms are obtained in the final step of the design procedure:

$$u = \frac{1}{\varphi_u(y)}\left(\alpha_{n-m}(y, \hat{\vartheta}, \hat{\kappa}_m, \varpi) - v_{m,n-m+1}\right) \tag{5.80}$$

and

$$\dot{\hat{\vartheta}} = \tau_{n-m}, \tag{5.81}$$

$$\dot{\hat{\kappa}}_m = -\gamma_2 \bar{\alpha}_1 z_1, \tag{5.82}$$

where $\gamma_2 > 0$ is the adaptation gain.

The control (5.70)–(5.82) results in the following closed-loop error model:

$$\dot{z} = A_z(z,t)z + W_\vartheta^\top(z,t)\tilde{\vartheta} - b_m \varsigma_1 \bar{\alpha}_1 \tilde{\kappa}_m + b_z(z,t)\epsilon, \tag{5.83}$$

$$\dot{\tilde{\vartheta}} = -\gamma_1 W_\vartheta z, \tag{5.84}$$

$$\dot{\tilde{\kappa}}_m = \gamma_2 \bar{\alpha}_1 \varsigma_1^\top z, \tag{5.85}$$

$$y = z_1, \tag{5.86}$$

where $z = [z_1, z_2, ..., z_{n-m}]^\top$ is the vector of the state errors, $\tilde{\vartheta} = \vartheta - \hat{\vartheta}$, $\tilde{\kappa}_m = \kappa_m - \hat{\kappa}_m$ are the parametric errors,

$$A_z = \begin{bmatrix} -b_m(c_1+s_1) & b_m & 0 & 0 & \cdots & 0 \\ -b_m & -c_2 - s_2 & 1+\sigma_{2,3} & \sigma_{2,4} & \ddots & \sigma_{2,n-m} \\ 0 & -1-\sigma_{2,3} & -c_3 - s_3 & 1+\sigma_{3,4} & \ddots & \sigma_{3,n-m} \\ 0 & -\sigma_{2,4} & -1-\sigma_{3,4} & \ddots & \ddots & \sigma_{4,n-m} \\ \vdots & \ddots & \ddots & \ddots & \ddots & \vdots \\ 0 & -\sigma_{2,n-m} & -\sigma_{3,n-m} & -\sigma_{4,n-m} & \cdots & -c_{n-m} - s_{n-m} \end{bmatrix},$$

$$b_z = \begin{bmatrix} 1 \\ -\frac{\partial \alpha_1}{\partial y} \\ -\frac{\partial \alpha_2}{\partial y} \\ \vdots \\ -\frac{\partial \alpha_{n-m-1}}{\partial y} \end{bmatrix}, \quad W_\vartheta^\top = \begin{bmatrix} \bar{\omega}^\top \\ \frac{\partial \alpha_1}{\partial y}\omega^\top \\ \frac{\partial \alpha_2}{\partial y}\omega^\top \\ \vdots \\ \frac{\partial \alpha_{n-m-1}}{\partial y}\omega^\top \end{bmatrix} + \varsigma_2 \varsigma_1^\top z_1,$$

$$\sigma_{i,j} = \begin{cases} 0 & \text{if } \ i = 1, \\ \gamma_1 \frac{\partial \alpha_{j-1}}{\partial y}\frac{\partial \alpha_1}{\partial \hat{\vartheta}}\bar{\omega} & \text{if } \ i = 2, \\ \frac{\partial \alpha_{j-1}}{\partial y}\gamma_1 \frac{\partial \alpha_{j-1}}{\partial \hat{\vartheta}}\omega & \text{if } \ i = 3, 4, \ldots, n-m, \end{cases}$$

$\varsigma_1 \in \mathbb{R}^{(n+m+1)(q+1)+q}$, $\varsigma_2 \in \mathbb{R}^n$ are the coordinate vectors. For the sake of compactness, we use the notations $A_z(z,t)=A_z(z,\hat{\vartheta},\hat{\kappa}_m,\varpi)$, $W_\vartheta^\top(z,t) = W_\vartheta^\top(z,\hat{\vartheta},\hat{\kappa}_m,\varpi)$, $b_z(z,t) = b_z(z,\hat{\vartheta},\hat{\kappa}_m,\varpi)$.

The model (5.83)–(5.86) is derived by taking the time derivative of z_i, $\tilde{\vartheta}$, and $\tilde{\kappa}_m$ in view of (5.65)–(5.67) and (5.70)–(5.82) (for more details, see [8, Sect. 10.2.1]).

Remark 5.1 The state error model (5.83) is not complete because it does not contain additional m equations of the zero dynamics. However, by virtue of Assumption 5.4.3, the stabilization of the model (5.83) implies the boundedness of all the closed-loop signals [6, Sect. 10.2.2]. ■

The properties of the closed-loop adaptive system are established as follows [18].

Proposition 5.4 *If Assumptions 2.1 and 5.4 hold, then the tuning functions adaptive controller (5.70)–(5.82) with the filters (5.58)–(5.62) when applied to the plant (5.37) ensures the boundedness of all the closed-loop signals and achievement of the control objective (5.39).*

Proof The proof of the proposition directly follows from the proof of Theorem 10.6 from [8]. First, from Proposition E.2 we obtain that z $\hat{\vartheta}$, and $\hat{\kappa}_m$ are bounded. Since z_1 is bounded, y is also bounded, and from (5.58), (5.59) we obtain the boundedness of ζ_i, $i = 0, 1, \ldots, n$. Then, analyzing the filters (5.60) and expressing u from y (with the use of the plant inverse dynamics and Assumption 5.4.3), we conclude that ν_i are bounded for $i = 0, 1, \ldots, m$. Therefore, β_1 is bounded, and, as a result, η_0 and $\bar{\varphi}$ are also bounded. Then from (5.62), we obtain the boundedness of η_i, $i = 1, 2, \ldots, n+m+1$ and, therefore, the boundedness of the regressor ω. By virtue of Assumption 5.4.3, we conclude that the zero dynamics of the closed-loop system is asymptotically stable and, as a result, the state of the plant x is bounded. In view of the boundedness of all the closed-loop signals, from Theorem B.4 we conclude that $z_1(t) \to 0$, and therefore, $y(t) \to 0$ as $t \to \infty$. □

Example 5.4 Consider the unstable plant of the form (5.37) with unknown A, b, c, d, and d_n given by

$$A = \begin{bmatrix} 0 & 1 \\ 6 & -1 \end{bmatrix}, \quad b = \begin{bmatrix} 0 \\ 2 \end{bmatrix}, \quad c = \begin{bmatrix} 3 \\ 4 \end{bmatrix}, \quad d = \begin{bmatrix} 1 \\ 0 \end{bmatrix}, \quad d_n = 0.25,$$

and with the inaccessible state x. The plant is affected by the disturbance $\upsilon(t) = 2\sin(t+1)$ with unknown amplitude, phase, and frequency.

Using the prior information about the plant and the form of disturbance, we construct the filters (5.59)–(5.62), where $n = 2$, $m = 0$, and $q = 2$. For the filters, we select

$$\bar{A} = \begin{bmatrix} -5 & 1 \\ -6 & 0 \end{bmatrix}, \quad \bar{l} = \begin{bmatrix} \bar{l}_1 \\ \bar{l}_0 \end{bmatrix} = \begin{bmatrix} 5 \\ 6 \end{bmatrix}, \quad G = \begin{bmatrix} 0 & 1 \\ -12 & -7 \end{bmatrix}, \quad l = \begin{bmatrix} 0 \\ 12 \end{bmatrix}.$$

For the controller design, we define the vector $\omega \in \mathbb{R}^{11}$ as

$$\omega = \left[-\bar{l}_1 \nu_{0,1} + \nu_{0,2}, -\bar{l}_1 \zeta_{1,1} + \zeta_{1,2} + y, -\bar{l}_1 \zeta_{0,1} + \zeta_{0,2}, (\eta_0 + ly)^\top, \eta_1^\top, \eta_2^\top, \eta_3^\top\right]^\top.$$

The vector $\bar{\omega} \in \mathbb{R}^{11}$ is obtained by zeroing of the first element of ω. Then, according to the backstepping procedure, we derive the following stabilizing functions:

$$\alpha_1 = -(c_1 + s_1)z_1 - \hat{\kappa}_0 \left(\beta_1 + \bar{\omega}^\top \hat{\vartheta}\right) + \bar{l}_1 \nu_{0,1},$$

$$\alpha_2 = -(c_2 + s_2)z_2 + \left(\frac{\partial \alpha_1}{\partial y} \omega^\top - \varsigma_1^\top z_1\right) \hat{\vartheta} + \frac{\partial \alpha_1}{\partial \hat{\vartheta}} \tau_2 + \beta_2,$$

where $z_1 = y$, $z_2 = \nu_{0,2} - \alpha_1$, $c_1 = c_2 = 1$,

$$s_1 = 1, \quad s_2 = \left(\frac{\partial \alpha_{i-1}}{\partial y}\right)^2,$$

$$\beta_1 = -\bar{l}_1 \zeta_{2,1} + \zeta_{2,2} + \bar{l}_1 y,$$

$$\beta_2 = \bar{l}_0 \nu_{0,1} + \frac{\partial \alpha_1}{\partial y} \beta_1 + \frac{\partial \alpha_1}{\partial \hat{\kappa}_0} \dot{\hat{\kappa}}_0 + \frac{\partial \alpha_1}{\partial \zeta_2} \left(\bar{A} \zeta_2 + \bar{l} y\right)$$
$$+ \sum_{r=0}^{1} \left(\frac{\partial \alpha_1}{\partial \zeta_r} \left(\bar{A} \zeta_r + \varsigma_{2-r} y\right)\right) + \frac{\partial \alpha_1}{\partial \nu_{0,1}} \left(-\bar{l}_1 \nu_{0,1} + \nu_{0,2}\right)$$
$$+ \frac{\partial \alpha_1}{\partial \eta_0} \left(G\eta_0 + Gly - l\beta_1\right) + \sum_{r=1}^{3} \left(\frac{\partial \alpha_1}{\partial \eta_r} \left(G\eta_r - l\bar{\omega}_r\right)\right),$$

$$\tau_2 = \gamma_1 \left(\bar{\omega} z_1 - \frac{\partial \alpha_1}{\partial y} \omega z_2 - \varsigma_1 z_1 z_2\right),$$

$\varsigma_1 \in \mathbb{R}^{11}$ is the coordinate vector. Then the actual control law is given by

$$u = \frac{1}{\varphi_u(y)} \alpha_2.$$

The adjustable parameters $\hat{\vartheta} \in \mathbb{R}^{11}$ and $\hat{\kappa}_0$ are generated by the adaptation algorithms

$$\dot{\hat{\vartheta}} = \tau_2, \qquad \dot{\hat{\kappa}}_0 = \gamma_2 \left(\beta_1 + \bar{\omega}^\top \hat{\vartheta}\right) z_1.$$

In simulation, we select $\gamma_1 = \gamma_2 = 10$ and zero initial conditions for x, ζ_1, ζ_2, ζ_3, ν_0, $\eta_0, \ldots, \eta_3$, $\hat{\vartheta}$, and $\hat{\kappa}_0$.

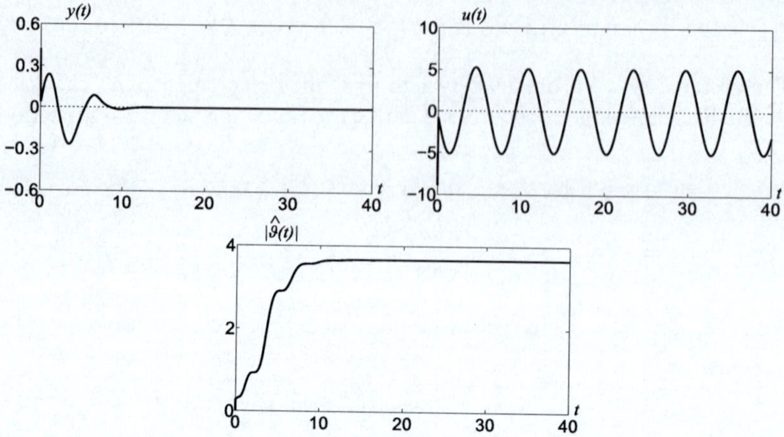

Fig. 5.4 Example 5.4: adaptive output-feedback compensation of the uncertain disturbance $\upsilon(t) = 2\sin(t+1)$ in the plant with unknown parameters (5.37)

Simulation results of the closed-loop system are shown in Fig. 5.4. It is seen from the presented plots that, despite the plant uncertainty, the control compensates the influence of the uncertain disturbance and drives the plant output to zero. ■

5.1.4 Regulation in LTI SISO Plants

Now, we combine the above approaches to simultaneously solve the problems of the adaptive reference tracking and the adaptive disturbance rejection.

Let us consider the plant (5.37) satisfying Assumption 5.3. The considered problem consists in design of a state-feedback adaptive controller providing the boundedness of all the closed-loop signals and achievement of the control objective

$$\lim_{t\to\infty} \varepsilon(t) = 0, \tag{5.87}$$

where $\varepsilon = y - g$ is the tracking error, g is the measured reference. In addition to Assumption 5.3, we also accept Assumption 4.3 concerning the properties of the exosignals.

Thus, we consider Problem 1.4 of the adaptive output regulation and solve it here for LTI SISO plants with measured state, unknown parameters, and affected by unmatched disturbance.[3]

[3] The case of LTI SISO plants with unknown parameters belonging to known bounded sets is considered in [10]. The case of partially known plant and known lower bounds on the disturbance frequencies is considered in [11]. In [12], the adaptive regulation problem is solved for unknown stable plant of unknown order.

In design of adaptive regulator, we will basically follow the approaches introduced in Sects. 4.1.4 and 5.1.3.1. Therefore, in this section, we outline the key steps of the design skipping sometimes intermediate arguments and calculations.

First, as in Sect. 4.1.4, we use Lemmas 1.1 and 1.2 to represent the plant in the following parameterized form (compare with (4.112)):

$$\dot{e} = A_{\Sigma} e + b_M \left(\bar{\psi}^{\top} x + \psi_u u + \theta_g^{\top} \bar{\xi}_g + \theta_v^{\top} \xi_v \right), \qquad \varepsilon = c^{\top} e, \tag{5.88}$$

where the state e is not accessible for measurements, $\bar{\psi} = \psi + k$, $k \in \mathbb{R}^n$ is a constant vector such that the matrix $A_{\Sigma} = A_M - b_M k^{\top}$ is Hurwitz, $\theta_g \in \mathbb{R}^{q_g}$ and $\theta_v \in \mathbb{R}^{q_v}$ are unknown constant vectors, $\bar{\xi}_g \in \mathbb{R}^{q_g}$ and $\xi_v \in \mathbb{R}^{q_v}$ are the states of the corresponding exosystems.

Second, using the observer

$$\dot{\hat{\xi}}_g = G_g \hat{\xi}_g + l_g g \tag{5.89}$$

with the pair (G_g, l_g) from the exosystem canonical form, we obtain

$$\bar{\xi}_g = \hat{\xi}_g + \epsilon_g, \tag{5.90}$$

where ϵ_g exponentially decays.

Third, using the observer (5.40)–(5.43), we represent the matched disturbance in the form

$$\theta_v^{\top} \xi_v = \vartheta_v^{\top} \omega_v + \epsilon_v, \tag{5.91}$$

where the extended vector of unknown parameters ϑ_v and the extended regressor ω_v are defined as (see Eqs. (5.45) and (5.46))

$$\vartheta_v = \left[\theta_v^{\top}, \bar{\psi}_1 \theta_v^{\top}, \bar{\psi}_2 \theta_v^{\top}, \ldots, \bar{\psi}_n \theta_v^{\top}, \psi_u \theta_v^{\top} \right]^{\top},$$
$$\omega_v = \left[(\eta_0 + N x)^{\top}, \eta_1^{\top}, \eta_2^{\top}, \ldots, \eta_n^{\top}, \eta_u^{\top} \right]^{\top},$$

the vectors η_0, η_i, η_u are generated by the filters (5.40)–(5.42), ϵ_v is the exponentially decaying term.

Then, substituting (5.90) and (5.91) into (5.88) and rearranging terms, we obtain the following error model:

$$\dot{e} = A_{\Sigma} e + \psi_u b_M (\vartheta^{\top} \omega + u + \epsilon), \qquad \varepsilon = c^{\top} e, \tag{5.92}$$

where $\vartheta \in \mathbb{R}^{n+q_g+(n+2)q_v}$ is a vector of unknown parameters given by

$$\vartheta = \frac{1}{\psi_u} \left[\bar{\psi}^{\top}, \theta_v^{\top}, \bar{\psi}_1 \theta_v^{\top}, \bar{\psi}_2 \theta_v^{\top}, \ldots, \bar{\psi}_n \theta_v^{\top}, \psi_u \theta_v^{\top}, \theta_g^{\top} \right]^{\top},$$

ω is the extended regressor defined as

$$\omega = \left[x^\top, (\eta_0 + Nx)^\top, \eta_1^\top, \eta_2^\top, \ldots, \eta_n^\top, \eta_u^\top, \hat{\xi}_g^\top\right]^\top,$$

ϵ is the exponentially decaying term.

Choosing the control law

$$u = -\hat{\vartheta}^\top \omega \tag{5.93}$$

with the vector of the adjustable parameters $\hat{\vartheta}$, we obtain the following closed-loop input–output error model:

$$\varepsilon = \psi_u W_\Sigma(s)\left[\tilde{\vartheta}^\top \omega\right] + \epsilon_\varepsilon, \tag{5.94}$$

where $\tilde{\vartheta} = \vartheta - \hat{\vartheta}$, $W_\Sigma(s) = c^\top(sI - A_\Sigma)^{-1}b_M$, ϵ_ε is the exponentially decaying term.

The model (5.94) is exactly in the form of (5.48), hence we can apply the algorithms of adaptation similar to the ones from Sect. 5.1.3.1:

$$\dot{\hat{\vartheta}} = \gamma_1 \frac{\bar{\omega}\bar{\varepsilon}}{1 + \bar{\omega}^\top \bar{\omega}}, \quad \dot{\hat{\psi}}_u = -\gamma_2 \frac{\zeta \bar{\varepsilon}}{1 + \bar{\omega}^\top \bar{\omega}}, \tag{5.95}$$

where $\gamma_1, \gamma_2 > 0$ are the adaptation gains, $\bar{\omega} = W_\Sigma(s)[\omega]$ is the filtered regressor, $\hat{\psi}_u$ is the additional adjustable parameter, $\bar{\varepsilon}$ is the augmented tracking error given by

$$\bar{\varepsilon} = \varepsilon + \hat{\psi}_u \zeta, \tag{5.96}$$

ζ is the swapping term defined as

$$\zeta = W_\Sigma(s)\left[\hat{\vartheta}^\top \omega\right] - \hat{\vartheta}^\top \bar{\omega}. \tag{5.97}$$

Then it can be proved that the adaptive controller (5.93) with the reference observer (5.89), the disturbance observer (5.40)–(5.43), and the adaptation algorithm (5.95)–(5.97) when applied to the plant (5.37) provides the boundedness of all the closed-loop signals and achievement of the control objective (5.87).

5.2 Nonlinear Plants

In this section, we consider the problems of adaptive tracking and adaptive disturbance compensation for nonlinear plants representable in the parametric strict-feedback and output-feedback canonical forms.

5.2.1 *Reference Tracking*

Consider the problem of adaptive reference tracking for nonlinear plants with unmatched parametric uncertainties. Let a nonlinear plant be represented in the parametric strict-feedback canonical form

$$\dot{x}_i = x_{i+1} + \psi^\top \phi_i(x_1, x_2, \dots, x_i), \quad i = 1, 2, \dots, n-1, \tag{5.98}$$

$$\dot{x}_n = \psi^\top \phi_n(x)\psi + \phi_x(x) + \phi_u(x)u, \tag{5.99}$$

$$y = x_1, \tag{5.100}$$

where x_i is the ith element of the measured state vector $x \in \mathbb{R}^n$, $u \in \mathbb{R}$ is the control, $y \in \mathbb{R}$ is the regulated variable, $\psi \in \mathbb{R}^k$ is a constant vector of unknown parameters, ϕ_x, ϕ_u, ϕ_i are the known smooth scalar functions. The function ϕ_u is bounded away from zero by some positive constant ϕ_0, i.e., $\phi_u(x) \geq \phi_0 > 0$. The control problem is to design a state-feedback controller providing the boundedness of all the closed-loop signals and the asymptotic reference tracking

$$\lim_{t\to\infty} \varepsilon(t) = 0, \tag{5.101}$$

where $\varepsilon = y - g$ is the tracking error, g is the measured reference satisfying Assumption 2.1 (with substitution $\upsilon = g$).

In order to get a solution to the problem, we first derive an appropriate parameterization of the reference g and then apply the modular backstepping with high-order tuner to cope with unmatched parametric uncertainties.

According to Lemma 2.3, the reference g can be represented in the form

$$g = \theta_\upsilon^\top \hat{\xi} + \epsilon,$$

where $\theta_\upsilon \in \mathbb{R}^q$ is an unknown constant vector, ϵ is the exponentially decaying term, and $\hat{\xi} \in \mathbb{R}^q$ is the regressor generated by the reference observer

$$\dot{\hat{\xi}} = G\hat{\xi} + lg \tag{5.102}$$

with an arbitrary Hurwitz matrix G and the controllable pair (G, l).

In the steps of the backstepping procedure, it will be required to use the HOTD of $\hat{\xi}$. To simplify analytical expressions, we introduce the filtered regressor

$$\hat{\xi}_\kappa = \frac{\kappa_0}{\kappa(s)} \left[\hat{\xi}\right], \tag{5.103}$$

where $\kappa(s)$ is a designer-chosen Hurwitz polynomial of an order no less than $n-1$, $\kappa_0 > 0$ is a constant. Then, according to Corollary A.5, we can write

$$g = \theta^\top \hat{\xi}_\kappa + \epsilon_\kappa, \tag{5.104}$$

where $\theta \in \mathbb{R}^q$ is a new vector of unknown parameters, ϵ_κ is the exponentially decaying term. It is worth noting that the HOTD of $\hat{\xi}_\kappa$

$$\hat{\xi}_\kappa^{(i)} = \frac{\kappa_0 s^i}{\kappa(s)} \left[\hat{\xi}\right], \qquad i = 1, 2, \ldots n-1 \tag{5.105}$$

are physically implementable.

We use the modular backstepping procedure with the high-order tuners that starts with evaluation of the time derivative $\dot{\varepsilon} = \dot{y} - \dot{g}$ in view of (5.98), (5.100) and (5.104). The procedure consists of design of the following virtual control laws (see Sect. E.1):

$$\alpha_1(x_1, \hat{\psi}, \hat{\theta}, \dot{\hat{\xi}}_\kappa, g) = -(c_1 + s_1)z_1 - \hat{\psi}^\top \phi_1 + \hat{\theta}^\top \dot{\hat{\xi}}_\kappa, \tag{5.106}$$

$$\alpha_i(\bar{x}_i, \bar{\psi}_{i-1}, \bar{\theta}_{i-1}, \bar{\xi}_i, g) = -z_{i-1} - (c_i + s_i)z_i \tag{5.107}$$
$$- \left(\phi_i^\top - \sum_{j=1}^{i-1} \frac{\partial \alpha_{i-1}}{\partial x_j} \phi_j^\top\right) \hat{\psi} + \frac{\partial \alpha_{i-1}}{\partial g} \hat{\theta}^\top \dot{\hat{\xi}}_\kappa$$
$$+ \sum_{j=1}^{i-1} \left(\frac{\partial \alpha_{i-1}}{\partial x_j} x_{j+1} + \frac{\partial \alpha_{i-1}}{\partial \hat{\psi}^{(j-1)}} \hat{\psi}^{(j)} + \frac{\partial \alpha_{i-1}}{\partial \hat{\theta}^{(j-1)}} \hat{\theta}^{(j)} + \frac{\partial \alpha_{i-1}}{\partial \hat{\xi}_\kappa^{(j)}} \hat{\xi}_\kappa^{(j+1)} \right),$$

where $i = 2, 3, \ldots, n$, $z_1 = \varepsilon = y - g$, $z_i = x_i - \alpha_{i-1}$ are the state errors, $c_i > 0$ are the design parameters, $\hat{\psi} \in \mathbb{R}^k$, $\hat{\theta} \in \mathbb{R}^q$ are the vectors of adjustable parameters,

$$s_1 = \mu_{1,1} |\phi_1|^2 + \mu_{1,2} \left|\dot{\hat{\xi}}_\kappa\right|^2, \tag{5.108}$$

$$s_i = \mu_{i,1} \left| \phi_i - \sum_{j=1}^{i-1} \frac{\partial \alpha_{i-1}}{\partial x_j} \phi_j \right|^2 + \mu_{i,2} \left| \frac{\partial \alpha_{i-1}}{\partial g} \dot{\hat{\xi}}_\kappa \right|^2 \tag{5.109}$$

are the damping terms with constant parameters $\mu_{i,1}, \mu_{i,2} > 0$, $\bar{x}_i = (x_1, x_2, \ldots, x_i)$, $\bar{\xi}_i = (\dot{\hat{\xi}}_\kappa, \ddot{\hat{\xi}}_\kappa, \ldots, \hat{\xi}_\kappa^{(i)})$, $\bar{\psi}^{(i)} = (\hat{\psi}, \dot{\hat{\psi}}, \ldots, \hat{\psi}^{(i)})$, and $\bar{\theta}^{(i)} = (\hat{\theta}, \dot{\hat{\theta}}, \ldots, \hat{\theta}^{(i)})$ are the sets of corresponding variables.

In the end of the procedure, we design the actual control law

$$u = \frac{1}{\phi_u(x)} \left(\alpha_n(x, \bar{\psi}_{n-1}, \bar{\theta}_{n-1}, \bar{\xi}_n, g) - \phi_x(x) \right). \tag{5.110}$$

The control law (5.106)–(5.110) when applied to the plant (5.98)–(5.100) results in the following closed-loop error model:

$$
\begin{aligned}
\dot{z} &= A_z(z, x, \bar{\vartheta}_{n-1}, \bar{\xi}_n)z + W_\vartheta^\top(z, x, \bar{\vartheta}_{n-1}, \bar{\xi}_n)\tilde{\vartheta} \\
&\quad + b_z(z, x, \bar{\vartheta}_{n-1}, \bar{\xi}_n)\dot{\epsilon}_\kappa, && (5.111)\\
\varepsilon &= z_1, && (5.112)
\end{aligned}
$$

where $z = [z_1, \; z_2, \ldots, \; z_n]^\top$ is the vector of the state errors, $\hat{\vartheta} = [\hat{\psi}^\top, \hat{\theta}^\top]^\top \in \mathbb{R}^{k+q}$ is the aggregated vector of adjustable parameters, $\tilde{\vartheta} = \vartheta - \hat{\vartheta}$ is the aggregated vector of parametric errors, $\vartheta = [\psi^\top, \theta^\top]^\top \in \mathbb{R}^{k+q}$ is the aggregated vector of unknown parameters,

$$
A_z = \begin{bmatrix} -c_1 - s_1 & 1 & 0 & \cdots & 0 \\ -1 & -c_2 - s_2 & 1 & \ddots & \vdots \\ 0 & -1 & \ddots & \ddots & 0 \\ \vdots & \ddots & \ddots & \ddots & 1 \\ 0 & \cdots & 0 & -1 & -c_n - s_n \end{bmatrix}, \quad W_\vartheta^\top = \left[W_\psi^\top, \; b_z \dot{\xi}_\kappa^\top \right],
$$

$$
W_\psi^\top = \begin{bmatrix} \phi_1^\top \\ \phi_2^\top - \frac{\partial \alpha_1}{\partial x_1}\phi_1^\top \\ \phi_3^\top - \frac{\partial \alpha_2}{\partial x_1}\phi_1^\top - \frac{\partial \alpha_2}{\partial x_2}\phi_2^\top \\ \vdots \\ \phi_n^\top - \sum_{j=1}^{n-1} \frac{\partial \alpha_{n-1}}{\partial x_j}\phi_j^\top \end{bmatrix}, \quad b_z = - \begin{bmatrix} 1 \\ \frac{\partial \alpha_1}{\partial g} \\ \frac{\partial \alpha_2}{\partial g} \\ \vdots \\ \frac{\partial \alpha_{n-1}}{\partial g} \end{bmatrix},
$$

$\bar{\vartheta}_{n-1} = [\bar{\psi}_{n-1}^\top, \bar{\theta}_{n-1}^\top]^\top$.

In order to derive an adaptation algorithm, we apply the nonlinear swapping scheme [8, Chap. 9] (see Lemma D.9 in Appendix D). To this end, we introduce the swapping filters

$$
\begin{aligned}
\dot{\zeta}_0 &= A_z\zeta_0 + W_\vartheta^\top \hat{\vartheta}, && (5.113)\\
\dot{\Omega}_\vartheta^\top &= A_z\Omega_\vartheta^\top + W_\vartheta^\top && (5.114)
\end{aligned}
$$

with the states $\zeta_0 \in \mathbb{R}^n$ and $\Omega_\vartheta \in \mathbb{R}^{n\times q+k}$ and define the augmented state error as

$$
\bar{z} = z + \zeta_0. \tag{5.115}
$$

Then, in view of Lemma D.9, we have

$$
\bar{z} = \Omega_\vartheta^\top \vartheta + \epsilon_z. \tag{5.116}
$$

The matrix Ω_ϑ is bounded since the matrix A_z is of special structure with damping terms. The term ϵ_z exponentially decays according to the equation $\dot{\epsilon}_z = A_z\epsilon_z + b_z\dot{\epsilon}_\kappa$.

Following Remark 4.6, we select the initial conditions of the filters (5.113), (5.114) as follows: $\zeta_0(0) = -z(0)$, $\Omega_\vartheta(0) = 0$. Hence, $\epsilon_z(0) = z(0) + \zeta_0(0) - \Omega_\vartheta(0)\vartheta = 0$.

Now, we apply a special realization of the adaptation algorithm with MRE to calculate the HOTD of the adjustable parameters $\hat{\vartheta}^{(i)}$ up to the order of $n-1$ used in the control laws (5.107), (5.110). The algorithm is given by (see Sect. 3.2.3)

$$\dot{\hat{\vartheta}} = \gamma\left(\bar{z}_\Omega - \bar{\Omega}_\vartheta\hat{\vartheta}\right), \tag{5.117}$$

where $\bar{z}_\Omega = L(s)\left[\Omega_\vartheta \bar{z}\right]$, $\bar{\Omega}_\vartheta = L(s)\left[\Omega_\vartheta\Omega_\vartheta^\top\right]$, $L(s)$ is a designer-chosen transfer function of the form (3.44) with a relative degree greater or equal to $n-2$, $\gamma > 0$ is the adaptation gain. In view of (5.116), this algorithm yields the closed-loop parametric error model

$$\dot{\tilde{\vartheta}} = -\gamma\bar{\Omega}_\vartheta\tilde{\vartheta} - \gamma L(s)\left[\Omega_\vartheta\epsilon_z\right]. \tag{5.118}$$

The algorithm (5.117) is suitable for calculation of the HOTD of the adjustable parameters $\hat{\vartheta}$ up to the $(n-1)$st order via the expressions

$$\hat{\vartheta}^{(i)} = \gamma\left(\bar{z}_\Omega^{(i-1)} - \sum_{j=0}^{i-1} C_j^{i-1}\bar{\Omega}_\vartheta^{(i-j-1)}\hat{\vartheta}^{(j)}\right), \quad i = 2, 3, \ldots, n-1, \tag{5.119}$$

where C_j^i are the binomial coefficients. The HOTD $\bar{z}_\Omega^{(i-1)}$, $\bar{\Omega}_\vartheta^{(i-j-1)}$ are implementable due to sufficiently large relative degree of the transfer function $L(s)$.

The main properties of the closed-loop system are formulated as follows.

Proposition 5.5 *If Assumption 2.1 holds, then the adjustable controller (5.106)–(5.110) with the reference observer (5.102), (5.103), the swapping scheme (5.113)–(5.115), and the adaptation algorithm (5.117), (5.119) when applied to the plant (5.98)–(5.100) provides the boundedness of all the closed-loop signals and achievement of the control objective (5.101).*

Proof Since $\hat{\xi}^{(i)}$ are bounded ($i = 0, 1, \ldots, n-1$), the proposition directly follows from Proposition E.1. □

Example 5.5 Consider the third-order unstable plant

$$\dot{x}_1 = x_2 + \phi_1^\top(x_1)\psi, \tag{5.120}$$

$$\dot{x}_2 = x_3, \tag{5.121}$$

$$\dot{x}_3 = u, \tag{5.122}$$

$$y = x_1 \tag{5.123}$$

with the vector of unknown parameters $\psi = [2, 1]^\top$ and the vector of nonlinear functions $\phi_1(x_1) = \left[x_1^2, \sin(x_1)\right]^\top$.

Reference signal to be tracked is given by the switching function

$$g(t) = \begin{cases} 1 & \text{if} \quad 0 \le t \le 17 \text{ sec}, \\ \sin\left(\frac{\pi t}{2}\right) & \text{if} \quad t > 17 \text{ sec} \end{cases} \tag{5.124}$$

with a priori unknown constant, amplitude, phase, and frequency.

Applying the backstepping procedure consisting of $n = 3$ steps, we design the virtual control laws α_1, α_2, α_3, and the actual control u:

$$\begin{aligned}
\alpha_1 &= -(c_1 + s_1)z_1 - \phi_1^\top \hat{\psi} + \dot{\hat{\xi}}_\kappa^\top \hat{\theta}, \\
\alpha_2 &= -z_1 - (c_2 + s_2)z_2 + \frac{\partial \alpha_1}{\partial x_1}\phi_1^\top \hat{\psi} + \frac{\partial \alpha_1}{\partial g}\dot{\hat{\xi}}_\kappa^\top \hat{\theta} \\
&\quad + \left(\frac{\partial \alpha_1}{\partial x_1}x_2 + \frac{\partial \alpha_1}{\partial \hat{\psi}}\dot{\hat{\psi}} + \frac{\partial \alpha_1}{\partial \hat{\theta}}\dot{\hat{\theta}} + \frac{\partial \alpha_1}{\partial \dot{\hat{\xi}}_\kappa}\ddot{\hat{\xi}}_\kappa\right), \\
\alpha_3 &= -z_2 - (c_3 + s_3)z_3 + \frac{\partial \alpha_2}{\partial x_1}\phi_1^\top \hat{\psi} + \frac{\partial \alpha_2}{\partial g}\dot{\hat{\xi}}_\kappa^\top \hat{\theta} \\
&\quad + \frac{\partial \alpha_2}{\partial x_1}x_2 + \frac{\partial \alpha_2}{\partial x_2}x_3 + \frac{\partial \alpha_2}{\partial \hat{\psi}}\dot{\hat{\psi}} + \frac{\partial \alpha_2}{\partial \dot{\hat{\psi}}}\ddot{\hat{\psi}} \\
&\quad + \frac{\partial \alpha_2}{\partial \hat{\theta}}\dot{\hat{\theta}} + \frac{\partial \alpha_2}{\partial \dot{\hat{\theta}}}\ddot{\hat{\theta}} + \frac{\partial \alpha_2}{\partial \dot{\hat{\xi}}_\kappa}\ddot{\hat{\xi}}_\kappa + \frac{\partial \alpha_2}{\partial \ddot{\hat{\xi}}_\kappa}\dddot{\hat{\xi}}_\kappa, \\
u &= \alpha_3,
\end{aligned}$$

where $z = [z_1, z_2, z_3]^\top = [y - g, x_2 - \alpha_1, x_3 - \alpha_2]^\top$,

$$s_1 = \mu_{1,1}\,|\phi_1|^2 + \mu_{1,2}\left|\dot{\hat{\xi}}_\kappa\right|^2, \quad s_2 = \mu_{2,1}\left|\frac{\partial \alpha_1}{\partial x_1}\phi_1\right|^2 + \mu_{2,2}\left|\frac{\partial \alpha_1}{\partial g}\dot{\hat{\xi}}_\kappa\right|^2,$$

$$s_3 = \mu_{3,1}\left|\frac{\partial \alpha_2}{\partial x_1}\phi_1\right|^2 + \mu_{3,2}\left|\frac{\partial \alpha_2}{\partial g}\dot{\hat{\xi}}_\kappa\right|^2,$$

$c_1 = c_2 = c_3 = 5$, $\mu_{1,1} = \mu_{1,2} = \mu_{2,1} = \mu_{2,2} = \mu_{3,1} = \mu_{3,2} = 0.0001$.

The regressor $\dot{\hat{\xi}}_\kappa$ and its time derivatives are calculated via the equalities

$$\hat{\xi}_\kappa^{(i)} = \frac{8s^i}{s^3 + 6s^2 + 12s + 8}\left[\hat{\xi}\right], \quad i = 1, 2, 3,$$

where $\hat{\xi}$ is generated by the reference observer (5.102) with G and l given by

$$G = \begin{bmatrix} 0 & 1 \\ -20 & -9 \end{bmatrix}, \quad l = \begin{bmatrix} 0 \\ 20 \end{bmatrix}.$$

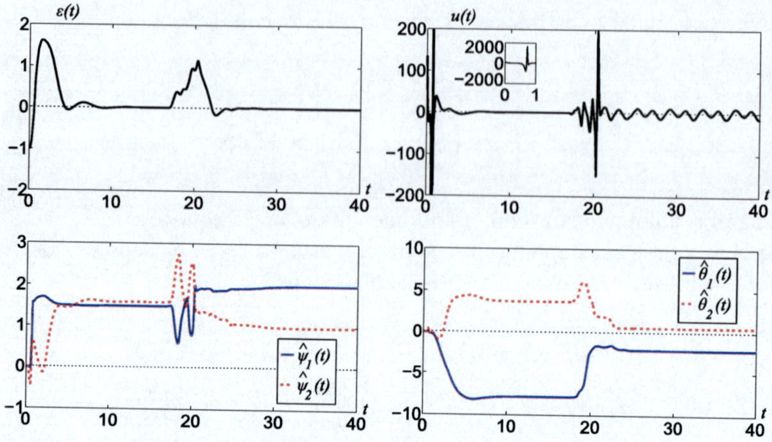

Fig. 5.5 Example 5.5: adaptive state-feedback tracking of the uncertain reference signal (5.124) in the nonlinear plant (5.120)–(5.123) with unmatched unknown parameters

To generate the adjustable parameters $\hat{\vartheta} = [\hat{\psi}^\top, \hat{\theta}^\top]^\top$ and their HOTD $\dot{\hat{\vartheta}} = [\dot{\hat{\psi}}^\top, \dot{\hat{\theta}}^\top]^\top$, $\ddot{\hat{\vartheta}} = [\ddot{\hat{\psi}}^\top, \ddot{\hat{\theta}}^\top]^\top$ used in α_1, α_2, and α_3, we select $L(s) = \frac{1}{s+1}$ and obtain the adaptation algorithm

$$\dot{\hat{\vartheta}} = \gamma \left(\frac{1}{s+1} [\Omega_\vartheta (z + \zeta_0)] - \frac{1}{s+1} \left[\Omega_\vartheta \Omega_\vartheta^\top \right] \hat{\vartheta} \right)$$

together with the expression for the second-order time derivative

$$\ddot{\hat{\vartheta}} = -\left(I_4 + \gamma \frac{1}{s+1} \left[\Omega_\vartheta \Omega_\vartheta^\top \right] \right) \dot{\hat{\vartheta}} + \gamma \Omega_\vartheta \left(z + \zeta_0 - \Omega_\vartheta^\top \hat{\vartheta} \right),$$

where Ω_ϑ and ζ_0 are generated by the swapping filters (5.113), (5.114), $\gamma = 5$.

All simulations are carried out with zero initial conditions $x(0)$, $\hat{\xi}(0)$, $\Omega_\vartheta(0)$, $\zeta_0(0)$, and $\hat{\vartheta}(0)$.

It is seen from the simulation results presented in Fig. 5.5 that the designed controller drives the tracking error to zero according to (5.101) despite the unmatched unknown parameters of the nonlinear plant (5.120)–(5.123) and the uncertainties of the reference signal (5.124). ■

5.2.2 *Compensation of Unmatched Disturbance*

5.2.2.1 Plants with Measured State

Let us consider uncertain nonlinear plant represented in the *parametric strict-feedback* canonical form with the external disturbance

$$\dot{x}_i = x_{i+1} + \psi^\top \phi_i(x_1, x_2, \ldots, x_i) + \bar{d}_{n-i}^\top \upsilon, \quad i = 1, \ldots, n-1, \tag{5.125}$$
$$\dot{x}_n = \psi^\top \phi_n(x) + \phi_x(x) + \phi_u(x)u + \bar{d}_0^\top \upsilon, \tag{5.126}$$
$$y = x_1, \tag{5.127}$$

where x_i are the elements of the measured state vector $x \in \mathbb{R}^n$, $u \in \mathbb{R}$ is the control, $y \in \mathbb{R}$ is the output, $\psi \in \mathbb{R}^k$ is a constant vector of unknown parameters, ϕ_x, ϕ_u, ϕ_i are the known smooth scalar functions, $\bar{d}_i = \left[d_{i,1}, d_{i,2}, \ldots, d_{i,\rho}\right]^\top$ is the known constant vector ($i = 0, 1, \ldots, n-1$), $\upsilon \in \mathbb{R}^\rho$ is the multidimensional disturbance satisfying Assumption 2.2.

We also represent the plant (5.125)–(5.127) in the matrix form

$$\dot{x} = Ex + \varsigma_n \left(\phi_x(x) + \phi_u(x)u\right) + \Phi(x)\psi + D\upsilon, \quad y = \varsigma_1^\top x, \tag{5.128}$$

where

$$E = \begin{bmatrix} 0 & \\ \vdots & I_{n-1} \\ 0 & 0 \cdots 0 \end{bmatrix}, \quad \Phi = \begin{bmatrix} \phi_1^\top \\ \phi_2^\top \\ \vdots \\ \phi_n^\top \end{bmatrix}, \quad D = \begin{bmatrix} \bar{d}_{n-1}^\top \\ \vdots \\ \bar{d}_1^\top \\ \bar{d}_0^\top \end{bmatrix}.$$

The considered control problem is to design a state-feedback controller ensuring the boundedness of all the closed-loop signals and asymptotic convergence of the output to zero:

$$\lim_{t \to \infty} y(t) = 0. \tag{5.129}$$

Following Lemma 2.10, we introduce the filters

$$\dot{\eta} = G\eta + GNx - NEx - N\varsigma_n \left(\phi_x(x) + \phi_u(x)u\right), \tag{5.130}$$
$$\dot{\Lambda} = G\Lambda - N\Phi(x), \tag{5.131}$$

where the pair (G, L) is given by (2.25), (2.26), and the matrix N is selected so that

$$ND = L.$$

Then, according to Proposition 2.13, the plant affected by the external disturbance can be represented in the following parameterized form:

$$\dot{x} = Ex + \varsigma_n \left(\phi_x(x) + \phi_u(x)u\right) + \Phi(x)\psi + T\omega_\kappa + \epsilon, \tag{5.132}$$

where $T \in \mathbb{R}^{n\times q(k+1)}$ is an unknown constant matrix, ω_κ is the regressor given by

$$\omega_\kappa = \frac{\kappa_0}{\kappa(s)}[\omega], \tag{5.133}$$

$$\omega = \left[(\eta + Nx)^\top, \Lambda_1^\top, \Lambda_2^\top, \ldots, \Lambda_k^\top\right]^\top, \tag{5.134}$$

$\kappa(s)$ is a designer-chosen Hurwitz polynomial of degree no less than $n-1$, $\kappa_0 > 0$ is a constant, $\Lambda_i \in \mathbb{R}^q$ is the ith column of the matrix Λ, ϵ is the exponentially decaying term.

For the design purposes, we also represent the model (5.132) in the row-wise form

$$\dot{x}_i = x_{i+1} + \psi^\top \phi_i(x_1, x_2, \ldots, x_i) + T_i^\top \omega_\kappa + \epsilon_i, \quad i = 1, 2, \ldots, n-1, \tag{5.135}$$

$$\dot{x}_n = \phi_x(x) + \phi_u(x)u + \psi^\top \phi_n(x) + T_n^\top \omega_\kappa + \epsilon_n, \tag{5.136}$$

$$y = x_1, \tag{5.137}$$

where $T_i^\top$ is the ith row of T, ϵ_i is the ith element of ϵ.

Based on the structure of (5.135)–(5.137), the modular backstepping includes the following steps (see Sect. E.1):

$$\alpha_1\left(x_1, \hat{\psi}, \hat{T}_1, \omega_\kappa\right) = -(c_1 + s_1)z_1 - \hat{\psi}^\top \phi_1 - \hat{T}_1^\top \omega_\kappa, \tag{5.138}$$

$$\begin{aligned}\alpha_i\left(\bar{x}_i, \bar{\psi}_{i-1}, \bar{T}_{1,i-1}, \bar{T}_{2,i-2}, \ldots, \bar{T}_{i-1,2}, \hat{T}_i, \bar{\omega}_{i-1}\right) &= -z_{i-1} - (c_i + s_i)z_i \\ &- \left(\phi_i^\top - \sum_{j=1}^{i-1} \frac{\partial \alpha_{i-1}}{\partial x_j}\phi_j^\top\right)\hat{\psi} + \sum_{j=1}^{i-1} \frac{\partial \alpha_{i-1}}{\partial x_j}\hat{T}_j^\top \omega_\kappa - \hat{T}_i^\top \omega_\kappa \\ &+ \sum_{j=1}^{i-1}\left(\frac{\partial \alpha_{i-1}}{\partial x_j}x_{j+1} + \frac{\partial \alpha_{i-1}}{\partial \hat{\psi}^{(j-1)}}\hat{\psi}^{(j)} + \frac{\partial \alpha_{i-1}}{\partial \omega_\kappa^{(j-1)}}\omega_\kappa^{(j)} + \sum_{p=1}^{j} \frac{\partial \alpha_{i-1}}{\partial \hat{T}_{i-j}^{(p-1)}}\hat{T}_{i-j}^{(p)}\right),\end{aligned} \tag{5.139}$$

where $i = 2, 3, \ldots, n$, $z_1 = y$, $z_i = x_i - \alpha_{i-1}$ are the state errors, $c_i > 0$ are the design parameters, $\hat{\psi} \in \mathbb{R}^k$, $\hat{T}_j \in \mathbb{R}^{q(k+1)}$ are the vectors of adjustable parameters,

$$s_1 = \mu_{1,1}\left|\phi_1\right|^2 + \mu_{1,2}\left|\omega_\kappa\right|^2 + \mu_{1,3}, \tag{5.140}$$

$$\begin{aligned}s_i = \mu_{i,1}\left|\phi_i - \sum_{j=1}^{i-1}\frac{\partial \alpha_{i-1}}{\partial x_j}\phi_j\right|^2 + \sum_{j=1}^{i-1}\mu_{i,j+1}\left|\frac{\partial \alpha_{i-1}}{\partial x_j}\omega_\kappa\right|^2 \\ + \mu_{i,i+1}\left|\omega_\kappa\right|^2 + \sum_{j=1}^{i-1}\mu_{i,j+i+1}\left|\frac{\partial \alpha_{i-1}}{\partial x_j}\right|^2 + \mu_{i,2i+1}\end{aligned} \tag{5.141}$$

are the damping terms with constant parameters $\mu_{i,j}$ ($i = 1, 2, \ldots, n$, $j = 1, 2, \ldots, 2i + 1$), $\bar{x}_i = (x_1, x_2, \ldots, x_i)$, $\bar{\omega}_i = (\hat{\omega}_\kappa, \dot{\hat{\omega}}_\kappa, \ldots, \hat{\omega}_\kappa^{(i)})$, $\bar{\psi}^{(i)} = (\hat{\psi}, \dot{\hat{\psi}}, \ldots, \hat{\psi}^{(i)})$, and $\bar{T}_{i,j} = (\hat{T}_i, \dot{\hat{T}}_i, \ldots, \hat{T}_i^{(j)})$ are the sets of corresponding variables.

The procedure is finished by design of the actual control law

$$u = \frac{1}{\phi_u(x)} \left(\alpha_n(x, \bar{\vartheta}, \bar{\omega}_{n-1}) - \phi_x(x) \right), \tag{5.142}$$

where $\bar{\vartheta} = (\bar{\psi}_{n-1}, \bar{T}_{1,n-1}, \bar{T}_{2,n-2}, \ldots, \bar{T}_{n-1,2}, \hat{T}_n)$ is the set of the corresponding functions and their time derivatives.

The closed-loop error model including the control law (5.138)–(5.142) when applied to the plant (5.98)–(5.100) takes the form

$$\dot{z} = A_z(z, x, \bar{\vartheta}, \bar{\omega}_{n-1})z + W_\vartheta^\top(z, x, \bar{\vartheta}, \bar{\omega}_{n-1})\tilde{\vartheta} + B_z(z, x, \bar{\vartheta}, \bar{\omega}_{n-1})\epsilon, \tag{5.143}$$

$$y = z_1, \tag{5.144}$$

where $z = [z_1,\ z_2, \ldots,\ z_n]^\top$ is the state errors vector, $\hat{\vartheta} = [\hat{\psi}^\top, \hat{T}_1^\top, \hat{T}_2^\top, \ldots, \hat{T}_n^\top]^\top \in \mathbb{R}^{nq(k+1)+k}$ is the aggregated vector of adjustable parameters, $\tilde{\vartheta} = \vartheta - \hat{\vartheta}$ is the aggregated vector of parametric errors, $\vartheta = [\psi^\top, T_1^\top, T_2^\top, \ldots, T_n^\top]^\top$ is the aggregated $(nq(k+1)+k)$-dimensional vector of unknown parameters,

$$A_z = \begin{bmatrix} -c_1 - s_1 & 1 & 0 & \cdots & 0 \\ -1 & -c_2 - s_2 & 1 & \ddots & \vdots \\ 0 & -1 & \ddots & \ddots & 0 \\ \vdots & \ddots & \ddots & \ddots & 1 \\ 0 & \cdots & 0 & -1 & -c_n - s_n \end{bmatrix}, \quad W_\vartheta^\top = \left[W_\psi^\top,\ B_z \otimes \omega_\kappa^\top \right],$$

$$W_\psi^\top = \begin{bmatrix} \phi_1^\top \\ \phi_2^\top - \frac{\partial \alpha_1}{\partial x_1}\phi_1^\top \\ \phi_3^\top - \frac{\partial \alpha_2}{\partial x_1}\phi_1^\top - \frac{\partial \alpha_2}{\partial x_2}\phi_2^\top \\ \vdots \\ \phi_n^\top - \sum_{j=1}^{n-1} \frac{\partial \alpha_{n-1}}{\partial x_j}\phi_j^\top \end{bmatrix}, \quad B_z = \begin{bmatrix} 1 & 0 & 0 & \cdots & 0 & 0 \\ -\frac{\partial \alpha_1}{\partial x_1} & 1 & 0 & \ddots & 0 & 0 \\ -\frac{\partial \alpha_2}{\partial x_1} & -\frac{\partial \alpha_2}{\partial x_2} & 1 & \ddots & 0 & 0 \\ \vdots & \ddots & \ddots & \ddots & \ddots & \vdots \\ -\frac{\partial \alpha_{n-1}}{\partial x_1} & -\frac{\partial \alpha_{n-1}}{\partial x_2} & -\frac{\partial \alpha_{n-1}}{\partial x_3} & \cdots & -\frac{\partial \alpha_{n-1}}{\partial x_{n-1}} & 1 \end{bmatrix}.$$

Now, to design an adaptation algorithm used for parameters tuning, we introduce the swapping filters [8, Chap. 9] (see Lemma D.9 in Appendix D)

$$\dot{\zeta}_0 = A_z\zeta_0 + W_\vartheta^\top \hat{\vartheta}, \tag{5.145}$$

$$\dot{\Omega}_\vartheta^\top = A_z\Omega_\vartheta^\top + W_\vartheta^\top \tag{5.146}$$

with the states $\zeta_0 \in \mathbb{R}^n$, $\Omega_\vartheta \in \mathbb{R}^{n \times q}$ and define the augmented state error

$$\bar{z} = z + \zeta_0. \tag{5.147}$$

By Lemma D.9, the signal $\bar{z}$ can be represented in the form of linear regression

$$\bar{z} = \Omega_\vartheta^\top \vartheta + \epsilon_z, \tag{5.148}$$

where ϵ_z is the exponentially decaying term satisfying the equation $\dot{\epsilon}_z = A_z\epsilon_z + B_z\epsilon$. Note that the matrix Ω_ϑ is bounded independently of the boundedness of W_ϑ due to the structure of the matrix A_z with the nonlinear damping terms s_i.

The effect of initial conditions $\epsilon_z(0)$ can be reduced by selecting the following initial conditions in the filters (5.145), (5.146) (see Remark 4.6): $\zeta_0(0) = -z(0)$, $\Omega_\vartheta(0) = 0$. As a result, $\epsilon_z(0) = z(0) + \zeta_0(0) - \Omega_\vartheta(0)\vartheta = 0$.

Now, based on the linear regression (5.148), we design a special implementation of the adaptation algorithm with MRE used for calculation of the HOTD of the adjustable parameters $\hat{\vartheta}$ (see Sect. 3.2.3) appearing in the control laws (5.139), (5.142). The algorithm of adaptation is of the following form:

$$\dot{\hat{\vartheta}} = \gamma\left(\bar{z}_\Omega - \bar{\Omega}_\vartheta\hat{\vartheta}\right), \tag{5.149}$$

where $\bar{z}_\Omega = L(s)\left[\Omega_\vartheta\bar{z}\right]$, $\bar{\Omega}_\vartheta = L(s)\left[\Omega_\vartheta\Omega_\vartheta^\top\right]$, $L(s)$ is a designer-chosen transfer function of the form (3.44) with a relative degree greater or equal to $n-2$, $\gamma > 0$ is the adaptation gain. In view of (5.148), this algorithm yields the closed-loop parametric error model

$$\dot{\tilde{\vartheta}} = -\gamma\bar{\Omega}_\vartheta\tilde{\vartheta} - \gamma L(s)\left[\Omega_\vartheta\epsilon_z\right]. \tag{5.150}$$

The HOTD of the adjustable parameters $\hat{\vartheta}$ up to the $(n-1)$st order can be calculated directly via the expressions

$$\hat{\vartheta}^{(i)} = \gamma\left(\bar{z}_\Omega^{(i-1)} - \sum_{j=0}^{i-1} C_j^{i-1}\bar{\Omega}_\vartheta^{(i-j-1)}\hat{\vartheta}^{(j)}\right), \quad i = 2, 3, \ldots, n-1, \tag{5.151}$$

where C_j^i are the binomial coefficients. The HOTD $\bar{z}_\Omega^{(i-1)}$, $\bar{\Omega}_\vartheta^{(i-j-1)}$ are accessible due to the appropriate relative degree of the transfer function $L(s)$.

As a conclusion, we formulate the main properties of the closed-loop system.

Proposition 5.6 *If Assumption 2.1 holds, then the adjustable controller (5.138)–(5.142) with the filters (5.130), (5.131), (5.133), (5.134), the swapping scheme (5.145)–(5.147), and the adaptation algorithm (5.149), (5.151) when applied to the plant (5.125)–(5.127) guarantees the boundedness of all the closed-loop signals and achievement of the control objective (5.129).*

Proof The proposition directly follows from Proposition E.1. □

Example 5.6 Let us design the adaptive disturbance compensator for the third-order unstable plant

$$\dot{x}_1 = x_2 + \phi_1(x_1)\psi + \bar{d}_2^\top \upsilon, \tag{5.152}$$
$$\dot{x}_2 = x_3 + \bar{d}_1^\top \upsilon, \tag{5.153}$$
$$\dot{x}_3 = u + \bar{d}_0^\top \upsilon, \tag{5.154}$$
$$y = x_1, \tag{5.155}$$

with the unknown parameter $\psi = -2$, the nonlinear function $\phi_1(x_1) = x_1^2$, the known vectors

$$\bar{d}_0 = \begin{bmatrix} 0 \\ 1 \end{bmatrix}, \quad \bar{d}_1 = \begin{bmatrix} 1 \\ 0 \end{bmatrix}, \quad \bar{d}_2 = \begin{bmatrix} 1 \\ 1 \end{bmatrix},$$

and the disturbance signal $\upsilon(t) = [\upsilon_1(t), \upsilon_2(t)]^\top = [2, \sin(t+3)]^\top$ with a priori unknown constant, amplitude, phase, and frequency.

After $n = 3$ steps of the backstepping design procedure we obtain the virtual control laws

$$\alpha_1 = -(c_1 + s_1)z_1 - \phi_1^\top \hat{\psi} - \omega_\kappa^\top \hat{T}_1,$$
$$\alpha_2 = -z_1 - (c_2 + s_2)z_2 + \frac{\partial \alpha_1}{\partial x_1}\phi_1\hat{\psi} + \omega_\kappa^\top \left(\frac{\partial \alpha_1}{\partial x_1}\hat{T}_1 - \hat{T}_2\right)$$
$$+ \frac{\partial \alpha_1}{\partial x_1}x_2 + \frac{\partial \alpha_1}{\partial \hat{\psi}}\dot{\hat{\psi}} + \frac{\partial \alpha_1}{\partial \hat{T}_1}\dot{\hat{T}}_1 + \frac{\partial \alpha_1}{\partial \omega_\kappa}\dot{\omega}_\kappa,$$
$$\alpha_3 = -z_2 - (c3 + s3)z_3 + \frac{\partial \alpha_2}{\partial x_1}\phi_1\hat{\psi} + \omega_\kappa^\top \left(\frac{\partial \alpha_2}{\partial x_1}\hat{T}_1 + \frac{\partial \alpha_2}{\partial x_2}\hat{T}_2 - \hat{T}_3\right)$$
$$+ \frac{\partial \alpha_2}{\partial x_1}x_2 + \frac{\partial \alpha_2}{\partial x_2}x_3 + \frac{\partial \alpha_2}{\partial \hat{\psi}}\dot{\hat{\psi}} + \frac{\partial \alpha_2}{\partial \dot{\hat{\psi}}}\ddot{\hat{\psi}} + \frac{\partial \alpha_2}{\partial \hat{T}_1}\dot{\hat{T}}_1 + \frac{\partial \alpha_2}{\partial \dot{\hat{T}}_1}\ddot{\hat{T}}_1 + \frac{\partial \alpha_2}{\partial \hat{T}_2}\dot{\hat{T}}_2$$
$$+ \frac{\partial \alpha_2}{\partial \omega_\kappa}\dot{\omega}_\kappa + \frac{\partial \alpha_2}{\partial \dot{\omega}_\kappa}\ddot{\omega}_\kappa,$$

and the actual control law $u = \alpha_3$ with $z = [z_1, z_2, z_3]^\top = [y, x_2 - \alpha_1, x_3 - \alpha_2]^\top$,

$$s_1 = \mu_{1,1}\phi_1^2 + \mu_{1,2}\,|\omega_\kappa|^2 + \mu_{1,3},$$
$$s_2 = \mu_{2,1}\left|\frac{\partial \alpha_1}{\partial x_1}\phi_1\right|^2 + \left(\mu_{2,2}\left|\frac{\partial \alpha_1}{\partial x_1}\right|^2 + \mu_{2,3}\right)|\omega_\kappa|^2 + \mu_{2,4}\left|\frac{\partial \alpha_1}{\partial x_1}\right|^2 + \mu_{2,5},$$
$$s_3 = \mu_{3,1}\left|\frac{\partial \alpha_1}{\partial x_1}\phi_1\right|^2 + \left(\mu_{3,2}\left|\frac{\partial \alpha_2}{\partial x_1}\right|^2 + \mu_{3,3}\left|\frac{\partial \alpha_2}{\partial x_2}\right|^2 + \mu_{3,4}\right)|\omega_\kappa|^2$$
$$+ \mu_{3,5}\left|\frac{\partial \alpha_2}{\partial x_1}\right|^2 + \mu_{3,6}\left|\frac{\partial \alpha_2}{\partial x_2}\right|^2 + \mu_{3,7},$$

$c_1 = c_2 = c_3 = 1$, $\mu_{1,1} = \mu_{1,2} = \mu_{1,3} = \mu_{2,1} = \mu_{2,2} = \mu_{2,3} = \mu_{2,4} = \mu_{2,5} = \mu_{3,1} = \mu_{3,2} = \mu_{3,3} = \mu_{3,4} = \mu_{3,5} = \mu_{3,6} = \mu_{3,7} = 0.00001$.

The regressor ω_κ and its time derivatives are calculated using the equalities

$$\omega_\kappa^{(i)} = \frac{6s^i}{s^2+5s+6}[\omega], \quad i = 0, 1, 2,$$
$$\omega = \left[(\eta + Nx)^\top, \Lambda_1^\top\right]^\top,$$

where η and Λ_1 are generated by the filters (5.130), (5.131) with the matrices

$$G = \begin{bmatrix} 0 & 1 & 0 \\ 0 & 0 & 1 \\ -60 & -47 & -12 \end{bmatrix}, \quad L = \begin{bmatrix} 0 & 0 \\ 0 & 0 \\ 60 & 60 \end{bmatrix}, \quad N = \begin{bmatrix} 0 & 0 & 0 \\ 0 & 0 & 0 \\ 0 & 60 & 60 \end{bmatrix}.$$

To calculate the adjustable parameters $\hat{\vartheta} = [\hat{\psi}^\top, \hat{T}_1^\top, \hat{T}_2^\top, \hat{T}_3^\top]^\top$ and the time derivatives $\dot{\hat{\psi}}$, $\ddot{\hat{\psi}}$, $\dot{\hat{T}}_1$, $\ddot{\hat{T}}_1$, $\dot{\hat{T}}_2$ used in α_1, α_2, and α_3, we choose $L(s) = \frac{1}{s+1}$ and obtain the adaptation algorithm

$$\dot{\hat{\vartheta}} = \gamma\left(\frac{1}{s+1}[\Omega_\vartheta(z+\zeta_0)] - \frac{1}{s+1}\left[\Omega_\vartheta\Omega_\vartheta^\top\right]\hat{\vartheta}\right)$$

together with the expression

$$\ddot{\hat{\vartheta}} = -\left(I_{19} + \gamma\frac{1}{s+1}\left[\Omega_\vartheta\Omega_\vartheta^\top\right]\right)\dot{\hat{\vartheta}} + \gamma\Omega_\vartheta\left(z + \zeta_0 - \Omega_\vartheta^\top\hat{\vartheta}\right),$$

where Ω_ϑ and ζ_0 are generated by the swapping filters (5.145), (5.146), and $\gamma = 2$.

All simulations are carried out with zero initial conditions $x(0)$, $\eta(0)$, $\Lambda(0)$, $\Omega_\vartheta(0)$, $\zeta_0(0)$, and $\hat{\vartheta}(0)$.

Simulation results are presented in Fig. 5.6 and demonstrate complete compensation of the unmatched uncertain disturbance $\upsilon(t) = [\upsilon_1(t), \upsilon_2(t)]^\top = [2, \sin(t+3)]^\top$ in the nonlinear plant (5.152)–(5.155) with unknown parameter. ■

5.2.2.2 Plants with Measured Output

Consider the nonlinear plant represented in the *parametric output-feedback* canonical form

$$\dot{x} = Ex + \phi_x(y) + \Phi(y)\psi + \bar{b}\phi_u(y)u + \bar{d}\upsilon, \tag{5.156}$$
$$y = \bar{c}^\top x + d_n\upsilon, \tag{5.157}$$

where $x \in \mathbb{R}^n$ is the inaccessible for measurement state, $\psi \in \mathbb{R}^k$ is a constant unknown vector,

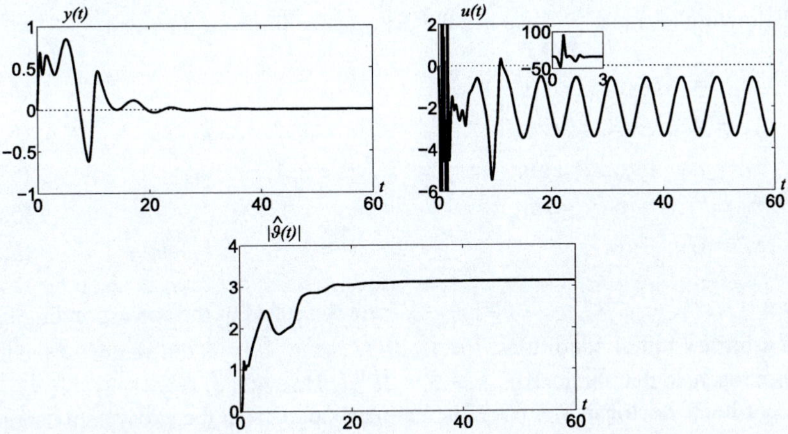

Fig. 5.6 Example 5.6: adaptive state-feedback compensation of the unmatched uncertain disturbance $\upsilon(t) = [\upsilon_1(t), \upsilon_2(t)]^\top = [2, \sin(t+3)]^\top$ in the nonlinear plant (5.152)–(5.155) with unknown parameter

$$E = \begin{bmatrix} 0 & \\ \vdots & I_{n-1} \\ 0 & \cdots 0 \end{bmatrix}, \ \bar{b} = \begin{bmatrix} 0 \\ \vdots \\ 0 \\ b_m \\ \vdots \\ b_1 \\ b_0 \end{bmatrix}, \ \bar{d} = \begin{bmatrix} d_{n-1} \\ \vdots \\ d_1 \\ d_0 \end{bmatrix}, \ \bar{c} = \begin{bmatrix} 1 \\ 0 \\ \vdots \\ 0 \end{bmatrix}, \ \phi_x(y) = \begin{bmatrix} \phi_{x1}(y) \\ \phi_{x2}(y) \\ \vdots \\ \phi_{xn}(y) \end{bmatrix},$$

$\Phi(y) = [\phi_1(y), \phi_2(y), \cdots, \phi_k(y)]$ is the nonlinear $n \times k$ dimensional matrix function, $\phi_x(y) \in \mathbb{R}^n$, $\phi_i(y) \in \mathbb{R}^n$, and $\phi_u(y) \in \mathbb{R}$ are the smooth functions, b_i and d_i are unknown constant parameters.

We accept the following additional assumptions regarding the plant (5.156), (5.157).

Assumption 5.5 *The plant* (5.156), (5.157) *is such that:*

5.5.1. *the polynomial* $\beta(s) = b_m s^m + b_{m-1} s^{m-1} + \cdots + b_1 s + b_0$ *is Hurwitz;*

5.5.2. *the sign of* b_m *is known. Without loss of generality, we assume that* $b_m > 0$*;*

5.5.3. *the function* $\phi_u(y)$ *is bounded away from zero by a constant* $\phi_0 > 0$*, i.e.,* $\phi_u(y) \geq \phi_0$.

The considered control problem is to design an output-feedback controller ensuring boundedness of all the closed-loop signals and asymptotic convergence of the output to zero in accordance with the control objective (5.129).[4]

[4] An alternative approach to the considered problem based on filtered transformation is presented in [9]. The case of SISO nonlinear plants represented in the normal form is considered in [23, 24].

Following Lemma 2.11, we use the disturbance filters of the form

$$\dot{\zeta}_{k+1} = \bar{A}\zeta_{k+1} + \bar{l}y + \phi_x(y), \tag{5.158}$$
$$\dot{\zeta}_i = \bar{A}\zeta_i + \phi_i(y), \qquad i = 1, 2, \ldots, k, \tag{5.159}$$
$$\dot{\nu}_i = \bar{A}\nu_i + \varsigma_{n-i}\phi_u(y)u, \qquad i = 0, 1, \ldots, m, \tag{5.160}$$
$$\dot{\eta}_0 = G\eta_0 + Gly - l\beta_1, \tag{5.161}$$
$$\dot{\eta}_i = G\eta_i - l\bar{\varphi}_i, \qquad i = 1, 2, \ldots, k+m+1, \tag{5.162}$$

where $\zeta_i, \zeta_{k+1}, \nu_j \in \mathbb{R}^n$, $\eta_0 \in \mathbb{R}^q$, $\eta_i \in \mathbb{R}^q$ are the states of the corresponding filters with arbitrary initial conditions, $\bar{l} = \left[\bar{l}_{n-1}, \bar{l}_{n-2}, \ldots, \bar{l}_0\right]^\top$ is the vector of feedback gains chosen so that the matrix $\bar{A} = E - \bar{l}\bar{c}^\top$ is Hurwitz, ς_i $(i = 1, 2, \ldots, n)$ is the ith coordinate vector in $\mathbb{R}^n$, (G, l) is the matrix pair from the exosystem canonical form (2.15),

$$\beta_1 = -\bar{l}_{n-1}\zeta_{k+1,1} + \zeta_{k+1,2} + \bar{l}_{n-1}y + \phi_{x1}(y) \tag{5.163}$$

is the measured function, $\bar{\varphi}_i$ $(i = 1, 2, \ldots, k+m+1)$ are the elements of the vector

$$\bar{\varphi} = \left[-\bar{l}_{n-1}\nu_{m,1} + \nu_{m,2}, -\bar{l}_{n-1}\nu_{m-1,1} + \nu_{m-1,2}, \ldots, -\bar{l}_{n-1}\nu_{0,1} + \nu_{0,2}, \right. \tag{5.164}$$
$$\left. -\bar{l}_{n-1}\zeta_{k,1} + \zeta_{k,2} + \phi_{k,1}(y), \ldots, -\bar{l}_{n-1}\zeta_{1,1} + \zeta_{1,2} + \phi_{1,1}(y)\right]^\top .$$

Then, in accordance with Proposition 2.14, a reduced-order plant model can be represented in the form

$$\dot{y} = \beta_1 + \vartheta^\top\bar{\omega} + b_m\left(-\bar{l}_{n-1}\nu_{m,1} + \nu_{m,2}\right) + \epsilon, \tag{5.165}$$
$$\dot{\nu}_m = \bar{A}\nu_m + \varsigma_{n-m}\phi_u(y)u, \tag{5.166}$$

where $\vartheta \in \mathbb{R}^{(k+m+1)(q+1)+q}$ is a vector of unknown parameters related to the parameters of the plant and the disturbance exosystem via the equalities

$$\vartheta = \left[\bar{\psi}^\top, \theta^\top, \bar{\psi}_1\theta^\top, \ldots, \bar{\psi}_{k+m+1}\theta^\top\right]^\top,$$
$$\bar{\psi} = [b_m, b_{m-1}, \ldots, b_0, \psi_k, \psi_{k-1}, \ldots, \psi_1]^\top,$$

$\bar{\omega} = \left[\bar{\varphi}_0^\top, (\eta_0 + ly)^\top, \eta_1^\top, \ldots, \eta_{k+m+1}^\top\right]^\top$, $\bar{\varphi}_0$ is obtained from $\bar{\varphi}$ (5.164) by zeroing of the first element, i.e.,

$$\bar{\varphi}_0 = \left[0, -\bar{l}_{n-1}\nu_{m-1,1} + \nu_{m-1,2}, \ldots, -\bar{l}_{n-1}\nu_{0,1} + \nu_{0,2}\,, \right. \tag{5.167}$$
$$\left. -\bar{l}_{n-1}\zeta_{k,1} + \zeta_{k,2} + \phi_{k,1}(y), \ldots, -\bar{l}_{n-1}\zeta_{1,1} + \zeta_{1,2} + \phi_{1,1}(y)\right]^\top ,$$

Extensions to the case of nonlinear plants with unknown high-frequency sign are introduced in [4, 5]. A particular case of nonlinear plants with integral ISS inverse dynamics is considered in [25].

and ϵ is the exponentially decaying term.

We will also use the more compact representation of (5.165)

$$\dot{y} = \beta_1 + \vartheta^\top \omega + \epsilon \tag{5.168}$$

with $\omega = \left[\bar{\varphi}^\top, (\eta_0 + ly)^\top, \eta_1^\top, \ldots, \eta_{k+m+1}^\top\right]^\top$.

Based on the model (5.165), (5.166), we design the adaptive controller via the following backstepping with tuning functions (see Sect. E.2):

$$\alpha_1(y, \hat{\vartheta}, \hat{\kappa}_m, \varpi) = -(c_1 + s_1)z_1 + \hat{\kappa}_m \bar{\alpha}_1 + \bar{l}_{n-1} v_{m,1}, \tag{5.169}$$

$$\bar{\alpha}_1(y, \hat{\vartheta}, \varpi) = -\beta_1 - \bar{\omega}^\top \hat{\vartheta}, \tag{5.170}$$

$$\alpha_2(y, \hat{\vartheta}, \hat{\kappa}_m, \varpi) = -(c_2 + s_2)z_2 + \left(\frac{\partial \alpha_1}{\partial y}\omega^\top - \varsigma_1^\top z_1\right)\hat{\vartheta} + \frac{\partial \alpha_1}{\partial \hat{\vartheta}}\tau_2 + \beta_2, \tag{5.171}$$

$$\alpha_i(y, \hat{\vartheta}, \hat{\kappa}_m, \varpi) = -z_{i-1} - (c_i + s_i)z_i + \frac{\partial \alpha_{i-1}}{\partial y}\omega^\top \hat{\vartheta} + \frac{\partial \alpha_{i-1}}{\partial \hat{\vartheta}}\tau_i \tag{5.172}$$

$$+ \beta_i - \gamma_1 \frac{\partial \alpha_{i-1}}{\partial y}\left(\sum_{j=2}^{i-1} \frac{\partial \alpha_{j-1}}{\partial \hat{\vartheta}} z_j\right)\omega, \quad i = 3, 4, \ldots, n-m,$$

where $z_i = v_{m,i} - \alpha_{i-1}$ are the new state errors,

$$s_1 = \mu_1, \quad s_i = \mu_i \left(\frac{\partial \alpha_{i-1}}{\partial y}\right)^2 \tag{5.173}$$

are the damping terms with constant gains $\mu_i > 0$ $(i = 1, 2, \ldots, n-m)$, $c_i > 0$ and $\gamma_1 > 0$ are the design parameters,

$$\tau_1 = \gamma_1 \bar{\omega} z_1, \tag{5.174}$$

$$\tau_2 = \tau_1 - \gamma_1 \left(\frac{\partial \alpha_1}{\partial y}\omega - \varsigma_1 z_1\right) z_2, \tag{5.175}$$

$$\tau_i = \tau_{i-1} - \gamma_1 \frac{\partial \alpha_{i-1}}{\partial y}\omega z_i, \quad i = 3, 4, \ldots, n-m \tag{5.176}$$

are the tuning functions, $\hat{\vartheta} \in \mathbb{R}^{(k+m+1)(q+1)+q}$ is the vector of adjustable parameters, $\hat{\kappa}_m$ is the additional adjustable parameter corresponding to the unknown constant $\kappa_m = \frac{1}{b_m}$, $\varpi = (\zeta_{k+1}, \zeta_1, \zeta_2, \ldots, \zeta_k, v_0, v_1, \ldots, v_m, \eta_0, \eta_1, \ldots, \eta_{k+m+1})$ is the set of variables necessary for the control laws design,

$$\beta_2 = \bar{l}_{n-2}\nu_{m,1} + \frac{\partial \alpha_1}{\partial y}\beta_1 + \frac{\partial \alpha_1}{\partial \hat{\kappa}_m}\dot{\hat{\kappa}}_m \tag{5.177}$$
$$+ \frac{\partial \alpha_1}{\partial \zeta_{k+1}}\left(\bar{A}\zeta_{k+1} + \bar{l}y + \phi_x(y)\right) + \sum_{r=1}^{k}\left(\frac{\partial \alpha_1}{\partial \zeta_r}\left(\bar{A}\zeta_r + \phi_r(y)\right)\right)$$
$$+ \sum_{r=0}^{m-1}\sum_{j=1}^{2}\left(\frac{\partial \alpha_1}{\partial \nu_{r,j}}\left(-\bar{l}_{n-j}\nu_{r,1} + \nu_{r,j+1}\right)\right) + \frac{\partial \alpha_1}{\partial \nu_{m,1}}\left(-\bar{l}_{n-1}\nu_{m,1} + \nu_{m,2}\right)$$
$$+ \frac{\partial \alpha_1}{\partial \eta_0}\left(G\eta_0 + Gly - l\beta_1\right) + \sum_{r=1}^{k+m+1}\left(\frac{\partial \alpha_1}{\partial \eta_r}\left(G\eta_r - l\bar{\varphi}_r\right)\right),$$

$$\beta_i = \bar{l}_{n-i}\nu_{m,1} + \frac{\partial \alpha_{i-1}}{\partial y}\beta_1 + \frac{\partial \alpha_{i-1}}{\partial \hat{\kappa}_m}\dot{\hat{\kappa}}_m \tag{5.178}$$
$$+ \frac{\partial \alpha_{i-1}}{\partial \zeta_{k+1}}\left(\bar{A}\zeta_{k+1} + \bar{l}y + \phi_x(y)\right) + \sum_{r=1}^{k}\left(\frac{\partial \alpha_{i-1}}{\partial \zeta_r}\left(\bar{A}\zeta_r + \phi_r(y)\right)\right)$$
$$+ \sum_{r=0}^{m-1}\sum_{j=1}^{i}\left(\frac{\partial \alpha_{i-1}}{\partial \nu_{r,j}}\left(-\bar{l}_{n-j}\nu_{r,1} + \nu_{r,j+1}\right)\right) + \sum_{j=1}^{i-1}\left(\frac{\partial \alpha_{i-1}}{\partial \nu_{m,j}}\left(-\bar{l}_{n-j}\nu_{m,1} + \nu_{m,j+1}\right)\right)$$
$$+ \frac{\partial \alpha_{i-1}}{\partial \eta_0}\left(G\eta_0 + Gly - l\beta_1\right) + \sum_{r=1}^{k+m+1}\left(\frac{\partial \alpha_{i-1}}{\partial \eta_r}\left(G\eta_r - l\bar{\varphi}_r\right)\right), \quad i = 3, 4, \ldots, n-m$$

are the known functions.

The adjustable parameters $\hat{\vartheta}$ and $\hat{\kappa}_m$ are generated by the adaptation algorithms

$$\dot{\hat{\vartheta}} = \tau_{n-m}, \tag{5.179}$$
$$\dot{\hat{\kappa}}_m = -\gamma_2\bar{\alpha}_1 z_1, \tag{5.180}$$

where $\gamma_2 > 0$ is the adaptation gain.

The procedure is finished by design of the actual control law

$$u = \frac{1}{\varphi_u(y)}\left(\alpha_{n-m}(y, \hat{\vartheta}, \hat{\kappa}_m, \varpi) - \nu_{m,n-m+1}\right). \tag{5.181}$$

The closed-loop error model involving the adaptive control law (5.169)–(5.181) when applied to the plant (5.156), (5.157) takes the following form:

$$\dot{z} = A_z(z,t)z + W_\vartheta^\top(z,t)\tilde{\vartheta} - b_m \varsigma_1 \bar{\alpha}_1 \tilde{\kappa}_m + b_z(z,t)\epsilon, \tag{5.182}$$

$$\dot{\hat{\vartheta}} = -\gamma_1 W_\vartheta z, \tag{5.183}$$

$$\dot{\hat{\kappa}}_m = \gamma_2 \bar{\alpha}_1 \varsigma_1^\top z, \tag{5.184}$$

$$y = z_1, \tag{5.185}$$

where $z = [z_1, z_2, ..., z_{n-m}]^\top$ is the vector of the state errors, $\tilde{\vartheta} = \vartheta - \hat{\vartheta}$, $\tilde{\kappa}_m = \kappa_m - \hat{\kappa}_m$ are the parametric errors,

$$A_z = \begin{bmatrix} -b_m(c_1+s_1) & b_m & 0 & 0 & \cdots & 0 \\ -b_m & -c_2 - s_2 & 1+\sigma_{2,3} & \sigma_{2,4} & \ddots & \sigma_{2,n-m} \\ 0 & -1-\sigma_{2,3} & -c_3 - s_3 & 1+\sigma_{3,4} & \ddots & \sigma_{3,n-m} \\ 0 & -\sigma_{2,4} & -1-\sigma_{3,4} & \ddots & \ddots & \sigma_{4,n-m} \\ \vdots & \ddots & \ddots & \ddots & \ddots & \vdots \\ 0 & -\sigma_{2,n-m} & -\sigma_{3,n-m} & -\sigma_{4,n-m} & \cdots & -c_{n-m} - s_{n-m} \end{bmatrix},$$

$$b_z = \begin{bmatrix} 1 \\ -\frac{\partial \alpha_1}{\partial y} \\ -\frac{\partial \alpha_2}{\partial y} \\ \vdots \\ -\frac{\partial \alpha_{n-m-1}}{\partial y} \end{bmatrix}, \quad W_\vartheta^\top = \begin{bmatrix} \bar{\omega}^\top \\ \frac{\partial \alpha_1}{\partial y}\omega^\top \\ \frac{\partial \alpha_2}{\partial y}\omega^\top \\ \vdots \\ \frac{\partial \alpha_{n-m-1}}{\partial y}\omega^\top \end{bmatrix} + \varsigma_2 \varsigma_1^\top z_1,$$

$$\sigma_{i,j} = \begin{cases} 0 & \text{if } i = 1, \\ \gamma_1 \frac{\partial \alpha_{j-1}}{\partial y} \frac{\partial \alpha_1}{\partial \hat{\vartheta}} \bar{\omega} & \text{if } i = 2, \\ \frac{\partial \alpha_{j-1}}{\partial y} \gamma_1 \frac{\partial \alpha_{j-1}}{\partial \hat{\vartheta}} \omega & \text{if } i = 3, 4, \ldots, n-m, \end{cases}$$

$\varsigma_1 \in \mathbb{R}^{(k+m+1)(q+1)+q}$, $\varsigma_2 \in \mathbb{R}^n$ are the coordinate vectors. For the sake of compactness, we use the notations $A_z(z,t) = A_z(z, \hat{\vartheta}, \hat{\kappa}_m, \varpi)$, $W_\vartheta^\top(z,t) = W_\vartheta^\top(z, \hat{\vartheta}, \hat{\kappa}_m, \varpi)$, $b_z(z,t) = b_z(z, \hat{\vartheta}, \hat{\kappa}_m, \varpi)$.

Proposition 5.7 *If Assumptions 2.1 and 5.5 hold, then the tuning functions adaptive controller* (5.169)–(5.181) *with the filters* (5.158)–(5.162) *when applied to the plant* (5.156), (5.157) *guarantees the boundedness of all the closed-loop signals and achievement of the control objective* (5.129).

Proof The proof of the proposition directly follows from the proof of Theorem 8.5 from [8]. Therefore, we just outline the main steps of the proof. First, from Proposition E.2, we obtain that z, $\hat{\vartheta}$, and $\hat{\kappa}_m$ are bounded. Since z_1 is bounded, y is also bounded, and from (5.158), (5.159) we obtain the boundedness of ζ_i, $i = 1, 2, \ldots, k+1$.

Then, analyzing the filters (5.160) and expressing u from y (with the use of the plant inverse dynamics and Assumption 5.5.1), we conclude that ν_i are bounded for $i = 0, 1, \dots, m$. Therefore, β_1 is bounded, and, as a result, η_0 and $\bar{\varphi}$ are also bounded. Then from (5.162), we obtain the boundedness of η_i, $i = 1, 2, \dots, k+m+1$ and, as a result, the boundedness of the regressor ω. By virtue of Assumption 5.5.1, we conclude that the zero dynamics of the closed-loop system is asymptotically stable and, hence, the state of the plant x is bounded. In view of the boundedness of all the closed-loop signals, from Theorem B.4 we conclude that $z_1(t) \to 0$, and therefore, $y(t) \to 0$ as $t \to \infty$. □

Example 5.7 Consider the plant

$$\dot{x}_1 = x_2 - y^2 + \psi_1 \sin(y) + \psi_2 y + 2\upsilon, \tag{5.186}$$

$$\dot{x}_2 = \psi_2 \cos(y) + (1 + y^2)u, \tag{5.187}$$

$$y = x_1 + \upsilon, \tag{5.188}$$

where $\psi = [3, 2]^\top$ is the vector of unknown parameters, $\upsilon(t) = \cos(2t + 3)$ is the disturbance with unknown amplitude, phase, and frequency. The state x is not accessible for measurements. The plant belongs to the class of plants (5.156), (5.157) with

$$\bar{b} = \begin{bmatrix} 0 \\ 1 \end{bmatrix}, \quad \bar{d} = \begin{bmatrix} 2 \\ 0 \end{bmatrix}, \quad d_2 = 1,$$

$$\phi_x(y) = \begin{bmatrix} \phi_{x1}(y) \\ \phi_{x2}(y) \end{bmatrix} = \begin{bmatrix} -y^2 \\ 0 \end{bmatrix},$$

$$\Phi(y) = \begin{bmatrix} \phi_1(y) \\ \phi_2(y) \end{bmatrix} = \begin{bmatrix} \sin(y) & y \\ 0 & \cos(y) \end{bmatrix}, \quad \phi_u(y) = 1 + y^2.$$

The vectors $\bar{b}$, $\bar{d}$, and the constant d_2 are unknown.

Based on the prior information about the plant and the form of the disturbance, we use $n = 2$, $k = 2$, $m = 0$, and $q = 2$ to introduce the filters (5.159)–(5.162). For the filters, we select

$$\bar{A} = \begin{bmatrix} -5 & 1 \\ -6 & 0 \end{bmatrix}, \quad \bar{l} = \begin{bmatrix} \bar{l}_1 \\ \bar{l}_0 \end{bmatrix} = \begin{bmatrix} 5 \\ 6 \end{bmatrix}, \quad G = \begin{bmatrix} 0 & 1 \\ -12 & -7 \end{bmatrix}, \quad l = \begin{bmatrix} 0 \\ 12 \end{bmatrix}.$$

Then, forming the vectors $\omega \in \mathbb{R}^{11}$ as

$$\omega = [-\bar{l}_1 \nu_{0,1} + \nu_{0,2}, -\bar{l}_1 \zeta_{2,1} + \zeta_{2,2} + \phi_{2,1}(y), -\bar{l}_1 \zeta_{1,1} + \zeta_{1,2} + \phi_{1,1}(y),$$
$$(\eta_0 + ly)^\top, \eta_1^\top, \eta_2^\top, \eta_3^\top]^\top$$

and $\bar{\omega} \in \mathbb{R}^{11}$ (obtained by zeroing of the first element of ω), we design two stabilizing functions

$$\alpha_1 = -(c_1 + s_1)z_1 - \hat{\kappa}_0\left(\beta_1 + \bar{\omega}^\top \hat{\vartheta}\right) + \bar{l}_1 v_{0,1},$$

$$\alpha_2 = -(c_2 + s_2)z_2 + \left(\frac{\partial \alpha_1}{\partial y}\omega^\top - \varsigma_1^\top z_1\right)\hat{\vartheta} + \frac{\partial \alpha_1}{\partial \hat{\vartheta}}\tau_2 + \beta_2,$$

where $z_1 = y$, $z_2 = v_{0,2} - \alpha_1$, $c_1 = c_2 = 1$,

$$s_1 = 1, \quad s_2 = \left(\frac{\partial \alpha_{i-1}}{\partial y}\right)^2,$$

$$\beta_1 = -\bar{l}_1 \zeta_{3,1} + \zeta_{3,2} + \bar{l}_1 y + \phi_{x1}(y),$$

$$\beta_2 = \bar{l}_0 v_{0,1} + \frac{\partial \alpha_1}{\partial y}\beta_1 + \frac{\partial \alpha_1}{\partial \hat{\kappa}_0}\dot{\hat{\kappa}}_0 + \frac{\partial \alpha_1}{\partial \zeta_3}\left(\bar{A}\zeta_3 + \bar{l}y + \phi_x(y)\right)$$

$$+ \sum_{r=1}^{2}\left(\frac{\partial \alpha_1}{\partial \zeta_r}\left(\bar{A}\zeta_r + \phi_r(y)\right)\right) + \frac{\partial \alpha_1}{\partial v_{0,1}}\left(-\bar{l}_1 v_{0,1} + v_{0,2}\right)$$

$$+ \frac{\partial \alpha_1}{\partial \eta_0}\left(G\eta_0 + Gly - l\beta_1\right) + \sum_{r=1}^{3}\left(\frac{\partial \alpha_1}{\partial \eta_r}\left(G\eta_r - l\bar{\omega}_r\right)\right),$$

$$\tau_2 = \gamma_1\left(\bar{\omega} z_1 - \frac{\partial \alpha_1}{\partial y}\omega z_2 - \varsigma_1 z_1 z_2\right),$$

$\varsigma_1 \in \mathbb{R}^{11}$ is the coordinate vector. Then the actual control law is given by

$$u = \frac{1}{\varphi_u(y)}\alpha_2.$$

In order to generate the adjustable parameters $\hat{\vartheta} \in \mathbb{R}^{11}$ and $\hat{\kappa}_0$, we apply the adaptation algorithms

$$\dot{\hat{\vartheta}} = \tau_2, \qquad \dot{\hat{\kappa}}_0 = \gamma_2\left(\beta_1 + \bar{\omega}^\top \hat{\vartheta}\right) z_1.$$

The adaptation gains are given by $\gamma_1 = \gamma_2 = 1$.

In simulation, the initial conditions for x, ζ_1, ζ_2, ζ_3, v_0, $\eta_0, \ldots, \eta_3$, $\hat{\vartheta}$, and $\hat{\kappa}_0$ are set to zero.

Simulation results of the closed-loop system are presented in Fig. 5.7 and demonstrate complete compensation of the uncertain sinusoidal disturbance in the nonlinear plant with unknown parameters. ■

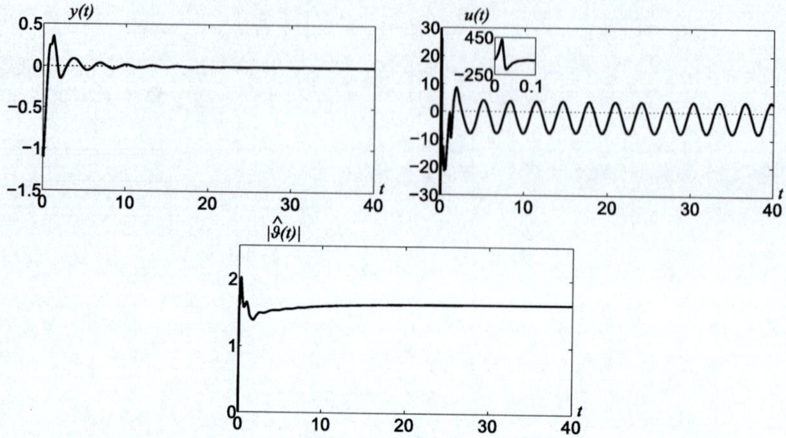

Fig. 5.7 Example 5.7: adaptive output-feedback compensation of uncertain disturbance $\upsilon(t) = \cos(2t + 3)$ in the nonlinear plant (5.186)–(5.188) with unknown parameters

References

1. Basturk, H.I., Krstić, M.: Adaptive cancelation of matched unknown sinusoidal disturbances for LTI systems by state derivative feedback. J. Dyn. Syst., Meas. Control **135**(1) (2013)
2. Basturk, H.I., Krstić, M.: State derivative feedback for adaptive cancellation of unmatched disturbances in unknown strict-feedback LTI systems. Automatica **50**, 2539–2545 (2014)
3. Bobtsov, A.A., Kolubin, S.A., Kremlev, A.S., Pyrkin, A.A.: An iterative algorithm of adaptive output control with complete compensation for unknown sinusoidal disturbance. Autom. Remote Control 1327–1363 (2012)
4. Chen, J., Ding, Z.: Adaptive output regulation of output feedback systems with unknown nonlinear exosystems and unknown high-frequency gain sign. In: Proceedings of the American Control Conference, pp. 4496–4501. Milwaukee, USA (2018)
5. Ding, Z.: Universal disturbance rejection for nonlinear systems in output feedback form. IEEE Trans. Autom. Control **48**(7), 1222–1227 (2003)
6. Fradkov, A.L., Miroshnik, I.V., Nikiforov, V.O.: Nonlinear and Adaptive Control of Complex Systems. Kluwer Academic Publishers (1999)
7. Ioannou, P., Sun, J.: Robust Adaptive Control. Prentice-Hall, Englewood Cliffs, NJ (1996)
8. Krstic, M., Kanellakopoulos, I., Kokotović, P.: Nonlinear and Adaptive Control Design. Wiley, NY (1995)
9. Marino, R., Tomei, P.: Adaptive tracking and disturbance rejection for uncertain nonlinear systems. IEEE Trans. Autom. Control **50**(1), 90–95 (2005)
10. Marino, R., Tomei, P.: An adaptive learning regulator for uncertain minimum phase systems with undermodeled unknown exosystems. Automatica **47**, 739–747 (2011)
11. Marino, R., Tomei, P.: Disturbance cancellation for linear systems by adaptive internal models. Automatica **49**, 1494–1500 (2013)
12. Marino, R., Tomei, P.: Output regulation for unknown stable systems. IEEE Trans. Autom. Control **60**(8), 2213–2218 (2015)
13. Monopoli, R.V.: Model reference adaptive control with an augmented signal. IEEE Trans. Autom. Control **19**(5), 474–484 (1974)
14. Morse, A.S.: High-order parameter tuners for adaptive control of nonlinear systems. In: Isidori, A., Tarn, T.J. (eds.) Systems, Models and Feedback: Theory and Applications, pp. 339–364. Birkhauser (1992)

15. Narendra, K.S., Annaswamy, A.M.: Stable Adaptive Systems. Prentice Hall (1989)
16. Nikiforov, V.O.: Adaptive servomechanism controller with an implicit reference model. Int. J. Control **68**(2), 277–286 (1997)
17. Nikiforov, V.O.: Adaptive stabilization of a linear plant subjected to external deterministic disturbances. J. Comput. Syst. Sci. Int. **36**(2), 258–261 (1997)
18. Nikiforov, V.O.: Adaptive controller rejecting uncertain deterministic disturbances in SISO systems. In: Proceedings of the 4th European Control Conference, pp. 1207–1212. Brussels, Belgium (1997)
19. Nikiforov, V.O.: Robust high-order tuner of simplified structure. Automatica **35**(8), 1409–1415 (1999)
20. Nikiforov, V.O.: Adaptive and Robust Control with Compensation of External Disturbance. Nauka, St.-Petersburg (2003). (in Russian)
21. Nikiforov, V.O.: Observers of external deterministic disturbances. Part II. Objects with unknown parameters. Autom. Remote. Control **65**(11) (2004)
22. Ortega, R.: On Morse's new adaptive controller: parameter convergence and transient performance. IEEE Trans. Autom. Control **38**(8), 1191–1202 (1993)
23. Priscoli, F.D., Marconi, L., Isidori, A.: A new approach to adaptive nonlinear regulation. SIAM J. Control Optim. **45**(3), 829–855 (2006)
24. Serrani, A., Isidori, A., Marconi, L.: Semiglobal nonlinear output regulation with adaptive internal model. IEEE Trans. Autom. Control **46**(8), 1178–1194 (2001)
25. Xu, D., Huang, J., Jiang, Z.-P.: Global adaptive output regulation for a class of nonlinear systems with iISS inverse dynamics using output feedback. Automatica **49**, 2184–2191 (2013)
26. Yilmaz, C.T., Basturk, H.I.: Output feedback control for unknown LTI systems driven by unknown periodic disturbances. Automatica **99**, 112–119 (2019)

Chapter 6
Adaptive Regulation in LTI Plants with Input Delay

Abstract The chapter presents controllers that provide adaptive reference tracking and adaptive disturbance rejection for LTI MIMO plants with known parameters and input delay. In other words, the chapter presents design methods applied to the plants with known parameters and input delay in order to solve control Problems 1.2, 1.3, and 1.4 formulated in Sect. 1.3.3. For methodological purposes, we first consider separately the problems of adaptive reference tracking (Sect. 6.1) and adaptive disturbance compensation (Sect. 6.2). Then combine the proposed approaches for solution to the general problem of adaptive output regulation (Sect. 6.3).

6.1 Reference Tracking

In this section, we consider Problem 1.2 of adaptive reference tracking and solve it for multidimensional reference and MIMO LTI plant with input delay. The plant is given by the equation

$$\dot{x} = Ax + Bu(t-\tau), \quad y = Cx, \tag{6.1}$$

where $x \in \mathbb{R}^n$ is the state, $y \in \mathbb{R}^\rho$ is the set of output variables ($\rho \le n$), $u \in \mathbb{R}^\rho$ is the vector of control signals, τ is the known constant delay, A, B, and C are the known matrices of appropriate dimensions. The plant satisfies Assumption 4.4.

The control problem is to design a state-feedback (or output feedback) controller that ensures the boundedness of all the closed-loop signals and achievement of the control objective

$$\lim_{t\to\infty} |y(t) - g(t)| = 0, \tag{6.2}$$

where $g \in \mathbb{R}^\rho$ is the measured reference satisfying Assumption 2.2 (with substitution $\upsilon = g$).

We introduce the controller (see [7])

$$u = -Kx(t+\tau) + u_t, \tag{6.3}$$

V. Nikiforov and D. Gerasimov, *Adaptive Regulation*, Lecture Notes in Control and Information Sciences 491, https://doi.org/10.1007/978-3-030-96091-9_6

where the matrix of feedback gains K is selected so that the matrix $A_\Sigma = A - BK$ is Hurwitz, u_t is the *tracking component* to be defined later, and $x(t+\tau)$ is the predicted state calculated as [1, 11–14]

$$x(t+\tau) = e^{A\tau}x(t) + \int_{t-\tau}^{t} e^{A(t-\mu)}Bu(\mu)d\mu. \tag{6.4}$$

Remark 6.1 The predictor (6.4) can be also represented in the following form:

$$x(t+\tau) = e^{A\tau}x + s_0, \tag{6.5}$$

where the integral

$$s_0(t) = \int_{t-\tau}^{t} e^{A(t-\mu)}Bu(\mu)d\mu \tag{6.6}$$

is calculated via the differential equation

$$\dot{s}_0 = As_0 + Bu - e^{A\tau}Bu(t-\tau), \quad s_0(0) = 0. \tag{6.7}$$

■

Remark 6.2 The control law (6.3) with the predictor (6.4) is not robust when the matrix A is not Hurwitz, and, as a result, the integral equation is not internally stable. In [14], for practical implementation of such a control law, it was proposed to approximate the integral term using one of the numerical quadrature rules [5]. However, as it was demonstrated in [6, 17, 23], such an approximation may not stabilize the system even for arbitrarily high accuracy of approximation. To overcome this practical limitation in [16], it was proposed to apply a low-pass filter to the approximated control and demonstrate that approximation of the integral by Riemann sum does not destroy the stability of the closed-loop system with the low-pass filter introduced. Later, in [9, 10, 24], this approach was successfully used for stabilization of systems with input as well as state delays.

In this book, for the sake of presentation simplicity, we will use predictors of the form (6.4) or (6.5)–(6.7) assuming that the problem of its practical implementation is resolved by the method mentioned above. ■

Remark 6.3 If the state x is not accessible for measurements, then in (6.4) instead of x we use the estimate $\hat{x}$ generated by the standard state observer

$$\dot{\hat{x}} = A\hat{x} + Bu(t-\tau) + L_o(y - C\hat{x}), \tag{6.8}$$

where the matrix of feedback gains $L_o \in \mathbb{R}^{n\times p}$ is such that the matrix $A_o = A - L_oC$ is Hurwitz [4, Sect. 8.8].

■

Substituting (6.3) into (6.1), we obtain

$$\dot{x} = A_\Sigma x + Bu_t(t-\tau), \quad y = Cx. \tag{6.9}$$

Then, in view of Lemmas 1.1 and 1.2, we conclude that there exists a solution pair (M_g, Θ_g) of the matrix equation (1.66) such that the plant (6.9) can be represented in the form

$$\dot{e} = A_\Sigma e + B\left(u_t(t-\tau) + \Theta_g\bar{\xi}\right), \quad \varepsilon = Ce, \tag{6.10}$$

where the state error $e = x - M_g\bar{\xi}$ is inaccessible for measurement, the output error $\varepsilon = y - g$ is measured, while the vector $\bar{\xi}$ is the state of the exosystem canonical form (2.23) or (2.24) (see Lemma 2.2).

Further, using Corollary A.7, we can write

$$\bar{\xi}(t) = \mathrm{e}^{(G+L\Theta)\tau}\bar{\xi}(t-\tau).$$

Hence,

$$\dot{e} = A_\Sigma e + B\left(u_t(t-\tau) + \bar{\Theta}\bar{\xi}(t-\tau)\right), \quad \varepsilon = Ce, \tag{6.11}$$

where $\bar{\Theta} = \Theta_g \mathrm{e}^{(G+L\Theta)\tau}$ is a matrix of unknown parameters.

Since $\bar{\xi}$ is not measured, it is necessary to apply an observer for calculation of the corresponding estimate. To this end, in view of (2.23), we directly extend the results of Lemma 2.3 to the case of multidimensional g and write

$$\bar{\xi} = \hat{\xi} + \epsilon_g,$$

where ϵ_g is the exponentially decaying term and $\hat{\xi}$ is the state of the reference observer

$$\dot{\hat{\xi}} = G\hat{\xi} + Lg \tag{6.12}$$

with an arbitrary initial condition $\hat{\xi}(0)$ and the matrices G, L defined by (2.25), (2.26).

Then we use the tracking component of the form

$$u_t = \hat{\Theta}\hat{\xi} \tag{6.13}$$

that yields the closed-loop model of the plant

$$\dot{e} = A_\Sigma e + B\left(\tilde{\Theta}(t-\tau)\hat{\xi}(t-\tau) + \epsilon_\tau\right), \quad \varepsilon = Cx \tag{6.14}$$

or

$$\varepsilon = W_\Sigma(s)\left[\tilde{\Theta}(t-\tau)\hat{\xi}(t-\tau)\right] + \epsilon, \tag{6.15}$$

where $W_\Sigma(s) = C(sI - A_\Sigma)^{-1}B$ is the asymptotically stable transfer function matrix, and ϵ_τ, ϵ exponentially decay.

At first glance, it seems that (6.14) is in the form of the canonical error model (3.90) that belongs to the class of linear dynamic error models with measured output. However, it is directly impossible to design conventional adaptation algorithms since the model (6.14) contains the delayed parametric error $\tilde{\Theta}(t-\tau)$. To overcome this problem, we employ a special swapping scheme. Below, we combine the results of Lemmas D.4 and D.7 from Appendix D to derive a suitable swapping scheme applicable to MIMO models with delay.

In view of Lemma D.4, we introduce the augmented error

$$\bar{\varepsilon} = \varepsilon + W_\Sigma(s)\left[\hat{\Theta}(t-\tau)\hat{\xi}(t-\tau)\right] - \Xi_w^\top(t-\tau)\hat{\vartheta}, \tag{6.16}$$

where $\hat{\vartheta} = [\hat{\Theta}_1, \hat{\Theta}_2, \ldots, \hat{\Theta}_\rho]^\top$ is the vector of adjustable parameters, $\hat{\Theta}_i$ is the ith row of the matrix $\hat{\Theta}$, and Ξ_w is the matrix regressor given by

$$\Xi_w^\top = \begin{bmatrix} W_{\Sigma 1,1}(s)\left[\hat{\xi}^\top\right] & \cdots & W_{\Sigma 1,\rho}(s)\left[\hat{\xi}^\top\right] \\ \vdots & \ddots & \vdots \\ W_{\Sigma \rho,1}(s)\left[\hat{\xi}^\top\right] & \cdots & W_{\Sigma \rho,\rho}(s)\left[\hat{\xi}^\top\right] \end{bmatrix}. \tag{6.17}$$

Then, applying Lemma D.7, we obtain the following model of the augmented error:

$$\bar{\varepsilon}(t) = \Xi_w^\top(t-\tau)\tilde{\vartheta}(t) + \epsilon_w(t), \tag{6.18}$$

where $\tilde{\vartheta} = \vartheta - \hat{\vartheta}$ is the vector of parametric errors, $\vartheta = \left[\bar{\Theta}_1, \bar{\Theta}_2, \ldots, \bar{\Theta}_\rho\right]^\top$ is a vector of unknown parameters, Θ_i is the ith row of the matrix Θ, and ϵ_w is the exponentially decaying term.

It is seen from (6.18) that the swapping scheme (6.16) enables one to remove the influence of delay from the vector of the parametric errors $\tilde{\vartheta}(t)$ and to obtain the error model for which the basic algorithms of adaptation from Chap. 3 are applicable. At the same time, since the regressor matrix Ξ_w is delayed, the adaptation process will naturally start with time delay τ.

The gradient algorithm of adaptation based on the model (6.18) is given by

$$\dot{\hat{\vartheta}} = \gamma\Xi_w(t-\tau)\bar{\varepsilon}, \quad \gamma > 0 \tag{6.19}$$

and yields the parametric error

$$\dot{\tilde{\vartheta}} = -\gamma \Xi_w(t-\tau)\Xi_w^\top(t-\tau)\tilde{\vartheta} - \gamma \Xi_w(t-\tau)\epsilon_w.$$

Then, with the use of standard arguments (see Proposition 3.10), we can prove the following statement.

Proposition 6.1 *Under Assumptions 2.2 and 4.4, the adjustable controller (6.3), (6.4), and (6.13) with the swapping scheme (6.16), (6.17) and with the adaptation algorithm (6.19) when applied to the plant (6.1) provides the boundedness of all the closed-loop signals and achievement of the control objective (6.2).*

Now, as shown in Sect. 4.2.1.1, we can derive adaptation algorithms with improved parametric convergence.

The algorithm with DRE. Define the dynamically extended augmented error $\bar{\varepsilon}_H \in \mathbb{R}^{\rho(p+1)}$ and the dynamically extended regressor $\bar{\xi}_w \in \mathbb{R}^{q\rho\times\rho(p+1)}$:

$$\bar{\varepsilon}_H = \begin{bmatrix} \hat{\varepsilon} - \Xi_w^\top(t-\tau)\hat{\vartheta} \\ H_1(s)\left[\hat{\varepsilon}\right] - \bar{\xi}_{w1}^\top(t-\tau)\hat{\vartheta} \\ \vdots \\ H_p(s)\left[\hat{\varepsilon}\right] - \bar{\xi}_{wp}^\top(t-\tau)\hat{\vartheta} \end{bmatrix}, \quad \bar{\xi}_w^\top = \begin{bmatrix} \Xi_w^\top \\ \bar{\xi}_{w1}^\top \\ \vdots \\ \bar{\xi}_{wp}^\top \end{bmatrix},$$

where $p \geq (q-1)$, $H_i(s)$ are distinct designer-chosen PASMP transfer functions,

$$\hat{\varepsilon} = \varepsilon + W_\Sigma(s)\left[\hat{\Theta}(t-\tau)\hat{\xi}(t-\tau)\right], \tag{6.20}$$

$$\bar{\xi}_{wi}^\top = H_i(s)\left[\Xi_w^\top\right], \qquad i = 1, 2, \ldots, p, \tag{6.21}$$

$\Xi_w^\top$ is defined by (6.17).

Then the adaptation algorithm with DRE takes the form

$$\dot{\hat{\vartheta}} - \gamma\bar{\xi}_w(t-\tau)\bar{\varepsilon}_H \tag{6.22}$$

and yields the closed-loop parametric error model

$$\dot{\tilde{\vartheta}} = -\gamma\bar{\xi}_w(t-\tau)\bar{\xi}_w^\top(t-\tau)\tilde{\vartheta} - \gamma\bar{\xi}_w(t-\tau)\epsilon_H,$$

where ϵ_H exponentially decays.

The algorithm with MRE is defined as

$$\dot{\hat{\vartheta}} = \gamma\left(L(s)\left[\Xi_w(t-\tau)\hat{\varepsilon}\right] - \Omega_\Xi(t-\tau)\hat{\vartheta}\right), \tag{6.23}$$

where $L(s)$ is a designer-chosen transfer function of the form (3.44), $\hat{\varepsilon}$ is given by (6.20), $\Omega_\Xi = L(s)\left[\Xi_w \Xi_w^\top\right]$. The algorithm yields the parametric error model

$$\dot{\tilde{\vartheta}} = -\gamma\Omega_\Xi(t-\tau)\tilde{\vartheta} - \gamma L(s)\left[\Xi_w(t-\tau)\epsilon_\Omega\right],$$

where ϵ_Ω exponentially decays.

In addition to Proposition 6.1, the algorithm of adaptation (6.22) provides the asymptotic parametric convergence if $\lambda_\Xi(t) \notin \mathcal{L}_1$, where $\lambda_\Xi(t)$ is the minimum eigenvalue of $\bar{\xi}_w(t)\bar{\xi}_w^\top(t)$, while the algorithm (6.23) provides the asymptotic parametric convergence if $\lambda_\Omega(t) \notin \mathcal{L}_1$, where $\lambda_\Omega(t)$ is the minimum eigenvalue of $\Omega_\Xi(t)$.

Example 6.1 Let us consider the two-input and two-output unstable plant (6.1) of the third order with the delay $\tau = 1$sec, the matrices

$$A = \begin{bmatrix} 0 & 1 & 0 \\ 0.1 & -0.9 & 0 \\ 0 & 0 & -1 \end{bmatrix}, \quad B = \begin{bmatrix} 0 & 0 \\ 1 & 0 \\ 0 & 1 \end{bmatrix}, \quad C = \begin{bmatrix} 1 & 0 & 0 \\ 1 & 0 & 1 \end{bmatrix},$$

and the transfer function matrix

$$W_u(s) = C\,(sI - A)^{-1}\,B = \begin{bmatrix} \frac{1}{s^2+0.9s-0.1} & 0 \\ \frac{1}{s^2+0.9s-0.1} & \frac{1}{s+1} \end{bmatrix}. \tag{6.24}$$

The reference signal is given by the function

$$g(t) = \begin{bmatrix} g_1(t) \\ g_2(t) \end{bmatrix} = \begin{bmatrix} \beta_2 \sin(wt + \varphi) \\ \beta_1 \end{bmatrix}$$

with the unknown constant parameters $\beta_1 = 2$, $\beta_2 = 3$, $w = 4$, and $\varphi = 5$. We design the observer (6.12) with

$$G = \begin{bmatrix} 0 & 1 & 0 \\ 0 & 0 & 0 \\ -24 & -26 & -9 \end{bmatrix}, \quad L = \begin{bmatrix} 0 & 0 \\ 0 & 0 \\ 24 & 24 \end{bmatrix}$$

and use the stabilizing component with the matrix

$$K = \begin{bmatrix} 5.685 & 8.246 & 20.732 \\ -1.197 & -1.049 & -0.146 \end{bmatrix}.$$

The latter ensures for the matrix $A_\Sigma = A - BK$ the set of eigenvalues $\lambda\{A_\Sigma\} = \{-2, -3, -4\}$ and together with the triple (A, B, C) gives the transfer function matrix

$$W_\Sigma(s) = \frac{1}{s^3 + 9s^2 + 26s + 24} \begin{bmatrix} s + 0.745 & -36.17 \\ 2.043s + 1.776 & s^2 + 8.255s - 54.043 \end{bmatrix}. \tag{6.25}$$

It is worth noting that the both transfer function matrices (6.24) and (6.25) have no transmission zeros.

The vector of adjustable parameters $\hat{\vartheta}$ is evaluated via one of the following three algorithms of adaptation:

- gradient algorithm (6.19) with $\gamma = 4$;
- algorithm (6.22) with DRE and $p = 2$, $H_1(s) = \frac{2}{s+1}$, $H_2(s) = \frac{2}{0.2s+1}$, $\gamma = 200$;
- algorithm (6.23) with MRE and $L(s) = \frac{1}{s+1}$, $\gamma = 200$.

All simulations are carried out with zero initial conditions $x(0)$, $s_0(0)$, $\hat{\xi}(0)$, and $\hat{\vartheta}(0)$. Simulation results for the closed-loop systems employing the adaptation algorithm with DRE (6.22) and the algorithm of adaptation with MRE (6.23) are presented in Fig. 6.1 and Fig. 6.2, respectively. We do not present the simulation results for the system closed by the gradient adaptation algorithm since the transients achieved after fitting of the gain γ take more than 5000 s of simulation time and cannot be significantly accelerated.

The results demonstrate the asymptotic tracking of a priori uncertain multidimensional reference g. In contrast to the gradient adaptation algorithm, the algorithms with DRE and MRE demonstrate fast transient performance. ■

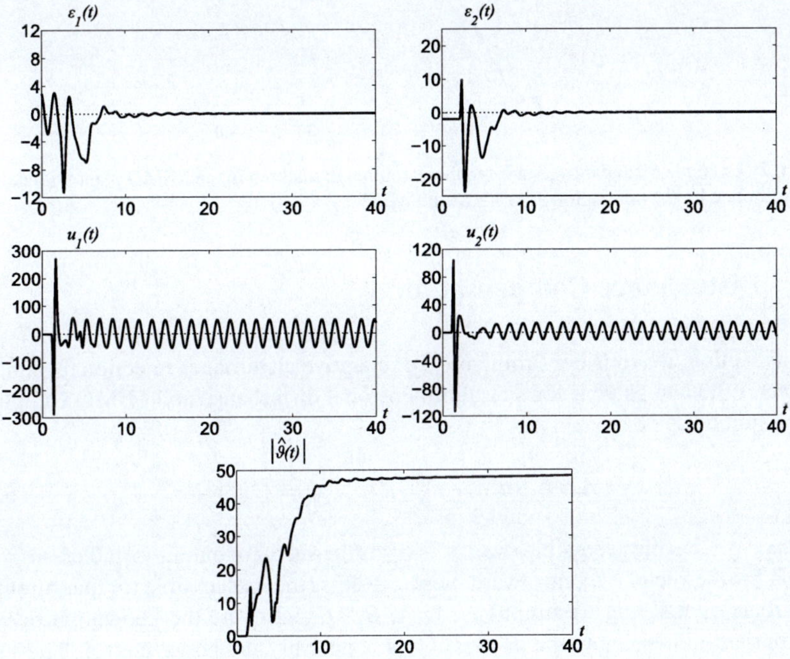

Fig. 6.1 Example 6.1: adaptive state-feedback reference tracking in the MIMO plant with delayed control tuned by the adaptation algorithm with DRE

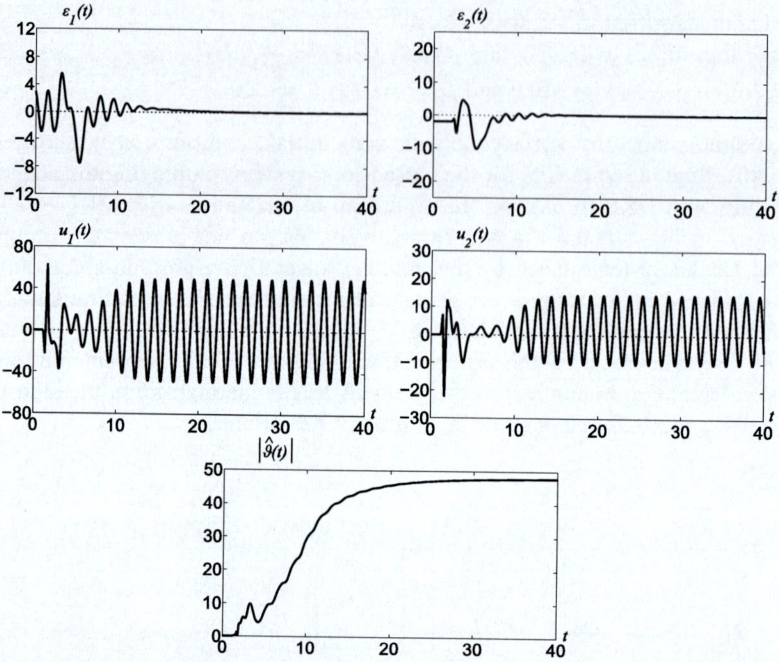

Fig. 6.2 Example 6.1: adaptive state-feedback reference tracking in the MIMO plant with delayed control tuned by the adaptation algorithm with MRE

6.2 Disturbance Compensation

In this section, we consider Problem 1.3 of adaptive disturbance rejection formulated in Sect. 1.3.3 and solve it for a multidimensional disturbance and MIMO LTI plants with input delay

$$\dot{x} = Ax + Bu(t-\tau) + D\upsilon, \quad y = Cx + Rx, \tag{6.26}$$

where $x \in \mathbb{R}^n$ is the accessible state, $y \in \mathbb{R}^m$ is the vector of output variables ($m \le n$), $u \in \mathbb{R}^m$ is the vector of control variables, $\upsilon \in \mathbb{R}^\rho$ is the inaccessible for measurement disturbance satisfying Assumption 2.2, A, B, D, and C are the known matrices of appropriate dimensions, and $R \in \mathbb{R}^{m\times\rho}$ is a possibly unknown matrix. The plant (6.1) satisfies Assumption 4.4.

The considered control problem consists in design of a state-feedback controller that provides the control objective

$$\lim_{t\to\infty} |y(t)| = 0 \tag{6.27}$$

with all the closed-loop signals bounded.[1]

First, in view of Lemmas 1.1 and 1.2, we conclude that there exists a solution pair $(M_\upsilon, \Theta_\upsilon)$ of the matrix equation (1.67) such that the plant (6.26) can be represented in the form

$$\dot{e} = Ae + B\left(u(t-\tau) + \Theta_\upsilon \bar{\xi}\right), \quad y = Ce, \tag{6.28}$$

where $e = x - M_\upsilon \bar{\xi}$ is the inaccessible for measurement state error, $\bar{\xi}$ is the state of the exosystem canonical form (2.23) or (2.24) (see Lemma 2.2).

Then, using Corollary A.7, we can write

$$\bar{\xi}(t) = \mathrm{e}^{(G+L\Theta)\tau}\bar{\xi}(t-\tau),$$

and, hence,

$$\dot{e} = Ae + B\left(u(t-\tau) + \bar{\Theta}\bar{\xi}(t-\tau)\right), \quad y = Ce, \tag{6.29}$$

where $\bar{\Theta} = \Theta_\upsilon \mathrm{e}^{(G+L\Theta)\tau}$ is a new matrix of unknown parameters.

Since $\bar{\xi}$ is unmeasured, we construct its estimate $\hat{\xi}$. In view of Lemma 2.2, we use Lemma 2.5 and define the estimate as

$$\hat{\xi} = \eta + Nx, \tag{6.30}$$

where $\eta \in \mathbb{R}^q$ is the state of the filter

$$\dot{\eta} = G\eta + (GN - NA)x - NBu(t-\tau) \tag{6.31}$$

with an arbitrary initial condition $\hat{\xi}(0)$, matrices G and L given by (2.25), (2.26), and the matrix $N \in \mathbb{R}^{q\times n}$ selected so that[2]

$$ND = L. \tag{6.32}$$

Then the following equality holds:

$$\bar{\xi} = \hat{\xi} + \epsilon_\xi, \tag{6.33}$$

where ϵ_ξ exponentially decays.

Replacing (6.33) in (6.29), we obtain

---

[1] Alternative identification-based approaches to the considered problem are presented in [3, 20–22]. The problem of direct adaptive compensation of matched disturbances affecting LTI SISO plant with input delay is solved in [8]. Extension to a certain class of unknown LTI systems is proposed in [2].

[2] See Remark 2.11.

$$\dot{e} = Ae + B\left(u(t-\tau) + \bar{\Theta}\hat{\xi}(t-\tau) + \epsilon\right), \quad y = Ce, \tag{6.34}$$

where $\epsilon(t) = \bar{\Theta}\epsilon_{\xi}(t-\tau)$ exponentially decays.

Motivated by the latter expression, we design the adjustable control law

$$u = u_s - \hat{\Theta}\hat{\xi}, \tag{6.35}$$

where u_s is a stabilizing component defined later, and $\hat{\Theta} \in \mathbb{R}^{m\times q}$ is the matrix of adjustable parameters generated by an adaptation algorithm. Substituting (6.35) into (6.34), we have

$$\dot{e} = Ae + B\left(u_s(t-\tau) + \tilde{\Theta}(t-\tau)\hat{\xi}(t-\tau) + \epsilon\right), \quad y = Ce, \tag{6.36}$$

where $\tilde{\Theta} = \bar{\Theta} - \hat{\Theta}$ is the matrix of parametric errors.

Since the regressor $\hat{\xi}$ is known to be bounded a priori, we first design an adaptation algorithm calculating the adjustable parameters $\hat{\Theta}$ and then stabilize the plant by an appropriate component u_s.

The model (6.36) cannot be directly used for design of an adaptation algorithm since the matrix A is possibly unstable and the matrix of parametric errors $\tilde{\Theta}(t-\tau)$ is delayed. To overcome these two obstacles, we apply a special swapping scheme obtained by combination of the results of Lemmas D.6 and D.8.

Lemma 6.1 *Let Assumptions 2.2 and 4.4 hold. Together with the plant (6.26), consider the filter*

$$\dot{\bar{e}} = A\bar{e} - L_e(y + C\bar{e}) - Bu(t-\tau), \tag{6.37}$$

where the matrix L_e is such that $A_e = A - L_eC$ is Hurwitz, and the initial condition $\bar{e}(0)$ is arbitrary. Define the augmented output as

$$\bar{y} = y + C\bar{e} - \Xi^{\top}(t-\tau)\hat{\vartheta}(t), \tag{6.38}$$

where $\hat{\vartheta} = [\hat{\Theta}_1, \hat{\Theta}_2, \ldots, \hat{\Theta}_m]^{\top}$ is the vector of adjustable parameters, $\hat{\Theta}_i$ is the ith row of the matrix $\hat{\Theta}$, and $\Xi \in \mathbb{R}^{mq\times m}$ is the matrix regressor given by

$$\Xi^{\top} = \begin{bmatrix} W_{e1,1}(s)\left[\hat{\xi}^{\top}\right] & \cdots & W_{e1,m}(s)\left[\hat{\xi}^{\top}\right] \\ \vdots & \ddots & \vdots \\ W_{em,1}(s)\left[\hat{\xi}^{\top}\right] & \cdots & W_{em,m}(s)\left[\hat{\xi}^{\top}\right] \end{bmatrix}. \tag{6.39}$$

Then:

L.6.1.1. *for the augmented output, we have*

$$\bar{y}(t) = \Xi^{\top}(t-\tau)\tilde{\vartheta}(t) + \epsilon_y(t), \tag{6.40}$$

where $\tilde{\vartheta} = \vartheta - \hat{\vartheta}$ is the vector of the parametric errors, $\vartheta = [\bar{\Theta}_1, \ldots, \bar{\Theta}_\rho]^\top$ is the vector of unknown parameters, $\bar{\Theta}_i$ is the ith row of $\bar{\Theta}$, and ϵ_y is the exponentially decaying term;

L.6.1.2. *if $|\dot{\hat{\vartheta}}(t)| \to 0$, then $\left|\Xi^\top(t-\tau)\tilde{\vartheta}(t) - W_e(s)[\tilde{\Theta}(t-\tau)\hat{\xi}(t-\tau)]\right| \to 0$ as $t \to \infty$.*

Proof Define the auxiliary vectors $e^* = e + \bar{e}$ and $y^* = y + C\bar{e} = C(e + \bar{e})$. Evaluating the time derivative of e^*, in view of (6.36) and (6.37), we have

$$\dot{e}^* = A_e e^* + B\left(\bar{\Theta}\hat{\xi}(t-\tau) + \epsilon\right), \quad y^* = Ce^*$$

or

$$y^* = y + C\bar{e} = W_e(s)\left[\bar{\Theta}\hat{\xi}(t-\tau)\right] + \epsilon_y. \tag{6.41}$$

By virtue of the properties of linear systems, we have

$$\begin{aligned} W_e(s)\left[\bar{\Theta}\hat{\xi}(t-\tau)\right] &= \begin{bmatrix} W_{e1,1}(s) & \cdots & W_{e1,m}(s) \\ \vdots & \ddots & \vdots \\ W_{em,1}(s) & \cdots & W_{em,m}(s) \end{bmatrix} \begin{bmatrix} \bar{\Theta}_1\hat{\xi}(t-\tau) \\ \vdots \\ \bar{\Theta}_m\hat{\xi}(t-\tau) \end{bmatrix} \\ &= \begin{bmatrix} W_{e1,1}(s)\left[\hat{\xi}^\top(t-\tau)\bar{\Theta}_1^\top\right] + \ldots + W_{e1,m}(s)\left[\hat{\xi}^\top(t-\tau)\bar{\Theta}_m^\top\right] \\ \vdots \\ W_{em,1}(s)\left[\hat{\xi}^\top(t-\tau)\bar{\Theta}_1^\top\right] + \ldots + W_{em,m}(s)\left[\hat{\xi}^\top(t-\tau)\bar{\Theta}_m^\top\right] \end{bmatrix} \\ &= \Xi^\top(t-\tau)\begin{bmatrix} \bar{\Theta}_1^\top \\ \vdots \\ \bar{\Theta}_m^\top \end{bmatrix} = \Xi^\top(t-\tau)\vartheta. \end{aligned}$$

Hence,

$$W_e(s)\left[\bar{\Theta}\hat{\xi}(t-\tau)\right] = \Xi^\top(t-\tau)\vartheta.$$

Substituting (6.41) into (6.38), in view of the latter expression, we directly obtain (6.40).

From Corollary D.1 we obtain that $\left|\Xi^\top(t-\tau)\tilde{\vartheta}(t) - W_e(s)[\tilde{\Theta}(t)\hat{\xi}(t-\tau)]\right| \to 0$ if $|\dot{\hat{\vartheta}}(t)| \to 0$ as $t \to \infty$. Since $|\dot{\hat{\vartheta}}(t)| \to 0$, we also have $|\tilde{\Theta}(t) - \tilde{\Theta}(t-\tau)| \to 0$ as $t \to \infty$. This completes the proof of L.6.1.2. □

Model (6.40) belongs to the class of standard static error models and enables one to employ the following basic algorithms of adaptation from Sect. 3.2:

The gradient algorithm of adaptation

$$\dot{\hat{\vartheta}} = \gamma \Xi(t-\tau)\bar{y}, \quad \gamma > 0, \tag{6.42}$$

where $\bar{y}$ and Ξ are defined by (6.38) and (6.39), respectively;

The adaptation algorithm with DRE

$$\dot{\hat{\vartheta}} = \gamma \bar{\xi}(t-\tau)\bar{y}_H, \quad \gamma > 0, \tag{6.43}$$

where

$$\begin{aligned} \bar{y}_H &= \begin{bmatrix} \hat{y} - \Xi^\top(t-\tau)\hat{\vartheta} \\ H_1(s)\left[\hat{y}\right] - \bar{\xi}_1^\top(t-\tau)\hat{\vartheta} \\ \vdots \\ H_p(s)\left[\hat{y}\right] - \bar{\xi}_p^\top(t-\tau)\hat{\vartheta} \end{bmatrix}, \quad \bar{\xi}^\top = \begin{bmatrix} \Xi^\top \\ \bar{\xi}_1^\top \\ \vdots \\ \bar{\xi}_p^\top \end{bmatrix}, \\ \hat{y} &= y + C\bar{e}, \end{aligned} \tag{6.44}$$

$\bar{H}(s) = \left[1, H_1(s), \ldots, H_p(s)\right]^\top$, $H_i(s)$ are distinct designer-chosen PASMP transfer functions, $p \geq (q-1)$,

$$\bar{\xi}_j = H_j(s)\left[\Xi\right], \quad j = 1, 2, \ldots, p,$$

and Ξ is defined by (6.39) ;

The adaptation algorithm with MRE

$$\dot{\hat{\vartheta}} = \gamma \left(L(s)\left[\Xi(t-\tau)\hat{y}\right] - \Omega_\Xi(t-\tau)\hat{\vartheta} \right), \quad \gamma > 0, \tag{6.45}$$

where $L(s)$ is a designer-chosen transfer function of the form (3.44), $\hat{y}$ is given by (6.44), and $\Omega_\Xi = L(s)\left[\Xi\Xi^\top\right]$.

Regardless of the stability properties of the closed-loop system, these adaptation algorithms have the following properties.

Lemma 6.2 *Let Assumptions 2.2 and 4.4 hold. Then the adaptation algorithm (6.42) (or (6.43), or (6.45)) with the swapping scheme (6.37)–(6.39) provides:*

L.6.2.1 *the boundedness of $\bar{y}$, $\hat{\vartheta}$, and $\hat{\Theta}$;*

L.6.2.2 *the asymptotic convergence $|\tilde{\Theta}(t)\hat{\xi}(t)| \to 0$ as $t \to \infty$.*

In addition to Lemma 6.2, the algorithm of adaptation (6.43) provides the asymptotic parametric convergence if $\lambda_\Xi(t) \notin \mathcal{L}_1$, where $\lambda_\Xi(t)$ is the minimum eigenvalue of $\bar{\xi}(t)\bar{\xi}^\top(t)$, while the algorithm (6.45) provides the asymptotic parametric convergence if $\lambda_\Omega(t) \notin \mathcal{L}_1$, where $\lambda_\Omega(t)$ is the minimum eigenvalue of $\Omega_\Xi(t)$.

Proof Algorithm (6.42) yields the closed-loop parametric error model

$$\dot{\tilde{\vartheta}} = -\gamma \Xi(t-\tau)\Xi^\top(t-\tau)\tilde{\vartheta}(t) - \gamma \Xi(t-\tau)\epsilon_y.$$

Since $\hat{\xi}$ is bounded, from Proposition 3.1 we directly obtain L.6.2.1 and the convergence $|\bar{y}(t)|$, $|\dot{\hat{\vartheta}}(t)| \to 0$ as $t \to \infty$. The convergence $|\bar{y}(t)| \to 0$ implies that $|\Xi^\top(t-\tau)\tilde{\vartheta}(t)| \to 0$ as $t \to \infty$. In view of Property L.6.1.2 of Lemma 6.1, Property L.2.2.2 follows from the convergence $|\dot{\hat{\vartheta}}(t)| \to 0$.

The proofs of Properties L.6.2.1 and L.6.2.2 for the adaptation algorithms (6.43) and (6.45) are carried out in a similar way by applying instead of Proposition 3.1, Propositions 3.3 and 3.4, respectively. □

Now, we construct the stabilizing component u_s. Since the state error e is not accessible for measurements, we apply the state observer defined by the following lemma.

Lemma 6.3 *Let Assumptions 2.2 and 4.4 hold. Together with the adaptation algorithm (6.42) (or (6.43), or (6.45)), we consider the following state error observer:*

$$\dot{\hat{e}} = A\hat{e} + L_e(y - C\hat{e}) + Bu_s(t-\tau), \tag{6.46}$$

where $\hat{e}$ is the state error estimate, L_e is the matrix selected so that $A_e = A - L_eC$ is Hurwitz. Then the estimation error $\tilde{e}(t) = e(t) - \hat{e}(t)$ approaches zero as $t \to \infty$:

$$\lim_{t\to\infty} |\tilde{e}(t)| = 0. \tag{6.47}$$

Proof Evaluation of the time derivative of $\tilde{e}$, in view of (6.36) and (6.46), gives

$$\dot{\tilde{e}} = A_e\tilde{e} + B\left(\tilde{\Theta}(t-\tau)\hat{\xi}(t-\tau) + \epsilon\right).$$

Then, by virtue of L.6.2.2, we directly obtain (6.47). □

The stabilizing component is defined as

$$u_s = -K\left(e^{A\tau}\hat{e}(t) + \int_{t-\tau}^{t} e^{A(t-\mu)}Bu_s(\mu)d\mu\right), \tag{6.48}$$

where the matrix of feedback gains K is such that $A_\Sigma = A - BK$ is Hurwitz, and $\hat{e}$ is the estimate generated by the state error observer (6.46).

The properties of the closed-loop system are determined as follows [18, 19].

Proposition 6.2 *Under Assumptions 2.2 and 4.4, the adaptive controller (6.35) with the adaptation algorithm (6.42) (or (6.43), or (6.45)), the swapping scheme (6.37)–(6.39), the state error observer (6.46), and the stabilizing component (6.48) when applied to the plant (6.26) provides the boundedness of all the closed-loop signals and achievement of the control objective (6.27).*

Proof Based on the expression (6.46), we calculate the prediction $\hat{e}(t+\tau)$:

$$\hat{e}(t+\tau) = \mathrm{e}^{A\tau}\hat{e}(t) + \int_{t-\tau}^{t} \mathrm{e}^{A(t-\mu)} \left(L_e C\tilde{e}(\mu+\tau) + Bu_s(\mu)\right) d\mu.$$

Hence, the stabilizing component (6.48) can be represented in the form

$$u_s(t) = -K\left(e(t+\tau) - \tilde{e}(t+\tau) - \Delta(t+\tau)\right), \tag{6.49}$$

where

$$\Delta(t+\tau) = \int_{t-\tau}^{t} \mathrm{e}^{A(t-\mu)} L_e C\tilde{e}(\mu+\tau) d\mu.$$

Substituting (6.49) into (6.36), we obtain

$$\dot{e} = A_\Sigma e + B\left(K\tilde{e}(t) + K\Delta(t) + \tilde{\Theta}(t-\tau)\hat{\xi}(t-\tau) + \epsilon\right).$$

In the latter equation, $|\tilde{\Theta}(t-\tau)\hat{\xi}(t-\tau)| \to 0$ as $t \to \infty$ by virtue of L.6.2.2, while $|\tilde{e}(t)|$, $|\Delta(t)| \to 0$ as $t \to \infty$ by virtue of (6.47). Therefore, $|e(t)| \to 0$ as $t \to \infty$, and the control objective (6.27) is achieved. Further signal analysis demonstrates the boundedness of all the closed-loop signals. □

Example 6.2 Consider the plant (6.26) with $\tau = 1$sec, the matrices

$$A = \begin{bmatrix} 0 & 1 & 0 \\ 0.1 & -0.9 & 0 \\ 0 & 0 & -1 \end{bmatrix},\ B = \begin{bmatrix} 0 & 0 \\ 1 & 0 \\ 0 & 1 \end{bmatrix},\ C = \begin{bmatrix} 1 & 0 & 0 \\ 1 & 0 & 1 \end{bmatrix}, \tag{6.50}$$

$$D = \begin{bmatrix} 1 & 1 \\ 0 & 1 \\ 1 & 0 \end{bmatrix},\ R = \begin{bmatrix} 1 & 2 \\ 2 & 3 \end{bmatrix},$$

and with the corresponding transfer function matrices

$$W_u(s) = C(sI-A)^{-1}B = \begin{bmatrix} \frac{1}{s^2+0.9s-0.1} & 0 \\ \frac{1}{s^2+0.9s-0.1} & \frac{1}{s+1} \end{bmatrix}, \tag{6.51}$$

$$W_\upsilon(s) = R + C(sI-A)^{-1}D = \begin{bmatrix} \frac{s^2+1.9s+0.8}{s^2+0.9s-0.1} & \frac{2s^2+2.8s+1.7}{s^2+0.9s-0.1} \\ \frac{2s^2+3.8s+0.6}{s^2+0.9s-0.1} & \frac{3s^2+3.7s+1.6}{s^2+0.9s-0.1} \end{bmatrix}. \tag{6.52}$$

Let the disturbance be given by

$$\upsilon(t) = \begin{bmatrix} \upsilon_1(t) \\ \upsilon_2(t) \end{bmatrix} = \begin{bmatrix} \beta_1 \\ \beta_{2,1} + \beta_{2,2}\sin(wt+\varphi) \end{bmatrix}, \tag{6.53}$$

where $\beta_1 = 1$, $\beta_{2,1} = 2$, $\beta_{2,2} = 3$, $w = 1$, and $\varphi = 0$ are unknown. We use the third-order observer (6.30), (6.31) with

$$G = \begin{bmatrix} 0 & 1 & 0 \\ 0 & 0 & 1 \\ -24 & -26 & -9 \end{bmatrix}, \quad L = \begin{bmatrix} 0 & 0 \\ 0 & 0 \\ 24 & 24 \end{bmatrix}, \quad N = \begin{bmatrix} 0 & 0 & 0 \\ 0 & 0 & 0 \\ 24 & 0 & 0 \end{bmatrix}.$$

For the swapping scheme (6.37) and for the state error observer (6.46), we select the matrix L_e as

$$L_e = \begin{bmatrix} 17.505 & 14.141 \\ -4.160 & -3.221 \\ -30 & -24.546 \end{bmatrix}.$$

For the stabilizing component (6.48), we select the matrix

$$K = \begin{bmatrix} -17.772 & 7.355 & 36.170 \\ -1.032 & -1.043 & -0.255 \end{bmatrix}.$$

The matrices L_e and K provide the eigenvalues sets of the matrices $A_e = A - L_eC$, $A_\Sigma = A - BK$ given by $\lambda\{A_e\} = \lambda\{A_\Sigma\} = \{-2, -3, -4\}$.

To implement the expression (6.39) of the swapping scheme, we obtain the transfer function matrix

$$\begin{aligned} W_e(s) &= C(sI - A_e)^{-1}B \\ &= \frac{1}{s^3 + 9s^2 + 26s + 24} \begin{bmatrix} s - 23.546 & -14.141s - 9.505 \\ s + 31 & s^2 + 18.405s + 11.495 \end{bmatrix}. \end{aligned}$$

To generate the adjustable parameters $\hat{\vartheta}$, we apply the adaptation algorithm with MRE and $L(s) = \frac{1}{2s+1}$, $\gamma = 20$. All simulations are carried out with zero initial conditions $x(0)$, $\eta(0)$, $\bar{e}(0)$, $\hat{e}(0)$, and $\hat{\vartheta}(0)$. Simulation results of the adaptive closed-loop system are presented in Fig. 6.3 and demonstrate the asymptotic rejection of a priori uncertain multidimensional disturbance υ affecting the system with input delay. ■

6.3 Regulation

In this section, we demonstrate how the above techniques can be combined to solve Problem 1.4 of adaptive output regulation formulated in Sect. 1.3.3. The problem consists in simultaneous reference tracking and disturbance rejection.

We consider the disturbed plant (6.26) satisfying Assumption 4.4. The control problem is to design a state-feedback controller ensuring the boundedness of all the closed-loop signals and achievement of the tracking objective

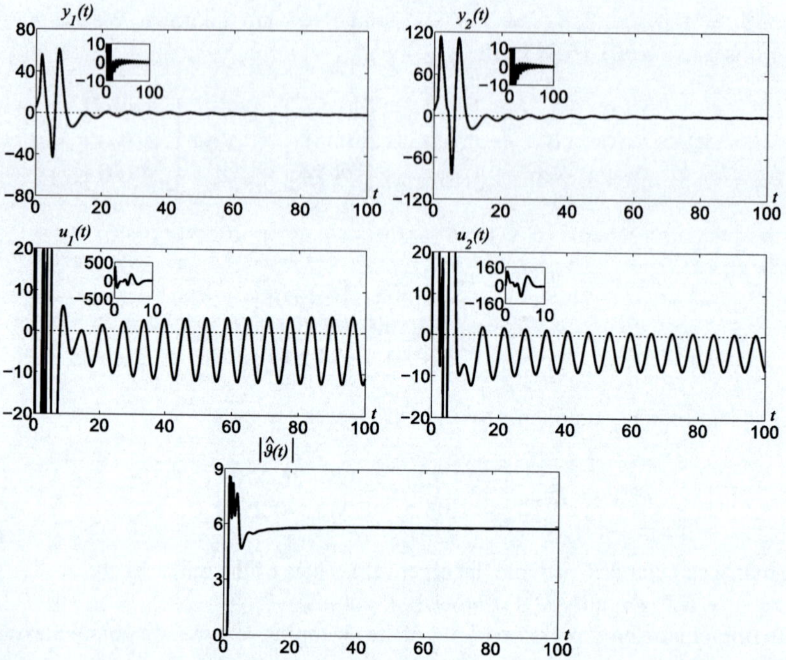

Fig. 6.3 Example 6.2: adaptive state-feedback disturbance compensation in the MIMO plant with delayed control tuned by the adaptation algorithm with MRE

$$\lim_{t\to\infty} |y(t) - g(t)| = 0, \tag{6.54}$$

where $g \in \mathbb{R}^m$ is the measured reference. Regarding the exosignals υ and g we make the following assumption.

Assumption 6.1[3] The disturbance $\upsilon \in \mathbb{R}^\rho$ and the reference $g \in \mathbb{R}^m$ are such that:

A.6.1.1. they can be modelled as the outputs of the exosystems

$$\dot{\xi}_g = \Gamma_g \xi_g, \qquad g = Q_g \xi_g, \tag{6.55}$$
$$\dot{\xi}_\upsilon = \Gamma_\upsilon \xi_\upsilon, \qquad \upsilon = H_\upsilon \xi_\upsilon, \tag{6.56}$$

where $\xi_g \in \mathbb{R}^{q_g}$, $\xi_\upsilon \in \mathbb{R}^{q_\upsilon}$ and the constant matrices Γ_g, Γ_υ, Q_g, and H_υ have appropriate dimensions;

A.6.1.2. both exosystems (6.55) and (6.56) satisfy Assumption 2.2.

Under Assumption 2.2, we can rewrite the exosystems (6.55) and (6.56) in the canonical forms (see Lemma 2.2 with appropriate change of notations):

---

[3] This is Assumption 4.6 repeated here for the sake of presentation convenience.

$$\dot{\bar{\xi}}_g = G_g\bar{\xi}_g + L_g g, \quad g = \bar{Q}_g\bar{\xi}_g,$$
$$\dot{\bar{\xi}}_\upsilon = G_\upsilon\bar{\xi}_\upsilon + L_\upsilon \upsilon, \quad \upsilon = \bar{H}_\upsilon\bar{\xi}_\upsilon,$$

where $\bar{\xi}_g \in \mathbb{R}^{q_g}$ and $\bar{\xi}_\upsilon \in \mathbb{R}^{q_\upsilon}$ are the states of the canonical forms, the Hurwitz matrices $G_g \in \mathbb{R}^{q_g \times q_g}$, $G_\upsilon \in \mathbb{R}^{q_\upsilon \times q_\upsilon}$ and the matrices $L_g \in \mathbb{R}^{q_g \times m}$, $L_\upsilon \in \mathbb{R}^{q_\upsilon \times \rho}$ are in the form of (2.25), (2.26), while the matrices $\bar{Q}_g \in \mathbb{R}^{m \times q_g}$ and $\bar{H}_\upsilon \in \mathbb{R}^{\rho \times q_\upsilon}$ are unknown.

According to Lemmas 1.1 and 1.2, there exist the solution pairs (M_g, Θ_g) and $(M_\upsilon, \Theta_\upsilon)$ of the matrix equations (1.66) and (1.67) such that in the new coordinates $e = x - M\bar{\xi}$ and $\varepsilon = y - g$ the plant (6.26) can be represented in the form

$$\dot{e} = Ae + B\left(\Theta_g\bar{\xi}_g + \Theta_\upsilon\bar{\xi}_\upsilon + u(t-\tau)\right), \quad \varepsilon = Ce, \tag{6.57}$$

where $M = \left[M_g,\ M_\upsilon\right]$, $\bar{\xi} = \left[\bar{\xi}_g^\top, \bar{\xi}_\upsilon^\top\right]^\top$.

In view of Corollary A.7, we can write

$$\bar{\xi}_g(t) = e^{(G_g + L_g\bar{Q}_g)\tau}\bar{\xi}_g(t-\tau), \quad \bar{\xi}_\upsilon(t) = \mathrm{e}^{(G_\upsilon + L_\upsilon\bar{H}_\upsilon)\tau}\bar{\xi}_\upsilon(t-\tau).$$

Replacing the latter expressions in (6.57), we have

$$\dot{e} = Ae + B(\bar{\Theta}_g\bar{\xi}_g(t-\tau) + \bar{\Theta}_\upsilon\bar{\xi}_\upsilon(t-\tau) + u(t-\tau)), \quad \varepsilon = e, \tag{6.58}$$

where

$$\bar{\Theta}_g = \Theta_g\mathrm{e}^{(G_g + L_g\bar{Q}_g)\tau}, \quad \bar{\Theta}_\upsilon = \Theta_\upsilon\mathrm{e}^{(G_\upsilon + L_\upsilon\bar{H}_\upsilon)\tau}$$

are unknown matrices.

Since the vectors $\bar{\xi}_g$ and $\bar{\xi}_\upsilon$ are inaccessible for measurements, we will construct the estimates $\hat{\xi}_g$ and $\hat{\xi}_\upsilon$ of the states $\bar{\xi}_g$ and $\bar{\xi}_\upsilon$, respectively.

To estimate $\bar{\xi}_g$, we use the reference observer

$$\dot{\hat{\xi}}_g = G_g\hat{\xi}_g + L_g g, \tag{6.59}$$

where the designer-chosen matrices $G_g \in \mathbb{R}^{q_g \times q_g}$ and $L_g \in \mathbb{R}^{q_g \times m}$ are given by (2.25), (2.26). The observer (6.59) ensures that

$$\bar{\xi}_g = \hat{\xi}_g + \epsilon_g, \tag{6.60}$$

where ϵ_g exponentially decays.

To estimate $\bar{\xi}_\upsilon$, we use the observer

$$\hat{\xi}_\upsilon = \eta + Nx, \quad \dot{\eta} = G_\upsilon\eta + (G_\upsilon N - NA)x - NBu(t-\tau) \tag{6.61}$$

with an arbitrary initial condition $\hat{\xi}(0)$, the matrix $N \in \mathbb{R}^{q\times n}$ satisfying the equality[4]

$$ND = L_{\upsilon}, \tag{6.62}$$

and the matrices $G_{\upsilon} \in \mathbb{R}^{q_{\upsilon}\times q_{\upsilon}}$ and $L_{\upsilon} \in \mathbb{R}^{q_{\upsilon}\times p}$ given by (2.25), (2.26). The observer (6.61), (6.62) ensures that

$$\bar{\xi}_{\upsilon} = \hat{\xi}_{\upsilon} + \epsilon_{\upsilon}, \tag{6.63}$$

where ϵ_{ξ} exponentially decays.

Replacement of (6.60) and (6.63) in (6.58) gives

$$\dot{e} = Ae + B\left(u(t-\tau) + \bar{\Theta}\hat{\xi}(t-\tau) + \epsilon\right), \quad \varepsilon = Ce, \tag{6.64}$$

where $\bar{\Theta} = [\bar{\Theta}_g,\ \bar{\Theta}_{\upsilon}]$, $\hat{\xi} = [\hat{\xi}_g^{\top},\ \hat{\xi}_{\upsilon}^{\top}]^{\top}$, ϵ is the exponentially decaying term.

The latter model is of the form (6.34) (taking into account the difference in definitions of $\hat{\xi}$ and $\bar{\Theta}$). Hence, we can directly use the results of Sect. 6.2 and employ:

- the adjustable controller

$$u = u_s - \hat{\Theta}\hat{\xi}, \tag{6.65}$$

  where $\hat{\Theta}$ is the matrix of the adjustable parameters, u_s is the stabilizing component;
- the state error observer

$$\dot{\hat{e}} = A\hat{e} + L_e(y - C\hat{e}) + Bu_s(t-\tau); \tag{6.66}$$

- the stabilizing component

$$u_s = -K\left(\mathrm{e}^{A\tau}\hat{e} + \int_{t-\tau}^{t} \mathrm{e}^{A(t-\mu)}Bu_s(\mu)d\mu\right), \tag{6.67}$$

  where the matrix of feedback gains K is such that $A_{\Sigma} = A - BK$ is Hurwitz, $\hat{e}$ is the estimate generated by the state error observer (6.66);
- the swapping scheme

$$\dot{\bar{e}} = A\bar{e} - L_e(\varepsilon + C\bar{e}) - Bu(t-\tau), \tag{6.68}$$

$$\bar{\varepsilon} = \varepsilon + C\bar{e} - \Xi^{\top}(t-\tau)\hat{\vartheta}, \tag{6.69}$$

  where the matrix L_e is such that $A_e = A - L_eC$ is Hurwitz, $\hat{\vartheta} = [\hat{\Theta}_1, \hat{\Theta}_2, \ldots, \hat{\Theta}_m]^{\top}$ is the vector of the adjustable parameters, $\hat{\Theta}_i$ is the ith row of the matrix $\hat{\Theta}$, and $\Xi \in \mathbb{R}^{m(q_g+q_{\upsilon})\times m}$ is the matrix regressor defined as

[4] See Remark 2.11.

$$\Xi^\top = \begin{bmatrix} W_{e1,1}(s)\left[\hat{\xi}^\top\right] \cdots W_{e1,m}(s)\left[\hat{\xi}^\top\right] \\ \vdots \quad \ddots \quad \vdots \\ W_{em,1}(s)\left[\hat{\xi}^\top\right] \cdots W_{em,m}(s)\left[\hat{\xi}^\top\right] \end{bmatrix}, \; W_e(s) = C(sI - A_e)^{-1}b. \quad (6.70)$$

The adjustable controller is tuned by one of the following adaptation algorithms:

- **the gradient algorithm of adaptation**

$$\dot{\hat{\vartheta}} = \gamma \Xi(t-\tau)\bar{\varepsilon}, \quad \gamma > 0; \tag{6.71}$$

- **the algorithm with DRE**

$$\dot{\hat{\vartheta}} = \gamma \bar{\xi}(t-\tau)\bar{\varepsilon}_H, \quad \gamma > 0, \tag{6.72}$$

where

$$\bar{\varepsilon}_H = \begin{bmatrix} \hat{\varepsilon} - \Xi^\top(t-\tau)\hat{\vartheta} \\ H_1(s)\left[\hat{\varepsilon}\right] - \bar{\xi}_1^\top(t-\tau)\hat{\vartheta} \\ \vdots \\ H_p(s)\left[\hat{\varepsilon}\right] - \bar{\xi}_p^\top(t-\tau)\hat{\vartheta} \end{bmatrix}, \quad \bar{\xi}^\top = \begin{bmatrix} \Xi^\top \\ \bar{\xi}_1^\top \\ \vdots \\ \bar{\xi}_p^\top \end{bmatrix},$$

$$\hat{\varepsilon} = \varepsilon + C\bar{e}, \tag{6.73}$$

$\bar{H}(s) = \left[1, H_1(s), \ldots, H_p(s)\right]^\top$, $H_i(s)$ are distinct designer-chosen PASMP transfer functions, $p \geq (q_g + q_v - 1)$,

$$\bar{\xi}_j = H_j(s)\left[\Xi\right], \quad j = 1, 2, \ldots, p,$$

Ξ is defined by (6.70);

- **the algorithm with MRE**

$$\dot{\hat{\vartheta}} = \gamma \left(L(s)\left[\Xi(t-\tau)\hat{\varepsilon}\right] - \Omega_\Xi(t-\tau)\hat{\vartheta} \right), \quad \gamma > 0, \tag{6.74}$$

where $L(s)$ is a designer-chosen transfer function of the form (3.44), $\hat{\varepsilon}$ is given by (6.73), $\Omega_\Xi = L(s)\left[\Xi\Xi^\top\right]$.

Now, using the same arguments as in the proof of Proposition 6.2, we can prove the following properties of the closed-loop system [18].

Proposition 6.3 *Under Assumptions 2.2 and 6.1, the adaptive controller (6.65) with the reference observer (6.59), the disturbance observer (6.61), (6.62), the state error observer (6.66), the stabilizing component (6.67), the swapping scheme (6.68)–(6.70), and the adaptation algorithm (6.71), (or (6.72), or (6.74)) when applied to the* *plant* (6.26) *provides the boundedness of all the closed-loop signals and achievement of the control objective* (6.54).

In addition to Proposition 6.3, the algorithm of adaptation (6.72) provides the asymptotic parametric convergence if $\lambda_{\Xi}(t) \notin \mathcal{L}_1$, where $\lambda_{\Xi}(t)$ is the minimum eigenvalue of $\bar{\xi}_w(t)\bar{\xi}_w^{\top}(t)$, while the algorithm (6.74) provides the asymptotic parametric convergence if $\lambda_{\Omega}(t) \notin \mathcal{L}_1$, where $\lambda_{\Omega}(t)$ is the minimum eigenvalue of $\Omega_{\Xi}(t)$.

Example 6.3 Consider the plant from Example 6.2. Let the reference and disturbance signals be given by

$$g(t) = \begin{bmatrix} g_1(t) \\ g_2(t) \end{bmatrix} = \begin{bmatrix} \beta_{g1,1} + \beta_{g1,2}\sin(w_g t + \varphi_g) \\ \beta_{g2,1} \end{bmatrix},$$
$$\upsilon(t) = \begin{bmatrix} \upsilon_1(t) \\ \upsilon_2(t) \end{bmatrix} = \begin{bmatrix} \beta_{\upsilon1,1} \\ \beta_{\upsilon2,1} + \beta_{\upsilon2,2}\sin(w_\upsilon t + \varphi_\upsilon) \end{bmatrix},$$

respectively. The parameters of these functions $\beta_{g1,1} = 6, \beta_{g1,2} = 5, w_g = 1, \varphi_g = 4$, $\beta_{g2,1} = 3$, $\beta_{\upsilon1,1} = 4$, $\beta_{\upsilon2,1} = -4$, $\beta_{\upsilon2,2} = 5$, $w_\upsilon = 2$, and $\varphi_\upsilon = 1$ are unknown.

To implement the adjustable control, we design the reference observer (6.59) and the disturbance observer (6.61) with the matrices

$$G_g = G_\upsilon = \begin{bmatrix} 0 & 1 & 0 \\ 0 & 0 & 1 \\ -24 & -26 & -9 \end{bmatrix}, \quad L_g = L_\upsilon = \begin{bmatrix} 0 & 0 \\ 0 & 0 \\ 24 & 24 \end{bmatrix}, \quad N = \begin{bmatrix} 0 & 0 & 0 \\ 0 & 0 & 0 \\ 24 & 0 & 0 \end{bmatrix},$$

design the state error observer (6.66) with the matrix

$$L_e = \begin{bmatrix} 17.505 & 14.141 \\ -4.160 & -3.221 \\ -30.000 & -24.546 \end{bmatrix}$$

and obtain the stabilizing component (6.67) with

$$K = \begin{bmatrix} -17.772 & 7.355 & 36.170 \\ -1.032 & -1.043 & -0.255 \end{bmatrix}.$$

The matrices L_e and K provide the eigenvalues sets of the matrices $A_e = A - L_e C$, $A_\Sigma = A - BK$ given by $\lambda\{A_e\} = \lambda\{A_\Sigma\} = \{-2, -3, -4\}$.

For expression (6.70) of the swapping scheme, we obtain the transfer function matrix

$$W_e(s) = C(sI - A_e)^{-1}B$$
$$= \frac{1}{s^3 + 9s^2 + 26s + 24}\begin{bmatrix} s - 23.546 & -14.141s - 9.505 \\ s + 31 & s^2 + 18.405s + 11.495 \end{bmatrix}.$$

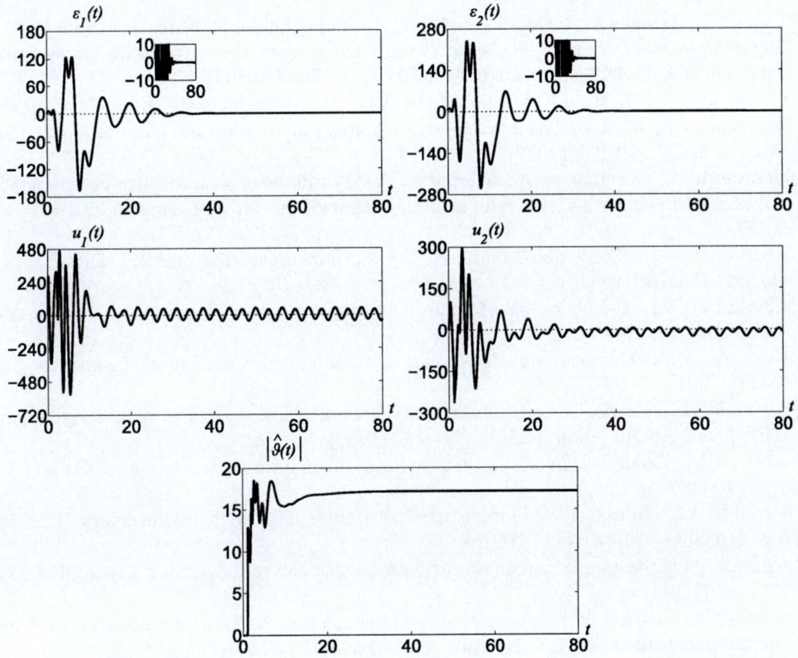

Fig. 6.4 Example 6.3: adaptive output-feedback reference tracking with compensation of the unmatched disturbance in the system with delayed control tuned by the adaptation algorithm with MRE

The vector of adjustable parameters $\hat{\vartheta}$ is generated by the algorithms of adaptation with MRE and $L(s) = \frac{1}{2s+1}$, $\gamma = 100$. All simulations are carried out with zero initial conditions $x(0)$, $\hat{\xi}_g(0)$, $\eta(0)$, $\bar{e}(0)$, $\hat{e}(0)$, and $\hat{\vartheta}(0)$.

It is shown by the simulation results presented in Fig. 6.4 that despite the influence of input delay and uncertain multidimensional disturbance $\upsilon(t)$ the plant output y asymptotically follows the uncertain multidimensional reference g. ■

References

1. Artstein, Z.: Linear systems with delayed controls: a reduction. IEEE Trans. Autom. Control **27**(4), 869–879 (1982)
2. Basturk, H.I., Krstić, M.: Adaptive sinusoidal disturbance cancellation for unknown LTI systems despite input delay. Automatica **58**, 131–138 (2015)
3. Bobtsov, A.A., Pyrkin, A.A.: The compensation of a harmonic perturbation under conditions of a delay in control. J. Comput. Syst. Sci. Int. **47**(4), 513–517 (2008)
4. Chen, C.-T.: Linear Systems. Theory and Design, 3rd edn. Oxford University Press, Oxford, New York (1999)

5. Davis, P.J., Rabinowitz, P.: Methods of Numerical Integration, 2nd edn. Academic (1984)
6. Engelborghs, K., Dambrine, M., Rose, D.: Limitations of a class of stabilization methods for delay systems. IEEE Trans. Autom. Control **46**(2), 336–339 (2001)
7. Gerasimov, D.N., Miliushin, A.S., Nikiforov, V.O.: Algorithms of adaptive tracking of unknown multisinusoidal signals in linear systems with arbitrary input delay. Int. J. Adapt. Control Signal Process. **33**(6), 900–912 (2019)
8. Gerasimov, D., Paramonov, A., Nikiforov, V.: Algorithms of fast adaptive compensation of disturbance in linear systems with arbitrary input delay. Int. J. Control **93**(7), 1596–1604 (2020)
9. Kharitonov, V.L.: Prediction-based controls: the implementation problem. Differ. Equ. **51**, 1675–1682 (2015)
10. Kharitonov, V.L: Prediction-based control for systems with state and several input delays. Automatica **79**, 11–16 (2017)
11. Krstić, M.: Delay Compensation for Nonlinear. Adaptive and PDE Systems. Birkhauser, Boston (2009)
12. Kwon, W.H., Pearson, A.E.: Feedback stabilization of linear systems with delayed control. IEEE Trans. Autom. Control **25**(2), 266–269 (1980)
13. Lewis, R.M.: Control-delayed system properties via an ordinary model. Int. J. Control **30**(3), 477–490 (1979)
14. Manitius, A.Z., Olbrot, A.W.: Finite spectrum assignment for systems with delays. IEEE Trans. Autom. Control **24**(4), 54–553 (1979)
15. Mirkin, L.: On the approximation of distributed-delay control laws. Syst. Control Lett. **51**(5), 331–342 (2004)
16. Mondie, S., Michiels, W.: Finite spectrum assignment of unstable time-delay systems with a safe implementation. IEEE Trans. Autom. Control **48**(12), 2207–2212 (2003)
17. Mondie, S., Dambrine, M., Santos, O.: Approximation of control laws with distributed delays: a necessary condition for stability. Kybernetika **38**(5), 541–551 (2002)
18. Nikiforov, V.O., Paramonov, A.V., Gerasimov, D.N.: Adaptive control algorithms in MIMO linear systems with control delay. Autom. Remote. Control. **81**(6), 1091–1106 (2020)
19. Nikiforov, V.O., Paramonov, A.V., Gerasimov, D.N.: Adaptive servocompensation of unmatched disturbances in MIMO LTI plants with input delay. In: Proceedings of American Control Conference, New Orleans, USA, pp. 2426–2431 (2021)
20. Pyrkin, A.A.: Adaptive algorithm to compensate parametrically uncertain biased disturbance of a linear plant with delay in the control channel. Autom. Remote. Control. **71**(8), 1562–1577 (2010)
21. Pyrkin, A.A., Bobtsov, A.A., Nikiforov, V.O., Kolyubin, S.A., Vedyakov, A.A., Borisov, O.I., Gromov, V.S.: Compensation of polyharmonic disturbance of state and output of a linear system with delay in the control channel. Autom. Remote. Control. **76**(12), 2124–2142 (2015)
22. Pyrkin, A., Smishlyaev, A., Bekiaris-Liberis, N., Krstić, M.: Rejection of sinusoidal disturbance of unknown frequency for linear system with input delay. In: American Control Conference, Baltimore, USA (2010)
23. Van Assche, V., Dambrine, M., Lafay, J.-F., Richard, J.-P.: Some problems arising in the implementation of distributed-delay control law. In: Proceedings of 38th Conference on Decision and Control, Phoenix, USA (1999)
24. Zhou, B.: Input delay compensation of linear systems with both state and input delay by nested prediction. Automatica **50**, 1434–1443 (2014)

Chapter 7
Robust Adaptive Regulation

Abstract The chapter presents solutions to Problems 1.5–1.7 of robust adaptive control formulated in Sect. 1.3.3. In this chapter, we suppose that the exosignal contains irregular component and demonstrate how the adaptive controllers must be modified in order to provide desirable properties of the closed-loop control systems—the boundedness of all the closed-loop signals in the presence of the irregular component and asymptotic achievement of the control objective when the irregular component approaches zero. In Sect. 7.1, the robustness properties of the reference and disturbance observers introduced in Chap. 2 are defined and analyzed. In Sect. 7.2, Problem 1.5 of robust adaptive tracking is solved for LTI plants, while Sect. 7.3 presents solutions of Problem 1.6 developed for some classes of linear and nonlinear plants. Problem 1.7 of robust adaptive output regulation is solved in Sect. 7.4 for a class of LTI SISO plants.

7.1 Robustness of Exosignal Observers

In this section, we demonstrate that the reference and the disturbance observers introduced in Chap. 2 are robust and applicable in the case when the exosignal υ_δ contains both the regular component υ exactly modelled by (2.1), (2.2) and the irregular component δ represented by an unknown bounded function of time, i.e., when

$$\upsilon_\delta = \upsilon + \delta. \tag{7.1}$$

We introduce the regressor estimation error $\tilde{\xi}_\delta$ as the difference between unmeasured state $\bar{\xi}$ of the exosystem canonical form (2.15) and its estimate $\hat{\xi}$ generated by an observer in the presence of the irregular component:

$$\tilde{\xi}_\delta = \bar{\xi} - \hat{\xi}. \tag{7.2}$$

V. Nikiforov and D. Gerasimov, *Adaptive Regulation*, Lecture Notes in Control and Information Sciences 491, https://doi.org/10.1007/978-3-030-96091-9_7

Definition 7.1 An observer is called *robust* if the regressor estimation error $\tilde{\xi}_\delta$ provided by this observer has the following properties:

D.7.1.1. if $\delta(t) \not\equiv 0$, then $\tilde{\xi}_\delta$ is bounded;
D.7.1.2. if $\delta(t) \to 0$ as $t \to \infty$, then $|\tilde{\xi}_\delta(t)| \to 0$ as $t \to \infty$;
D.7.1.3. if $\delta(t) \equiv 0$, then $\tilde{\xi}_\delta$ exponentially decays.

Definition 7.2 The regressor estimation error $\tilde{\xi}_\delta$ is called *robust with respect to* δ if it has Properties D.7.1.1–D.7.1.3.[1]

Properties D.7.1.1–D.7.1.3 enable one to use the robust modifications of the basic algorithms of adaptation (see Sect. 3.5) in order to solve Problems 1.5–1.7 of robust adaptive control formulated in Sect. 1.3.3. These properties are based on the fact that for all plant parameterizations (including parameterizations for nonlinear plants) behavior of the regressor estimation error is described by a linear model with the irregular component δ considered as an input. Below, we rigorously analyze the robustness of some observers introduced in Chap. 2. The obtained results can be straightforwardly extended to all observers from Sects. 2.4 and 2.5.

7.1.1 Reference Observer

Consider the reference observer

$$\dot{\hat{\xi}} = G\hat{\xi} + l\upsilon_\delta, \tag{7.3}$$

where G and l are from the exosystem canonical form (2.15). Then, evaluating the time derivative of $\tilde{\xi}_\delta = \bar{\xi} - \hat{\xi}$, in view of (2.15) and (7.3), we obtain

$$\dot{\tilde{\xi}}_\delta = G\tilde{\xi}_\delta - l\delta, \quad \tilde{\xi}_\delta(0) = \bar{\xi}(0) - \hat{\xi}(0). \tag{7.4}$$

Therefore, in view of Lemma 2.3, for the regular component of the reference, we have

$$\upsilon = \theta^\top \hat{\xi} + \delta_\upsilon, \tag{7.5}$$

where θ is a vector of unknown parameters, and $\delta_\upsilon = \theta^\top \tilde{\xi}_\delta$ is robust in the sense of Definition 7.2.

[1] This notion will be also used for any variable depending on the irregular component δ.

7.1.2 Disturbance Observers: Plants with Known Parameters

In this section, we demonstrate the robustness of the disturbance observers designed in Sect. 2.4 for plants with known parameters. We do not analyze all observers from Sect. 2.4 but consider only two ones since the extensions to the other observers are almost straightforward.

7.1.2.1 LTI SISO Plants with Unmatched Disturbance and Measured State

Let us consider the LTI plant (2.48) from Sect. 2.4.1.2 affected by the disturbance υ_δ given by (7.1):

$$\dot{x} = Ax + bu + d\upsilon_\delta, \quad y = c^\top x + d_n\upsilon_\delta. \tag{7.6}$$

Then we employ the observer (2.42), (2.43), (2.49) and define the properties of the regressor estimate $\hat{\xi}$ as follows.

Lemma 7.1 *If Assumption 2.1 holds and there exists a matrix $N \in \mathbb{R}^{q\times n}$ satisfying (2.49), then for the state $\bar{\xi}$ of the exosystem canonical form (2.15) the following equality is valid: $\bar{\xi} = \hat{\xi} + \tilde{\xi}_\delta$, where the bounded regressor estimate $\hat{\xi}$ is generated by the observer (2.42), (2.43), and the regressor estimation error $\tilde{\xi}_\delta$ is robust in the sense of Definition 7.2.*

Proof Taking the time derivative of $\tilde{\xi}_\delta = \bar{\xi} - \hat{\xi}$, in view of (2.15), (7.6), (2.42), (2.43), and (2.49), we obtain

$$\begin{aligned}\dot{\tilde{\xi}}_\delta &= G\bar{\xi} + l\upsilon - G\eta - GNx + NAx + Nbu - NAx - Nbu - Nd\upsilon - Nd\delta\\ &= G\left(\bar{\xi} - \eta - Nx\right) - l\delta = G\tilde{\xi}_\delta - l\delta.\end{aligned}$$

The latter means that $\tilde{\xi}_\delta$ satisfies Definition 7.2. □

Now, using the same approach as in Sect. 2.4.1.2, we can prove the following statement defining the plant parameterization.

Proposition 7.1 *If Assumptions 2.1 and 2.4 hold, then the plant (7.6) can be represented by the model*

$$\dot{e} = Ae + b(u + \theta_\upsilon^\top\hat{\xi} + \delta_e), \qquad y = c^\top e + d_n\delta, \tag{7.7}$$

where $\theta_\upsilon \in \mathbb{R}^q$ is a vector of unknown parameters, $\hat{\xi}$ is the regressor estimate generated by the observer (2.42), (2.43), (2.49), the state e is not accessible for measurement, and $\delta_e = \theta_\upsilon^\top\tilde{\xi}_\delta$ is robust in the sense of Definition 7.2.

Extension of these results to the plant with measured output is presented in [3] (see Sect. 2.2.2).

7.1.2.2 Nonlinear Plant in General Form

Consider the nonlinear plant (2.83) from Sect. 2.4.2.1 affected by the disturbance υ_δ given by (7.1):

$$\dot{x} = f(x) + g(x)u + d(x)\upsilon_\delta. \tag{7.8}$$

Then the properties of the regressor estimate $\hat{\xi}$ generated by the observer (2.84)–(2.86) are established as follows.

Lemma 7.2 *If Assumption 2.1 holds and there exists a vector function $\pi(x)$ satisfying (2.86), then for the state $\bar{\xi}$ of the exosystem canonical form (2.15) the following equality holds: $\bar{\xi} = \hat{\xi} + \tilde{\xi}_\delta$, where the bounded regressor estimate $\hat{\xi}$ is generated by the observer (2.84)–(2.86), and the regressor estimation error $\tilde{\xi}_\delta$ is robust in the sense of Definition 7.2.*

Proof Evaluating the time derivative of $\tilde{\xi}_\delta = \bar{\xi} - \hat{\xi}$, in view of (2.15), (7.8) and (2.84)–(2.86), we obtain

$$\begin{aligned}\dot{\tilde{\xi}}_\delta &= G\bar{\xi} + l\upsilon - G\eta - G\pi(x) + \frac{\partial \pi(x)}{\partial x}(f(x) + g(x)u) \\ &\quad - \frac{\partial \pi(x)}{\partial x}(f(x) + g(x)u + d(x)\upsilon + d(x)\delta) \\ &= G\left(\bar{\xi} - \eta - \pi(x)\right) - l\delta = G\tilde{\xi}_\delta - l\delta.\end{aligned}$$

The latter means that $\tilde{\xi}_\delta$ satisfies Definition 7.2. □

The above result can be directly extended to the observers designed for particular classes of nonlinear plants including the plants represented in the normal canonical form and the plants represented in the strict-feedback and output-feedback canonical forms.

7.1.3 *Disturbance Observers: Plants with Unknown Parameters*

In this section, we demonstrate the robustness of the disturbance observers designed in Sect. 2.5 for plants with unknown parameters. We illustrate the main approach on the following three examples of disturbance observers.

7.1.3.1 LTI SISO Plants with Matched Disturbance and Measured State

Consider the LTI plant (2.117) from Sect. 2.5.1.1 with unknown parameters satisfying Assumption 2.7 and affected by the disturbance υ_δ (7.1), i.e.,

$$\dot{x} = Ax + b_M(\psi_u u + \upsilon_\delta) \tag{7.9}$$

and use the observer (2.119)–(2.122). Then the properties of the regressor estimate $\hat{\xi}$ are defined as follows.

Lemma 7.3 *If Assumptions 2.1 and 2.7 hold and there exists a matrix $N \in \mathbb{R}^{q\times n}$ satisfying (2.122), then the state $\bar{\xi}$ of the exosystem canonical form (2.15) can be represented in the form $\bar{\xi} = \hat{\xi} + \tilde{\xi}_\delta$, where the bounded regressor estimate $\hat{\xi}$ is given by*

$$\hat{\xi} = \eta_0 + Nx + \sum_{i=1}^{n} \psi_i \eta_i + \psi_u \eta_u, \tag{7.10}$$

ψ_i is the ith element of the vector ψ, η_0, η_i, and η_u are the states of the filters (2.119)–(2.122), and the estimation error $\tilde{\xi}_\delta$ is robust in the sense of Definition 7.2.

Proof Define the regressor estimation error as

$$\tilde{\xi}_\delta = \bar{\xi} - \hat{\xi} = \bar{\xi} - \eta_0 - Nx - \sum_{i=1}^{n} \psi_i \eta_i - \psi_u \eta_u,$$

where $\bar{\xi}$ is the state of the exosystem canonical form (2.23), and calculate its time derivative in view of (2.23), (7.9), and (2.119)–(2.122):

$$\begin{aligned}
\dot{\tilde{\xi}}_\delta &= G\bar{\xi} + l\upsilon - G\eta_0 - GNx + NA_M x \\
&\quad - NA_M x - Nb_M\psi^\top x - Nb_M\psi_u u - Nb_M\upsilon \\
&\quad - Nb_M\delta - G\sum_{i=1}^{n} \psi_i \eta_i + l\sum_{i=1}^{n} \psi_i x_i - G\psi_u\eta_u + \psi_u l u \\
&= G(\bar{\xi} - \eta_0 - Nx - \sum_{i=1}^{n} \psi_i \eta_i - \psi_u \eta_u) - l\delta = G\tilde{\xi}_\delta - l\delta.
\end{aligned}$$

The latter means that $\tilde{\xi}_\delta$ satisfies Definition 7.2. □

Now, using the same approach as in Sect. 2.5.1.1, we can prove the statement defining the plant parameterization.

Proposition 7.2 *If Assumptions 2.1 and 2.7 hold and there exists a matrix $N \in \mathbb{R}^{q\times n}$ satisfying (2.122), then the plant (7.9) can be represented in the form*

$$\dot{x} = A_M x + b_M(\vartheta^\top \omega + \psi_u u + \delta_x), \tag{7.11}$$

where $\vartheta \in \mathbb{R}^{n+(n+2)q}$ is an extended vector of unknown parameters given by (2.125), $\omega \in \mathbb{R}^{n+(n+2)q}$ is the extended regressor given as

$$\omega = \left[x^\top, (\eta_0 + Nx)^\top, \eta_1^\top, \eta_2^\top, \dots, \eta_n^\top, \eta_u^\top\right]^\top, \tag{7.12}$$

η_0, η_i and η_u are generated by the filters (2.119)–(2.121), and $\delta_x = \theta^\top \tilde{\xi}_\delta$ is robust in the sense of Definition 7.2.

7.1.3.2 LTI SISO Plants with Unmatched Disturbance and Measured State

The results of the above section can be directly extended to the case of the LTI plants (2.128) satisfying Assumption 2.8 and affected by the disturbance υ_δ (7.1):

$$\dot{x} = Ax + bu + d\upsilon_\delta, \quad y = c^\top x + d_n \upsilon_\delta. \tag{7.13}$$

If we use the observer (2.133)–(2.136), then the parameterized plant model is defined by the following statement.

Proposition 7.3 *If Assumptions 2.1 and 2.8 hold and there exists a matrix $N \in \mathbb{R}^{q\times n}$ satisfying (2.136), then the plant (7.13) can be represented in the form*

$$\dot{e} = A_\Sigma e + b_M(\vartheta^\top \omega + \psi_u u + \delta_e), \quad y = c^\top e + d_n \delta, \tag{7.14}$$

where the state e is not accessible for measurements, $\vartheta \in \mathbb{R}^{n+(n+2)q}$ is an extended vector of unknown parameters given by (2.138), $\omega \in \mathbb{R}^{n+(n+2)q}$ is the extended regressor defined as

$$\omega = \left[x^\top, (\eta_0 + Nx)^\top, \eta_1^\top, \eta_2^\top, \dots, \eta_n^\top, \eta_u^\top\right]^\top, \tag{7.15}$$

η_0, η_i, η_u are generated by the filters (2.133)–(2.135), and δ_e is robust in the sense of Definition 7.2.

This proposition is proved in the same way as Proposition 2.10. It can be also shown that $\delta_e = \theta^\top \tilde{\xi}_\delta$, where $\tilde{\xi}_\delta$ satisfies the equation $\dot{\tilde{\xi}}_\delta = G\tilde{\xi}_\delta - l\delta$.

7.1.3.3 Nonlinear Plants with Measured State and Unmatched Disturbances

Consider the uncertain nonlinear plant (2.177) from Sect. 2.5.2.2 affected by the multidimensional disturbance $\upsilon_\delta \in \mathbb{R}^\rho$ given by (7.1) with $\upsilon, \delta \in \mathbb{R}^\rho$, i.e.,

$$\dot{x} = Ex + \varsigma_n\left(\phi_x(x) + \phi_u(x)u\right) + \Phi(x)\psi + D\upsilon_\delta, \quad y = \varsigma_1^\top x \tag{7.16}$$

and use the observer (2.182), (2.183), and (2.178). Then the properties of the regressor estimate $\hat{\xi}$ are defined as follows.

Lemma 7.4 *If Assumption 2.2 holds and there exists a matrix $N \in \mathbb{R}^{q\times n}$ satisfying (2.178), then the state $\bar{\xi}$ of the exosystem canonical form (2.23) can be represented in the form $\bar{\xi} = \hat{\xi} + \tilde{\xi}_\delta$, where the bounded regressor estimate $\hat{\xi}$ is given by*

$$\hat{\xi} = \eta + Nx + \Lambda\psi,$$

η and Λ are the states of the filters (2.182) and (2.183), respectively, and $\tilde{\xi}_\delta$ is robust in the sense of Definition 7.2.

Proof Define the regressor estimation error as

$$\tilde{\xi}_\delta = \bar{\xi} - \eta - Nx - \Lambda\psi,$$

where $\bar{\xi}$ is the state of the exosystem canonical form (2.23), and calculate its time derivative in view of (2.23), (7.16), (2.182), (2.183), and (2.178):

$$\begin{aligned}\dot{\tilde{\xi}}_\delta &= G\bar{\xi} + L\upsilon - G\eta - GNx + NEx + N\varsigma_n\left(\phi_x(x) + \phi_u(x)u\right)\\ &\quad - NEx - N\varsigma_n(\phi_x(x) + \phi_u(x)u) - N\Phi(x)\psi - ND\upsilon - ND\delta\\ &\quad - G\Lambda\psi + N\Phi(x)\psi = G(\bar{\xi} - \eta - Nx - \Lambda\psi) - L\delta = G\tilde{\xi}_\delta - L\delta.\end{aligned}$$

The latter means that $\tilde{\xi}_\delta$ satisfies Definition 7.2. □

Then, in view of Lemma 2.2, we can write

$$\upsilon = T_0\omega + \bar{\delta}_\upsilon,$$

where T_0 is defined by (2.180), ω is defined by (2.181), while $\bar{\delta}_\upsilon = \Theta\tilde{\xi}_\delta$ is robust in the sense of Definition 7.2. Therefore, the plant (7.16) can be represented in the parameterized form

$$\dot{x} = Ex + \varsigma_n\left(\phi_x(x) + \phi_u(x)u\right) + \Phi(x)\psi + T_0\omega + D\bar{\delta}_\upsilon, \quad y = \varsigma_1^\top x$$

or, in view of Proposition 2.13, can be defined as

$$\dot{x} = Ex + \varsigma_n\left(\phi_x(x) + \phi_u(x)u\right) + \Phi(x)\psi + T\omega_\kappa + \delta_x, \quad y = \varsigma_1^\top x, \tag{7.17}$$

where T is a new matrix of unknown parameters, $\omega_\kappa = \frac{\kappa_0}{\kappa(s)}[\omega]$ is the filtered regressor, $\kappa(s)$ is a designer-chosen Hurwitz polynomial of degree no less than $n-1$, $\kappa_0 > 0$ is a constant, $\delta_x = D(\bar{\delta}_\upsilon + \epsilon_\kappa)$, and ϵ_κ is the exponentially decaying term. Signal δ_x is robust in the sense of Definition 7.2.

7.2 Reference Tracking

Now, we illustrate how Problem 1.5 of the robust adaptive reference tracking can be solved taking into account robustness of the reference observer established in Sect. 7.1.1.

Consider the plant (4.1) satisfying Assumption 4.1. It is also assumed that the signal of the form

$$g_\delta = g + \delta \tag{7.18}$$

is accessible for measurements, where the regular component g satisfies Assumption 2.1 (with substitution $\upsilon = g$), and the irregular component δ is an unknown bounded function of time, i.e., $|\delta(t)| \le \bar{\delta}$ for all $t \ge 0$. In this case, the regular component g represents the reference to be tracked, while the irregular component δ is considered as harmful noise of measurements (in particular, high-frequency noise).

For the robust adaptive tracking, we use the same controller as in Sect. 4.1.1.2 with only the difference that the algorithm of adaptation contains the continuous switching term (see Sect. 3.5). Thus, as in Sect. 4.1, we use stabilizing controller

$$u = -k^\top x + u_t, \tag{7.19}$$

where the vector of feedback gains k is selected so that $A_\Sigma = A - bk^\top$ is Hurwitz, and u_t is the tracking component. We also define the closed-loop transfer function

$$W_\Sigma(s) = c^\top(sI - A_\Sigma)^{-1}b = \frac{\beta(s)}{\alpha(s)} = \frac{b_m s^m + b_{m-1}s^{m-1} + \dots + b_0}{s^n + a_{n-1}s^{n-1} + \dots + a_0},$$

where the polynomial $\beta(s)$ is Hurwitz by Assumption A.4.1.2.

Further, we employ the tracking component

$$u_t = \frac{1}{\beta(s)}\left[a_0 g + \sum_{i=1}^{n} a_i \hat{g}_i\right], \tag{7.20}$$

where $\hat{g}_i$ is the estimate of the HOTD $g^{(i)}$ calculated by the estimator

$$\dot{\hat{\xi}} = G\hat{\xi} + lg_\delta, \tag{7.21}$$

$$\hat{g}_i = \hat{\theta}^\top \left(G + l\hat{\theta}^\top\right)^i \hat{\xi}, \quad i = 1, 2, \ldots, n \tag{7.22}$$

with the pair (G, l) from (2.15) and the vector of adjustable parameters $\hat{\theta} \in \mathbb{R}^q$ generated by an adaptation algorithm.

To calculate $\hat{\theta}$, we can use the following algorithms of adaptation (see Sect. 3.5):

The robust gradient algorithm of adaptation

$$\dot{\hat{\theta}} = \gamma\hat{\xi}\left(g_\delta - \hat{\theta}^\top\hat{\xi}\right) - \gamma\sigma(\hat{\theta})\hat{\theta}, \tag{7.23}$$

where $\gamma > 0$ is the adaptation gain, and $\sigma(\hat{\theta})$ is the continuous switching term defined as

$$\sigma(\hat{\theta}) = \begin{cases} 0, & |\hat{\theta}| < \theta^*, \\ \frac{|\hat{\theta}|}{\theta^*} - 1, & \theta^* \leq |\hat{\theta}| \leq 2\theta^*, \\ 1, & |\hat{\theta}| > 2\theta^* \end{cases} \tag{7.24}$$

with a large positive constant θ^*;

The robust adaptation algorithm with DRE

$$\dot{\hat{\theta}} = \gamma\Xi\left(\bar{g}_\delta - \Xi^\top\hat{\theta}\right) - \gamma\sigma(\hat{\theta})\hat{\theta}, \tag{7.25}$$

where Ξ is defined by (4.30), $\sigma(\hat{\theta})$ is defined by (7.24), and $\bar{g}_\delta$ is defined as

$$\bar{g}_\delta = \bar{H}(s)[g_\delta] = \left[g_\delta, H_1(s)[g_\delta], \ldots, H_p(s)[g_\delta]\right]^\top, \quad \bar{g}_\delta \in \mathbb{R}^{p+1}; \tag{7.26}$$

The robust adaptation algorithm with MRE

$$\dot{\hat{\theta}} = \gamma\left(L(s)\left[\hat{\xi}g_\delta\right] - \Omega\hat{\theta}\right) - \gamma\sigma(\hat{\theta})\hat{\theta}, \tag{7.27}$$

where $L(s)$ is a designer-chosen transfer function of the form (3.44), $\Omega = L(s)\left[\hat{\xi}\hat{\xi}^\top\right]$, and $\sigma(\hat{\theta})$ is the function defined by (7.24).

We first establish preliminary result concerning the properties of the HOTD estimates.

Lemma 7.5 *Under Assumption 2.1 (with $\upsilon = g$), the HOTD estimator (7.21), (7.21) with the robust adaptation algorithm (7.23) (or (7.25), or (7.24)) ensures the following properties of the closed-loop system:*

L.7.5.1. *$\hat{\theta}$, $\hat{g}_i$ and the estimation errors $\tilde{g}_i = g^{(i)} - \hat{g}_i$ are bounded ($i \in \mathbb{N}_0$);*

L.7.5.2. *if $\theta^* > |\theta|$, then $\tilde{g}_i$ are of the order of $\bar{\delta}$ in the mean square sense, i.e., $\tilde{g}_i \in \mathcal{MS}(\bar{\delta})$;*

L.7.5.3. *if $\theta^* > |\theta|$ and $\delta(t) \equiv 0$, then*

$$\lim_{t\to\infty} \left(g^{(i)}(t) - \hat{g}_i(t)\right) = 0, \quad i \in \mathbb{N}_0. \tag{7.28}$$

Proof Evaluating the reference estimation error, in view of (7.18), we obtain

$$\tilde{g}_0 = g_\delta - \hat{g}_0 = \theta^\top \bar{\xi} + \delta - \hat{\theta}^\top \hat{\xi} = \theta^\top \hat{\xi} + \theta^\top \tilde{\xi}_\delta + \delta - \hat{\theta}^\top \hat{\xi} = \tilde{\theta}^\top \hat{\xi} + \bar{\delta},$$

where $\bar{\xi}$ is the state of the exosystem canonical model, $\tilde{\xi}_\delta$ satisfies the equation (7.4), and $\bar{\delta} = \delta + \theta^\top \tilde{\xi}_\delta$. The obtained equality is in the form of (3.109). Therefore, Properties L.7.5.1 and L.7.5.2 directly follow from Proposition 3.12, while Property L.7.5.3 is established by Lemma 4.1. □

Remark 7.1 The estimates (7.21) involve the signals $\hat{\xi}$ and $\hat{\theta}$ that are the outputs of the low-pass filters. Therefore, the influence of the high-frequency noise δ is attenuated in these signals, and we can suppose that for $\theta^* > |\theta|$ the steady-state estimation errors $\tilde{g}_i$ are small for small δ (not necessarily in the mean square sense). ■

Example 7.1 Consider the problem of the HOTD estimation of the reference $g(t) = \sin(3t)$ solved in Example 4.2. In contrast to Example 4.2, we now assume that the noisy signal $g_\delta = g + \delta$ is measured instead of the actual reference g. We apply the second-order observer (7.21) with G and l defined by (4.24). Then the HOTD estimates (7.21) are defined by the expressions (4.33).

We choose $\theta^* = 5$ for (7.24), $p = 1$ and $H_1(s) = \frac{1}{0.25s+1}$ for the adaptation algorithm with DRE (7.25) and $L(s) = \frac{1}{0.2s+1}$ for the adaptation algorithm with MRE (7.27). The adaptation gain for all the algorithms is given by $\gamma = 10$. Figure 7.1 demonstrates the measured signal g_δ corrupted by noise. In the simulation, the irregular component δ is generated by Simulink block "Band-Limited White Noise" with the following parameters: noise power 10^{-6}, sample time 10^{-4} and seed 23341.

Simulation results are presented in Fig. 7.2 and demonstrate the boundedness of $\hat{\theta}$ and almost asymptotic estimation of the HOTD of the reference g despite the measurement of the noisy signal g_δ. ■

Now, we formulate the main result establishing the properties of the closed-loop control system.

Proposition 7.4 *If Assumption 2.1 (with $\upsilon = g$) and Assumption 4.1 hold, then the adaptive controller (7.19), (7.20) with the reference observer (7.21), the HOTD*

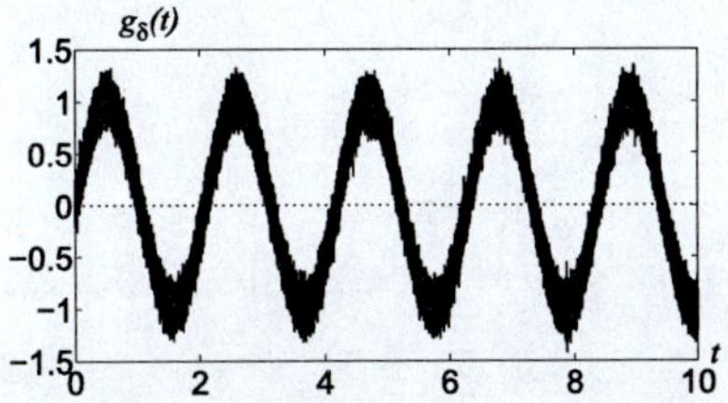

Fig. 7.1 Evolution of the reference signal $g_\delta(t)$

estimator (7.21) *and the adaptation algorithm* (7.23) *(or* (7.25)*, or* (7.27)*) when applied to the plant* (7.18) *ensures the following properties of the closed-loop system:*

P.7.4.1. *all the signals in the system including the tracking error $\varepsilon = y - g$ are bounded, and if $\theta^* > |\theta|$, then ε is of the order of $\bar{\delta}$ in the mean square sense, i.e., $\varepsilon \in \mathcal{MS}(\bar{\delta})$;*

P.7.4.2. *if $\delta(t) \equiv 0$ and $\theta^* > |\theta|$, then the control objective* (4.2) *is achieved.*

This proposition directly follows from (4.37) (where ϵ is replaced by a variable that is robust in the sense of Definition 7.2) and Lemma 7.5.

In view of Remark 7.1, we can suppose that the steady-state tracking error $\varepsilon = y - g$ is small for small δ not in the mean square sense only.

Example 7.2 Consider the plant (4.1) with the matrices (4.22) and the reference $g(t) = \sin(3t)$. We use the second-order reference observer (7.21) with G and l defined by (4.24). Selecting $k = [-30, -11]^\top$, we design the control law

$$u = -30x_1 - 11x_2 + \frac{1}{30}\left(30g_\delta + 11\hat{g}_1 + \hat{g}_2\right),$$

where the estimates $\hat{g}_1$ and $\hat{g}_2$ are defined by (7.21). Again, as in Example 7.1, the irregular component δ is simulated with the use of the Simulink block "Band-Limited White Noise" with the following parameters: noise power 10^{-6}, sample time 10^{-4} and seed 23341.

Simulation results of the closed-loop systems employing the algorithms of adaptation (7.23), (7.25) and (7.27) with the filter parameters taken from Example 7.1 and different gains γ are presented in Fig. 7.3. It is shown that, despite the influence of the noise, the plant output tracks a priori unknown reference g with high accuracy at steady state. At the same time, an increase of the adaptation gain γ does not cause noise amplification but enables one to increase the rate of parametric convergence in the algorithms with DRE and MRE. ■

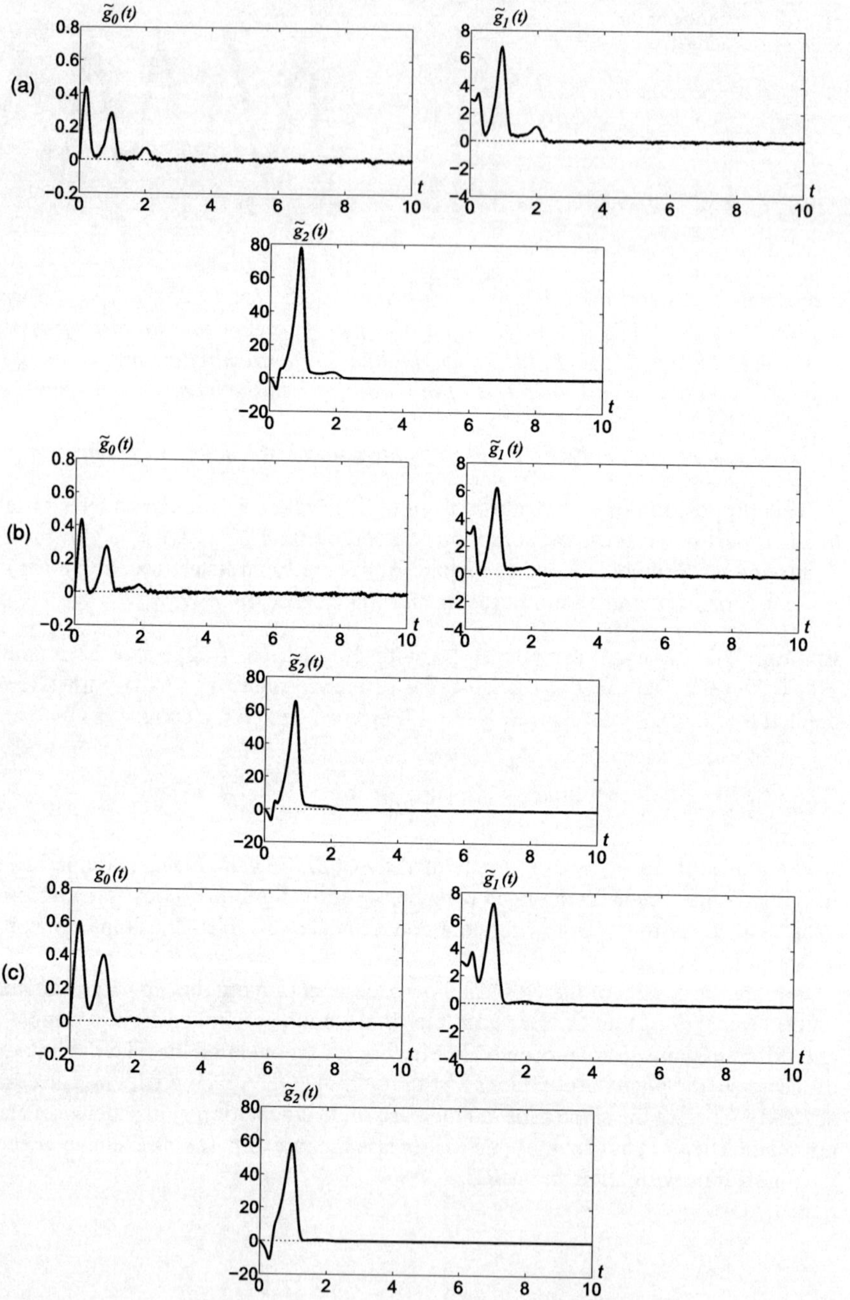

Fig. 7.2 Example 7.1: robust adaptive estimation of the HOTD of the reference g by noisy measurements g_δ and applying: **a** the gradient-based algorithm; **b** the algorithm with DRE; **c** the algorithm with MRE

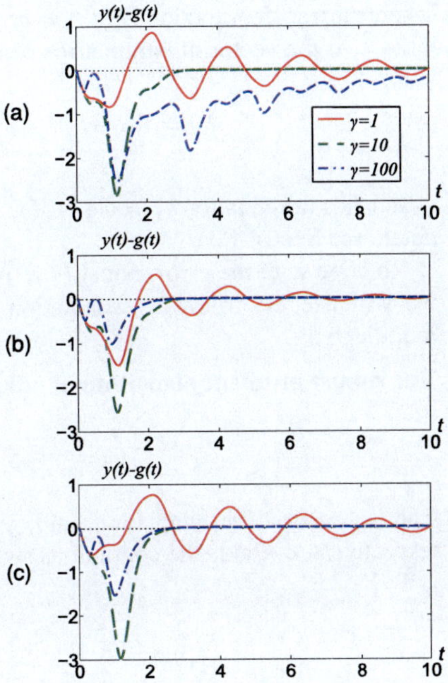

Fig. 7.3 Example 7.2: adaptive tracking of the reference $g(t) = \sin(3t)$ corrupted by noise in the system with: **a** the gradient algorithm of adaptation; **b** the algorithm of adaptation with DRE; **c** the algorithm of adaptation with MRE

7.3 Disturbance Rejection

In this section, we illustrate how the robustness of the disturbance observers established in Sects. 7.1.2 and 7.1.3 allows one to solve Problem 1.5 of robust adaptive disturbance rejection.

7.3.1 LTI SISO Plants with Known Parameters

Consider the LTI plant (7.6) satisfying Assumption 4.2. For the robust adaptive disturbance rejection, we use the same controller as in Sect. 4.1.3 with only the difference that the algorithm of adaptation contains the continuous switching term (see Sect. 3.5). Thus, as in Sect. 4.1.3, we use the controller (4.53), (4.57) and the disturbance observer (4.43), (4.44), (4.55).

Then, in view of Proposition 7.1, it can be shown that the closed-loop error model in the input–output form takes the form

$$y = W_{\Sigma}(s)\left[\tilde{\theta}_{\upsilon}^{\top}\hat{\xi}\right] + \delta_y, \tag{7.29}$$

where the transfer function $W_\Sigma(s) = c^\top(sI - A_\Sigma)^{-1}b$ is asymptotically stable, $\tilde{\theta}_\upsilon = \theta_\upsilon - \hat{\theta}_\upsilon$ is the vector of parametric errors, θ_υ is a vector of unknown parameters,

$$\delta_y = W_\Sigma(s)\left[\theta_\upsilon^\top \tilde{\xi}_\delta\right] + c^\top e^{A_\Sigma t} e(0) + d_n \delta, \tag{7.30}$$

$\tilde{\xi}_\delta$ satisfies the equation $\dot{\tilde{\xi}}_\delta = G\tilde{\xi}_\delta - l\delta$, and e is the inaccessible state error (for more detail, see Sect. 4.1.3).

To cope with the error model (7.29) containing the bounded disturbance (7.30), we will use the following adaptation algorithms with continuous switching σ-modification.

The robust gradient algorithm of adaptation is defined as

$$\dot{\hat{\theta}}_\upsilon = \gamma \xi_w \bar{y} - \gamma\sigma(\hat{\theta}_\upsilon)\hat{\theta}_\upsilon, \tag{7.31}$$

where $\gamma > 0$ is the adaptation gain, $\bar{y}$ and ξ_w are defined by (4.61) and (4.62), respectively, $\sigma(\hat{\theta}_\upsilon)$ is the continuous switching term given by

$$\sigma(\hat{\theta}_\upsilon) = \begin{cases} 0, & |\hat{\theta}_\upsilon| < \theta^*, \\ \frac{|\hat{\theta}_\upsilon|}{\theta^*} - 1, & \theta^* \le |\hat{\theta}_\upsilon| \le 2\theta^*, \\ 1, & |\hat{\theta}_\upsilon| > 2\theta^*, \end{cases} \tag{7.32}$$

and θ^* is a large positive constant.

The robust algorithm with DRE is defined as

$$\dot{\hat{\theta}}_\upsilon = \gamma \Xi_y y_{H} - \gamma\sigma(\hat{\theta}_\upsilon)\hat{\theta}_\upsilon, \tag{7.33}$$

where Ξ_y and y_H are determined by (4.65) and (4.66), respectively. The function $\sigma(\hat{\theta}_\upsilon)$ is given by (7.32).

The robust algorithm with MRE is defined as

$$\dot{\hat{\theta}}_\upsilon = \gamma\left(L(s)\left[\xi_w\left(y + W_\Sigma(s)\left[\hat{\theta}_\upsilon^\top \xi\right]\right)\right] - \Omega\hat{\theta}_\upsilon\right) - \gamma\sigma(\hat{\theta}_\upsilon)\hat{\theta}_\upsilon, \tag{7.34}$$

where $L(s)$ is a designer-chosen transfer function of the form (3.44), $\Omega = L(s)\left[\xi_w \xi_w^\top\right]$, and $\sigma(\hat{\theta}_\upsilon)$ is given by (7.32).

The properties of the closed-loop system are defined as follows.

Proposition 7.5 *If Assumptions 2.1 and 4.2 hold, then the adaptive controller* (4.53), (4.57) *with the disturbance observer* (4.43), (4.44), (4.55) *and the adaptation algorithm* (7.31) *(or* (7.33)*, or* (7.34)*) when applied to the plant* (7.6) *ensures the following properties of the closed-loop system:*

P.7.5.1. *all the signals in the system are bounded, and if $\theta^* > |\theta_\upsilon|$, then y is of the order of $\bar{\delta} > |\delta(t)|$ in the mean square sense, i.e., $y \in \mathcal{MS}(\bar{\delta})$;*

P.7.5.2. *if* $\theta^* > |\theta_\upsilon|$ *and* $\delta(t) \equiv 0$, *then* $y(t) \to 0$ *as* $t \to \infty$.

Proof Consider the gradient algorithm of adaptation (7.31). In view of (7.29), (7.30) and (4.61), we have $\bar{y} = \tilde{\theta}_\upsilon^\top \hat{\xi} + \delta_y$. Since the latter equality is in the form of (3.109), from Property P.3.12.2 of Proposition 3.12 ($\hat{\xi}$ is bounded), we obtain that $\hat{\theta}_\upsilon$ is bounded, and $\bar{y}$ and $\dot{\hat{\theta}}_\upsilon$ are of the order of δ_y in the mean square sense. In view of the boundedness of $\hat{\xi}$ and $\hat{\theta}_\upsilon$, we can conclude that all the closed-loop signals are bounded. From (7.30), we obtain that δ_y is of the order of $\bar{\delta}$ in the mean square sense. Further, from Corollary 3.4, we conclude that the difference between y and $\bar{y}$ is of the order of δ_y and, hence, of $\bar{\delta}$ in the mean square sense. This completes the proof of P.7.5.1. Property P.7.5.2 directly follows from Property P.3.12.3 of Proposition 3.12. Using similar arguments, we can extend these results to the algorithms (7.33) and (7.34). □

Example 7.3 Let the plant be described by (7.6) with

$$A = \begin{bmatrix} 0 & 1 \\ 0 & 0 \end{bmatrix}, \; b = \begin{bmatrix} 0 \\ 1 \end{bmatrix}, \; c = \begin{bmatrix} 30 \\ 0 \end{bmatrix}, \; d = \begin{bmatrix} 1 \\ 2 \end{bmatrix}, \; d_n = 0,$$

and the regular component of the disturbance υ be modelled by the exosystem (2.1), (2.2) of the second order. We use the second-order disturbance observer (4.43), (4.44), (4.55) with G and l defined by (4.24) and select

$$N = \begin{bmatrix} 0 & 0 \\ 0 & 10 \end{bmatrix}, \; k = \begin{bmatrix} 30 \\ 11 \end{bmatrix}.$$

For simulations, we use the initial condition $x(0) = [0.02, 0]^\top$ and zero initial conditions for $\hat{\xi}(0)$ and $\hat{\theta}_\upsilon(0)$. The adaptation gain is selected as $\gamma = 100$.

Consider first the experiments when $\upsilon(t) \equiv 0$ and the system is closed by the gradient algorithm of adaptation (7.31) with: (a) $\sigma(\hat{\theta}_\upsilon) \equiv 0$; (b) $\sigma(\hat{\theta}_\upsilon)$ defined by (7.32) and $\theta^* = 1.5$. Let the irregular component of the disturbance be represented by the function $\delta(t) = 0.1 \sin(t) \sin(10t)$. Then, as shown by the right plot of Fig. 7.4, the system with $\sigma(\hat{\theta}_\upsilon) \equiv 0$ suffers from unbounded parametric drift. As seen from the right plot in Fig. 7.5, the drift can be avoided by applying the σ-modification (7.31) with the switching rule (7.32).

Now, we investigate the case, when $\upsilon(t) = \sin\left(\frac{\pi t}{2}\right)$ and

$$\delta(t) = \begin{cases} 0.1 \sin(t) \sin(10t) & \text{if } \; 0 \le t \le 20 \text{ s}, \\ 0 & \text{if } \; t > 20 \text{ s}. \end{cases} \tag{7.35}$$

The vector of adjustable parameters $\hat{\theta}_\upsilon$ is calculated using one of the following three adaptation algorithms with $\theta^* = 15$:

- the gradient-based algorithm (7.31);
- the algorithm (7.33) with DRE and $p = 1$, $H_1(s) = \frac{1}{0.25s+1}$;

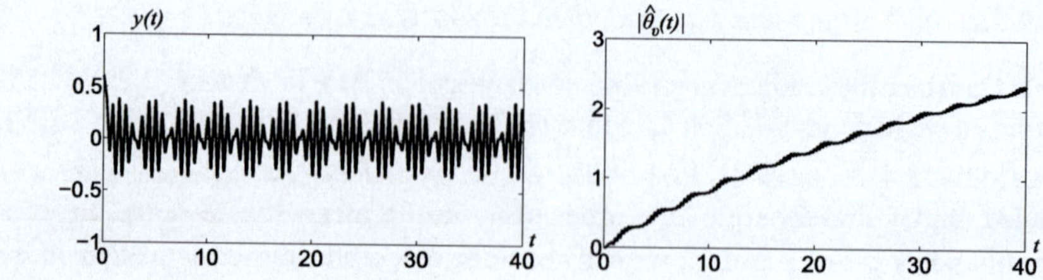

Fig. 7.4 Example 7.3: the response of the adaptive system to the disturbance $\upsilon_\delta(t) = 0.1\sin(t)\sin(10t)$ for $\sigma(\hat{\theta}_\upsilon) \equiv 0$

- the algorithm (7.34) with MRE and $L(s) = \frac{1}{0.2s+1}$.

Simulation results for the closed-loop systems employing the robust modifications of the gradient algorithm of adaptation, the adaptation algorithm with DRE and the algorithm of adaptation with MRE are presented in Fig. 7.6. It is seen from the plots that, in presence of nonzero irregular component δ, the plant outputs approach small steady-state values. If $\delta(t) = 0$, then the disturbance υ is completely rejected and the output y is driven to zero. ■

7.3.2 *LTI SISO Plants with Unknown Parameters*

Consider the LTI plant (7.9) satisfying Assumption 5.2. For the robust adaptive rejection of the disturbance, we use the controller (5.30) with the disturbance observer (5.24)–(5.27) (see Sect. 5.1.2) employing the robust modification of the algorithm of adaptation.

In view of Proposition 7.2, it can be shown that the closed-loop error model takes the form

$$\dot{x} = A_\Sigma x + \psi_u b_M(\tilde{\vartheta}^\top \omega + u + \bar{\delta}_x), \tag{7.36}$$

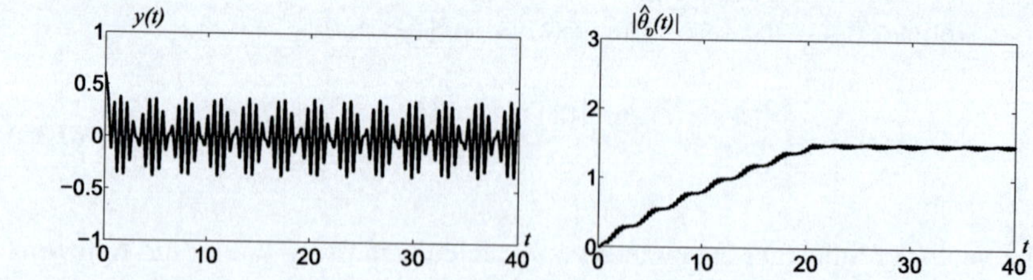

Fig. 7.5 Example 7.3: the response of the robust adaptive system of disturbance compensation to the disturbance $\upsilon_\delta(t) = 0.1\sin(t)\sin(10t)$ for $\sigma(\hat{\theta}_\upsilon)$ defined by (7.32) and $\theta^* = 1.5$

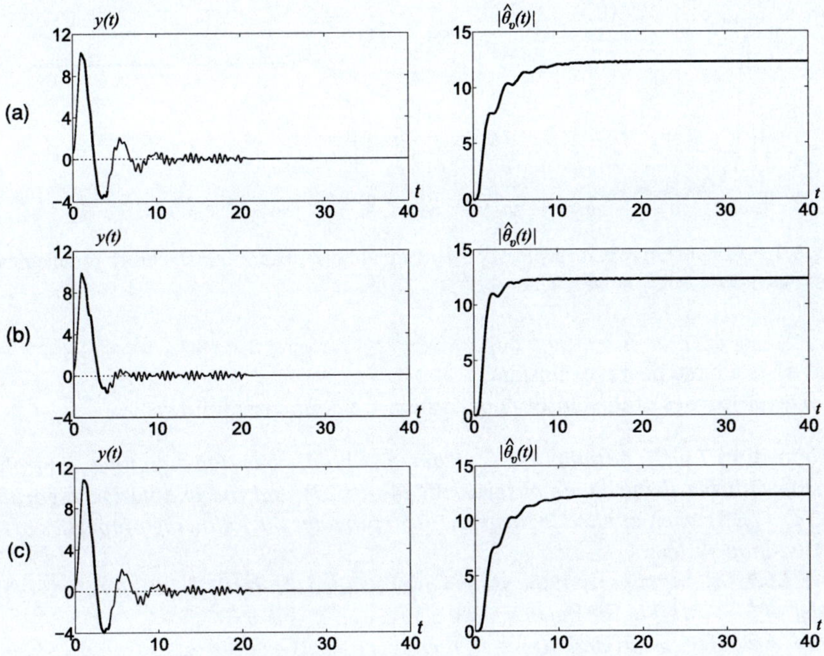

Fig. 7.6 Example 7.3: robust adaptive compensation of the disturbance υ_δ containing the irregular component by applying: **a** the gradient algorithm of adaptation; **b** the algorithm of adaptation with DRE; **c** the algorithm of adaptation with MRE

where $\tilde{\vartheta} = \bar{\vartheta} - \hat{\vartheta}$ is the vector of parametric errors, $\bar{\vartheta} \in \mathbb{R}^{n+(n+2)q} = \vartheta/\psi_u$ is a vector of unknown parameters, $\hat{\vartheta}$ is the vector of the adjustable parameters generated by an adaptation algorithm, ω is the regressor defined by (5.29) with the elements η_u and η_i $(i = 0, 1, \ldots, n)$ generated by the filters (5.24)–(5.26), $\bar{\delta}_x = \delta_x/\psi_u$, and δ_x is given in Proposition 7.2.

The model (7.36) is a particular case of the model (3.121). Therefore, we use Proposition 3.14 to construct the robust algorithm of adaptation

$$\dot{\hat{\vartheta}} = \gamma \omega b_M^\top P x - \gamma \sigma(\hat{\vartheta})\hat{\vartheta}, \tag{7.37}$$

where $\gamma > 0$ is the adaptation gain, $P = P^\top \succ 0$ is the solution of the Lyapunov matrix equation $A_\Sigma^\top P + P A_\Sigma = -Q$ with $Q = Q^\top \succ 0$, $\sigma(\hat{\vartheta})$ is the continuous switching term given by

$$\sigma(\hat{\vartheta}) = \begin{cases} 0, & |\hat{\vartheta}| < \vartheta^* \\ \frac{|\hat{\vartheta}|}{\vartheta^*} - 1, & \vartheta^* \le |\hat{\vartheta}| \le 2\vartheta^*, \\ 1, & |\hat{\vartheta}| > 2\vartheta^*, \end{cases} \tag{7.38}$$

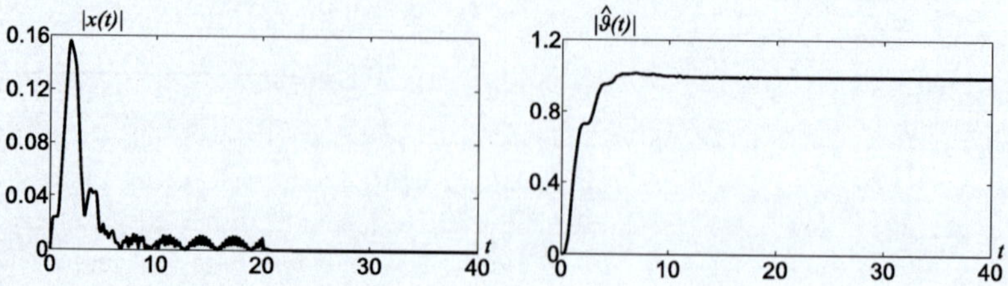

Fig. 7.7 Example 7.4: robust adaptive compensation of disturbance υ in the linear parametrically uncertain plant from Example 5.2

and ϑ^* is a large positive constant.

The properties of the closed-loop system are defined as follows.

Proposition 7.6 *If Assumptions 2.1 and 5.2 hold, then the adaptive controller* (5.30) *with the disturbance observer* (5.24)–(5.27) *and the adaptation algorithm* (7.37), (7.38) *when applied to the plant* (7.9) *provides the following properties of the closed-loop system:*

P.7.6.1. *all the signals in the system are bounded, and if* $\theta^* > |\theta|$, *then* x *is of the order of* $\bar{\delta} > |\delta(t)|$ *in the mean square sense, i.e.,* $x \in \mathcal{MS}(\bar{\delta})$;

P.7.6.2. *if* $\theta^* > |\theta|$ *and* $\delta(t) \equiv 0$, *then* $|x(t)| \to 0$ *as* $t \to \infty$.

This proposition directly follows from Proposition 3.14.

Example 7.4 Consider the plant, controller and disturbance observer from Example 5.2. However, to tune the parameters $\hat{\vartheta}$, we apply the adaptation algorithm (7.37), (7.38) with the matrix P defined in Example 5.2 and $\vartheta^* = 15$. In simulation, we investigate the case when $\upsilon = \sin(2t)$, and the irregular component of the disturbance is defined by (7.35).

Simulation results presented in Fig. 7.7 demonstrate a small steady-state error in presence of the irregular component δ and complete rejection of a priori unknown disturbance with asymptotic convergence of the norm $|x(t)|$ to zero if $\delta(t) \equiv 0$. ■

7.3.3 Nonlinear Plant with Unmatched Disturbance and Measured State

Now, consider the nonlinear plant (7.16) with unknown parameters. For the robust adaptive disturbance rejection, we apply the adjustable controller (5.138)–(5.142) from Sect. 5.2.2.1. In view of (7.17), this controller yields the following closed-loop error model:

$$\dot{z} = A_z z + W_\vartheta^\top \tilde{\vartheta} + B_z \delta_x, \qquad y = z_1,$$

where z, A_z, W_ϑ, and B_z are defined in Sect. 5.2.2.1, while δ_x is defined in Sect. 7.1.3.3. Then, with the use of nonlinear swapping scheme (5.145)–(5.147), we obtain the following model of augmented state error:

$$\bar{z} = \Omega_\vartheta^\top \vartheta + \epsilon_z, \tag{7.39}$$

where ϵ_z satisfies the equation $\dot{\epsilon}_z = A_z\epsilon_z + B_z\delta_x$ and, hence, is robust in the sense of Definition 7.2. The matrix Ω_ϑ is bounded due to the structure of the matrix A_z with the nonlinear damping terms s_i.

To adjust parameters $\hat{\vartheta}$, we can use robust modification of the adaptation algorithm with MRE

$$\dot{\hat{\vartheta}} = \gamma\left(\bar{z}_\Omega - \bar{\Omega}_\vartheta\hat{\vartheta}\right) - \gamma\bar{\sigma}\hat{\vartheta}, \tag{7.40}$$

where $\bar{z}_\Omega = L(s)\,[\Omega_\vartheta \bar{z}]$, $\bar{\Omega}_\vartheta = L(s)\left[\Omega_\vartheta\Omega_\vartheta^\top\right]$, $L(s)$ is a designer-chosen transfer function of the form (3.44) with a relative degree greater or equal to $n-2$, $\gamma > 0$ is the adaptation gain,

$$\bar{\sigma} = L(s)\left[\sigma(\hat{\vartheta})\right],$$

and $\sigma(\hat{\vartheta})$ is the continuous switching term defined by (7.38).

The HOTD of the adjustable parameters $\hat{\vartheta}$ up to the $(n-1)$st order can be calculated directly via the expressions

$$\hat{\vartheta}^{(i)} = \gamma\left(\bar{z}_\Omega^{(i-1)} - \sum_{j=0}^{i-1} C_j^{i-1}\left(\bar{\Omega}_\vartheta + \bar{\sigma}I\right)^{(i-j-1)}\hat{\vartheta}^{(j)}\right),$$

where $i = 2, 3, \ldots, n-1$, C_j^i are the binomial coefficients. The HOTD $\bar{z}_\Omega^{(i-1)}$, $\bar{\Omega}_\vartheta^{(i-j-1)}$ and $\bar{\sigma}^{(i-j-1)}$ are accessible due to the appropriate relative degree of the transfer function $L(s)$.

Then, in view of Proposition 3.13, we can formulate the following properties of the closed-loop system.

Proposition 7.7 *If Assumption 2.1 holds, then the adjustable controller (5.138)–(5.142) with the filters (5.130), (5.131), (5.133), (5.134), the swapping scheme (5.145)–(5.147), and the adaptation algorithm (7.40) when applied to the plant (7.16) guarantees that*

P.7.7.1. *all the closed-loop signals are bounded, and if $\theta^* > |\theta|$, then y is of the order of $\bar{\delta} > |\delta(t)|$ in the mean square sense, i.e., $y \in \mathcal{MS}(\bar{\delta})$;*

P.7.7.2. *if $\theta^* > |\theta|$ and $\delta(t) \equiv 0$, then $y(t) \to 0$ as $t \to \infty$.*

7.4 Output Regulation

Now, we demonstrate that, with a minor modification of the adaptive controller, the approach introduced in Sect. 5.1.4 can be extended for robust adaptive regulation in order to cope with the exosignals containing irregular components.

Thus, we consider the plant (7.13) satisfying Assumption 5.3.[2] Additionally, we make the following assumption regarding the properties of the exosignals.

Assumption 7.1 The disturbance υ_δ and the measured reference g_δ are such that:
A.7.1.1. they can be represented as

$$\upsilon_\delta = \upsilon + \delta_\upsilon, \qquad g_\delta = g + \delta_g, \tag{7.41}$$

where υ and g are the regular components of the disturbance and reference, respectively, while δ_υ and δ_g are the irregular components of these exosignals;
A.7.1.2. the regular components υ and g satisfy Assumption 4.3;
A.7.1.3. the irregular components δ_υ and δ_g are unknown but bounded time functions.

The considered control problem consists in design of a state-feedback adaptive controller providing (i) the boundedness of all the closed-loop signals; (ii) small (in a certain sense) steady-state tracking error $\varepsilon = y - g$ for small irregular components; (iii) achievement of the control objective

$$\lim_{t\to\infty} \varepsilon(t) = 0 \tag{7.42}$$

if $\delta_\upsilon(t) \equiv 0$ and $\delta_g(t) \equiv 0$.

Now, following the first step of the design procedure from Sect. 5.1.4, in view of (7.5) and Proposition 7.3, we can write the parameterized form of the plant as

$$\dot{e} = A_\Sigma e + b_M(\vartheta^\top \omega + \theta^\top \hat{\xi}_g + \psi_u u + \delta_e), \quad \varepsilon = c^\top e + d_n \delta_\upsilon - \delta_g, \tag{7.43}$$

where $\vartheta \in \mathbb{R}^{n+(n+2)q_\upsilon}$ and $\theta \in \mathbb{R}^{q_g}$ are vectors of unknown parameters, $\hat{\xi}_g$ is the state of the reference filter

$$\dot{\hat{\xi}}_g = G_g \hat{\xi}_g + l_g g_\delta, \tag{7.44}$$

ω is the regressor defined by (7.15) with η_0, η_i, *and* η_u generated by the filters (2.133)–(2.135), and δ_e is the term which is robust in the sense of Definition 7.2.

Finally, we can write

$$\dot{e} = A_\Sigma e + \psi_u b_M(\vartheta_\psi^\top \omega_\psi + u + \delta_\psi), \quad \varepsilon = c^\top e + d_n \delta_\upsilon - \delta_g \tag{7.45}$$

---

[2] A particular case of an asymptotically stable plant affected by a noisy sinusoidal disturbance with known lower and upper bounds on its frequency is considered in [2].

or

$$\varepsilon = \psi_u W_\Sigma(s)[\vartheta_\psi^\top \omega_\psi + u] + \delta_\varepsilon, \tag{7.46}$$

where $\vartheta_\psi = \frac{1}{\psi_u}[\vartheta^\top, \theta^\top]^\top$ is the aggregated vector of unknown parameters, $\omega_\psi = [\omega^\top, \hat{\xi}_g^\top]^\top$ is the aggregated regressor, $W_\Sigma(s) = c^\top(sI - A_\Sigma)^{-1}b_M$ is the asymptotically stable transfer function, and δ_ε is the term which is robust in the sense of Definition 7.2.

If we choose the adjustable controller as

$$u = -\hat{\vartheta}_\psi^\top \omega_\psi, \tag{7.47}$$

where $\hat{\vartheta}_\psi$ is the vector of adjustable parameters, then we obtain the closed-loop error model

$$\varepsilon = \psi_u W_\Sigma(s)\left[\tilde{\vartheta}_\psi^\top \omega_\psi\right] + \delta_\varepsilon,$$

where $\tilde{\vartheta}_\psi = \vartheta_\psi - \hat{\vartheta}_\psi$.

For the tuning of the controller (7.47), one can use the following robust algorithm of adaptation with augmented error [1, Sect. 9.3.3]:

$$\dot{\hat{\vartheta}}_\psi = \gamma_1 \frac{\bar{\omega}_\psi \bar{\varepsilon}}{1 + \bar{\omega}_\psi^\top \bar{\omega}_\psi} - \gamma_1 \sigma(\hat{\vartheta}_\psi)\vartheta_\psi, \tag{7.48}$$

$$\dot{\hat{\psi}}_u = -\gamma_2 \frac{\zeta \bar{\varepsilon}}{1 + \bar{\omega}_\psi^\top \bar{\omega}_\psi} - \gamma_2 \sigma(\hat{\psi}_u)\hat{\psi}_u, \tag{7.49}$$

where $\gamma_1, \gamma_2 > 0$ are the adaptation gains, $\bar{\omega}_\psi = W_\Sigma(s)[\omega_\psi]$ is the filtered regressor, $\hat{\psi}_u$ is the additional adjustable parameter, $\bar{\varepsilon}$ and ζ are the augmented tracking error and the swapping term given by

$$\bar{\varepsilon} = \varepsilon + \hat{\psi}_u \zeta, \qquad \zeta = W_\Sigma(s)\left[\hat{\vartheta}_\psi^\top \omega_\psi\right] - \hat{\vartheta}_\psi^\top \bar{\omega}_\psi, \tag{7.50}$$

$$\sigma(\hat{\vartheta}_\psi) = \begin{cases} 0, & |\hat{\vartheta}_\psi| < \vartheta^*, \\ \frac{|\hat{\vartheta}_\psi|}{\vartheta^*} - 1, & \theta^* \le |\hat{\vartheta}_\psi| \le 2\vartheta^*, \\ 1, & |\hat{\vartheta}_\psi| > 2\vartheta^*, \end{cases} \quad \sigma(\hat{\psi}_u) = \begin{cases} 0, & |\hat{\psi}_u| < \psi^*, \\ \frac{|\hat{\psi}_u|}{\psi^*} - 1, & \theta^* \le |\hat{\psi}_u| \le 2\psi^*, \\ 1, & |\hat{\psi}_u| > 2\psi^*, \end{cases}$$

and $\vartheta^* > 0$ and $\psi_u^* > 0$ are some large constants.

Then the properties of the closed-loop system are as follows (see [1, Theorem 9.3.2]).

Proposition 7.8 *If Assumptions 5.3 and 7.1 hold, then the adaptive controller (7.47) with the reference observer (7.44), disturbance observer (2.133)–(2.136), and the* *adaptation algorithm* (7.48)–(7.50) *when applied to the plant* (7.13) *provides the following properties of the closed-loop system:*

P.7.8.1. *all the signals in the system are bounded, and if* $\vartheta^* > |\vartheta_\psi|$ *and* $\psi^* > |\psi_u|$, *then* ε *is of the order of* $\bar{\delta} > |\delta_\upsilon(t)| + |\delta_g(t)|$ *in the mean square sense, i.e.,* $\varepsilon \in \mathcal{MS}(\bar{\delta})$;

P.7.8.2. *if* $\vartheta^* > |\vartheta_\psi|$, $\psi^* > |\psi_u|$ *and* $\delta_\upsilon(t) \equiv 0$, $\delta_g(t) \equiv 0$, *then* $\varepsilon(t) \to 0$ *as* $t \to \infty$.

References

1. Ioannou, P., Sun, J.: Robust Adaptive Control. Prentice-Hall, Enlewood Cliffs (1996)
2. Marino, R., Santosuosso, G.L., Tomei, P.: Robust adaptive compensation of biased sinusoidal disturbances with unknown frequency. Automatica **39**, 1755–1761 (2003)
3. Nikiforov, V.O.: Adaptive and Robust Control with Disturbance Compensation. Nauka, St. Petersburg (2003). (in Russian)

Appendix A
Exosignals: Parameterizations and Properties

In this appendix, we present several technical corollaries of Lemmas 2.1 and 2.2 used in the book. We first consider the case of scalar exosignals and then extend it for multidimensional ones.

For convenience, we repeat Lemma 2.1.

Lemma A.1 *The exosignal $\upsilon \in \mathbb{R}$ satisfying Assumption A.2.1 can be represented in the form of linear regression*

$$\upsilon = \theta^{\top}\bar{\xi}, \tag{A.1}$$

where $\theta \in \mathbb{R}^q$ is a vector of unknown parameters, while the bounded regressor $\bar{\xi} \in \mathbb{R}^q$ is the state of the model

$$\dot{\bar{\xi}} = G\bar{\xi} + l\upsilon \tag{A.2}$$

or

$$\dot{\bar{\xi}} = \left(G + l\theta^{\top}\right)\bar{\xi} \tag{A.3}$$

with a certain initial condition $\bar{\xi}(0)$, an arbitrary Hurwitz matrix $G \in \mathbb{R}^{q\times q}$ and an arbitrary constant vector $l \in \mathbb{R}^q$ forming the controllable pair (G, l).

Below, by ϵ_i $(i \in \mathbb{N})$ we will denote different exponentially decaying terms of appropriate dimensions.

Corollary A.1 *The ith time derivative $\upsilon^{(i)}$ of the exosignal υ can be represented in the form*

$$\upsilon^{(i)} = \bar{\theta}_i^{\top}\bar{\xi}, \quad i \in \mathbb{N}, \tag{A.4}$$

where $\bar{\theta}_i^{\top} = \theta^{\top}(G + l\theta^{\top})^i$ is a vector of unknown parameters.

V. Nikiforov and D. Gerasimov, *Adaptive Regulation*, Lecture Notes in Control and Information Sciences 491, https://doi.org/10.1007/978-3-030-96091-9

Remark A.1 If υ is a function with jump discontinuity at $t = 0$, e.g.,

$$\upsilon(t) = \begin{cases} 0, & \text{for} \quad t < 0, \\ \theta^\top \xi(t), & \text{for} \quad t \geq 0, \end{cases}$$

then a term with appropriately scaled ith time derivative of the unit step function $1(t)$ must be added to the expression (A.4). For example, for $i = 1$, we have

$$\dot{\upsilon}(t) = \bar{\theta}_1^\top \xi(t) + \upsilon(0) D_\delta(t), \tag{A.5}$$

where $D_\delta(t) = \frac{d}{dt} 1(t)$ is the Dirac delta function defined as

$$D_\delta(t) = \begin{cases} \infty, & \text{for} \quad t = 0, \\ 0, & \text{for} \quad t > 0 \end{cases}$$

and

$$\int_0^\infty D_\delta(\tau) d\tau = 1.$$

A term with the time derivative of the unit step function must be taken into account for correct analysis of the transients in dynamic systems with exosignals applied at the instance $t = 0$ (see the proof of Proposition 2.11). However, since such derivatives are zero everywhere but at the origin, one can use the simplified expression (A.4) to analyze steady-state regimes in asymptotically stable systems even if the exosignal has jump discontinuity at $t = 0$. ■

Corollary A.2 *For any proper transfer function* $W(s) = \frac{\beta(s)}{\alpha(s)}$, *where* $\alpha(s) = \sum_{i=0}^n a_i s^i$, $\beta(s) = \sum_{i=0}^m b_i s^i$, a_i, b_i *are constants, and* $\alpha(s)$ *is Hurwitz, the filtered exosignal*

$$\upsilon_f = W(s)[\upsilon] \tag{A.6}$$

can be represented in the form

$$\upsilon_f = \theta^\top \xi_w \tag{A.7}$$

with the filtered regressor ξ_w *given by*

$$\xi_w = W(s)\left[\xi\right]. \tag{A.8}$$

Proof By the properties of LTI systems, we have

$$\upsilon_f = W(s)[\upsilon] = W(s)\left[\theta^\top \xi\right] = \theta^\top W(s)\left[\xi\right] = \theta^\top \xi_w.$$

□

Corollary A.3 *The filtered regressor $\bar{\xi}_w$ (A.8) can be represented in the form*

$$\bar{\xi}_w = \bar{\xi}_f + \epsilon_1, \tag{A.9}$$

where $\bar{\xi}_f$ is the state of the filter

$$\dot{\bar{\xi}}_f = G\bar{\xi}_f + l\upsilon_f \tag{A.10}$$

with an arbitrary initial condition $\bar{\xi}_f(0)$ and

$$\epsilon_1(t) = W(s)\left[\mathrm{e}^{Gt}\bar{\xi}(0)\right] - \mathrm{e}^{Gt}\bar{\xi}_f(0). \tag{A.11}$$

Then the following expression holds:

$$\upsilon_f = \theta^\top\bar{\xi}_f + \epsilon_2, \tag{A.12}$$

where $\epsilon_2 = \theta^\top\epsilon_1$ exponentially decays.

Proof We represent the solutions $\bar{\xi}$ of (A.2) and $\bar{\xi}_f$ of (A.10) as the sums of the zero-state response and the zero-input response ([2], p. 116):

$$\bar{\xi} = (sI - G)^{-1}l\,[\upsilon] + \mathrm{e}^{Gt}\bar{\xi}(0), \tag{A.13}$$
$$\bar{\xi}_f = (sI - G)^{-1}l\left[\upsilon_f\right] + \mathrm{e}^{Gt}\bar{\xi}_f(0). \tag{A.14}$$

Therefore,

$$\begin{aligned}\bar{\xi}_w &= W(s)\left[\bar{\xi}\right] = W(s)\left[(sI - G)^{-1}l\,[\upsilon] + \mathrm{e}^{Gt}\bar{\xi}(0)\right] \\ &= (sI - G)^{-1}l\,[W(s)[\upsilon]] + W(s)\left[\mathrm{e}^{Gt}\bar{\xi}(0)\right] \\ &= (sI - G)^{-1}l\left[\upsilon_f\right] + W(s)\left[\mathrm{e}^{Gt}\bar{\xi}(0)\right].\end{aligned} \tag{A.15}$$

Then, calculating $\epsilon_1 = \bar{\xi}_w - \bar{\xi}_f$, in view of (A.15) and (A.14), we directly obtain (A.11) and, as a result, (A.9). Substituting (A.9) into (A.7), we get (A.12). □

Corollary A.4 *Define $\bar{\Gamma} = G + l\theta^\top$ and $M_w = \alpha^{-1}(\bar{\Gamma})\beta(\bar{\Gamma})$, where the matrices $\alpha(\bar{\Gamma})$ and $\beta(\bar{\Gamma})$ are calculated by replacement of $\bar{\Gamma}$ instead of s in $\alpha(s)$ and $\beta(s)$ (scalar terms $\alpha(0)$ and $\beta(0)$ are replaced by $\alpha(0)I$ and $\beta(0)I$), respectively. Then, for any Hurwitz polynomial $\alpha(s)$, the filtered regressor $\bar{\xi}_w$ (A.8) can be represented in the form*

$$\bar{\xi}_w = \tilde{\xi}_f + \epsilon_3, \tag{A.16}$$

where $\tilde{\xi}_f$ is the state of the autonomous model

$$\dot{\tilde{\xi}}_f = \bar{\Gamma}\tilde{\xi}_f \tag{A.17}$$

with the initial condition

$$\bar{\xi}_f(0) = M_w \bar{\xi}(0), \tag{A.18}$$

and $\epsilon_3 \in \mathbb{R}^q$ *is the exponentially decaying term satisfying the differential equation*

$$\sum_{i=0}^{n} a_i \epsilon_3^{(i)} = 0 \tag{A.19}$$

with the initial conditions $\epsilon_3^{(i)}(0)$ $(i = 0, 1, \ldots, n-1)$ *given by*

$$\epsilon_3^{(i)}(0) = \begin{cases} -\bar{\Gamma}^i M_w \bar{\xi}(0), & \text{for } i < n-m, \\ \left(\frac{b_m}{a_n} I - \bar{\Gamma}^i M_w\right) \bar{\xi}(0), & \text{for } i = n-m, \\ \frac{1}{a_n}\left(\sum\limits_{j=n-m}^{i} \bar{\Gamma}^{j-n+m} b_{m+j-i} - \bar{\Gamma}^i M_w\right) \bar{\xi}(0) & \\ \quad -\frac{1}{a_n} \sum\limits_{j=n-i}^{n-1} a_j \left(\bar{\Gamma}^{j-n+i} M_w \bar{\xi}(0) + \epsilon_3^{(j-n+i)}(0)\right), & \text{for } i > n-m. \end{cases} \tag{A.20}$$

Therefore, the following expression holds:

$$\upsilon_f = \theta^\top \bar{\xi}_f + \epsilon_4, \tag{A.21}$$

where $\epsilon_4 = \theta^\top \epsilon_3$ *exponentially decays.*

Before we start the proof, we note that the matrix $\alpha(\bar{\Gamma})$ is not singular (and, therefore, the inverse matrix $\alpha^{-1}(\bar{\Gamma})$ exists) since the roots of the Hurwitz polynomial $\alpha(s)$ do not coincide with the eigenvalues of $\bar{\Gamma}$.

Proof The filtered regressor $\bar{\xi}_w$ defined by (A.8) can be considered as the solution of the differential equation

$$\sum_{i=0}^{n} a_i \bar{\xi}_w^{(i)} = \sum_{i=0}^{m} b_i \bar{\xi}^{(i)} \tag{A.22}$$

with the initial values[1] $\bar{\xi}^{(i)}(0)$, $i = 0, 1, \ldots, n-1$ given by[2]

[1] We distinguish *initial conditions* of a differential equation defined at time $t = 0^-$ and *initial values* of a solution and its HOTD defined at time $t = 0^+$.

[2] Equations (A.23) are derived by successive calculation of the Laplace transforms of $\bar{\xi}_w^{(i)}$ and applying the Laplace initial-value theorem.

$$\bar{\xi}_w^{(i)}(0) = \begin{cases} 0, & \text{for } i < n-m, \\ \frac{b_m}{a_n}\bar{\xi}(0), & \text{for } i = n-m, \\ \frac{1}{a_n}\Bigg(\sum\limits_{j=n-m}^{i} \bar{\Gamma}^{j-n+m} b_{m+j-i}\bar{\xi}(0) & \\ \qquad - \sum\limits_{j=n-i}^{n-1} a_j \bar{\xi}_w^{(j-n+i)}(0) \Bigg), & \text{for } i > n-m. \end{cases} \tag{A.23}$$

Now, we demonstrate that the solution of (A.17) $\tilde{\xi}_f(t) = \mathrm{e}^{\bar{\Gamma}t}\tilde{\xi}_f(0)$ with the initial condition (A.18) and with the initial values

$$\tilde{\xi}_f^{(i)}(0) = \bar{\Gamma}^i M_w \bar{\xi}(0), \quad i = 0, 1, \ldots, n-1 \tag{A.24}$$

obeys the differential equation

$$\sum_{i=0}^{n} a_i \tilde{\xi}_f^{(i)} = \sum_{i=0}^{m} b_i \bar{\xi}^{(i)}. \tag{A.25}$$

Substituting $\tilde{\xi}_f(t) = \mathrm{e}^{\bar{\Gamma}t}\tilde{\xi}_f(0)$ and $\bar{\xi}(t) = \mathrm{e}^{\bar{\Gamma}t}\bar{\xi}(0)$ into (A.25), we have[3]

$$\alpha(\bar{\Gamma})\mathrm{e}^{\bar{\Gamma}t}\tilde{\xi}_f(0) = \beta(\bar{\Gamma})\mathrm{e}^{\bar{\Gamma}t}\bar{\xi}(0).$$

In view of (A.18), the latter equality is reduced to the identity $\beta(\bar{\Gamma})\mathrm{e}^{\bar{\Gamma}t}\bar{\xi}(0) = \beta(\bar{\Gamma})\mathrm{e}^{\bar{\Gamma}t}\bar{\xi}(0)$ that proves (A.25). Further, introducing $\epsilon_3 = \bar{\xi}_w - \tilde{\xi}_f$ and subtracting (A.25) from (A.22), we obtain (A.19). Finally, subtracting (A.24) from (A.23), in view of (A.16) and (A.18), we obtain (A.20).[4] □

Remark A.2 Comparing two exosystems (A.3) and (A.17) generating different exosignals due to different initial conditions, we establish the equivalent state transformation

$$\tilde{\xi}_f = M_w \bar{\xi} \tag{A.26}$$

with the state transformation matrix $M_w = \alpha^{-1}(\bar{\Gamma})\beta(\bar{\Gamma})$. ■

Remark A.3 The solution $\tilde{\xi}_f(t) = \mathrm{e}^{\bar{\Gamma}t}\tilde{\xi}_f(0)$ with the initial condition (A.18) can be considered as a steady-state component of the trajectory $\bar{\xi}_w(t)$. ■

Corollary A.5 *If the polynomial $\beta(s)$ from Corollary A.2 is Hurwitz, then the exosignal υ satisfying Assumption A.2.1 can be represented in the form of linear regression*

---

[3] The case of a scalar exponential input is considered in [2], page 6.

[4] Since (A.19) is a homogeneous equation, its initial conditions are equal to the initial values of its solution.

Table A.1 Regression models for the exosignals υ and $\upsilon_f = W(s)[\upsilon]$

№	Regressor	Regressor formula	Regression	
			υ	υ_f
1	$\bar{\xi}$	(A.2)	$\theta^\top \bar{\xi}$	$\theta^\top M_w \bar{\xi} + \theta^\top \epsilon_3$
2	$\bar{\xi}_w$	(A.8)	$\bar{\theta}^\top \bar{\xi}_w + \epsilon_5$	$\theta^\top \bar{\xi}_w$
3	ξ_f	(A.10)	$\bar{\theta}^\top \xi_f + \epsilon_6$	$\theta^\top \xi_f + \epsilon_2$
4	$\bar{\xi}_f$	(A.17)	$\bar{\theta}^\top \bar{\xi}_f$	$\theta^\top \bar{\xi}_f + \epsilon_4$

$$\upsilon = \bar{\theta}^\top \bar{\xi}_w + \epsilon_5 \tag{A.27}$$

or

$$\upsilon = \bar{\theta}^\top \xi_f + \epsilon_6, \tag{A.28}$$

where $\bar{\theta}^\top = \theta^\top M_w^{-1} \in \mathbb{R}^q$ is a new vector of unknown parameters, $\bar{\xi}_w$ is the filtered regressor defined by (A.8), ξ_f is the state of (A.10), $\epsilon_5 = -\bar{\theta}^\top \epsilon_3$ and $\epsilon_6 = \bar{\theta}^\top(\epsilon_1 - \epsilon_3)$ exponentially decay, while ϵ_1 and ϵ_3 are defined by (A.11) and (A.19), respectively.

Proof Since $\beta(s)$ is Hurwitz, then the matrix $\beta(\bar{\Gamma})$ is nonsingular, and the inverse matrix M_w^{-1} exists. Then the equality (A.27) is proved by replacement of $\bar{\xi} = M_w^{-1}\bar{\xi}_f$ in (A.1) and, further, by substitution of $\bar{\xi}_f = \bar{\xi}_w - \epsilon_3$ given by (A.16). The equality (A.28) is obtained by substitution of (A.9) into (A.27). □

Taking into account the transformation (A.26) and the relation (A.9), we summarize the results of Corollaries A.2–A.5 in Table A.1.

Example A.1 Consider the second-order exosystem (2.1), (2.2) with

$$\Gamma = \begin{bmatrix} 0 & 1 \\ -1 & 0 \end{bmatrix}, \quad h = \begin{bmatrix} 1 \\ 0 \end{bmatrix},$$

its canonical form (2.16), (2.14) with

$$G = \begin{bmatrix} 0 & 1 \\ -2 & -3 \end{bmatrix}, \quad l = \begin{bmatrix} 0 \\ 1 \end{bmatrix}$$

and

$$\theta = \begin{bmatrix} 1 \\ 3 \end{bmatrix}, \quad \bar{\Gamma} = G + l\theta^\top = \begin{bmatrix} 0 & 1 \\ -1 & 0 \end{bmatrix}$$

calculated in Example 2.7. For the initial condition $\bar{\xi}(0) = [1,\ 3]^\top$, we get $\upsilon(t) = 10\cos t$. Consider the filtered exosignal $\upsilon_f = W(s)[\upsilon]$ with

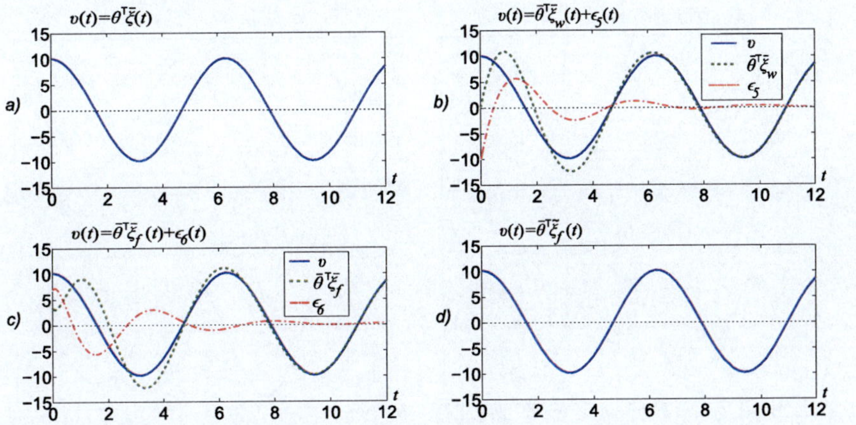

Fig. A.1 Exosignal $\upsilon(t) = 10\cos t$ calculated using regression models with the regressor: **a** $\bar{\xi}$; **b** $\bar{\xi}_w$; **c** $\bar{\xi}_f$; **d** $\bar{\xi}_f$

$$W(s) = \frac{\beta(s)}{\alpha(s)} = \frac{6s^2 + 7s + 8}{s^3 + 3s^2 + 4s + 5}.$$

In this case, we have

$$M_w = \left(\bar{\Gamma}^3 + 3\bar{\Gamma}^2 + 4\bar{\Gamma} + 5I\right)^{-1}\left(6\bar{\Gamma}^2 + 7\bar{\Gamma} + 8I\right) = \begin{bmatrix} 1.923 & 0.615 \\ -0.615 & 1.923 \end{bmatrix},$$

$$\epsilon_3(0) = -M_w\bar{\xi}_f(0) = \begin{bmatrix} -3.769 \\ -5.154 \end{bmatrix},$$

$$\dot{\epsilon}_3(0) = 6\bar{\xi}(0) - \bar{\Gamma} M_w\bar{\xi}(0) = \begin{bmatrix} 0.846 \\ 21.769 \end{bmatrix},$$

$$\ddot{\epsilon}_3(0) = 7\bar{\xi}(0) - 4M_w\bar{\xi}(0) - 4\epsilon_3(0) + 6\bar{\Gamma}\bar{\xi}(0) - 3\bar{\Gamma} M_w\bar{\xi}(0) - 3\dot{\epsilon}_3(0) - \bar{\Gamma}^2 M_w\bar{\xi}(0)$$
$$= \begin{bmatrix} 10.769 \\ -33.846 \end{bmatrix}.$$

For the regressions (A.27), (A.28), the vector $\bar{\theta}$ is calculated as

$$\bar{\theta}^\top = \theta^\top M_w^{-1} = \left[0.925,\ 1.264\right].$$

Note that, for the control problems considered in the book, the vector $\bar{\theta}$ is unknown. However, we calculate it here to illustrate Corollaries A.2–A.5 by simulation. Figure A.1 demonstrates the plots of υ calculated according to (A.1), (A.27) and (A.28). Figure A.2 demonstrates the plots of υ_f calculated according to (A.6), (A.7), (A.12), and (A.21). ∎

The corollary is directly proved via the fundamental solution of (A.3).

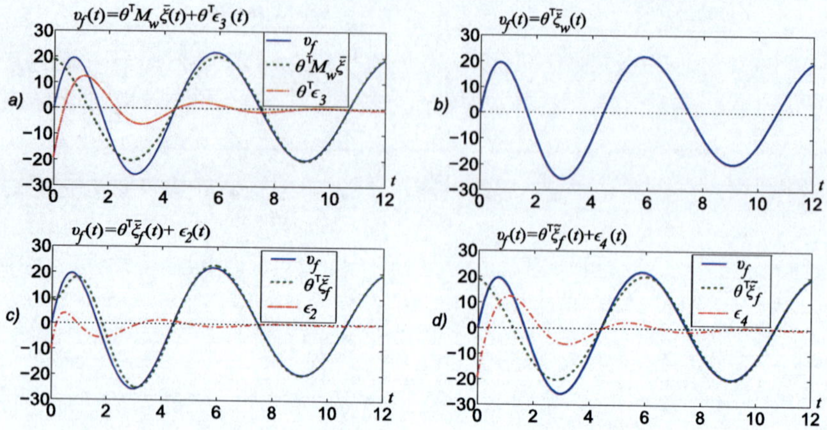

Fig. A.2 Filtered exosignal $\upsilon_f(t) = W(s)[10\cos t]$ calculated using regression models with the regressor: **a** $\bar{\xi}$; **b** $\bar{\xi}_w$; **c** $\bar{\xi}_f$; **d** $\bar{\xi}_f$

Now, consider the case of multidimensional exosignal $\upsilon \in \mathbb{R}^q$ given by (2.21) and repeat Lemma 2.2.

Lemma A.2 *The exosignal* $\upsilon \in \mathbb{R}^\rho$ *satisfying Assumption* 2.2 *can be represented in the form of linear regression*

$$\upsilon = \Theta\bar{\xi} \tag{A.29}$$

with a matrix of unknown parameters $\Theta \in \mathbb{R}^{\rho\times q}$ *and the bounded regressor* $\bar{\xi} \in \mathbb{R}^q$ *generated by the model*

$$\dot{\bar{\xi}} = G\bar{\xi} + L\upsilon \tag{A.30}$$

or

$$\dot{\bar{\xi}} = (G + L\Theta)\,\bar{\xi} \tag{A.31}$$

with a certain initial state $\bar{\xi}(0)$, *where* $G \in \mathbb{R}^{q\times q}$ *is a Hurwitz matrix given by*

$$G = \begin{bmatrix} 0 & 1 & 0 & \cdots & 0 \\ 0 & 0 & 1 & \cdots & 0 \\ \vdots & \vdots & \vdots & \ddots & \vdots \\ 0 & 0 & 0 & \cdots & 1 \\ -\gamma_0 & -\gamma_1 & -\gamma_2 & \cdots & -\gamma_{q-1} \end{bmatrix}, \tag{A.32}$$

$L \in \mathbb{R}^{q\times\rho}$ *is a matrix defined as*

$$L = \begin{bmatrix} 0 & 0 & \cdots & 0 \\ 0 & 0 & \cdots & 0 \\ \vdots & \vdots & \ddots & \vdots \\ 0 & 0 & \cdots & 0 \\ \bar{\lambda} & \bar{\lambda} & \cdots & \bar{\lambda} \end{bmatrix}, \tag{A.33}$$

γ_i ($i = 0, 1, \ldots, q-1$) and $\bar{\lambda} > 0$ are constants.

Below, by $\bar{\epsilon}_i$ ($i \in \mathbb{N}$) we denote different exponentially decaying terms of appropriate dimensions.

Corollary A.6 *The ith time derivative $\upsilon^{(i)}$ of the exosignal $\upsilon \in \mathbb{R}^q$ can be represented in the form*

$$\upsilon^{(i)} = \bar{\Theta}_i \bar{\xi}, \qquad i \in \mathbb{N}, \tag{A.34}$$

where $\bar{\Theta}_i = \Theta(G + L\Theta)^i$ is a matrix of unknown parameters.

Corollary A.7 *The predicted values of the regressor $\bar{\xi}(t+\tau)$ and the exosignal $\upsilon(t+\tau)$ with the prediction time $\tau > 0$ are defined as*

$$\bar{\xi}(t+\tau) = \mathrm{e}^{(G+L\Theta)\tau}\bar{\xi}(t), \tag{A.35}$$

$$\upsilon(t+\tau) = \bar{\Theta}\bar{\xi}(t), \tag{A.36}$$

where $\mathrm{e}^{(G+L\Theta)\tau}$ and $\bar{\Theta} = \Theta \mathrm{e}^{(G+L\Theta)\tau}$ are matrices with unknown parameters.

The corollary is directly proved via the fundamental solution of (A.31).

References

1. Chen, C.-T.: Linear Systems. Theory and Design, 3rd edn. Oxford University Press, NY, Oxford (1999)
2. Lewin, W.S.(Ed.): Control Handbook. (In two volumes), Jaico Publishing House, CRC Press in Cooperation with IEEE Press (1999)

Appendix B
Stability, Lyapunov Functions and Robustness

In this appendix, some stability notions and theorems used in this book are presented. Starting with the stability definitions and some basic theorems of Lyapunov's direct method, we briefly discuss the robustness of stability properties with respect to different classes of perturbations. Finally, we introduce the notion of the ISS. More detailed information on these subjects can be found in vast literature devoted to the stability theory and Lyapunov's direct method [2, 4, 7, 8, 10, 12, 13] as well as in the books devoted to the problems of adaptive control [1, Sect. 2.1], [3, Sect. 3.4], [5, Sect. 2.1], [6, Sect. 2.4]. In the presentation of the robustness issues, we basically follow [1, Sect. 2.1.2.]. The notion of the ISS is due to [5, Appendix C].

Variable notations used in this appendix are local. Initial conditions for $x(t)$ are denoted by $x(0)$, $x_0 = x(0)$ or $x(t_0)$ depending on a context.

B.1 Definitions of Stability

Consider a nonlinear system described by the ordinary differential equation

$$\dot{x} = f(x, t), \tag{B.1}$$

where $x \in \mathbb{R}^n$ is the state, t is time, $f\colon \mathbb{R}^n \times \mathbb{R}_+ \to \mathbb{R}^n$, and $f(0, t) = 0$ for all $t \geq t_0$. We assume that $f(x, t)$ is such that for defined initial conditions $x_0 = x(t_0)$ and $t = t_0$ the system (B.1) possesses a unique solution $x(t, x_0, t_0)$.

Definition B.1 A state $x^* \in \mathbb{R}^n$ is called an *equilibrium state of the system* (B.1) if $f(x^*, t) \equiv 0$ for all $t \geq t_0$.

For the adaptive systems considered in this book, the primary interest is in the stability of the origin $x^* = 0$ that is an equilibrium state of (B.1). Therefore, we consider the origin as an equilibrium state, the stability properties of which are to be investigated.

V. Nikiforov and D. Gerasimov, *Adaptive Regulation*, Lecture Notes in Control and Information Sciences 491, https://doi.org/10.1007/978-3-030-96091-9

Definition B.2 The equilibrium state $x^* = 0$ of (B.1) is called:

D.B.2.1. *stable* if for arbitrary t_0 and $\epsilon > 0$ there exists $\delta(\epsilon, t_0) > 0$ such that $|x_0| \le \delta(\epsilon, t_0)$ implies that $|x(t, x_0, t_0)| \le \epsilon$ for all $t \ge t_0$;

D.B.2.2. *uniformly stable* if in Definition D.B.2.1 δ is independent of the initial time t_0;

D.B.2.3. *asymptotically stable* if it is stable and there exists $\delta(t_0) > 0$ such that $|x_0| \le \delta(t_0)$ implies that

$$\lim_{t\to\infty} |x(t, x_0, t_0)| = 0;$$

D.B.2.4. *uniformly asymptotically stable* if it is uniformly stable and for every $\epsilon > 0$ and $t_0 \ge 0$ there exist $\delta > 0$ independent of t_0 and ϵ and $T(\epsilon) > 0$ independent of t_0 such that $|x(t, t_0, x_0)| \le \epsilon$ for all $t \ge t_0 + T(\epsilon)$ and $|x_0| \le \delta$;

D.B.2.5. *exponentially stable* if there exists $c > 0$ and for any $\epsilon > 0$ there exists $\delta(\epsilon)$ such that for each $|x_0| \le \delta(\epsilon)$ it holds that

$$|x(t, x_0, t_0,)| \le \epsilon e^{-c(t-t_0)} \quad \text{for all} \quad t \ge t_0;$$

D.B.2.6. *unstable* if it is not stable.

As well known, exponential stability always implies uniform asymptotic stability. If a system is autonomous (i.e., $f = f(x)$), all stability properties are uniform. For linear systems, uniform asymptotic stability and exponential stability are equivalent.

Definition B.3 If the system (B.1) has a unique equilibrium state $x^* = 0$ and:

D.B.3.1. $\lim_{\epsilon\to\infty} \delta(\epsilon) = \mathbb{R}^n$ in Definitions D.B.2.1 and D.B.2.2, then the equilibrium state and the system itself are called *globally stable* and *globally uniformly stable*, respectively;

D.B.3.2. $\delta = \mathbb{R}^n$ in Definitions D.B.2.3 and D.B.2.4, then the equilibrium state and the system itself are called *globally asymptotically stable* and *globally uniformly asymptotically stable*, respectively;

D.B.3.3. $\delta = \mathbb{R}^n$ in Definition D.B.2.5, then the equilibrium state and the system itself are called *globally exponentially stable.*

Global uniform asymptotic stability and global exponential stability are most desirable stability properties of closed-loop control systems. For linear systems, all the stability properties hold globally.

B.2 Lyapunov Functions and Lyapunov's Direct Method

Lyapunov's direct method allows us to determine the stability of (B.1) from the properties of $f(x, t)$ and a specially selected *Lyapunov function* $V(x, t)$ without finding the trajectories $x(t, x_0, t_0)$. In this book, under a Lyapunov function $V(x, t)$: $\mathbb{R}^n \times \mathbb{R}_+ \to \mathbb{R}_+$ we will mean a continuously differentiable scalar function the time derivative of which along the solutions of an investigated system possesses some desirable properties. The *time derivative of* $V(x,t)$ *along the solutions of* (B.1) is defined as

$$\dot{V}(x,t) = \frac{\partial V(x,t)}{\partial t} + \frac{\partial V(x,t)}{\partial x} f(x,t).$$

Remark B.1 Sometimes in literature, before calculation of the time derivative $\dot{V}(x,t)$, the selected function $V(x,t)$ is called the *Lyapunov function candidate*. In this book, the function $V(x,t)$ is called the *Lyapunov function* immediately after its first introduction before calculating its derivative. ■

In order to formulate the main stability theorems, we introduce the following definitions.

Definition B.4 A continuous function $w(|x|)$: $[0, c) \to \mathbb{R}_+$ is said to belong to a class $\mathcal{K}$ (i.e., $w \in \mathcal{K}$) if it is strictly increasing and $w(0) = 0$. It is said to belong to a class $\mathcal{K}_\infty$ (i.e., $w \in \mathcal{K}_\infty$) if $c = \infty$ and $w(|x|) \to \infty$ as $|x| \to \infty$.

The following two theorems establish sufficient conditions for stability and asymptotic stability of the equilibrium state [5, Theorem A.5].

Theorem B.1 (Stability) *Let* V: $\mathbb{R}^n \times \mathbb{R}_+ \to \mathbb{R}_+$ *be a continuously differentiable function such that for all* $x \in \mathcal{X} \subset \mathbb{R}^n$, $t \geq 0$

$$\dot{V}(x,t) = \frac{\partial V(x,t)}{\partial t} + \frac{\partial V(x,t)}{\partial x} f(x,t) \leq 0,$$

where $\mathcal{X} = \{x \in \mathbb{R}^n\colon |x| < c\}$, $c > 0$. *Then the equilibrium state* $x^* = 0$ *of the system* (B.1) *is:*

T.B.1.1. *stable if* $w_1(|x|) \leq V(x,t)$ *and* $w_1 \in \mathcal{K}$ *for all* $x \in \mathcal{X}$, $t \geq 0$;

T.B.1.2. *uniformly stable if* $w_1(|x|) \leq V(x,t) \leq w_2(|x|)$ *and* $w_1, w_2 \in \mathcal{K}$ *for all* $x \in \mathcal{X}$, $t \geq 0$;

T.B.1.3. *globally uniformly stable if* $w_1(|x|) \leq V(x,t) \leq w_2(|x|)$ *and* $w_1, w_2 \in \mathcal{K}_\infty$ *for all* $x \in \mathcal{X} = \mathbb{R}^n$, $t \geq 0$.

Theorem B.2 (Asymptotic stability) *Let* V: $\mathbb{R}^n \times \mathbb{R}_+ \to \mathbb{R}_+$ *be a continuously differentiable function such that for all* $x \in \mathcal{X} \subset \mathbb{R}^n$, $t \geq 0$

$$\dot{V}(x,t) = \frac{\partial V(x,t)}{\partial t} + \frac{\partial V(x,t)}{\partial x} f(x,t) \leq -w_3(|x|),$$

where $\mathcal{X} = \{x \in \mathbb{R}^n\colon |x| < c\}$, $c > 0$. *Then the equilibrium state* $x^* = 0$ *of the system* (B.1) *is:*

T.B.2.1. *uniformly asymptotically stable if* $w_1(|x|) \leq V(x,t) \leq w_2(|x|)$ *and* $w_1, w_2, w_3 \in \mathcal{K}$ *for all* $x \in \mathcal{X}$, $t \geq 0$;

T.B.2.2. *globally uniformly asymptotically stable if* $w_1(|x|) \leq V(x,t) \leq w_2(|x|)$ *and* $w_1, w_2 \in \mathcal{K}_\infty$, $w_3 \in \mathcal{K}$ *for all* $x \in \mathcal{X} = \mathbb{R}^n$, $t \geq 0$.

If in Theorem B.2 we set $w_i(|x|) = c_i|x|^2, c_i > 0, i = 1, 2, 3$, then we obtain sufficient conditions of exponential and global exponential stability. The next theorem (known as the Krasovskii theorem) establishes necessary and sufficient conditions for global exponential stability for systems with continuously differentiable function $f(x, t)$ [1, Theorem 2.7].

Theorem B.3 *(Exponential stability) The equilibrium state $x^* = 0$ of the system* (B.1) *with continuously differentiable right-hand side is globally exponentially stable iff there exists a continuously differentiable function $V(x, t)$: $\mathbb{R}^n \times \mathbb{R}_+ \to \mathbb{R}_+$ such that*

$$c_1|x|^2 \leq V(x, t) \leq c_2|x|^2, \qquad \left|\frac{\partial V(x, t)}{\partial x}\right| \leq c_3|x|^2,$$

$$\dot{V}(x, t) = \frac{\partial V(x, t)}{\partial t} + \frac{\partial V(x, t)}{\partial x} f(x, t) \leq -c_4|x|^2,$$

where $c_i > 0$, $i = 1, 2, 3, 4$.

In general, the methods of adaptive control ensure asymptotic stability with respect to a part of the state variables of the closed-loop system represented by the plant state, output or tracking error. Usually, the origin of the remaining part of the states represented by the parametric errors is stable only. Therefore, the case of *partial asymptotic stability* (or *partial asymptotic convergence*) plays important role in the theory of adaptive control. For the partial convergence analysis, the following theorem (known as the LaSalle–Yoshizawa theorem) can be efficiently used [9, 13].

Theorem B.4 *(Partial asymptotic convergence) Assume that $x^* = 0$ is the equilibrium state of* (B.1) *and $f(x, t)$ is locally Lipschitz in x uniformly in t. If there exists a continuously differentiable function $V(x)$: $\mathbb{R}^n \to \mathbb{R}_+$, $V(x) \geq w(|x|)$ such that*

$$\dot{V} = \frac{\partial V(x)}{\partial x} f(x, t) \leq -\bar{w}(x) \leq 0, \quad \text{for all } t \geq 0,\ x \in \mathbb{R}^n,$$

where $w(|x|) \in \mathcal{K}_\infty$ and $\bar{w}(x)$:$\mathbb{R}^n \to \mathbb{R}_+$ is a continuous function, then all solutions of the system (B.1) *are globally uniformly bounded and satisfy the equality*

$$\lim_{t\to\infty} \bar{w}(x(t)) = 0.$$

B.3 Robustness of Stability Properties

The notion of robustness plays important role in analysis and design of real-world control systems because it answers the question: is a desirable property of a system preserved for a certain class of perturbations? Let us consider the perturbed system

$$\dot{x} = f(x, t) + \Delta(x, t), \tag{B.2}$$

where the term $\Delta(x, t)$ represents the *perturbations* of the *nominal* system (B.1).[5] Under *perturbation* we mean uncertainties of the system mathematical model that are not taken into account in design of the closed-loop control.

Definition B.5 If a property of the nominal system (B.1) is preserved for sufficiently small (in some sense) perturbations of its mathematical model $\Delta(x, t)$ (i.e., it is preserved for the perturbed system (B.2) with small $\Delta(x, t)$), then this property is called *robust with respect to this class of perturbations*. If the property can be destroyed by arbitrarily small (in some sense) perturbation $\Delta(x, t)$, then it is called *nonrobust with respect to this class of perturbations*.

Now, we briefly review the robustness of the stability properties with respect to different classes of perturbations.

The perturbation $\Delta(x, t)$ is called *multiplicative* if it satisfies the equality

$$|\Delta(x, t)| \leq c_5|x|, \qquad c_5 > 0. \tag{B.3}$$

Definition B.6 A stability property of the equilibrium state $x^* = 0$ of the system (B.1) is called robust with respect to multiplicative perturbations if there exists a number $c_5 > 0$ such that this property also holds for the perturbed system (B.2) with any $\Delta(x, t)$ satisfying the inequality (B.3).

As it is known, stability, asymptotic stability and even uniform asymptotic stability are not robust with respect to multiplicative perturbations.

Example B.1 Consider the scalar system

$$\dot{x} = -\frac{1}{1+t}x \tag{B.4}$$

the equilibrium state $x^* = 0$ of which is asymptotically (but not *uniformly* asymptotically) stable. It is obvious that the origin of the perturbed system

$$\dot{x} = -\frac{1}{1+t}x + c_5 x$$

is unstable for any arbitrarily small $c_5 > 0$ since $\left(c_5 - \frac{1}{1+t}\right) > 0$ for all $t > \left(\frac{1}{c_5} - 1\right)$. ■

Example B.2 Consider the scalar system $\dot{x} = -x^3$ the equilibrium state $x^* = 0$ of which is uniformly asymptotically stable. However, the origin of the perturbed system $\dot{x} = -x^3 + c_5 x$ is unstable for any arbitrarily small $c_5 > 0$. ■

---

[5] We assume that $\Delta(x, t)$ is such that the system (B.2) has a unique solution for any initial conditions x_0 and t_0.

For the Lyapunov function $V(x,t)$ from Theorem B.3, we have

$$\dot{V} \leq -c_4|x|^2 + \left|\frac{\partial V}{\partial x}\right| |\Delta| \leq -(c_4 - c_3 c_5)|x|^2.$$

If $c_5 \leq \frac{c_4}{c_3}$, then $(c_4 - c_3 c_5) > 0$ and the equilibrium state $x^* = 0$ of the perturbed system (B.2) is exponentially stable. Therefore, the property of exponential stability is robust with respect to multiplicative perturbations.

Now, we consider another class of perturbations that may not vanish for $|x(t)| \to 0$, but satisfies the boundedness condition

$$|\Delta(x,t)| \leq c_6 \tag{B.5}$$

for all $x \in \mathbb{R}^n$, $t \geq 0$ and some $c_6 > 0$. We will call such perturbations *additive*. Usually, additive perturbations represent external disturbances affecting controlled plants.

Definition B.7 A stability property of the equilibrium state $x^* = 0$ of the system (B.1) is called robust with respect to additive perturbations if there exists c_6, $\delta(c_6, t_0) > 0$ and $\epsilon(c_6, t_0)$, $T(c_6, t_0) \geq 0$ such that for the system (B.2) with the perturbation $\Delta(x,t)$ satisfying (B.5) the inequality $|x_0| \leq \delta$ implies that $x(t, x_0, t_0) \in \mathcal{L}_\infty$ and $|x(t, x_0, t_0)| \leq \epsilon$ for all $t \geq t_0 + T$. If $\delta = \mathbb{R}^n$, this property is robust in global.

The properties of stability, uniform stability, and asymptotic stability are not robust with respect to the additive perturbations.

Example B.3 Consider the system (B.4) and the constant perturbation $\Delta(x,t) \equiv c_6$. Then, for the perturbed system

$$\dot{x} = -\frac{1}{1+t}x + c_6,$$

we have

$$x(t, x_0, 0) = \frac{1}{t+1}x_0 + \frac{c_6}{2}\left(t + 1 - \frac{1}{t+1}\right).$$

Hence, $x(t, x_0, 0) \to \infty$ as $t \to \infty$ for any arbitrarily small $c_6 > 0$. ■

As it is known, the properties of uniform asymptotic stability and exponential stability are robust with respect to the additive perturbations. However, the property of global exponential stability enables us to derive simple trajectory bounds for the transients in perturbed systems. These bounds are used in Sect. 3.5 devoted to robust adaptation algorithms and can be defined via the following lemma.

Lemma B.1 *If* $V\colon \mathbb{R}_+ \to \mathbb{R}_+$ *is such that*

$$\dot{V}(t) \le -\alpha V(t) + c, \tag{B.6}$$

where $\alpha, c > 0$ *are constants, then*

$$V(t) \le V(0)\mathrm{e}^{-\alpha t} + \frac{c}{\alpha}. \tag{B.7}$$

If $c_1|x(t)|^2 \le V(t) \le c_2|x(t)|^2$, *where* $x\colon\mathbb{R}_+ \to \mathbb{R}^n$ *and* $c_1, c_2 > 0$ *are constants, then for* $|x(t)|$ *we have*

$$|x(t)| \le \sqrt{\frac{c_2}{c_1}}|x(0)|\mathrm{e}^{-\frac{\alpha}{2}t} + \sqrt{\frac{c}{c_1\alpha}}. \tag{B.8}$$

Proof Solving (B.6), we have

$$V(t) \le V(0)\mathrm{e}^{-\alpha t} + c\int_0^t \mathrm{e}^{-\alpha(t-\tau)}d\tau \le V(0)\mathrm{e}^{-\alpha t} + \frac{c}{\alpha}(1-\mathrm{e}^{-\alpha t}) \le V(0)\mathrm{e}^{-\alpha t} + \frac{c}{\alpha}.$$

This completes the proof of (B.7).

Further, we obtain

$$|x(t)|^2 \le \frac{1}{c_1}V(t) \le \frac{c_2}{c_1}|x(0)|^2\mathrm{e}^{-\alpha t} + \frac{c}{c_1\alpha}$$

and

$$|x(t)| \le \sqrt{\frac{c_2}{c_1}|x(0)|^2\mathrm{e}^{-\alpha t} + \frac{c}{c_1\alpha}} \le \sqrt{\frac{c_2}{c_1}}|x(0)|\mathrm{e}^{-\frac{\alpha}{2}t} + \sqrt{\frac{c}{c_1\alpha}}.$$

This completes the proof. □

Calculating the time derivative of the Lyapunov function $V(x, t)$ from Theorem B.3 along the solutions of (B.2) with (B.5), we have

$$\dot{V} \le -c_4|x|^2 + \left|\frac{\partial V}{\partial x}\right||\Delta| \le -c_4|x|^2 + c_3c_6|x|.$$

Completing the squares and strengthening the inequality, we can write

$$\dot{V} \le -\frac{c_4}{2}|x|^2 - \left(\sqrt{\frac{c_4}{2}}|x| - \frac{c_3c_6}{\sqrt{2c_4}}\right)^2 + \frac{c_3^2c_6^2}{2c_4} \le -\frac{c_4}{2c_2}V + \frac{c_3^2c_6^2}{2c_4}. \tag{B.9}$$

Applying the result of Lemma B.1 (with substitutions $\alpha = \frac{c_4}{2c_2}$ and $c = \frac{c_3^2c_6^2}{2c_4}$), we have

$$|x(t, x(0), 0)| \leq \sqrt{\frac{c_2}{c_1}} \left(|x(0)| e^{-\frac{c_4}{4c_2} t} + \frac{c_3 c_6}{c_4} \right).$$

The latter means that $x(t, x_0, t_0)$ is bounded and exponentially converges to the residual set

$$D_x = \left\{ x\colon |x| \leq \sqrt{\frac{c_2}{c_1} \frac{c_3 c_6}{c_4}} \right\}.$$

Therefore, the property of global exponential stability is globally robust with respect to additive disturbances in the sense of Definition B.7.

B.4 Input-To-State Stability

For some design methods, it is important to prove that bounded input (e.g., external disturbance affecting the plant) causes bounded state response. In some cases, more strengthened behavior is needed: the bound on the state response must be proportional to the bound on the input. This property is known as ISS-property, and the controllers providing this property are called *ISS-controllers* [11]. In this book, we use the ISS-controllers for modular adaptive backstepping (see Appendix E).

In order to formulate the ISS-property, we introduce the following instrumental definition [5, Definition A.3].

Definition B.8 A continuous function $w(|x|, t)\colon [0, c] \times [0, \infty) \to \mathbb{R}_+$ is said to belong to a class $\mathcal{KL}$ (i.e., $w \in \mathcal{KL}$) if, for each fixed t, the function $w(|x|, t)$ belongs to the class $\mathcal{K}$ with respect to $|x|$ and, for each fixed $|x|$, the function $w(|x|, t)$ is decreasing, i.e., $w(|x|, t) \to 0$ as $t \to \infty$.

Now, consider the nonlinear system

$$\dot{x} = f(x, \upsilon, t), \tag{B.10}$$

where f is locally Lipschitz in x and υ uniformly in t.

Definition B.9 ([5], *Definition C.1*) The system (B.10) is ISS with respect to input υ if there exist a function $\alpha \in \mathcal{KL}$ and a function $\beta \in \mathcal{K}$ such that for all $t_0 \geq 0$, any initial condition $x(t_0)$ and any bounded piecewise continuous input υ the solution of (B.10) exists and satisfies the inequality

$$|x(t)| \leq \alpha\left(|x(t_0)|, t - t_0\right) + \beta\left(\sup_{\tau \in [t_0, t]} |\upsilon(\tau)| \right).$$

For the system (B.10), the ISS means that:

1. if $\upsilon \in \mathcal{L}_\infty$, then $x \in \mathcal{L}_\infty$;
2. the norm $|x(t)|$ is ultimately bounded by the function $\beta\left(\sup_{\tau \in [t_0, t]} |\upsilon(\tau)| \right)$;

3. if $\upsilon(t) \to 0$ as $t \to \infty$, then $x(t) \to 0$ as $t \to \infty$;
4. the origin of the system $\dot{x} = f(x, 0, t)$ is globally asymptotically stable.

Now, we formulate a sufficient condition for ISS [5,Theorem C.2].

Theorem B.5 *If there exists a continuously differentiable function $V(x,t)$: $\mathbb{R}^n \times \mathbb{R}_+ \to \mathbb{R}_+$*[6] *such that for all $x \in \mathbb{R}^n$ and $\upsilon \in \mathbb{R}^m$*

$$w_1(|x|) \le V(x,t) \le w_2(|x|),$$
$$\frac{\partial V}{\partial t} + \frac{\partial V}{\partial x} f(x,u) \le -w_3(|x|), \quad \text{whenever} \quad |x| \ge w_4(|\upsilon|),$$

where $w_1,\ w_2,\ w_4 \in \mathcal{K}_\infty$, and $w_3 \in \mathcal{K}$, then the system (B.10) is ISS.

Example B.4 Consider the plant

$$\dot{x} = -x + (x^2 + 1)\upsilon. \tag{B.11}$$

It is seen that the equilibrium $x^* = 0$ of the autonomous system $\dot{x} = -x$ is globally uniformly asymptotically stable. However, if $\upsilon(t) \not\equiv 0$ but bounded, then $x(t)$ may grow to infinity. Let $\upsilon(t) \equiv 1$ and $x_0 = 0.5$. Then the solution of (B.11) is given by

$$x(t) = \frac{1}{2} + \frac{\sqrt{3}}{2} \tan\left(\frac{\sqrt{3}}{2} t\right).$$

It is seen that, despite the bounded input of the system (B.11), its trajectory grows unbounded at the instance $t_1 = \frac{\pi}{\sqrt{3}}$ (*finite escape time* phenomenon). ∎

Example B.5 Consider the plant

$$\dot{x} = -x^3 + x\upsilon. \tag{B.12}$$

The autonomous system $\dot{x} = -x^3$ is uniformly asymptotically stable. To analyze the ISS-property of this system, we select the function $V = \frac{1}{2}x^2$ and evaluate its time derivative in view of (B.12):

$$\dot{V} = -x^4 + x^2\upsilon = -(1-\mu)x^4 - \mu x^4 + x^2\upsilon,$$

where $0 < \mu < 1$ is a constant. Then we can get the ISS condition of Theorem B.5

$$\dot{V} \le -(1-\mu)x^4 \quad \text{if} \quad |x| \ge \sqrt{\frac{\bar{\upsilon}}{\mu}},$$

[6] Sometimes in literature, $V(x,t)$ is called *ISS-Lyapunov* function.

where $\bar{\upsilon} = \sup_{\tau \in [t_0, t]} |\upsilon(\tau)|$. Therefore, the system (B.12) is ISS with respect to the input υ. ∎

References

1. Fradkov, A.L., Miroshnik, I.V., Nikiforov, V.O.: Nonlinear and Adaptive Control of Complex Systems. Kluwer Academic Publishers (1999)
2. Hahn, W.: Stability of Motion. Springer, NY (1967)
3. Ioannou, P., Sun, J.: Robust Adaptive Control. Prentice-Hall, Enlewood Cliffs, NJ (1996)
4. Khalil, H.K.: Nonlinear Control. Prentice-Hall, NJ (2015)
5. Krstić, M., Kanellakopoulos, I., Kokotović, P.: Nonlinear and Adaptive Control Design. Wiley, Inc., NY (1995)
6. Lakshmikantham, V., Matrosov, V.M., Sivasundaram, S.: Vector Lyapunov Functions and Stability Analysis of Nonlinear Systems. Springer Netherlands (1991)
7. LaSalle, J., Lefshetz, S.: Stability by Direct Lyapunov Method. Academic, NY (1961)
8. LaSalle, J.: Stability theory for ordinary differential equations. J. Differ. Equ. **4**, 57–65 (1968)
9. Narendra, K.S., Annaswamy, A.M.: Stable Adaptive Systems. Prentice Hall (1989)
10. Rouche, N.P., Habets, P., Laloy, M.: Stability Theory by Lyapunov's Direct Method. Springer, NY (1977)
11. Sontag, E.D.: Smooth stabilization implies coprime factorization. IEEE Trans. Autom. Control **34**(4), 435–443 (1989)
12. Yoshizawa, T.: Stabilty Theory by Lyapunov's Second Method. The Mathematical Society of Japan (1966)
13. Zubov, V.I.: Lyapunov's Methods and Their Applications. Pergamon Press, Oxford (1964)

Appendix C
Norms and Properties of Functions

In this book, we refer to the following classes of real-valued functions.

Definition C.1 (*Continuity*) A scalar function $f(t)$: $\mathbb{R}_+ \to \mathbb{R}$ is called *continuous* if for any given c_f there exists a $c_0(c_f, t_0)$ such that for all $t, t_0 \geq 0$ for which $|t - t_0| \leq c_0$ we have $|f(t) - f(t_0)| \leq c_f$.

Definition C.2 (*Uniform Continuity*) A scalar function $f(t)$: $\mathbb{R}_+ \to \mathbb{R}$ is called *uniformly continuous* if for any given c_f there exists a $c_0(c_f)$ such that for all t, $t_0 \geq 0$ for which $|t - t_0| \leq c_0$ we have $|f(t) - f(t_0)| \leq c_f$.

Definition C.3 (*Piecewise Continuity*) A scalar function $f(t)$: $\mathbb{R}_+ \to \mathbb{R}$ is called *piecewise continuous* (i.e., $f \in \mathcal{PC}$) if f is continuous on any finite interval $[t_0, t_1] \subset [0, \infty)$ except for a finite number of points.

Definition C.4 (*Smoothness*) A scalar function $f(t)$: $\mathbb{R}_+ \to \mathbb{R}$ is called *smooth* if its partial derivatives on t of any order are defined and continuous.

Definition C.5 (*Boundedness*) A scalar function $f(t)$: $\mathbb{R}_+ \to \mathbb{R}$ is called *bounded on a set* $\mathcal{T}$ if there exists some positive $c(\mathcal{T}) < \infty$ such that $|f(t)| \leq c$ for all $t \in \mathcal{T}$. If $\mathcal{T} = [0, \infty)$, then the function $f(t)$ is *globally bounded* (or, simply, *bounded*).

A vector function $f(t) = [f_1(t), f_2(t), \ldots, f_n(t)]^\top$ is called continuous (uniformly continuous, piecewise continuous, smooth, bounded on a set or globally bounded) if all the scalar functions $f_i(t)$ are continuous (uniformly continuous, piecewise continuous, smooth, bounded on a set or globally bounded), respectively.

Definition C.6 (*Lipshitz property*) A function $f(x, t)$: $\mathbb{R}^n \times \mathbb{R}_+ \to \mathbb{R}^m$ is called *locally Lipschitz in x uniformly in t* (i.e., $f \in \mathcal{L}$) if for any compact set $\mathcal{X} \in \mathbb{R}^n$ there exists a constant $L = L(\mathcal{X})$ such that for all x, $x^* \in \mathcal{X}$ the following inequality holds:

$$|f(x, t) - f(x^*, t)| < L|x - x^*|.$$

V. Nikiforov and D. Gerasimov, *Adaptive Regulation*, Lecture Notes in Control and Information Sciences 491, https://doi.org/10.1007/978-3-030-96091-9

Definition C.7 (*Persistent excitation*) A bounded piecewise continuous vector function $f(t)$: $\mathbb{R}_+ \to \mathbb{R}^n$ is said to be *persistently exciting* (i.e., $f \in \mathcal{PE}$) if there exist constants $T > 0$, $\alpha > 0$ and $t_0 > 0$ such that

$$\int_t^{t+T} f(\tau)f^\top(\tau)d\tau \succ \alpha I \text{ for all } t \geq t_0.$$

Definition C.8 (*The same order in the mean square sense*) [2, Definition 3.3.1] A vector function $f(t)$: $\mathbb{R}_+ \to \mathbb{R}^n$ is said to be *of the order of c in the mean square sense* (i.e., $f \in \mathcal{MS}(c)$) if for a given constant $c \geq 0$ there exist some finite constants $c_0,\ c_1 \geq 0$, where c_0 is independent of c, such that

$$\int_t^{t+T} f^\top(\tau)f(\tau)d\tau \leq c_0 c^2 T + c_1, \quad \forall t, T \geq 0.$$

For a vector function $f(t)$: $\mathbb{R}_+ \in \mathbb{R}^n$, the $\mathcal{L}_p$ norm, where $1 \leq p < \infty$, is defined as

$$\| f(t) \|_p = \left(\int_0^\infty |f(t)|^p dt \right)^{\frac{1}{p}},$$

while the $\mathcal{L}_\infty$ norm is defined as

$$\| f(t) \|_\infty = \sup |f(t)|.$$

Definition C.9 ($\mathcal{L}_p$ *norms*) A vector function $f(t)$: $\mathbb{R}_+ \in \mathbb{R}^n$ is said to belong to the function space $\mathcal{L}_p$ (i.e., $f \in \mathcal{L}_p$) if its $\mathcal{L}_p$ norm exists and is finite.

The notations $f \in \mathcal{L}_\infty$ and $f \in \mathcal{L}_2$ define globally bounded and square-integrable functions, respectively.

The following properties are useful in stability analysis of the closed-loop control systems [1]:

1. $f \in \mathcal{L}_1 \cap \mathcal{L}_\infty \;\Rightarrow\; f \in \mathcal{L}_p, \quad 1 \leq p < \infty$;
2. $\dot{f} \in \mathcal{L}_\infty$, $f \in \mathcal{L}_p$, $1 \leq p < \infty$, $\Rightarrow |f(t)| \to 0$ as $t \to \infty$;
3. $\dot{f} \in \mathcal{L}_\infty$, $f \in \mathcal{L}_2 \Rightarrow f \in \mathcal{L}_\infty$ and $|f(t)| \to 0$ as $t \to \infty$;
4. $\dot{f} \in \mathcal{L}_1 \Rightarrow$ there exists $\lim_{t\to\infty} f(t) < \infty$.

Lemma C.1 (*Barbalat's Lemma*) [3] *If a function* $f(t)$: $\mathbb{R}_+ \to \mathbb{R}$ *is uniformly continuous and the limit of the integral* $\lim_{t\to\infty} \int_0^t f(\tau)d\tau$ *exists and is finite, then* $\lim_{t\to\infty} f(t) = 0$.

References

1. Desoer, C.A., Vidyasagar, M.: Feedback Systems: Input-Output Properties. Academic, New York (1975)
2. Ioannou, P., Sun, J.: Robust Adaptive Control. Prentice-Hall, Englewood Cliffs, NJ (1996)
3. Popov, V.M.: Hyperstability of Control Systems. Springer, New York (1973)

Appendix D
Schemes of Swapping

The concept of *augmented error* and *swapping* technique, initially proposed by Monopoli in [7] to overcome the structural obstacle caused by high-order relative degree, now is widely used in adaptive control with applications to different control problems including MRAC of SISO plants [3, 5, 8, 9, 10], adaptive control of MIMO systems [11], adaptive modular backstepping in nonlinear systems [6, 12], and adaptive control with input delay [1, 2, 11]. In addition to the main lemmas introduced in Sect. 3.4.2, we now present different swapping schemes used in this book. For the sake of completeness and convenience of presentation, we repeat (without proofs) Lemmas 3.1 and 3.2.

D.1 Swapping in LTI SISO Systems

D.1.1 Stable Error Model

Consider the state error model

$$\dot{e} = Ae + b\tilde{\theta}^\top \omega, \qquad \varepsilon = c^\top e, \tag{D.1}$$

where $e \in \mathbb{R}^n$ is the unmeasured state error with initial condition $e(0)$, $\varepsilon \in \mathbb{R}$ is the output error, $\tilde{\theta} = \theta - \hat{\theta}$ is the vector of parametric errors, $\theta \in \mathbb{R}^q$ is a vector of unknown constant parameters, $\hat{\theta} \in \mathbb{R}^q$ is the vector of adjustable parameters generated by an adaptation algorithm, $\omega = \omega(t)$: $\mathbb{R}_+ \to \mathbb{R}^q$ is the measured regressor, A, b, and c are known and have appropriate dimensions, and A is Hurwitz. The triplet (A, b, c) is assumed to be controllable and observable and results in the asymptotically stable transfer function $W(s) = c^\top (sI - A)^{-1} b$. In this case, the model (D.1) can be also represented in the input–output form

$$\varepsilon = W(s)\left[\tilde{\theta}^\top \omega\right] + \epsilon, \tag{D.2}$$

V. Nikiforov and D. Gerasimov, *Adaptive Regulation*, Lecture Notes in Control and Information Sciences 491, https://doi.org/10.1007/978-3-030-96091-9

where $\epsilon(t) = c^\top e^{At} e(0)$ exponentially decays.

Lemma D.1 (*Transfer function swapping*)[7] *[7, 9]. Along with (D.1), we define the* swapping term

$$\zeta = W(s)\left[\hat{\theta}^\top \omega\right] - \hat{\theta}^\top \omega_w, \tag{D.3}$$

where

$$\omega_w = W(s)[\omega] \tag{D.4}$$

is the filtered regressor, and introduce the augmented error

$$\bar{\varepsilon} = \varepsilon + \zeta. \tag{D.5}$$

Then the following equality holds:

$$\bar{\varepsilon} = \tilde{\theta}^\top \omega_w + \epsilon. \tag{D.6}$$

Lemma D.2 (*State-space swapping*)[8] *[4, 8]. Along with (D.1), we consider the* swapping filters

$$\dot{\Omega}_w = A\Omega_w + b\omega^\top, \qquad \omega_w^\top = c^\top \Omega_w,$$
$$\dot{\zeta}_w = A\zeta_w - \Omega_w \dot{\hat{\theta}}, \qquad \zeta = c^\top \zeta_w,$$

with arbitrary initial conditions $\Omega_w(0)$ and $\zeta_w(0)$. Define the augmented error *as*

$$\bar{\varepsilon} = \varepsilon + \zeta.$$

Then the following equality holds:

$$\bar{\varepsilon} = \tilde{\theta}^\top \omega_w + \bar{\epsilon},$$

where $\bar{\epsilon} = c^\top \epsilon_w$ exponentially decays, and ϵ_w is the exponentially decaying term satisfying the equation $\dot{\epsilon}_w = A\epsilon_w$ with $\epsilon_w(0) = e(0) + \zeta_w(0) - \Omega_w(0)\hat{\theta}(0)$.

Corollary D.1[9] *The difference between the* output error ε *and the* augmented error *$\bar{\varepsilon}$ is given by*

$$\zeta = \bar{\varepsilon} - \varepsilon = -W_\zeta(s)\left[W_\Omega(s)\left[\omega^\top\right]\dot{\hat{\theta}}\right]$$

or

---

[7] Lemma 3.1.

[8] Lemma 3.2.

[9] Corollary 3.4 on p. 128.

$$\tilde{\theta}^\top W(s)\,[\omega] - W(s)\left[\tilde{\theta}^\top \omega\right] = -W_\zeta(s)\left[W_\Omega(s)\,[\omega^\top]\,\dot{\hat{\theta}}\right] + \bar{\epsilon}_w,$$

where $W_\zeta(s) = c^\top (sI - A)^{-1}$, $W_\Omega(s) = (sI - A)^{-1} b$,

$$\zeta = -W_\zeta(s)\left[W_\Omega(s)\,[\omega^\top]\,\dot{\hat{\theta}}\right]$$

is the swapping term, $\bar{\epsilon}_w = \epsilon - \bar{\epsilon}$.

D.1.2 Unstable Error Model

Consider the state error model

$$\dot{e} = Ae + b\left(\tilde{\theta}^\top \omega + u_s\right), \qquad \varepsilon = c^\top e, \tag{D.7}$$

where A is possibly unstable, and u_s is a stabilizing component of control design of which is postponed for some reasons. All other notations are defined in the previous section as for the model (D.1).

Lemma D.3 (*SISO observer-based swapping*) *Along with* (D.7), *we consider the filter*

$$\dot{\hat{e}} = A\hat{e} - l_e(\varepsilon + c^\top \hat{e}) + b(\hat{\theta}^\top \omega - u_s), \tag{D.8}$$

where the vector of feedback gains l_e *is selected so that the matrix* $A_e = A - l_e c^\top$ *is Hurwitz. Define the* augmented error

$$\bar{\varepsilon} = \varepsilon + c^\top \hat{e} - \hat{\theta}^\top \omega_e, \tag{D.9}$$

where $\omega_e = W_e(s)[\omega]$ *is the filtered regressor, and* $W_e(s) = c^\top (sI - A_e)^{-1} b$ *is the asymptotically stable transfer function. Then the following equality holds:*

$$\bar{\varepsilon} = \tilde{\theta}^\top \omega_e + \epsilon_e, \tag{D.10}$$

where $\epsilon_e(t) = c^\top e^{A_e t}(e(0) + \hat{e}(0))$ *exponentially decays.*

Proof Define auxiliary vector $\tilde{e} = e + \hat{e}$. Then, evaluating the time derivative of $\tilde{e}$, in view of (D.7) and (D.8), we obtain

$$\dot{\tilde{e}} = A_e \tilde{e} + b\theta^\top \omega.$$

Therefore,

$$
\begin{aligned}
\bar{\varepsilon} &= \varepsilon + c^\top(\tilde{e} - e) - \hat{\theta}^\top \omega_e = c^\top \tilde{e} - \hat{\theta}^\top \omega_e \\
&= W_e\left[\theta^\top \omega\right] + c^\top \mathrm{e}^{A_e} \tilde{e}(0) - \hat{\theta}^\top \omega_e^\top = \tilde{\theta}^\top \omega_e + \epsilon_e.
\end{aligned}
$$

□

D.2 Swapping in LTI MIMO Systems

D.2.1 Stable Error model

Consider the state error model

$$\dot{e} = Ae + B\tilde{\Theta}\omega, \qquad \varepsilon = Ce, \tag{D.11}$$

where $e \in \mathbb{R}^n$ is the unmeasured state error with the initial condition $e(0)$, $\varepsilon \in \mathbb{R}^m$ is the output error vector, $\tilde{\Theta} = \Theta - \hat{\Theta}$ is the matrix of parametric errors, $\Theta \in \mathbb{R}^{m \times q}$ is a matrix of unknown constant Parameters, $\hat{\Theta} \in \mathbb{R}^{m \times q}$ is the matrix of adjustable parameters generated by an adaptation algorithm, $\omega = \omega(t)$: $\mathbb{R}_+ \to \mathbb{R}^q$ is the measured regressor, A, B, and C are known and have appropriate dimensions, and A is Hurwitz. The triplet (A, B, C) is assumed to be controllable and observable and results in the asymptotically stable transfer function matrix $\bar{W}(s) = C(sI - A)^{-1}B$. In this case, the model (D.11) can be also represented in the input–output form

$$\varepsilon = \bar{W}(s)\left[\tilde{\Theta}\omega\right] + \epsilon, \tag{D.12}$$

where $\epsilon(t) = C\mathrm{e}^{At}e(0)$ exponentially decays.

Before proceeding, we rewrite the state error model in a form which will be more suitable for derivation of swapping schemes and adaptation algorithms. Denote the ith row of $\tilde{\Theta}$ by $\tilde{\Theta}_i$ and consider

$$\tilde{\Theta}\omega = \begin{bmatrix} \tilde{\Theta}_1\omega \\ \tilde{\Theta}_2\omega \\ \vdots \\ \tilde{\Theta}_m\omega \end{bmatrix} = \begin{bmatrix} \omega^\top & 0 & \cdots & 0 \\ 0 & \omega^\top & \cdots & 0 \\ \vdots & \vdots & \ddots & \vdots \\ 0 & 0 & \cdots & \omega^\top \end{bmatrix} \begin{bmatrix} \tilde{\Theta}_1^\top \\ \tilde{\Theta}_2^\top \\ \vdots \\ \tilde{\Theta}_m^\top \end{bmatrix} = \left(I_m \otimes \omega^\top\right) \begin{bmatrix} \tilde{\Theta}_1^\top \\ \tilde{\Theta}_2^\top \\ \vdots \\ \tilde{\Theta}_m^\top \end{bmatrix}.$$

Then, denoting the vector of the parametric errors by

$$\tilde{\vartheta} = [\tilde{\Theta}_1, \tilde{\Theta}_2, \ldots, \tilde{\Theta}_m]^\top \in \mathbb{R}^{mq} \tag{D.13}$$

and the matrix regressor by

$$\Omega^\top = I_m \otimes \omega^\top \in \mathbb{R}^{m \times mq}, \tag{D.14}$$

we rewrite the model (D.11) in the form

$$\dot{e} = Ae + B\bar{\Omega}^\top \tilde{\vartheta}, \qquad \varepsilon = Ce, \tag{D.15}$$

where

$$\tilde{\vartheta} = \vartheta - \hat{\vartheta}, \;\; \vartheta = [\Theta_1, \Theta_2, \ldots, \Theta_m]^\top, \;\; \hat{\vartheta} = [\hat{\Theta}_1, \hat{\Theta}_2, \ldots, \hat{\Theta}_m]^\top, \tag{D.16}$$

and $\hat{\Theta}_i$ is the ith row of the matrix $\hat{\Theta}$.

Lemma D.4 (*Transfer function matrix swapping*) *Along with* (D.11), *we define the* swapping term

$$\bar{\zeta} = \bar{W}(s)\left[\hat{\Theta}\omega\right] - \bar{\Omega}_w^\top \hat{\vartheta} = \bar{W}(s)\left[\bar{\Omega}^\top \hat{\vartheta}\right] - \bar{\Omega}_w^\top \hat{\vartheta},$$

where

$$\bar{\Omega}_w^\top = \bar{W}(s)\left[\bar{\Omega}^\top\right] = \begin{bmatrix} \bar{W}_{11}(s)\left[\omega^\top\right] & \ldots & \bar{W}_{1m}[\omega^\top] \\ \vdots & \ddots & \vdots \\ \bar{W}_{m1}(s)\left[\omega^\top\right] & \ldots & \bar{W}_{mm}\left[\omega^\top\right] \end{bmatrix},$$

$\bar{W}_{ij}$ *are the entries of* $\bar{W}(s)$, $\bar{\Omega}$, *and* $\hat{\vartheta}$ *are defined by* (D.14) *and* (D.16), *respectively, and introduce the* augmented error

$$\bar{\varepsilon} = \varepsilon + \bar{\zeta}. \tag{D.17}$$

Then the following equality holds:

$$\bar{\varepsilon} = \bar{\Omega}_w^\top \vartheta + \epsilon, \tag{D.18}$$

where ϵ *exponentially decays.*

Proof The lemma is proved with the use of the same arguments as for the case of scalar output error (see the proof of Lemma 3.1 on p. 127). □

Lemma D.5 (*MIMO state-space swapping*) *Along with* (D.11), *we consider the* swapping filters

$$\dot{\Omega}_A^\top = A\Omega_A^\top + B\bar{\Omega}^\top, \quad \bar{\Omega}_w^\top = C\Omega_A^\top,$$
$$\dot{\zeta}_A = A\zeta_A - \Omega_A^\top \dot{\hat{\vartheta}}, \qquad \bar{\zeta} = C\zeta_A$$

with arbitrary initial conditions $\Omega_A(0)$ *and* $\zeta_A(0)$, *where* $\bar{\Omega}$ *and* $\hat{\vartheta}$ *are defined by* (D.14) *and* (D.16). *Define the* augmented error *as*

$$\bar{\varepsilon} = \varepsilon + \bar{\zeta}.$$

Then the following equality holds:

$$\bar{\varepsilon} = \bar{\Omega}_w^\top \tilde{\vartheta} + \bar{\epsilon},$$

where $\bar{\epsilon} = C\epsilon_w$ *exponentially decays, and* ϵ_w *satisfies the equation* $\dot{\epsilon}_w = A\epsilon_w$ *with* $\epsilon_w(0) = e(0) + \zeta_A(0) - \Omega_A^\top(0)\hat{\vartheta}(0)$.

Proof The lemma is proved with the use of the same arguments as for the case of scalar output error (see the proof of Lemma 3.2 on p. 127). □

Corollary D.2 *The difference between the* output error ε *and the* augmented error $\bar{\varepsilon}$ *satisfies the equalities*

$$\bar{\varepsilon} - \varepsilon = -\bar{W}_\zeta(s)\left[\bar{W}_\Omega(s)\left[\bar{\Omega}^\top\right]\dot{\hat{\vartheta}}\right]$$

and

$$\bar{W}(s)\left[\bar{\Omega}^\top\right]\tilde{\vartheta} - \bar{W}(s)\left[\bar{\Omega}^\top\tilde{\vartheta}\right] = -W_\zeta(s)\left[\bar{W}_\Omega(s)[\bar{\Omega}^\top]\dot{\hat{\vartheta}}\right] + \bar{\epsilon}_w,$$

where $\bar{W}_\zeta(s) = C(sI - A)^{-1}$, $\bar{W}_\Omega(s) = (sI - A)^{-1}B$,

$$\bar{\zeta} = -\bar{W}_\zeta(s)\left[\bar{W}_\Omega(s)[\bar{\Omega}^\top]\dot{\hat{\vartheta}}\right]$$

is the swapping term, $\bar{\epsilon}_w = \epsilon - \bar{\epsilon}$.

D.2.2 Unstable Error Model

Consider the error model

$$\dot{e} = Ae + B\left(\tilde{\Theta}\omega + u_s\right), \quad \varepsilon = Ce, \tag{D.19}$$

where A is possibly unstable, and u_s is a stabilizing component of control design of which is postponed for some reasons. All other notations are defined in the previous section as for the plant (D.11). With the use of notations (D.13) and (D.14), we rewrite the model (D.19) in the form

$$\dot{e} = Ae + B\left(\bar{\Omega}^\top\tilde{\vartheta} + u_s\right), \quad \varepsilon = Ce. \tag{D.20}$$

Lemma D.6 (*MIMO observer-based swapping* [11]) *Along with* (D.19) *consider the filter*

$$\dot{\hat{e}} = A\hat{e} - L_e\left(\varepsilon + C\hat{e}\right) + B\left(\bar{\Omega}^\top\hat{\vartheta} - u_s\right), \tag{D.21}$$

where the matrix of feedback gains L_e is selected so that the matrix $A_e = A - L_e C$ is Hurwitz, and $\bar{\Omega}$ and $\hat{\vartheta}$ are defined by (D.14) *and* (D.16). *Define the* augmented error

$$\hat{\varepsilon} = \varepsilon + C\hat{e} - \bar{\Omega}_e^\top \hat{\vartheta}, \tag{D.22}$$

where $\bar{\Omega}_e^\top = \bar{W}_e\left[\bar{\Omega}^\top\right]$, and $\bar{W}_e(s) = C(sI - A_e)^{-1}B$. Then the following equality holds:

$$\bar{\varepsilon} = \bar{\Omega}_e^\top \tilde{\vartheta} + \epsilon, \tag{D.23}$$

where $\epsilon \in \mathbb{R}^m$ exponentially decays.

Proof The lemma is proved with the use of the same arguments as for the case of scalar output error (see the proof of Lemma D.3). □

D.3 Swapping in SISO LTI Systems with Input Delay

D.3.1 Stable Error Model

Consider the state error model

$$\dot{e} = Ae + b\tilde{\theta}^\top(t-\tau)\omega, \qquad \varepsilon = c^\top e, \tag{D.24}$$

where $\tilde{\theta}(t-\tau) = \theta - \hat{\theta}(t-\tau)$, τ is a known time delay, A is Hurwitz, and all other notations are defined as for the model (D.1). The following lemma allows us to remove the delay from the augmented error.

Lemma D.7 (*Transfer function swapping with delay* [2]) *Along with* (D.24), *we define the* swapping term

$$\zeta = W(s)\left[\hat{\theta}^\top(t-\tau)\omega\right] - \hat{\theta}^\top(t)\omega_w, \tag{D.25}$$

where $\omega_w = W(s)[\omega]$, and introduce the augmented error

$$\bar{\varepsilon} = \varepsilon + \zeta. \tag{D.26}$$

Then the following equality holds:

$$\bar{\varepsilon} = \tilde{\theta}^\top(t)\omega_w + \epsilon, \tag{D.27}$$

where $\tilde{\theta}(t) = \theta - \hat{\theta}(t)$, and $\epsilon(t) = c^\top e^{A_e t} e(0)$ is the exponentially decaying term.

Proof Taking into account the properties of linear systems and the fact that θ is constant, we get

$$\begin{aligned}\bar{\varepsilon} = \varepsilon + \zeta &= W(s)\left[\left(\theta - \hat{\theta}^{\top}(t-\tau)\right)\omega\right] + \epsilon + W(s)\left[\hat{\theta}^{\top}(t-\tau)\omega\right] - \hat{\theta}^{\top}(t)W(s)[\omega] \\ &= \theta^{\top}W(s)[\omega] - W(s)\left[\hat{\theta}^{\top}(t-\tau)\omega\right] + W(s)\left[\hat{\theta}^{\top}(t-\tau)\omega\right] - \hat{\theta}^{\top}(t)W(s)[\omega] + \epsilon \\ &= \tilde{\theta}^{\top}(t)\omega_w + \epsilon.\end{aligned}$$

□

D.3.2 *Unstable Error Model*

Consider the state error model

$$\dot{e} = Ae + b\left(\tilde{\theta}^{\top}(t-\tau)\omega + u_s\right), \qquad \varepsilon = c^{\top}e, \tag{D.28}$$

where A is possibly unstable, and u_s is a stabilizing component of control design of which is postponed for some reasons. All other notations are defined in the previous section as for the plant (D.24).

Lemma D.8 (*SISO observer-based swapping with delay* [11]) *Along with* (D.28), *we consider the filter*

$$\dot{\hat{e}} = A\hat{e} - l_e\left(\varepsilon + c^{\top}\hat{e}\right) + b\left(\hat{\theta}^{\top}(t-\tau)\omega - u_s\right),$$

where the vector of feedback gains l_e is selected so that the matrix $A_e = A - l_e c^{\top}$ is Hurwitz. Define the augmented error

$$\bar{\varepsilon} = \varepsilon + c^{\top}\hat{e} - \hat{\theta}^{\top}(t)\omega_e, \tag{D.29}$$

where $\omega_e = W_e(s)[\omega]$ is the filtered regressor, and $W_e(s) = c^{\top}(sI - A_e)^{-1}b$ is the asymptotically stable transfer function. Then the following equality holds:

$$\bar{\varepsilon} = \tilde{\theta}^{\top}(t)\omega_e + \epsilon_e,$$

where $\epsilon_e(t) = c^{\top}e^{A_e t}\left(e(0) + \hat{e}(0)\right)$ exponentially decays.

The lemma is proved in the same way as Lemma D.3.

Remark D.1 The results of Lemmas D.7 and D.8 can be directly extended to the case of MIMO systems [11]. ■

D.4 Nonlinear Swapping

Consider the nonlinear time-varying state error model

$$\dot{e} = A(e,t)e + B(e,t)\Omega^\top(e,t)\tilde{\theta} + D(e,t)\delta, \qquad \varepsilon = C(e,t)e, \tag{D.30}$$

where $e \in \mathbb{R}^n$ is the state error, $\varepsilon \in \mathbb{R}^m$ is the output error, $\delta \in \mathbb{R}^n$ is the external disturbance, $A(e,t)\colon \mathbb{R}^n \times \mathbb{R}_+ \to \mathbb{R}^{n\times n}$, $B(e,t)\colon \mathbb{R}^n \times \mathbb{R}_+ \to \mathbb{R}^{n\times n}$, $\Omega(e,t)\colon \mathbb{R}^n \times \mathbb{R}_+ \to \mathbb{R}^{q\times n}$ and $D(e,t)\colon \mathbb{R}^n \times \mathbb{R}_+ \to \mathbb{R}^{n\times n}$ are locally Lipschitz in e and continuous and bounded in t, $C(e,t)\colon \mathbb{R}^n \times \mathbb{R}_+ \to \mathbb{R}^{m\times n}$ is bounded in e and t, and $\tilde{\theta} = \theta - \hat{\theta} \in \mathbb{R}^q$ is the vector of parametric errors. We also assume that the model (D.30) is such that for any $e(0)$ there exists a unique solution $e(t)$ defined for $t \in [0,\infty)$. As in [6, Appendix F], we accept the following assumption.

Assumption D.1 There exists a continuously differentiable function $V(\chi,t)\colon \mathbb{R}^n \times \mathbb{R}_+ \to \mathbb{R}_+$ such that

$$c_1|\chi|^2 \le V(\chi,t) \le c_2|\chi|^2 \tag{D.31}$$

and for each $e(t) \in C^0$

$$\frac{\partial V}{\partial \chi}A(e,t)\chi + \frac{\partial V}{\partial t} \le -c_3|\chi|^2 - c_4\left|\frac{\partial V}{\partial \chi}D(e,t)\right|^2 \tag{D.32}$$

for all $t \ge 0$, $\chi \in \mathbb{R}^n$ and some $c_1, c_2, c_3, c_4 > 0$.

Lemma D.9 *(Nonlinear swapping [6, Appendix F]) Along with* (D.30)*, consider the* nonlinear swapping filters

$$\dot{\bar{\Omega}}_{_A}^\top = A(e,t)\bar{\Omega}_{_A}^\top + B(e,t)\Omega^\top(e,t), \quad \Omega_{_C}^\top = C(e,t)\bar{\Omega}_{_A}^\top, \tag{D.33}$$

$$\dot{\bar{\zeta}}_{_A} = A(e,t)\bar{\zeta}_{_A} - \bar{\Omega}_A^\top\dot{\hat{\vartheta}}, \qquad \zeta_c = C\bar{\zeta}_{_A} \tag{D.34}$$

with arbitrary initial conditions $\Omega_{_A}(0)$ and $\zeta_{_A}(0)$ and $\dot{\hat{\vartheta}}$ obtained from an adaptation algorithm. Define the augmented error *as*

$$\bar{\varepsilon} = \varepsilon + \zeta_c.$$

If Assumption D.1 holds and $|\delta(t)| \to 0$ as $t \to \infty$, then

$$\bar{\varepsilon} = \Omega_c^\top\tilde{\theta} + \tilde{\varepsilon}, \tag{D.35}$$

where $\tilde{\varepsilon}$ is uniformly bounded and $|\tilde{\varepsilon}(t)| \to 0$ as $t \to \infty$.

Proof Calculating the time derivative of $\tilde{e} = e + \bar{\zeta}_{_A} - \bar{\Omega}_{_A}^\top\tilde{\theta}$, in view of (D.30), (D.33) and (D.34), we obtain

$$\begin{aligned}\dot{\tilde{e}} &= A(e,t)e + B(e,t)\Omega^\top\tilde{\theta} + D(e,t)\delta + A(e,t)\bar{\zeta}_A - \bar{\Omega}_A^\top\dot{\hat{\theta}} - A(e,t)\bar{\Omega}_A^\top\tilde{\theta} \\ &- B(e,t)\Omega^\top\tilde{\theta} + \bar{\Omega}_A^\top\dot{\hat{\theta}} = A(e,t)(e + \bar{\zeta}_A - \bar{\Omega}_A^\top\tilde{\theta}) + D(e,t)\delta \\ &= A(e,t)\tilde{e} + D(e,t)\delta.\end{aligned}$$

Then, in view of (D.32), we get

$$\begin{aligned}\dot{V}(\tilde{e},t) &= \frac{\partial V}{\partial \tilde{e}}(A(e,t)\tilde{e} + D(e,t)\delta) + \frac{\partial V}{\partial t} \\ &\le -c_3|\tilde{e}|^2 - c_4\left|\frac{\partial V}{\partial \tilde{e}}D(e,t)\right|^2 + \left|\frac{\partial V}{\partial \tilde{e}}D(e,t)\right||\epsilon| \\ &= -c_3|\tilde{e}|^2 - c_4\left(\left|\frac{\partial V}{\partial \tilde{e}}D(e,t)\right| - \frac{1}{2c_4}|\tilde{e}|\right)^2 + \frac{1}{4c_4}|\tilde{e}|^2 \\ &\le -c_3|\tilde{e}|^2 + \frac{1}{4c_4}|\epsilon|^2.\end{aligned}$$

Therefore, according to (D.31),

$$\dot{V} \le -\frac{c_3}{c_2}V + \frac{1}{4c_4}|\epsilon|^2.$$

In view of Lemma B.2, the latter means that V is uniformly bounded and $V(t) \to 0$ as $t \to \infty$. From (D.31), we have that $|\tilde{e}| \le \sqrt{\frac{V}{c_1}}$, and, therefore, $|\tilde{e}(t)| \to 0$ as $t \to \infty$.

Then, from $e + \bar{\zeta}_A = \bar{\Omega}_A^\top\tilde{\theta} + \tilde{e}$ and $\bar{\varepsilon} = C(e,t)(e + \bar{\zeta}_A) = C(e,t)(\bar{\Omega}_A^\top\tilde{\theta} + \tilde{e})$, we directly obtain (D.35) with $\tilde{\varepsilon} = C(e,t)\tilde{e}$ (note that $C(e,t)$ is bounded in e and t). □

References

1. Gerasimov, D.N., Paramonov, A.V., Nikiforov, V.O.: Algorithms of adaptive disturbance compensation in linear Systems with Arbitrary Input Delay. Int. J. Control 1–9 (2018)
2. Gerasimov, D.N., Miliushin, A.S., Nikiforov, V.O.: Algorithms of adaptive tracking of unknown multisinusoidal signals in linear systems with arbitrary input delay. Int. J. Adapt. Control Signal Proces. **33**(6), 900–912 (2019)
3. Feuer, A., Morse, A.S.: Adaptive control of single-input single-output linear systems. IEEE Trans. Autom. Control **23**(4), 557–569 (1978)
4. Fradkov, A.L., Miroshnik, I.V., Nikiforov, V.O.: Nonlinear and Adaptive Control of Complex Systems. Kluwer Academic Publishers (s1999)
5. Kreisselmeier, G., Joos, D.: Rate of convergence in model reference adaptive control. IEEE Trans. Autom. Control **27**(3), 710–713 (1982)
6. Krstić, M., Kanellakopoulos, I., Kokotović, P.: Nonlinear and Adaptive Control Design. Wiley, NY (1995)
7. Monopoli, R.V.: Model reference adaptive control with an augmented signal. IEEE Trans. Autom. Control **19**(5), 474–484 (1974)

8. Morse, A.S.: Global stability of parameter-adaptive control systems. IEEE Trans. Autom. Control **25**(3), 433–439 (1980)
9. Narendra, K.S., Annaswamy, A.M.: Stable Adaptive Systems. Prentice Hall (1989)
10. Nikiforov, V.O., Fradkov, A.L.: Adaptive control schemes with augmented error. Survey. Autom. Remote Control **55**, 1239–1255 (1994)
11. Nikiforov, V.O., Paramonov, A.V., Gerasimov, D.N.: Adaptive control algorithms in MIMO systems with control delay. Autom. Remote Control. **81**, 1107–1124 (2020)
12. Nikiforov, V.O., Gerasimov, D.N., Pashenko, A.V.: Modular adaptive backstepping design with high-order tuner. IEEE Trans. Autom. Control. https://doi.org/10.1109/TAC.2021.3091442 (to appear)

Appendix E
Adaptive Backstepping

In this appendix, we present recursive design techniques called *adaptive backstepping* which are used in this book to overcome the structural obstacle caused by unmatched uncertainties. The problem of counteracting unmatched uncertainties with the use of iterative design technique was resolved in the work [1]. Later, this procedure received the name of *adaptive backstepping with overparameterization*. In [2, 3], the authors developed this approach and presented two modifications of the recursive design procedures without overparameterization: *adaptive backstepping with tuning functions* and *modular adaptive backstepping*. The first technique means step-by-step design of virtual control and virtual adaptation laws called *stabilizing* and *tuning* functions, respectively, while the second approach implies step-by-step design of virtual control laws (stabilizing functions) that ensures strong ISS-property and independent design of an adaptation algorithm. In [3], the ISS-controllers were combined with standard first-order adaptation laws (Lyapunov-based, gradient and least-square algorithms). In [4, 5], the ISS-controllers were combined with high-order tuners.

In this book, we use *modular adaptive backstepping with high-order tuners* applied to the plants with measured state (Sect. E.1) and *adaptive backstepping with tuning functions* applied to the plants with measured output and unknown high-frequency gain b_m (Sect. E.2).

In both cases, the main structural properties of the closed-loop systems are determined by the nonlinear system matrix A_z. In the case of modular backstepping, the diagonal entries of the matrix A_z contain nonlinear damping terms that provide strong ISS-property of the closed-loop system. In the case of backstepping with tuning functions, the matrix A_z is skew-symmetric with negative diagonal elements that is extremely useful in the proof of stability properties.

V. Nikiforov and D. Gerasimov, *Adaptive Regulation*, Lecture Notes in Control and Information Sciences 491, https://doi.org/10.1007/978-3-030-96091-9

E.1 Modular Adaptive Backstepping with High-Order Tuners

We consider an uncertain SISO nonlinear plant represented in the nonautonomous *parametric strict-feedback* form

$$\dot{x}_1 = x_2 + \phi_1^\top(x_1, \xi)\theta, \tag{E.1}$$
$$\dot{x}_i = x_{i+1} + \phi_i^\top(x_1, x_2, \cdots, x_i, \xi)\theta, \quad i = 2, 3, \ldots, n-1, \tag{E.2}$$
$$\dot{x}_n = \phi_n^\top(x, \xi)\theta + \phi_u(x)u, \tag{E.3}$$
$$y = x_1, \tag{E.4}$$

where $u \in \mathbb{R}$ is the control, $y \in \mathbb{R}$ is the regulated variable, x_i are the elements of the measured state vector, $\theta \in \mathbb{R}^q$ is a constant vector of unknown parameters, $\phi_u \in \mathbb{R}$ and $\phi_i \in \mathbb{R}^q$ are the smooth functions. The function $\phi_u(x)$ is bounded away from zero by a constant ϕ_0, i.e., $\phi_u(x) \geq \phi_0 > 0$. The function $\xi = \xi(t)$ is bounded and has bounded measured (or analytically calculated) time derivatives up to the order $n-1$.

Remark E.1 Even if $u(t) \equiv 0$, the model (E.1)–(E.4) is nonautonomous because the functions ϕ_i depend on the time function $\xi(t)$. In the control problems considered in this book, this function represents a filtered regressor (see, for example, Sect. 4.1.3.2). ■

The control objective is to design a state-feedback control law such that for any initial state all the signals in the closed-loop systems are bounded and the output y asymptotically converges to zero, i.e.,

$$\lim_{t\to\infty} y(t) = 0. \tag{E.5}$$

The modular backstepping procedure consists in recursive design of the stabilizing functions (virtual controls) α_i ($i = 1, 2, \ldots, n$) given by [4, 5]

$$\alpha_1\left(x_1, \hat{\theta}, \xi\right) = -\left(c_1 + s_1\right) z_1 - \phi_1^\top \hat{\theta}, \tag{E.6}$$

$$\alpha_i\left(\bar{x}_i, \bar{\theta}^{(i-1)}, \bar{\xi}^{(i-1)}\right) = -z_{i-1} - \left(c_i + s_i\right) z_i - \left(\phi_i^\top - \sum_{j=1}^{i-1} \frac{\partial \alpha_{i-1}}{\partial x_j}\phi_j^\top\right)\hat{\theta} \tag{E.7}$$
$$+ \sum_{j=1}^{i-1}\left(\frac{\partial \alpha_{i-1}}{\partial x_j} x_{j+1} + \frac{\partial \alpha_{i-1}}{\partial \hat{\theta}^{(j-1)}}\hat{\theta}^{(j)} + \frac{\partial \alpha_{i-1}}{\partial \xi^{(j-1)}}\xi^{(j)}\right), \quad i = 2, 3, \ldots, n,$$

where $z_1 = x_1$, $z_i = x_i - \alpha_{i-1}$ are the new state errors,

$$s_1(x_1, \xi) = \mu_1|\phi_1|^2, \quad s_i(\bar{x}_i, \bar{\theta}^{(i-2)}, \bar{\xi}^{(i-2)}) = \mu_i \left|\phi_i - \sum_{j=1}^{i-1} \frac{\partial \alpha_{i-1}}{\partial x_j}\phi_j\right|^2 \tag{E.8}$$

are the *damping terms*, $c_i > 0$ and $\mu_i > 0$ are the design parameters, $\bar{x}_i = (x_1, x_2, \ldots, x_i)$, and $\bar{\theta}^{(i)} = (\hat{\theta}, \dot{\hat{\theta}}, \ldots, \hat{\theta}^{(i)})$, and $\bar{\xi}^{(i)} = (\xi, \dot{\xi}, \ldots, \xi^{(i)})$ are the sets of corresponding variables.

The *actual control law* derived in the final step of the procedure is defined as

$$u = \frac{1}{\phi_u(x)}\alpha_n(x, \bar{\theta}^{(n-1)}, \bar{\xi}^{(n-1)}). \tag{E.9}$$

Lemma E.1 *The control law* (E.6)–(E.9) *applied to the plant* (E.1)–(E.4) *results in the closed-loop error model*

$$\dot{z} = A_z(z, x, \bar{\theta}^{(n-1)}, \bar{\xi}^{(n-1)})z + W_\theta^\top(z, x, \bar{\theta}^{(n-1)}, \bar{\xi}^{(n-1)})\tilde{\theta}, \tag{E.10}$$

$$y = z_1, \tag{E.11}$$

where $z = [z_1, z_2, \ldots, z_n]^\top$ *is the vector of the state errors,*

$$A_z = \begin{bmatrix} -c_1 - s_1 & 1 & 0 & \cdots & 0 \\ -1 & -c_2 - s_2 & 1 & \ddots & \vdots \\ 0 & -1 & \ddots & \ddots & 0 \\ \vdots & \ddots & \ddots & \ddots & 1 \\ 0 & \cdots & 0 & -1 & -c_n - s_n \end{bmatrix}, \quad W_\theta^\top = \begin{bmatrix} \phi_1^\top \\ \phi_2^\top - \frac{\partial \alpha_1}{\partial x_1}\phi_1^\top \\ \vdots \\ \phi_n^\top - \sum_{j=1}^{n-1} \frac{\partial \alpha_{n-1}}{\partial x_j}\phi_j^\top \end{bmatrix}.$$

The lemma is proved by calculation of the time derivatives $\dot{z}_i$ in view of the plant (E.1)–(E.4), and the control law (E.6)–(E.9).

An important feature of the control law (E.6)–(E.9) consists in the fact that it provides the ISS-property of the closed-loop system (E.10) with $\tilde{\theta}$ considered as inputs.

Lemma E.2 *If* $\bar{\theta}^{(n-1)}$ *is bounded, then in the closed-loop system x and z are bounded and the state z exponentially converges to a compact residual set satisfying the inequality*

$$|z(t)| \le \frac{1}{2\sqrt{c_0\mu_0}}\|\tilde{\theta}\|_\infty + |z(0)|e^{-c_0 t}, \tag{E.12}$$

where

$$c_0 = \min_{i=1,2,\ldots,n} \{c_i\}, \quad \mu_0 = \left(\sum_{i=1}^{n} \frac{1}{\mu_i}\right)^{-1}.$$

Lemma E.2 is proved by calculating the time derivative of $\frac{1}{2}|z|^2$ along the solutions of (E.10):

$$\frac{d}{dt}\left(\frac{1}{2}|z|^2\right) = \sum_{i=1}^{n}\left(-c_i z_i^2 - \mu_i |b_{z1}|^2 z_i^2 + z_i b_i^\top \tilde{\theta}\right)$$

$$\leq -c_0|z|^2 - \sum_{i=1}^{n}\left(\sqrt{\mu_i}|b_1|z_i - \frac{\| \tilde{\theta} \|_\infty}{2\sqrt{\mu_i}}\right)^2 + \sum_{i=1}^{n}\frac{\| \tilde{\theta} \|_\infty^2}{4\mu_i}$$

$$\leq -c_0|z|^2 + \frac{\| \tilde{\theta} \|_\infty^2}{4\mu_0}.$$

The latter means (E.12).

Lemma E.3 *Together with closed-loop model* (E.10)*, we consider the filters*

$$\dot{\zeta}_z = A_z \zeta_z + W_\theta^\top \hat{\theta}, \tag{E.13}$$

$$\dot{\Omega}_z^\top = A_z \Omega_z^\top + W_\theta^\top \tag{E.14}$$

and define the augmented state error

$$\bar{z} = z + \zeta_z. \tag{E.15}$$

Then the following equality holds:

$$\bar{z} = \Omega_z^\top \theta + \epsilon_z, \tag{E.16}$$

where ϵ_z exponentially decays.

Proof Calculating the time derivative of the error $\epsilon_z = z + \zeta_z + \Omega_z^\top \vartheta$, in view of (E.10), (E.13), and (E.14), we have

$$\dot{\epsilon}_z = A_z \epsilon_z$$

with the initial condition $\epsilon_z(0) = z(0) + \zeta_z(0) - \Omega_z(0)\theta$. Now, taking into account the structures of matrix A_z, we conclude that ϵ_z exponentially decays. □

Remark E.2 Without loss of generality, we set $\zeta_z(0) = -z(0)$ and $\Omega_z(0) = 0$. In this case, $\epsilon_z(0) = 0$ that simplifies further derivations. ■

It is worth noting a very important property of the filter (E.14). As it was proved by Lemma 6.1 in [3], due to the special structure of the matrix A_z, the filter state Ω_z is bounded irrespectively of the boundness of its input W_θ. Since the regressor Ω_z is bounded, it allows us to use both normalized and unnormalized adaptation laws on the basis of the regression model (E.16).

Following [4], we employ a high-order tuner which enables one to generate HOTD of the adjustable parameters required for implementation of the control (E.6)–(E.9). The high-order tuner is defined as

$$\dot{\hat{\theta}} = \gamma \left(\bar{z}_{\Omega} - \bar{\Omega}\hat{\theta} \right), \tag{E.17}$$

where $\bar{z}_{\Omega} = L(s) \left[\Omega_z \bar{z} \right]$, $\bar{\Omega} = L(s) \left[\Omega_z \Omega_z^\top \right]$, and $L(s)$ is a designer-chosen transfer function of the form (3.44) with $p \geq (n-1)$. The matrix $\bar{\Omega}$ is positive semi-definite by construction. In view of (E.16) and Remark E.2, it can be shown that in this case the closed-loop parametric error model takes the form

$$\dot{\tilde{\theta}} = -\gamma \bar{\Omega}\tilde{\theta}. \tag{E.18}$$

Now, we define $\bar{\lambda}(t) \geq 0$ as the minimum eigenvalue of $\bar{\Omega} = L(s) \left[\Omega_z \Omega_z^\top \right]$ and formulate the main stability properties of the system closed by the ISS-controller and the high-order tuner [4].

Proposition E.1 *The control law* (E.6)–(E.9) *together with the swapping scheme* (E.13)–(E.15) *and the adaptation algorithm* (E.17) *when applied to the plant* (E.1)–(E.4) *ensures the following properties in the closed-loop system:*

P.E.1.1. *the global uniform boundedness of all the signals in the system and global asymptotic convergence $z(t) \to 0$ as $t \to \infty$ and, as a result, achievement of the control objective* (E.5);

P.E.1.2. *if $\bar{\lambda}(t) \notin \mathcal{L}_1$, then $|\tilde{\theta}(t)|$ converges to zero asymptotically and the rate of convergence can be increased by increasing adaptation gain γ.*

Proof Since Property P.E.1.1 basically follows from Theorem 6.3 and Theorem 6.7 in [3], we just outline its proof. First, due to the specific structure of A_z involving the nonlinear damping terms, Ω_z and $\bar{\Omega}$ are uniformly bounded (see Lemma 6.1 and Lemma 6.5 in [3]). Then since $\bar{\Omega}$ is positive semi-definite, uniformly bounded and filtered with the transfer function $L(s)$, we immediately conclude from (E.18) that $\tilde{\theta}^{(n-1)} \in \mathcal{L}_\infty$ and, therefore, $z, \dot{z}, x \in \mathcal{L}_\infty$. Hence, all the closed-loop signals are globally uniformly bounded on $t \in [0, \infty)$. Further, integrating the time derivative of the Lyapunov function $V = \frac{\tilde{\theta}^\top \tilde{\theta}}{\gamma}$ with the use of its bounded initial $V(0)$ and asymptotical $V(\infty)$ values, we prove that $\tilde{z}_\Omega = \bar{\Omega}\tilde{\theta}, \dot{\hat{\theta}} \in \mathcal{L}_2$ and $\tilde{z}_\Omega(t), \dot{\hat{\theta}}(t) \to 0$ as $t \to \infty$. Since $\dot{\hat{\theta}} \in \mathcal{L}_2$, from Lemma F.4 of [3], we obtain that $z \in \mathcal{L}_2$ and, therefore, $z(t) \to 0$ as $t \to \infty$. Property P.E.1.1 is proved.

Property P.E.1.2 is proved by calculating the time derivative of $\frac{1}{2}|\tilde{\theta}|^2$ along the solutions of (E.18):

$$\frac{d}{dt}\left(\frac{1}{2}|\tilde{\theta}|^2 \right) = -\gamma \tilde{\theta}^\top \Omega \tilde{\theta} \leq -\gamma \bar{\lambda}(t) |\tilde{\theta}|^2. \tag{E.19}$$

Therefore,

$$|\tilde{\theta}(t)| \leq \mathrm{e}^{-\gamma \int_0^t \bar{\lambda}(\tau)d\tau} |\tilde{\theta}(0)|.$$

The latter means that $|\tilde{\theta}(t)|$ approaches zero asymptotically as time tends to infinity if $\bar{\lambda}(t) \notin \mathcal{L}_1$. The rate of convergence of the norm $|\tilde{\theta}(t)|$ to zero can be increased by increasing the gain γ. □

E.2 Tuning Functions Adaptive Backstepping: Unknown High-Frequency Gain b_m

As demonstrated in Sects. 2.5.1.3 and 2.5.2.3, in the case of output-feedback adaptive control, a SISO plant with unknown parameters admits the following parameterization:

$$\dot{y} = \beta_1 + \theta^\top \bar{\omega} + b_m(-l_{n-1}\nu_1 + \nu_2) + \epsilon, \tag{E.20}$$
$$\dot{\nu}_i = -l_{n-i}\nu_1 + \nu_{i+1}, \quad i = 1, 2, \ldots, n-1, \tag{E.21}$$
$$\dot{\nu}_n = -l_0\nu_1 + u, \tag{E.22}$$

where $y \in \mathbb{R}$ is the output (regulated) variable, ν_i are the measured variables, l_j $(j = 0, 1, \ldots, n-1)$ are some known constants, $\theta \in \mathbb{R}^{q-1}$ is a vector of unknown parameters, $b_m > 0$ is an unknown parameter, $\beta_1 \in \mathbb{R}$, $\bar{\omega} \in \mathbb{R}^{q-1}$ are bounded function and regressor with bounded and measured (or analytically calculated) time derivatives up to the order $n-1$, and ϵ is some exponentially decaying term.

For the sake of presentation simplicity, we will also use the more compact form of (E.20) given by

$$\dot{y} = \beta_1 + \vartheta^\top \omega + \epsilon, \tag{E.23}$$

where $\vartheta = \left[b_m, \theta^\top\right]^\top$ and $\omega = \left[-l_{n-1}\nu_1 + \nu_2, \bar{\omega}^\top\right]^\top$ are the q-dimensional vector of unknown parameters and regressor, respectively. We will also use the following notation $\omega_0 = \left[0, \bar{\omega}^\top\right]^\top$.

Remark E.3 In SISO systems, b_m represents so-called high-frequency gain, while ν_i represent the elements of the state of some auxiliary filter (see Sects. 2.5.1.3, 2.5.2.3, 5.1.3.2, and 5.2.2.2). ■

Based on the procedure of the tuning functions adaptive backstepping (see [3, Sect. 4.5]), we design the following stabilizing functions starting from ν_2 as the first virtual control:

$$\alpha_1(y, \nu_1, \beta_1, \bar{\omega}, \hat{\vartheta}, \hat{\kappa}_m,) = -(c_1 + s_1)z_1 + \hat{\kappa}_m\bar{\alpha}_1 + l_{n-1}\nu_1, \tag{E.24}$$

$$\bar{\alpha}_1(\beta_1, \bar{\omega}, \hat{\vartheta}) = -\beta_1 - \omega_0^\top\hat{\vartheta}, \tag{E.25}$$

$$\alpha_2(y, \nu_1, \nu_2, \beta_1, \dot{\beta}_1, \bar{\omega}, \dot{\bar{\omega}}, \hat{\vartheta}, \hat{\kappa}_m) = -(c_2 + s_2)z_2 + \left(\frac{\partial\alpha_1}{\partial y}\omega^\top - \varsigma_1^\top z_1\right)\hat{\vartheta} + \frac{\partial\alpha_1}{\partial\hat{\vartheta}}\tau_2 + \beta_2, \tag{E.26}$$

$$\alpha_i(y, \bar{\nu}_i, \varpi^{(i-1)}, \hat{\vartheta}, \hat{\kappa}_m) = -z_{i-1} - (c_i + s_i)z_i + \frac{\partial\alpha_{i-1}}{\partial y}\omega^\top\hat{\vartheta} + \frac{\partial\alpha_{i-1}}{\partial\hat{\vartheta}}\tau_i + \beta_i - \gamma_1\frac{\partial\alpha_{i-1}}{\partial y}\left(\sum_{j=2}^{i-1}\frac{\partial\alpha_{j-1}}{\partial\hat{\vartheta}}z_j\right)\omega, \tag{E.27}$$

where $i = 3, 4, \ldots, n$, $z_1 = x_1$, $z_i = x_i - \alpha_{i-1}$ are the new state errors,

$$s_1 = \mu_1, \quad s_i = \mu_i\left(\frac{\partial\alpha_{i-1}}{\partial y}\right)^2 \tag{E.28}$$

are the damping terms with constant gains $\mu_i > 0$ $(i = 1, 2, \ldots, n)$,

$$\tau_1 = \gamma_1\omega_0 z_1, \tag{E.29}$$

$$\tau_2 = \tau_1 - \gamma_1\left(\frac{\partial\alpha_1}{\partial y}\varpi - \varsigma_1 z_1\right)z_2, \tag{E.30}$$

$$\tau_i = \tau_{i-1} - \gamma_1\frac{\partial\alpha_{i-1}}{\partial y}\omega z_i, \quad i = 3, 4, \ldots, n \tag{E.31}$$

are the tuning functions, $c_i > 0$, $\gamma_1 > 0$ are the design parameters, $\hat{\vartheta} = [\hat{b}_m, \hat{\theta}^\top]^\top \in \mathbb{R}^q$ is the vector of adjustable parameters, $\hat{b}_m$, $\hat{\theta}$ are the adjustable parameters corresponding to unknown b_m and θ, respectively, $\hat{\kappa}_m$ is the additional adjustable parameter corresponding to the unknown constant $\kappa_m = \frac{1}{b_m}$, $\bar{\nu}_i = (\nu_1, \nu_2, \ldots, \nu_i)$, $\varpi^{(i)} = (\beta_1, \dot{\beta}_1, \ldots, \beta_1^{(i)}, \bar{\omega}, \dot{\bar{\omega}}; \ldots, \bar{\omega}^{(i)})$ are the sets of corresponding variables, and

$$\beta_2 = l_{n-2}x_1 + \frac{\partial\alpha_1}{\partial y}\beta_1 + \frac{\partial\alpha_{i-1}}{\partial\hat{\kappa}_m}\dot{\hat{\kappa}}_m + \frac{\partial\alpha_1}{\partial\nu_1}\dot{\nu}_1 + \frac{\partial\alpha_1}{\partial\beta_1}\dot{\beta}_1 + \frac{\partial\alpha_1}{\partial\bar{\omega}}\dot{\bar{\omega}}, \tag{E.32}$$

$$\beta_i = l_{n-i}x_1 + \frac{\partial\alpha_{i-1}}{\partial y}\beta_1 + \frac{\partial\alpha_{i-1}}{\partial\hat{\kappa}_m}\dot{\hat{\kappa}}_m + \sum_{j=1}^{i-1}\left(\frac{\partial\alpha_{i-1}}{\partial\nu_i}(-l_{n-j}\nu_1 + \nu_{j+1}) + \frac{\partial\alpha_{i-1}}{\partial\beta_1^{(j-1)}}\beta_1^{(j)} + \frac{\partial\alpha_{i-1}}{\partial\bar{\omega}^{(j-1)}}\bar{\omega}^{(j)}\right) \tag{E.33}$$

are the terms involving known functions.

The adjustable parameters $\hat{\vartheta}$ and $\hat{\kappa}_m$ are generated by the adaptation algorithms

$$\dot{\hat{\vartheta}} = \tau_n, \tag{E.34}$$

$$\dot{\hat{\kappa}}_m = -\gamma_2 \bar{\alpha}_1 z_1, \tag{E.35}$$

where $\gamma_2 > 0$ is the adaptation gain.

The procedure is finished by design of the actual control law

$$u = \alpha_n(y, \bar{v}_{n-1}, \varpi^{(n-1)}, \hat{\vartheta}, \hat{\kappa}_m). \tag{E.36}$$

Lemma E.4 *The control law* (E.24)–(E.36) *together with the adaptation algorithms* (E.34), (E.35) *when applied to the plant* (E.20)–(E.22) *results in the closed-loop error model*

$$\dot{z} = A_z(z,t)z + W_\theta^\top(z,t)\tilde{\vartheta} - b_m \varsigma_1 \bar{\alpha}_1 \tilde{\kappa}_m + b_z(z,t)\epsilon, \tag{E.37}$$

$$\dot{\tilde{\vartheta}} = -\gamma_1 W_\theta(z,t) z, \tag{E.38}$$

$$\dot{\tilde{\kappa}}_m = \gamma_2 \bar{\alpha}_1(\beta_1, \bar{\omega}, \hat{\vartheta}) \varsigma_1^\top z, \tag{E.39}$$

$$y = z_1, \tag{E.40}$$

where $z = [z_1, z_2, \dots, z_n]^\top$ *is the vector of the state errors,* $\tilde{\vartheta} = \vartheta - \hat{\vartheta}$, $\tilde{\kappa}_m = \kappa_m - \hat{\kappa}_m$ *are the parametric errors,*

$$A_z = \begin{bmatrix} -b_m(c_1+s_1) & b_m & 0 & 0 & \cdots & 0 \\ -b_m & -c_2 - s_2 & 1+\sigma_{2,3} & \sigma_{2,4} & \ddots & \sigma_{2,n} \\ 0 & -1-\sigma_{2,3} & -c_3 - s_3 & 1+\sigma_{3,4} & \ddots & \sigma_{3,n} \\ 0 & -\sigma_{2,4} & -1-\sigma_{3,4} & \ddots & \ddots & \sigma_{4,n} \\ \vdots & \ddots & \ddots & \ddots & \ddots & \vdots \\ 0 & -\sigma_{2,n} & -\sigma_{3,n} & -\sigma_{4,n} & \cdots & -c_n - s_n \end{bmatrix},$$

$$b_z = \begin{bmatrix} 1 \\ -\frac{\partial \alpha_1}{\partial y} \\ -\frac{\partial \alpha_2}{\partial y} \\ \vdots \\ -\frac{\partial \alpha_{n-1}}{\partial y} \end{bmatrix}, \quad W_\theta^\top = \begin{bmatrix} \omega_0^\top \\ \frac{\partial \alpha_1}{\partial y}\omega^\top \\ \frac{\partial \alpha_2}{\partial y}\omega^\top \\ \vdots \\ \frac{\partial \alpha_{n-1}}{\partial y}\omega^\top \end{bmatrix} + \varsigma_2 \varsigma_1^\top z_1,$$

$$\sigma_{i,j} = \begin{cases} 0 & \text{if } i = 1, \\ \gamma_1 \frac{\partial \alpha_{j-1}}{\partial y} \frac{\partial \alpha_1}{\partial \hat{\vartheta}} \omega_0 & \text{if } i = 2, \\ \frac{\partial \alpha_{j-1}}{\partial y} \gamma_1 \frac{\partial \alpha_{j-1}}{\partial \hat{\vartheta}} \omega & \text{if } i = 3, 4, \ldots, n, \end{cases}$$

and $\varsigma_1 \in \mathbb{R}^q$, $\varsigma_2 \in \mathbb{R}^n$ *are the coordinate vectors. In* (E.37), (E.38), *we use the notations* $A_z(z, t) = A_z(z, \hat{\vartheta}, \hat{\kappa}_m, \varpi)$, $W_{\vartheta}^{\top}(z, t) = W_{\vartheta}^{\top}(z, \hat{\vartheta}, \hat{\kappa}_m, \varpi)$, *and* $b_z(z, t) = b_z(z, \hat{\vartheta}, \hat{\kappa}_m, \varpi)$.

The lemma is proved by calculation of the time derivatives $\dot{z}_i$ in view of the plant (E.20)–(E.22), the control law (E.24)–(E.36), and the adaptation algorithms (E.34), (E.35).

Now, we are in a position to formulate the following statement.

Proposition E.2 *The tuning functions control law* (E.24)–(E.36) *when applied to the plant* (E.20)–(E.22) *ensures the global uniform boundedness of all the signals in the system and global asymptotic convergence* $z(t) \to 0$ *as* $t \to \infty$ *and, as a result, achievement of the control objective* (E.5).

Proof First of all, we note that due to the skew-symmetric property of the matrix A_z, we have

$$z^{\top}(A_z^{\top} + A_z)z = -b_m(c_1 + s_1)z_1^2 - \sum_{i=2}^{n}(c_i + s_i)z_i^2$$
$$\le -c_0|z|^2 - b_m\mu_1 z_1^2 - \mu_0 \sum_{i=2}^{n} \left(\frac{\partial \alpha_{i-1}}{\partial y}\right)^2 z_i^2,$$

where $c_0 = \min\{b_m c_1, c_2, \ldots, c_n\}$, $\mu_0 = \min\{\mu_2, \ldots, \mu_n\}$, and $\alpha_0 = 1$. Then the time derivative of the Lyapunov function

$$V(z, \tilde{\vartheta}, \tilde{\kappa}_m, \epsilon) = \frac{1}{2} z^{\top} z + \frac{1}{2\gamma_1} \tilde{\vartheta}^{\top} \tilde{\vartheta} + \frac{b_m}{2\gamma_2} \tilde{\kappa}_m^2 + \left(\frac{1}{4 b_m \mu_1} + \frac{n-1}{4\mu_0}\right) \int_t^{\infty} \epsilon^2(\tau) d\tau$$

along the solution of (E.37), (E.34), (E.35) takes the form

$$\dot{V} \le -c_0|z|^2 - b_m\mu_1 z_1^2 - \mu_0 \sum_{i=2}^{n} \left(\frac{\partial \alpha_{i-1}}{\partial y}\right)^2 z_i^2 + z^{\top} W_{\theta}^{\top} \tilde{\vartheta} - b_m z^{\top} \varsigma_1 \bar{\alpha}_1 \tilde{\kappa}_m + z^{\top} b_z \epsilon$$
$$- \tilde{\vartheta}^{\top} W_{\theta} z + b_m \tilde{\kappa}_m \bar{\alpha}_1 \varsigma_1^{\top} z - \left(\frac{1}{4 b_m \mu_1} + \frac{n-1}{4\mu_0}\right) \epsilon^2$$
$$\le -c_0|z|^2 - b_m\mu_1 z_1^2 - \mu_0 \sum_{i=2}^{n} \left(\frac{\partial \alpha_{i-1}}{\partial y}\right)^2 z_i^2 + |z_1||\epsilon| + \sum_{i=2}^{n} \left|\frac{\partial \alpha_{i-1}}{\partial y}\right| |z_i||\epsilon|$$
$$- \frac{1}{4 b_m \mu_1} \epsilon^2 - \frac{n-1}{4\mu_0} \epsilon^2 = -c_0|z|^2 - \left(\sqrt{b_m\mu_1}|z_1| - \frac{1}{2\sqrt{b_m\mu_1}}|\epsilon|\right)^2$$

$$
- \sum_{i=2}^{n} \left(\sqrt{\mu_0} \left| \frac{\partial \alpha_{i-1}}{\partial y} \right| |z_i| - \frac{1}{2\sqrt{\mu_0}} |\epsilon| \right)^2
$$
$$
\leq -c_0 |z|^2.
$$

Therefore, by virtue of Theorem B.1, z, $\hat{\vartheta}$, and $\hat{\kappa}_m$ are bounded. Since β_1, $\bar{\omega}$, and their time derivatives up to the necessary order are bounded, then, by Theorem B.4, $|z(t)| \to 0$ as $t \to \infty$. □

Remark E.4 In real-world control problems, we cannot assume a priori that the functions β_1 and $\bar{\omega}$ are bounded. Therefore, their boundedness must be rigorously proved by taking into account the structure of additional filters and the properties of the plant zero dynamics. ■

References

1. Kanellakopoulos, I., Kokotović, P.V., Morse, A.S.: Systematic design of adaptive controllers for feedback linearizable systems. IEEE Trans. Autom. Control **36**, 1241–1253 (1991)
2. Krstić, M., Kanellakopoulos, I., Kokotović, P.V.: Adaptive nonlinear control without overparametrization. Syst. Control Lett. **19**, 177–185 (1992)
3. Krstić, M., Kanellakopoulos, I., Kokotović, P.: Nonlinear and Adaptive Control Design. Wiley, NY (1995)
4. Nikiforov, V.O., Gerasimov, D.N., Pashenko, A.V.: Modular adaptive backstepping design with a high-order tuner. IEEE Trans. Autom. Control https://doi.org/10.1109/TAC.2021.3091442, (to appear)
5. Nikiforov, V.O., Voronov, K.V.: Adaptive backstepping with a high-order tuner. Automatica. **37**(12), 1953–1960 (2001)

Index

V. Nikiforov and D. Gerasimov, *Adaptive Regulation*, Lecture Notes in Control and Information Sciences 491, https://doi.org/10.1007/978-3-030-96091-9